T. G. BECKWITH
N. LEWIS BUCK
Department of Mechanical Engineering
University of Pittsburgh

MECHANICAL MEASUREMENTS

SECOND EDITION

ADDISON-WESLEY PUBLISHING COMPANY
Reading, Massachusetts · Menlo Park, California
London · Amsterdam · Don Mills, Ontario · Sydney

This book is in the
Addison-Wesley Series in Mechanics and Thermodynamics

Consulting editors
Howard W. Emmons and Bernard Budiansky

Second printing, June 1973

ISBN 0-201-00454-2
IJKLMNOPQ-AL-7987

PREFACE

Since the first edition of this text appeared, both the practical application of mechanical measurements and its inclusion in engineering curricula have expanded. The broad subject area of mechanical measurements, in its most general sense, has continued to replace the older mechanical-power laboratory course in many schools. Older methods and techniques have been improved and new procedures developed. In addition, the philosophy of "experimental design/development" has received even greater acceptance as a legitimate approach to the complex engineering design problem.

Hence the original purposes of the first edition remain unchanged. The book is specifically directed to the undergraduate mechanical-engineering student who receives his first industrial assignment as a member of some project group in an engineering development section. The acquisition of numerical data through experimental measurement is a fundamental objective of such efforts; and having a feel for the possibilities and the limitations of the methods of mechanical measurements is essential to obtaining meaningful results. This book is also intended as a reference source for the practicing engineer who wishes to use the text as a guide for further study in specific areas of measurement, or to obtain an overview of the subject as a whole.

Each and every chapter is a worthy topic for a complete treatise in itself. Therefore the book serves as an introduction—a jumping-off place for a more intensive study of specific areas of mechanical measurements. For this reason an extensive bibliography is included to assist both the student and the practitioner in seeking additional information in his particular area of interest.

The original division of the book has been retained, namely, Part 1, "Fundamentals of Mechanical Measurement," and Part 2, "Applied Mechanical Measurement." Some rearrangement of the subject matter has been made in order to present the material in a more logical sequence.

The section on basic principles, Chapter 2, has been considerably expanded, particularly in the area of dynamic response of systems. In spite of the bed-rock nature of fundamental standards, the basic units for both time and length have been redefined since 1961, and these changes are reflected in the revision. Although not fundamental to the basic purpose of the text, great strides have been taken in the hardware associated with mechanical measurement; e.g., most electron tubes have given way to the transistor, resulting in lighter, much more compact, and reliable gear. Where pertinent, recognition of these changes has been made. Solid-state techniques appear to have a very bright future in the field of strain measurement, and the current state of that art is covered in new sections. The chapters on pressure and flow measurement have been expanded.

Mechanical devices as sources of noise are coming under an increasingly greater attack both by private and governmental groups. Legal questions concerning "pollution of silence" are becoming more critical, and scientifically correct methods for properly measuring the characteristics of a sound or noise are quite important. Therefore a new chapter, "Acoustical Measurements," has been added.

An important part of the revision consists in the inclusion of problems at the end of each chapter. Many of the problems are stated in a form which suggests laboratory experiments. Specific, detailed experiments have been avoided, however, because of the wide diversity of equipment found in different college laboratories.

Certain errors which always, it seems, insidiously creep into work of this sort, have been corrected. However, the authors are not so optimistic as to believe that all such defects are nonexistent in this new edition. It is sincerely hoped, therefore, that readers noting discrepancies will be so kind as to bring them to the writers' attention so corrections may be made at the earliest opportunity.

Finally, the authors wish to recognize those colleagues and users of the original edition who either directly or by letter offered many constructive criticisms and suggestions. They were all gratefully received. Specifically, the writers are indebted to Dr. Gene Geiger and Dr. Roy Marangoni, University of Pittsburgh, for their most helpful assistance in reading and criticizing the manuscript, and to Mrs. Lillian Malin for her professional secretarial help.

Pittsburgh, Penn.   T. G. B.
March 1969   N. L. B.

CONTENTS

Chapter 10   Measurement of Dimension, Displacement and Linear Velocity

Chapter 11   Strain Measurement

PART 1

FUNDAMENTALS OF
MECHANICAL MEASUREMENT

CHAPTER 1

THE SIGNIFICANCE
OF MECHANICAL MEASUREMENTS

A basic function of all branches of engineering is design: design of manufactured goods of various kinds, design of machinery to perform the manufacturing operations, design of power sources, design of electrical and electronic apparatus, design of roads and waterways and sanitary systems, design of processes and process equipment, design of materials, and so on. Another function is to provide for proper operation and maintenance of such equipment or systems; for unless the intended purposes are reliably performed, both operationally and economically, the systems are of questionable practical value.* Both of these areas of engineering require *measurement* as a source of very important and necessary information—information without which the function could not be properly performed.

For measurement to be of greatest usefulness, all factors influencing the measuring process must be understood. For example, in practice, a measurement seldom provides the exact information desired, for the very act of measuring alters the condition to be determined. Often this effect is negligible; many times it is not. In any case, the more that is known regarding the characteristics of the measuring system, the more valuable will be the results.

Most measurements associated with mechanical engineering may be placed in either or both of two broad categories: (a) those of a mechanics type, or (b) those of a power type. The former are commonly applied to experimental or developmental programs, whereas the latter are generally used for monitoring or operational measurements, which are often a part of a control system.

In the latter case, provision for various measurements may be a required part of a processing system or power source. For example, proper and economical operation of a steam power plant requires continual monitoring of the various phases of the process. Steam pressure, temperature, and flow rates, along with various combustion measurements, and vibration

---

* Certain research or experimental apparatus excepted.

3

frequencies and amplitudes—all must be continuously evaluated, not for the purpose of developing the system further, but for the purpose of maintaining proper operation. The quantities measured are often both indicated and recorded, and, in addition, used to control some part of the process. In fact, *the whole area of automation or automatic control is based on measurements.* The very concept of control requires the *measured* discrepancy between actual and desired performance. The "controlling" portion of the system must know magnitude and direction of error to react intelligently.

On the other hand, measurement may be used as a design tool. Complete solutions to complex problems in mechanical design require the integration of three distinct methods or approaches: (a) the empirical method, (b) the rational method, and (c) the experimental method.

Empirical design is dependent on "good engineering judgment," based on knowledge of satisfactory previous performance, either personally observed or generally recognized as being "good practice." Information of this sort may be available in the form of "rule of thumb," handbook data, or perhaps incorporated in a design code.

Rational design, if strictly applied, would base each design step on accepted scientific theory or law. In the field of mechanical design such theory would, to a great extent, be in the areas of mechanics and thermodynamics. Of major importance would be such topics as statics, dynamics, strength of materials and elasticity, fluid flow, and heat transfer. Included would be all scientific knowledge applicable to the problem.

In practice, neither the empirical nor the strictly rational approach is very often capable of *singlehandedly* completely solving any but the simplest of design problems. For that matter, in many cases even in combination they may not lead to a direct solution.

In years past, new machines have been "laid out" on the drawing board, detailed, and a prototype produced, followed by a period of "ironing out the bugs." Test programs were seldom planned and only proceeded as necessity dictated. Often the development came after the machine was put on the job. In short, *development* was not accepted as an integral part of the design procedure, but only as a necessary evil.

At least for modern, complexly engineered devices, that period is past. *Experimental design* must be considered not only a necessary design step, but also *the* practical way of getting the job done. The foundation for this broad area of design is accurate, correlated *measurement* of all physical quantities involved.

In addition to experimental design, the term *developmental design* or if you insist, design by trial and error, may be applied. It is important to recognize, however, that such procedure must never be *random;* the trials must be made with intelligence and the errors wisely interpreted. Practically, this is what is meant by the phrase "research and development," which has recently become so common as to be abbreviated as R&D.

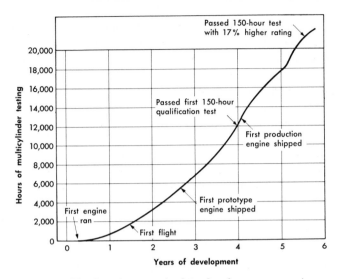

Fig. 1.1   Testing time required to develop a new engine.

As an example of what we are talking about, let us consider a relatively complex problem, the development of an automobile automatic transmission. Such a device may be assembled of mechanical components each of rather ordinary nature, presenting no particular individual design problems. However, in combined form, synchronization of the clutching, braking, or gear shifting is all-important to proper operation. Compatible response of the integrated components becomes a governing requirement. In short, satisfactory design of the individual elements is an important, but far from sufficient requirement for proper operation. Usually the question of proper performance cannot be finally answered until the prototype is produced and tested. After a few test runs (i.e., measurements) a much more realistic understanding of the whole problem is possible. Then, if the outlook appears promising, there follows an experimental period during which the transmission is *developed* (i.e., designed experimentally).

Figure 1.1 further illustrates the point being made. It shows the time required to develop a new aircraft engine [1].* Note that the first engine was running about five months after design was initiated. There followed approximately one year of testing before the engine actually powered an aircraft in flight, and the first production engine was shipped only after four years of development. Of course, the basis for the long period of developmental testing was careful measurement and intelligent interpretation of the measured quantities.

A statement widely credited to Lord Kelvin is as follows: "I often say that when you can measure what you are speaking about, and express it in

---

* Bracketed numerals identify the references at the end of the book.

numbers, you know something about it, but when you cannot express it in numbers, your knowledge is of a meager and unsatisfactory kind: it may be the beginning of knowledge, but you have scarcely, in your thoughts, advanced to the stage of science, whatever the matter may be." Development engineering depends heavily upon the determination of numbers—numbers obtained experimentally. Acquiring reliable measure, then correctly interpreting its meaning, invariably leads one nearer and nearer to the desired solution.

Developmental design probably requires the greatest application of real engineering ingenuity, except perhaps for the initial conception of the project. Programs of this nature require many ingenious test facilities, equipment, and techniques. Since each situation presents its own problems, the good development engineer is the one who through intelligent application of what is at hand knows how to get the answer. He may have access to an almost unlimited budget from which to draw his tools, or he may be limited to improvisation with what is immediately available. In either case, the *right* answer is supreme, and the equipment and technique are of importance only as they provide that answer. A procedure which produces the wrong answer is *many times worse than useless.*

An important basic area of measurement, often given only superficial attention, is the relation of a quantity to time. Measurement is often considered only in terms of static or steady-state conditions. *Modern mechanical measurement must give particular attention to determination of transient phenomena.* This is an important aim of the material which follows.

The results of most experiments must be reported in written form. Such reports may be placed in either of two broad categories, the *laboratory note* or the *formal report.* In most research and development organizations, many more lab notes than formal reports are written. Both usually include the following areas of information: (a) the purpose of the test or aim of the project, (b) pertinent theory, (c) a description of the equipment and statement of procedure, (d) a presentation of the experimentally determined and calculated results, and (e) a discussion of results, conclusions, and recommendations.

The primary difference between the lab note and the formal report is dictated by the end purposes of the two. The lab note is prepared with the understanding that the reader will be familiar with the specific problem, project, or assignment. Indeed, he is often the immediate supervisor or, in the case of educational courses, the instructor. It was he who made the assignment in the first place, spelled out the area of investigation, and probably discussed apparatus, procedures, and techniques. Why should he wade through pages of background materials which, although pertinent, are certainly not new or unfamiliar to him? He wants to know the results and to be alerted to those factors which may have affected the results. It is only necessary to include enough information so that the experimenter and those immediately associated with the project can at some later date recreate the setup and procedures. In its simplest form, it is conceivable that the lab note could be complete on a single page of graph paper: a short title covering the objective, a block diagram depicting the equipment and procedure, data plotted in the form of a curve, and a short statement of results. The date and the experimenter's name should always be included.

On the other hand, the formal report is written for circulation to people relatively unfamiliar with the immediate detailed problem—perhaps the vice president in charge of engineering for a very large company, interested members of the board of directors, or people outside the organization itself. Or it may be that the formal report is written as a final communication to some foundation or agency which supported the work. In this case, considerably more elaboration is required, and the result is a full-blown story of the complete project.

Personnel in most large industrial research organizations write many laboratory notes, but only an occasional all-inclusive formal report. In fact, most formal reports become detailed summations of many previously written lab notes, the formal report serving as the final wrap-up of a major project.

The chapters that follow are written with the idea of presenting a reasonably complete cross section of mechanical-measurement methods. They are written, however, with the full realization that the field is broadening at a tremendously accelerating rate. It is important, therefore, that the serious, experimentally minded engineer make every effort to keep abreast of developments in his field. For this purpose there are a number of excellent journals and reference sources which should be regularly reviewed (see Problem 1.1).

SUGGESTED READINGS

Ambrosius, E. E., R. D. Fellows, and A. D. Brickman, *Mechanical Measurement and Instrumentation*. New York: The Ronald Press Co., 1966.

Cook, N. H., and E. Rabinowicz, *Physical Measurement and Measurement Analysis*. Reading, Mass.: Addison-Wesley Publishing Co., 1963.

Doeblin, E. O., *Measurement Systems: Application and Design*. New York: McGraw-Hill Book Co., 1966.

Fribance, A. E., *Industrial Instrumentation Fundamentals*. New York: McGraw-Hill Book Co., 1962.

Holman, J. P., *Experimental Methods for Engineers*. New York: McGraw-Hill Book Co., 1966.

Messersmith, C. W., C. F. Warner, and R. A. Olsen, *Mechanical Engineering Laboratory*, 2nd Ed. New York: John Wiley & Sons, Inc., 1958.

National Bureau of Standards: *Precision Measurement and Calibration*, Handbook 77, U.S. Department of Commerce, 1961; Vol. I, *Electricity and Electronics*, Vol. II, *Heat and Mechanics*, Vol. III, *Optics, Metrology and Radiation*.

Sweeney, R. J., *Measurement Techniques in Mechanical Engineering*. New York: John Wiley & Sons, Inc., 1953.

Tuve, G. L., and L. C. Domholdt, *Engineering Experimentation*. New York: McGraw-Hill Book Co., 1966.

PROBLEMS

1.1 As a semester assignment, determine which of the following journals are available in your school library. Review one or more issues of each, and prepare a short statement covering the apparent editorial policies, including what seems to be the particular clientele to whom they are directed.

   1. *Control Engineering*
   2. *Electromechanical Design*
   3. *Experimental Mechanics* (Society for Experimental Stress Analysis)
   4. *Instrumentation Technology* (Journal of Instrument Society of America)
   5. *Instruments and Control Systems*
   6. *Instruments and Experimental Techniques* (USSR: English translation)
   7. *Instrument Society of America* (Proceedings)
   8. *Journal of the Acoustical Society of America*
   9. *Journal of the Optimal Society of America*
  10. *Journal of Research* (National Bureau of Standards)
  11. *Journal of Scientific Instruments* (British)
  12. *Measurement Techniques* (USSR: English translation)
  13. *Metrologia*
  14. *Review of Scientific Instruments*
  15. *Transactions and Journal* (Britain: Inst. of Measurement and Control)
  16. *Technical News Bulletin* (National Bureau of Standards)

1.2 As a semester assignment, write a short synopsis of one article selected from each of the periodicals listed in Problem 1.1. The articles selected for review should be from an issue published within the past year.

1.3 A list of general references is appended to each of the following chapters. Select two books from these lists and write a short book review for each.

1.4 Ingenuity is often a more important tool in experimental development work than is a piece or system of advanced gear.* Ingenious use of simple equipment can often produce results not usually obtainable from simple use of sophisticated equipment. The following project is assigned with the idea of providing an opportunity to demonstrate ingenuity.

---

* For good examples of ingenious experimental solutions to scientific problems, often obtained with a minimum of resources, the reader is referred to a regular monthly column, "The Amateur Scientist," which appears in *The Scientific American*.

a) Purchase a box of ordinary paper drinking straws. Do not use the cellophane or plastic type.

b) Make simple, but scientifically proper, test setups (employ materials and facilities easily obtained, such as cardboard, wood, glue, cord, etc.) and measure the following properties of drinking straws:

    1. tensile load to cause failure,

    2. compressive load to cause failure as a column,

    3. moment causing failure in bending,

    4. torque to cause failure,

    5. pressure to cause bursting.

c) Write a report covering the above tests, including

    1. methods of testing, including photos or sketches,

    2. data as requested,

    3. sketches or photos of typical modes of failure,

    4. discussion of results and conclusions, along with claimed accuracies.

1.5 Design and construct a simple truss employing drinking straws of the type tested in the above project.

a) Through supporting calculations, predict the theoretical collapse load of the truss.

b) Experimentally determine the collapse load.

c) Write a report discussing your findings.

CHAPTER 2

THE GENERALIZED MEASUREMENT SYSTEM

2.1 WHAT IS MEASUREMENT?

Fundamentally, measurement is the act, or the result, of a quantitative comparison between a predefined standard and an unknown magnitude. If the result is to be generally meaningful, two requirements must be met in the act of measurement: (a) the standard which is used for comparison must be accurately defined and commonly accepted, and (b) the procedure and apparatus employed for obtaining the comparison must be provable.

The first requirement is that there be an accepted standard of comparison. A weight cannot simply be *heavy*. It can only be proportionately as heavy as something else, namely, our standard. A comparison must be made, and unless it is made relative to something generally recognized as standard, the measurement can only have limited meaning. This holds for any quantitative measurement we may wish to make. In general, the comparison is one of magnitude, and a *numerical* result is presupposed.

2.2 FUNDAMENTAL METHODS OF MEASUREMENT

There are two basic methods of measurement: (a) *direct comparison* with either a primary or a secondary standard, and (b) *indirect comparison* with a standard through the use of a calibrated system. Both methods are discussed in this article.

a) Direct comparison. How do you measure the length of a cold-rolled bar? You probably use a steel tape. You compare the bar's length with a *standard*. The bar is so many feet long, because that many units on your standard have the same length as the bar. You have determined this by making a *direct comparison*. Although you do not have access to the primary standard defining the unit, you manage very well with a secondary standard. The standard which you use could, no doubt, trace its ancestry directly back, through only three or four generations, to the primary length standard, the specified wavelength from krypton 86 (Article 4.3).

While to measure by direct comparison is to strip the measurement problem to its barest essentials, the method is not always the most accurate or indeed the best. The human senses are not equipped to make direct comparisons of all quantities with equal facility. In many cases they are not sensitive enough. We can make direct length comparisons using a steel rule with a preciseness of, say, 0.01 inch. Often we wish for greater accuracy, in which case we must call for additional assistance from some form of calibrated measuring system.

While we can do a reasonable job through direct comparisons of length, how well can we compare masses, for example? Our senses enable us to make rough comparisons. We can "heft" a pound of butter and compare the effect with that of some unknown mass. If the unknown is about the same weight, we may be able to say that it is slightly heavier or perhaps not quite as heavy as our "standard" pound, but we could never be certain that the two masses were the same, even say within one ounce. Our ability to make this comparison is not as good as it is for displacement.

In making most engineering measurements, therefore, we require the assistance of some form of measuring *system*, and measurement by *direct* comparision is less general than is measurement by *indirect* comparison.

b) Using a calibrated system. Indirect comparison makes use of some form of transducing device coupled to a chain of connecting apparatus, which we shall call, *in toto*, the *measuring system*. This chain of devices converts the basic form of input into an analogous form [2] which it then processes and presents at the output as a known function of the input. Such a conversion is often necessary in order to make the desired information intelligible. The human senses are simply not equipped to detect the strain in a machine member, for instance. Assistance is required from a system which senses, converts, and finally presents an analogous output in the form of a displacement on a scale or chart.

Processing of the analogous signal may take many forms. Often it is necessary to increase an amplitude or a power through some form of amplification. Or in another case it may be necessary to extract the desired information from a mass of extraneous input requiring filtering. Then again, a remote reading or recording may be needed, such as ground recording of a temperature or pressure in a missile in flight. This would most certainly require that the pressure or temperature be combined with a radio-frequency signal for transmission to the ground.

In each of the various cases requiring amplification, or filtering, or remote recording, etc., electrical methods suggest themselves. In fact, the majority of transducers in use, *particularly for dynamic mechanical measurements*, convert the mechanical input into an analogous electrical form for processing.

TABLE 2.1

THE THREE STAGES OF A GENERALIZED MEASUREMENT SYSTEM

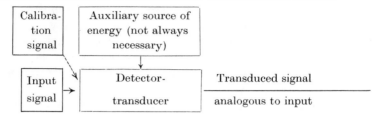

Typical mechanical input signals	Dimensions	Stage one Detector-transducer
Events or items	E	Senses desired input to the exclusion of all others and provides an analogous output.
Displacement	L	
Time	t	
Force	F	*Types and examples*
Temperature	T	
Frequency (events per unit time)	E/t	*Mechanical;* contacting spindle, spring mass, elastic devices (such as
Period (time per event)	t/E	Bourdon tube, proving ring, etc.),
Velocity	L/t	gyro.
Acceleration	L/t^2	
Mass	Ft^2/L	*Hydraulic-pneumatic;* buoyant float, orifice, venturi, vane, propeller.
Angle	L/L	
Unit strain	L/L	*Optical;* photographic film, photoelectric cell.
Stress	F/L^2	
Pressure	F/L^2	
Area	L^2	*Electrical;* contactor, resistance, capacitance, inductance, piezoelectric crystal, thermocouple, moving electrode, streaming potential.
Volume	L^3	
Moment	FL	
Torque	FL	
Volume flow	L^3/t	
Mass flow	Ft/L	

2.3 THE GENERALIZED SYSTEM

Most measurement systems fall within the framework of a generalized arrangement consisting of three phases or stages, as follows [3, 4]:

STAGE I A *detector-transducer stage.*

STAGE II An intermediate stage, which we shall call the *intermediate modifying stage.*

TABLE 2.1

THE THREE STAGES OF A GENERALIZED MEASUREMENT SYSTEM (*Continued*)

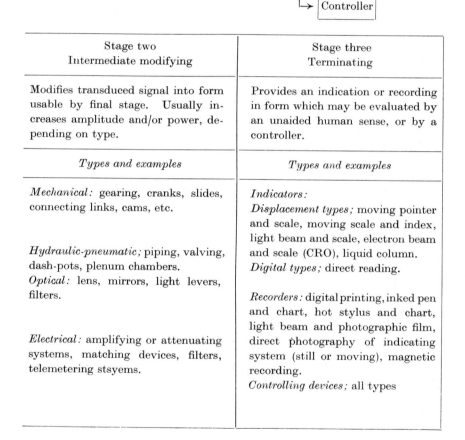

Stage two Intermediate modifying	Stage three Terminating
Modifies transduced signal into form usable by final stage. Usually increases amplitude and/or power, depending on type.	Provides an indication or recording in form which may be evaluated by an unaided human sense, or by a controller.
Types and examples	*Types and examples*
Mechanical: gearing, cranks, slides, connecting links, cams, etc. *Hydraulic-pneumatic;* piping, valving, dash-pots, plenum chambers. *Optical:* lens, mirrors, light levers, filters. *Electrical:* amplifying or attenuating systems, matching devices, filters, telemetering stsyems.	*Indicators:* *Displacement types;* moving pointer and scale, moving scale and index, light beam and scale, electron beam and scale (CRO), liquid column. *Digital types;* direct reading. *Recorders:* digital printing, inked pen and chart, hot stylus and chart, light beam and photographic film, direct photography of indicating system (still or moving), magnetic recording. *Controlling devices;* all types

STAGE III The *terminating stage*, consisting of one or a combination of the following: an indicator, a recorder, or some form of controller.

Each stage is made up of a distinct component or grouping of components which perform required and definite steps in the measurement. These may be termed *basic elements* [3], whose scope is determined by their functioning rather than their construction. Table 2.1 outlines the significance of each of these three stages.

a) First stage detector-transducer. The prime function of the first stage is to detect or to sense the input signal. At the same time, ideally it should be insensitive to every other possible input. For instance, if it is a pressure pickup, it should not be sensitive to, say, acceleration; if it is a strain gage, it should be insensitive to temperature; or if a linear accelerometer, it should be insensitive to angular acceleration, and so on. Unfortunately, it is very rare indeed to find a detecting device that is completely selective.

As an illustration of a very simple detector-transducer, let us consider the familiar *tire* gage used for checking automobile tire pressures. Such a device is shown in section in Fig. 2.1. It consists of a cylinder and piston, a spring resisting the piston movement, and a stem with scale divisions. As the air pressure bears against the piston, the resulting force compresses the spring until the spring and air forces are balanced. The calibrated stem, which remains in place after the spring returns the piston, indicates the applied pressure.

In this case the piston-cylinder combination along with the spring make up the detector-transducer. The piston and cylinder form one basic element, while the spring is another basic element. The piston-cylinder combination, serving as a force-summing device, senses the pressure effect, and the spring transduces it into displacement.

A function usually, though not always, required of the first stage is the changing or converting of the input signal into another type of quantity. Such a sensing device transduces, or transforms, or converts the basic input into an analogous variable. It changes the input from one alphabet to another. In recent years the analogous signal more often than not has been an electrical quantity. For another simple example, refer to the spring scale (Fig. 7.3). In this case force is transduced to displacement.

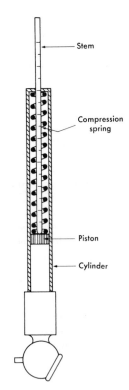

Fig. 2.1 Gage for measuring pressure in automobile tires.

b) Intermediate stage. The major purpose of the second stage of the generalized system is to modify the transduced information so that it is acceptable to the third, or terminating, stage. In addition, it may perform one or more basic operations, such as selective filtering, integration, differentiation, telemetering, etc., as may be required.

Probably the most common function of the second stage is to increase amplitude or power of the signal, or both, to the level required to drive the final terminating device. In addition, it must be designed for proper matching characteristics between first and second and between second and third stages.

c) *Third, or terminating, stage.* The third, or terminating, stage provides the information sought in a form comprehensible to one of the human senses or to a controller. If the output is intended for immediate human recognition, it is, with rare exception, presented in one of the following forms:

1. As a *relative displacement,* such as movement of an indicating hand, displacement of oscilloscope trace or oscillograph light beam, etc., or,

2. In *digital* form, as presented by a mechanical counter such as the familiar odometer portion of an automobile speedometer, or one of the modern digital voltmeters, etc.

As an example of a complete system, let us say that an acceleration is to be measured, as in Fig. 2.2. The *first-stage* device, the accelerometer, provides an analogous voltage. In addition to a voltage amplifier, the *second stage* may also include a filter which selectively attenuates unwanted high-frequency components. It may also integrate the analog signal with respect

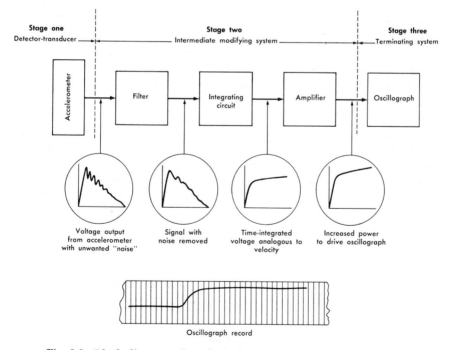

Fig. 2.2 Block diagram of a relatively complex measuring system.

to time, thereby providing a velocity-time relation rather than an acceleration-time signal. Finally, the signal power will probably be increased to the level necessary to drive the *third, or terminating, stage*, which may consist of a galvanometer-type oscillograph. The final record would then be in the form of a trace (a displacement-time plot) on photographic paper, and with proper calibration an intelligible velocity-time measurement should be the result.

2.4 CALIBRATION

Every measuring system must be *provable;* i.e., it must prove its ability to measure reliably. The procedure for this is called *calibration*. It consists of determining the system's scale. At some point during the preparation of the system for measurement, *known* magnitudes of the basic input quantity must be fed into the detector-transducer, and the system's behavior must be observed.

If the system has been proved linear, perhaps *single-point* calibration will suffice, wherein the effect of only a single value of the input is used. If the system is not linear, or if it has not been so proved, a number of values must be used and their results observed.

The input may be static or dynamic, depending on the application; however, quite often dynamic response must be based on static calibration, simply because a known dynamic source cannot be had. Naturally this is not optimum procedure; the more nearly the calibration standard corresponds to the unknown in all of its characteristics, the better the situation.

Once in a while the nature of the system or one of its components makes the introduction of a sample of the basic input quantity difficult or impossible. One of the important characteristics of the bonded resistance-type strain gage is the fact that through quality control at the time of manufacture, *spot* calibration may be applied to a complete lot of gages. As a result, an indirect calibration of a strain-measuring system may be provided through the gage factor supplied by the manufacturer. Instead of attempting to apply a known unit strain to the gage installed on the test structure, which if possible would often result in an ambiguous situation, a resistance change is substituted. Through the predetermined gage factor, the system's strain response may thereby be obtained (see Article 11.13).

2.5 RELATION BETWEEN SENSITIVITY AND PERIOD

Many measuring devices or system components involve elements constrained by gravity or spring force, whose deflection is analogous to the signal input. The ordinary balance scale is an example, as are the D'Arsonval meter movement and the mirror-type galvanometer (Article 8.9). The same is true of most pressure transducers, elastic-force transducers, and many other measuring devices.

If the system is a translational one, a spring-constrained mass may be involved. Our tire gage illustrates the case in which the piston and stem and a portion of the spring constitute the mass whose motion is controlled by the interaction of the applied pressure and the spring force. Other examples are the seismic-type instruments discussed in Chapter 16.

These devices depend on equilibrium* for correct indication. When equilibrium is disturbed by a change of input, the system requires time to readjust to the new equilibrium, and a number of oscillations may take place before the new output is correctly indicated. The rate at which the amplitude of such oscillations decrease is a function of system's damping. In addition, the frequency of oscillation is a function of both damping and sensitivity.

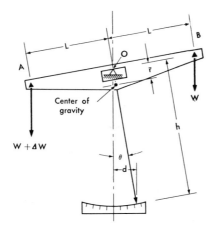

Fig. 2.3 Schematic of a beam balance.

Let us consider a symmetrical scale beam without damping (Fig. 2.3). For simplification, we will assume that the masses of the scale pans and the weights being compared are concentrated at points A and B, and that they are also included in the moment of inertia I, which is referred to the main pivot point O. Further, we will assume that a small difference, ΔW, exists between the two weights being compared, and that points A, B, and O lie along a straight line. We define sensitivity, η, as the ratio of the displacement of the end of the pointer to the length of the pointer, h, divided by ΔW, or

$$\eta = \frac{1}{\Delta W}\frac{d}{h} = \frac{1}{\Delta W}\tan\theta. \qquad (2.1)$$

---

* This assumes application of D'Alembert's principle to the accelerometer.

The system behaves like a compound pendulum, and the period of oscillation
is [5]

$$\mathscr{T} = 2\pi\sqrt{I/\bar{r}w_b},$$ (2.2)

where

$$I = \text{moment of inertia,}$$

$$\bar{r} = \text{distance between center of gravity}$$
$$\text{of the beam alone, and about pivot}$$
$$\text{point } O,$$

$$w_b = \text{weight of beam.}$$

With the weights applied,

$$w_b\bar{r}\sin\theta = \overline{\Delta W}L\cos\theta \quad\text{or}\quad \tan\theta = \frac{L\,\overline{\Delta W}}{w_b\bar{r}}.$$

Hence using Eq. (2.1), we find that

$$\eta = \frac{L}{w_b\bar{r}}.$$ (2.3)

Combining Eqs. (2.2) and (2.3), we have

$$\eta = \frac{L}{I}\left(\frac{\mathscr{T}}{2\pi}\right)^2 = \frac{L}{I}\left(\frac{1}{2\pi f}\right)^2,$$ (2.3a)

where f = natural frequency.

Equation (2.3a) indicates that the sensitivity is a function of $\mathscr{T}$, the
period of oscillation of the balance scale, with increased sensitivity corresponding to a long beam and low moment of inertia. Putting it another way,
the more sensitive instrument oscillates more slowly than the less sensitive
instrument. This is an important observation having significant bearing
on the dynamic response of most single-degree-of-freedom instruments.

2.6 IMPORTANCE OF DAMPING

Another factor having an important bearing on the usefulness of any general
instrument of this type is damping. Damping in this connection is usually
thought of as viscous rather than as Coulomb or frictional damping, and may
be obtained by use of fluids or by electrical means.

Viscous damping is a velocity function, and the force opposing the
motion may be expressed as

$$F = -c\frac{dS}{dt},$$

where

$$c = \text{damping coefficient,} \quad \frac{dS}{dt} = \text{velocity.}$$

The negative sign indicates that the resulting force opposes the velocity.

The effect of viscous damping on a free vibration is to reduce the vibrational amplitudes with respect to time according to a logarithmic relation.

Damping magnitude is conveniently thought of in terms of critical damping, which is the minimum damping that can be used to just prevent overshoot when a spring-mass system is deflected from equilibrium and then released. This limiting condition is illustrated in Fig. 2.4. The value of the critical damping coefficient c_c, for a simple spring-supported mass m, is expressed by the relation [6]

$$c_c = 2\sqrt{mk}, \qquad (2.4)$$

in which k is the spring constant.

The importance of proper damping to dynamic measurement may be understood by assuming that our scale beam in the previous example is a part of an instrument which is required to come to *different* equilibriums as rapidly as possible. A situation of this sort exists in the application of light-beam galvanometers to recording oscillographs. The galvanometer suspension is driven by varying frequency inputs, and its ability to follow is governed by its natural period and by damping.

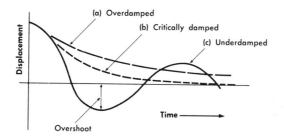

Fig. 2.4 Time-displacement relations for damped motion (a) for damping greater than critical, (b) for critical damping, (c) for damping less than critical.

Suppose that our scale beam has very low damping. When a disturbing force is applied, the scale will be caused to oscillate, and the oscillation will continue for a long period of time. A final balance will be obtained only by prolonged waiting, which limits the frequency with which the weighing process may be repeated.

On the other hand, suppose considerable damping is provided—well above critical. An extreme example of this would be to submerge the entire scale in a container of molasses. Balance would be approached at a very slow rate again, but in this case there would be no oscillation. Here, again, excessive time would be required before the next weighing operation could commence. Theoretically, viscous damping does not change the inherent sensitivity of the device; however, sensitivity and natural period are related, as discussed in the previous article.

It appears, therefore, that if we were to design a beam-type scale for quickly determining magnitudes of different masses, the final form would necessarily be a result of compromise. We would like equilibrium to be reached as quickly as possible in order to *get on with the job,* but in addition we would require a certain maximum sensitivity from our instrument, which is one of the factors determining accuracy.

It would seem that there might be an optimum damping value that should be used. Although this is not exactly the case, because of other factors involved, damping in the order of 60 to 75% of critical is provided in most instruments of this type (see Article 2.9b). In addition, it would appear that sensitivities greater than required by the application should be avoided because sensitivity is gained at the expense of frequency response.

2.7 DYNAMIC CHARACTERISTICS OF SIMPLIFIED MEASURING SYSTEMS

In addition to those characteristics made explicit by basic definitions, mechanical-input quantities may also have distinctive time-amplitude properties which may be classified as follows:

I. Static (unchanging with time)

II. Dynamic (varying with time)

    A. Steady-state periodic

        1. Simple harmonic (e.g., 60 Hz line voltage)

        2. Complex (e.g., many mechanical-vibration amplitudes)

    B. Transient

        1. Step (e.g., a sudden temperature change)

        2. Complex (e.g., stress variation caused by a single mechanical impact)

By making certain simplifying assumptions, the dynamic characteristics of *most* measuring systems may be placed in one or the other of two basic categories: first-order systems and second-order systems. The assumptions are that any restoring elements (springs, etc.) are linear, that damping is viscous, and that the system is single degree. Although these assumptions are seldom completely realized, a study of the performance of such idealized systems can be profitable.

It is the purpose of the next several articles to consider the performance of first- and second-order systems when excited by simple dynamic inputs.

2.8 CHARACTERISTICS OF FIRST-ORDER SYSTEMS

Figure 2.5 shows a model of a first-order system consisting of a spring and dashpot in parallel, along with a necessary connecting element AB, which is acted on by a force that is a function of time. The feature that distinguishes first- from second-order systems is the condition that for the former, mass is ignored, whereas for the latter, it is not. It will be seen later that there are other situations in mechanical measurements, which do not involve these same mechanical components but which have analogous behavior in terms of time.

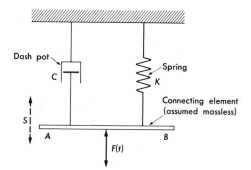

Fig 2.5 Mechanical model of a first-order system.

Although many different forcing functions $F(t)$ could be applied, we will consider only two: a stepped force and a harmonically varying force.

a) The step-forced first-order system. Let

$$F(t) = 0, \qquad \text{for} \qquad t < 0$$

and

$$F(t) = F_0, \qquad \text{for} \qquad t \geq 0.$$

For force equilibrium on the connecting element (which is assumed to be massless),

$$c \frac{ds}{dt} + ks = F_0, \tag{2.5}$$

where

$t = $ time, sec,

$s = $ displacement, in.,

$c = $ damping coefficient, lb · sec/in.,

$k = $ deflection constant, lb /in.,

$F_0 = $ amplitude of the constant input force, lb.

Then

$$\int_0^t dt = c \int_{s_A}^s \frac{ds}{(F_0 - ks)},$$

from which we obtain

$$\frac{F_0 - ks}{F_0 - ks_A} = e^{-kt/c} = e^{-t/\tau}. \tag{2.6}$$

The units of $\tau = c/k$ are seconds, and this quantity is known as the *time constant*.

Equation (2.6) may be written

$$s = s_\infty[1 - e^{-t/\tau}] + s_A e^{-t/\tau} = s_\infty + [s_A - s_\infty]e^{-t/\tau}, \tag{2.7}$$

where

$s_\infty = F_0/k =$ the limiting displacement of the system as $t \to \infty$,

$s_A =$ any initial displacement at $t = 0$.

We have assumed that the first-order system represents any dynamic condition wherein the elements are essentially massless, the displacement constraint is linear, and a significant viscous rate constraint is present. More generally, Eq. (2.7) may be written

$$P = P_\infty[1 - e^{-t/\tau}] + P_A e^{-t/\tau}$$

or

$$P = P_\infty + [P_A - P_\infty]e^{-t/\tau}, \tag{2.8}$$

where

$P =$ the magnitude of any first-order process at $t = t$,

$P_\infty =$ the limiting magnitude of the process as $t \to \infty$,

$P_A =$ initial magnitude of the process at $t = 0$.

Although the basic relationship was derived in terms of a spring-dashpot arrangement, other processes which behave in an analogous manner include (a) a heated (or cooled) bulk or mass, such as a temperature sensor subjected to a step-temperature change (Article 15.10c), (b) simple capacitive-resistive or inductive-resistive circuits (Article 7.25a), and (c) the decay of a radio-active source with time (Articles 18.3 and 18.4).

Figure 2.6 represents two different process-time conditions for the step-excited first-order system: (a) for a progressive process, wherein the action is an increasing function of time, and (b) the *decaying* process, wherein the magnitude decreases with time.

Significance of the time constant, τ. If we substitute the magnitude of one time constant for t in Eq. (2.8), we obtain

$$P = P_\infty + (P_A - P_\infty)(0.368),$$

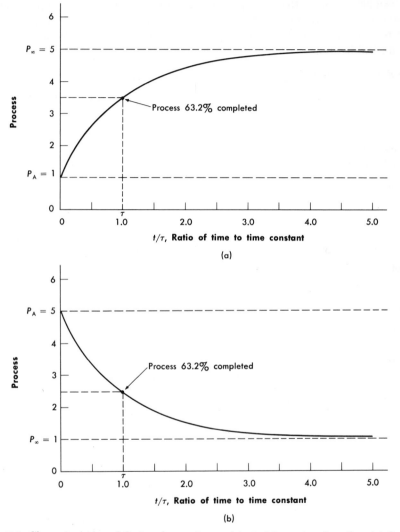

Fig. 2.6 Characteristics of first-order system subjected to a step function (a) for a progressive process, (b) for a decaying process.

from which we see that $(1 - 0.368)$, or 63.2%, of the dynamic portion of the process will have been completed. Two time constants yield 86.5%; three yield 95.0%; four yield 98.2%; and so on. These percentages of completed processes are important because they will always be the same regardless of the process, provided that the process is governed by the conditions of the step-excited first-order system. It is often assumed that a process is completed during a period of *five* time constants.

Example. Assume that the application of a temperature probe approximates first-order conditions,* that the probe has a time-constant of 6 sec, and that it is suddenly subjected to a temperature step of 75–300°F. What temperature will be indicated 10 sec after the process has been initiated?

Solution. Applying Eq. (2.8), we find that

$$P_\infty = 300°F, \qquad P_A = 75°F, \qquad t = 10 \text{ sec},$$

$$P = 300 + (75 - 300)e^{-10/6} \qquad = 257°F.$$

Example. Assume the same conditions as for the example above, except that the step is 300–75°F. Find the indicated temperature after 10 sec.

Solution.

$$P_\infty = 75°F, \qquad P_A = 300°F, \qquad t = 10 \text{ sec},$$

$$P = 75 + (300 - 75)e^{-10/6} \qquad = 117°F.$$

b) The harmonically excited first-order system. Again referring to Fig. 2.5, let us now consider the case for

$$F(t) = F_0 \cos \Omega t,$$

or

$$c \frac{ds}{dt} + ks = F_0 \cos \Omega t, \tag{2.9}$$

where

$F_0 = $ the amplitude of the forcing function, lb,

$\Omega = $ the circular frequency of the forcing function, rad/sec.

The solution of Eq. (2.9) yields

$$s = A_1 e^{-t/\tau} + \frac{F_0/k}{\sqrt{1 + (\tau\Omega)^2}} \cos (\Omega t - \phi), \tag{2.10}$$

where

$A_1 = $ constant whose value depends on the initial conditions,

$\tau = $ time constant $= \dfrac{c}{k}$,

$\phi = $ phase lag $= \tan^{-1} \dfrac{\Omega c}{k} = \tan^{-1} \dfrac{2\pi}{\mathscr{T}} \tau$, \tag{2.11}

$\mathscr{T} = \dfrac{2\pi}{\Omega} = $ period of excitation cycle, sec.

---

* This represents a first approximation for many practical temperature sensors. For further discussion of this matter, see Article 15.10c.

It is seen that the first term on the right side of Eq. (2.10), the complementary function, is *transient* and that after a period of several time constants, becomes very small. The second term is the *steady-state* relationship and, except for the short initial period, we may write

$$s = \frac{F_0/k}{\sqrt{1 + (\tau\Omega)^2}} \cos(\Omega t - \phi) \qquad (2.12)$$

or

$$\frac{s}{s_s} = \frac{\cos(\Omega t - \phi)}{\sqrt{1 + (\tau\Omega)^2}},$$

and

$$\frac{s_d}{s_s} = \frac{1}{\sqrt{1 + (\tau\Omega)^2}},$$

$$= \frac{1}{\sqrt{1 + (2\pi\tau/\mathscr{T})^2}}, \qquad (2.13)$$

where

s_d = amplitude of the periodic dynamic displacement, in.,

and

$$s_s = \frac{F_0}{k}.$$

The quantity s_s is the static deflection which would occur, should the force amplitude F_0 be applied as a *static* force. The ratio s_d/s_s is often termed the *amplification ratio*. For analogous situations, Eq. (2.13) may be written

$$\frac{P_d}{P_s} = \frac{1}{\sqrt{1 + (2\pi\tau/\mathscr{T})^2}}, \qquad (2.13a)$$

where P represents the magnitude of the applicable process.

Figures 2.7 and 2.8 illustrate the relationships of the phase angle and the magnification ratio described by Eqs. (2.11) and (2.13a), respectively.

To a great extent the harmonically excited first-order system is of academic interest only. In most situations involving motion of bodies, a moving mass exists and cannot be ignored. When this is the case, the problem becomes one of second order and will be discussed in subsequent paragraphs.

Although harmonically excited processes that do not involve moving mass are rare, for the purpose of an example we may assume a condition wherein a temperature probe is subjected to a harmonically varying input. Let the time constant of the probe be 10 sec. Assume that the probe is inserted into an environment for which the temperature varies harmonically between 75°F and 300°F, with a period of 20 sec. Describe the temperature readout in terms of input.

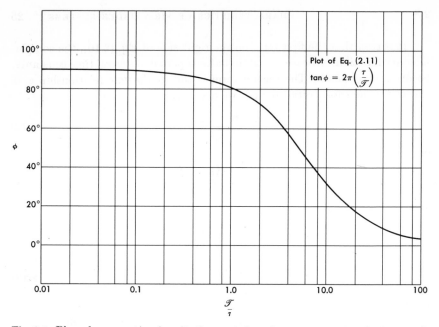

Fig. 2.7 Phase lag vs. ratio of excitation period to time constant for the harmonically excited first-order system.

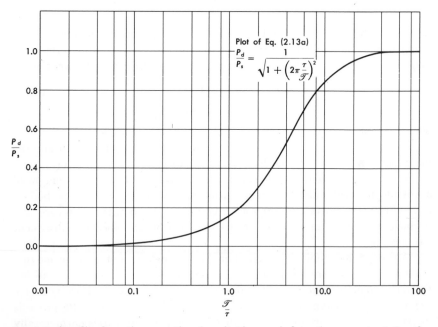

Fig. 2.8 Amplitude ratio vs. ratio of excitation period to time constant for the harmonically excited first-order system.

In this case the temperature input may be expressed as

$$T(t) = \left(\frac{300 + 75}{2}\right) + \left(\frac{300 - 75}{2}\right) \cos\left(\frac{2\pi}{20}\right)t$$

$$= 187.5 + 112.5 \cos\left(\frac{2\pi}{20}\right)t.$$

From Eq. (2.13a), we find that

$$\frac{T_d}{T_s} = \frac{1}{\sqrt{1 + \left(\dfrac{2\pi}{20} \times 10\right)^2}},$$

$$T_d = \frac{112.5}{3.3} = 34°F.$$

From Eq. (2.11), we find that the phase lag is

$$\phi = \tan^{-1} 2\pi(\tau/\mathscr{T}) = \tan^{-1}\left(\frac{2\pi \times 10}{20}\right) = \tan^{-1} \pi = 72\tfrac{1}{2}° \text{ (Angle)}$$

or

$$\text{time lag} = \frac{72\tfrac{1}{2}}{360} \times 20 = 4 \text{ sec.}$$

A graphical representation of the situation is shown in Fig. 2.9. For further discussion of the response of temperature-measuring probes, see Article 15.10c.

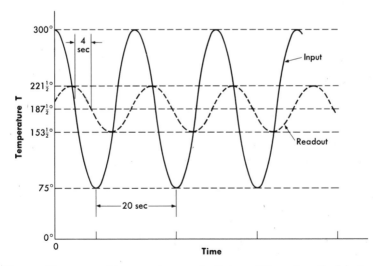

Fig. 2.9 Response of temperature probe for conditions described in text.

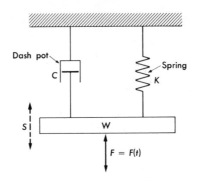

Fig. 2.10 Mechanical model of force-excited second-order system.

2.9 CHARACTERISTICS OF SECOND-ORDER SYSTEMS

Figure 2.10 illustrates the essentials of a second-order system. This arrange-ment approximates many actual mechanical arrangements including simple weighing systems, such as elastic-type load cells supporting mass, D'Arsonval meter movements, including the ordinary galvanometers, and many force-excited mechanical-vibration systems.

As with the first-order system, many excitation modes are possible, ranging from the simple step-function to a whole multitude of more complex possibilities. We shall investigate the step and the harmonic forms. These two forms approximate many actual situations, and inasmuch as all periodic inputs may be reduced to combinations of simple harmonic components (Article 3.4), the latter can give us an insight into system performance when subject to most forms of dynamic input.

a) Step-excited second-order system. Referring to Fig. 2.10, we let

$$F = 0, \quad \text{when} \quad t < 0$$

and

$$F = F_0, \quad \text{when} \quad t \geq 0.$$

Application of Newton's second law yields

$$\frac{d^2s}{dt^2} + \frac{c}{m}\frac{ds}{dt} + \frac{k}{m}s = \frac{F}{m}. \tag{2.14}$$

If we assume *under-damping*, that is, $(kg/W) > (cg/2W)^2$, the general solution for Eq. (2.14) may be written as follows:

$$s = e^{(-cg/2W)t}[A \cos \omega_{nd}t + B \sin \omega_{nd}t] + F/k, \tag{2.15}$$

where A and B are constants governed by boundary conditions, and

$$\omega_{nd} = \text{damped natural frequency}$$
$$= \sqrt{(kg/W) - (cg/2W)^2}. \tag{2.15a}$$

Note that the exponential multiplier may be written $e^{-t/\tau}$, where the time constant $\tau = 2W/cg$. If we let $s = 0$ and $v = 0$ at $t = 0$, and evaluate A and B, then by rearrangement and substitution of terms, Eq. (2.15) may be put into the following form:

$$\frac{s_d}{s_s} = \left\{ 1 - e^{-(c/c_c)\omega_n t}\left[\frac{c/c_c}{\sqrt{1 - (c/c_c)^2}} \sin \sqrt{1 - (c/c_c)^2}\,\omega_n t - \cos\sqrt{1 - (c/c_c)^2}\,\omega_n t \right] \right\},$$

(2.16)

where

$$\omega_n = \sqrt{kg/W} = \text{natural frequency, rad/sec,}$$

$$c_c = \text{critical damping coefficient, lb-sec/in.}$$

Here again we may write

$$\frac{P_d}{P_s} = \frac{S_d}{S_s},$$

where P represents the magnitude of any applicable process.

For the overdamped condition, $c/c_c > 1$, the solution of Eq. (2.14) may be written as follows:

$$\frac{s_d}{s_s} = \left[\frac{-c/c_c - \sqrt{(c/c_c)^2 - 1}}{2\sqrt{(c/c_c)^2 - 1}} e^{[-c/c_c + \sqrt{(c/c_c)^2 - 1}]\omega_n t} \right.$$

$$\left. + \frac{c/c_c - \sqrt{(c/c_c)^2 - 1}}{2\sqrt{(c/c_c)^2 - 1}} e^{[-c/c_c - \sqrt{(c/c_c)^2 - 1}]\omega_n t} + 1 \right]. \qquad (2.17)$$

Figure 2.11 shows the plots for Eqs. (2.16) and (2.17) for various damping ratios.

b) *Harmonically excited second-order systems.* Referring to Fig. 2.10, when

$$F(t) = F_0 \cos \Omega t,$$

we may write

$$\frac{W}{g}\frac{d^2s}{dt^2} + c\frac{ds}{dt} + ks = F_0 \cos \Omega t, \qquad (2.18)$$

for which the solution becomes

$$s = e^{-(cg/2W)t}[A \cos \omega_{nd}t + B \sin \omega_{nd}t]$$

$$+ \frac{(F_0/k)\cos(\Omega t - \phi)}{\sqrt{[1 - (W\Omega^2/kg)]^2 + (c\Omega/k)^2}} \qquad (2.19a)$$

$$= e^{-t/\tau}[A \cos\sqrt{1 - (c/c_c)^2}\,\omega_n t + B \sin\sqrt{1 - (c/c_c)^2}\,\omega_n t]$$

$$+ \frac{s_s \cos(\Omega t - \phi)}{\sqrt{[1 - (\Omega/\omega_n)^2]^2 + [2(c/c_c)(\Omega/\omega_n)]^2}}, \qquad (2.19b)$$

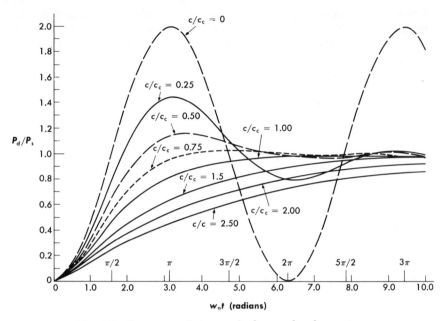

Fig. 2.11 Response of step-excited second-order system.

where A and B are constants which depend on particular boundary conditions, and

$$\Omega = \text{frequency of excitation, rad/sec,}$$

$$\phi = \tan^{-1}\left[\frac{2(c/c_c)(\Omega/\omega_n)}{1 - (\Omega/\omega_n)^2}\right] = \text{phase angle.} \tag{2.20}$$

It is seen that the first term on the right side of Eq. (2.19b) is transient and after several time constants will disappear. The second term is the steady-state relationship, for which we may write

$$\frac{s_d}{s_s} = \frac{P_d}{P_s} = \frac{1}{\sqrt{[1 - (\Omega/\omega_n)^2]^2 + [2(c/c_c)(\Omega/\omega_n)]^2}}$$

$$= \text{the magnification ratio.} \tag{2.21}$$

Figures 2.12 and 2.13 are plots of Eqs. (2.21) and (2.20), respectively, for various values of the damping ratio. The ratio s_d/s_s is a measure of the system response to the frequency of input. Normally, it is desired that this relation be constant with frequency; i.e., we would like the system to be insensitive to changes in the frequency of input $F(t)$. Inspection of Fig. 2.12 shows that the amplitude ratio is reasonably constant for only a limited frequency range and then only for certain damping ratios. We see that for a

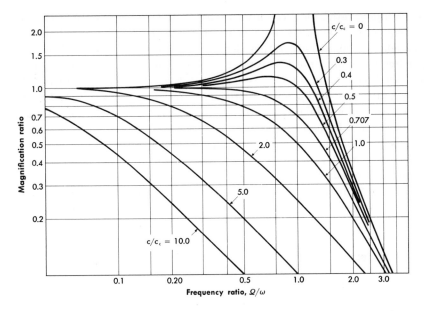

Fig. 2.12 Plot of Eq. (2.21) illustrating the frequency response to harmonic excitation of the system shown in Fig. 2.9.

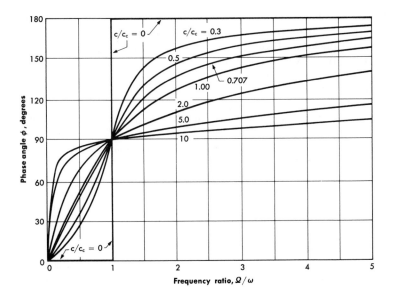

Fig. 2.13 Plot of Eq. (2.20) illustrating the phase response of the system shown in Fig. 2.9.

given damping ratio, ideal conditions $(s_d/s_s = 1)$ may occur only at one or two frequencies. If the system is to be employed for general dynamic measurement applications, rather definite damping must be used and an upper frequency limit must be established. Practically, if a damping ratio in the neighborhood of 65–75% is employed, then the amplitude ratio will approximate unity over a range of frequency ratios of about 0–40%. Even for these conditions, inherent error $(1 - s_d/s_s)$ exists, and a usable system can be had only through compromise.

Inspection of Fig. 2.13 indicates that damping ratios in the order of 65–75% of critical provide an approximately linear phase shift for the frequency-ratio range of 0–40%. This is desirable if a proper time relationship is to be maintained between the harmonic components of a complex input (see Articles 3.4, 3.8, 16.6, and 16.7 for further discussion of phase relationships).

From this discussion it is apparent that if simple measurements are to be made as rapidly as possible, or more importantly, if the input signal is continuous and complex, rather definite limitations are imposed. For a more extensive discussion of system dynamics, the reader is referred to Chapter 3 and to references listed below.

SUGGESTED READINGS

Cook, N. H., and E. Rabinowicz, *Physical Measurement and Analysis.* Reading, Mass.: Addison-Wesley Publishing Co., 1963.

Doeblin, E. O., *Measurement Systems; Application and Design.* New York: McGraw-Hill Book Co., 1966.

Draper, C. S., W. McKay, and S. Lees, *Instrument Engineering.* 3 vols. New York: McGraw-Hill Book Co., Inc., 1952.

Haberman, C. M., *Engineering Systems Analysis.* Columbus, Ohio: Charles E. Merrill Books, Inc., 1965.

Pearson, E. B., *Technology of Instrumentation.* Princeton, N.J.: D. Van Nostrand Co., 1957.

Reswick, J. B., and C. K. Taft, *Introduction to Dynamic Systems.* Englewood Cliffs, N.J.: Prentice-Hall, Inc. 1967.

Shearer, J. L., A. T. Murphy, and H. H. Richardson, *Introduction to System Dynamics.* Reading, Mass.: Addison-Wesley Publishing Co., 1967.

PROBLEMS

2.1 Review current literature (see the list of periodicals at the end of Chapter 1 for possible sources) and prepare a short resume of some article or paper pertinent to the subjects discussed in this chapter. Be prepared to make an oral presentation of your report.

2.2 Set up and conduct tests to evaluate your ability to estimate, with use of nothing more than your judgment and experience, the magnitudes of the common quantities listed below. You may also wish to determine some way of measuring improvement with practice.

a) Linear distances within specified ranges
b) Weight of small objects of various densities
c) Time intervals, either predetermined intervals or the elapsed time corresponding to some event
d) Frequency of a pure sound
e) Velocity

What limits of accuracy in each category do you think might be developed with practice?

2.3 Each of the following devices include the three basic stages of the generalized measuring system. Make a general analysis by breaking each system into its fundamental stages.

a) Bourdon pressure gage
b) Aneroid barometer
c) Mercury-column barometer
d) Automobile speedometer
e) Pendulum clock
f) Tuning-fork watch [14]
g) Dial indicator (Fig. 6.1)
h) Pendulum scale (Fig. 12.5)
i) Dimensional comparator (Fig. 10.6)
j) Tuckerman strain gage (Fig. 11.4)
k) Pressure thermometer (Fig. 15.2)
l) Engine indicator (Fig. 7.1)
m) Temperature-measuring system (Fig. 15.15)
n) Pyrometer (Fig. 15.22)

2.4 The diameter of an automobile tire may vary depending on several factors such as (a) wear, (b) inflation, and (c) type of tire, e.g., "ordinary" or "winter" tread. Investigate the error (see Articles 5.2 and 5.3) that may be introduced in the calibration of both the speedometer and the odometer by this variation. If you were responsible for the original design of the speed- and distance-measuring systems, what would you employ as the "standard" tire diameter? List reasons for your choice.

2.5 Obtain a laboratory chemical-balance scale and determine its free unloaded period. Displace the rider or apply a small known weight to one of the pans and determine the sensitivity. Apply weights to the pans, balance the scale, and determine the period. What conclusions do you draw from the latter experiment?

2.6 A process behaves according to a decreasing exponential form with respect to time. If the time constant for the process is 1.2 sec, what times, measured from initiation of the process, will be required for 30%, 50%, and 80% completion of the process? [*Note:* If 30% of the process has been completed, then 70% remains and the magnitude of the process becomes

$$P = P_\infty + 0.70(P_A - P_\infty).$$

2.7 A process behaves according to an increasing exponential form. If the time constant for the process is 0.128 sec, what times, measured from initial conditions, will be required for 25%, 50%, and 75% completion of the process?

2.8 A given output increases from zero in an exponential manner so that it reaches 20% of its maximum value in 2 sec.
   a) What is the time constant of the function?
   b) Approximately how long will it take to reach "maximum" value?

2.9 Suppose that a temperature probe exhibiting single-time constant behavior and having a time constant of $3\frac{1}{2}$ sec is quickly taken from water at 32°F and plunged into water at 212°F. What temperature should be indicated $1\frac{1}{2}$ sec after the process is initiated?

2.10 A thermometer behaving in accordance with single-time constant theory has a time constant of 2.5 secs. If it is subjected to a step temperature change from 212 to 32°F, what temperature will be indicated at 1, $2\frac{1}{2}$, and 5 sec?

2.11 A temperature-sensitive transducer is subjected to a sudden temperature change. It takes 10.0 sec for the transducer to reach equilibrium (five time constants). How long, in seconds, will it take for the transducer to read half of the temperature difference?

2.12 The time constant for a first-order process is 0.35 sec. If the process is subjected to a step input from 65 units to 240 units:
   a) What value will be indicated after one time constant?
   b) After two time constants?
   c) After three time constants?
   d) After two seconds?

2.13 What are the values for Problem 2.12 if the step is from 240 units to 65 units?

2.14 What is the time constant of an RC circuit,
   a) if $R = 10$ megohms and $C = 0.5$ mfd.?
   b) if $R = 500$ ohms and $C = 0.5$ mfd.? (See Article 7.25a.)

2.15 What is the time constant of an RL circuit
   a) if $R = 100k$ ohms and $L = 75$ mh?
   b) if $R = 2$ megohms and $L = 3$ henrys? (See Article 7.25a.)

2.16 What is the time constant for a single-degree-of-freedom viscous-damped system for which $W = 5$ lb and $c = 1.2$ lb-sec/in.? (See Eq. [2.19].)

2.17 Radioactive decay is normally given in terms of "half-life." Helium 6 has a half-life of 0.85 sec. What is the corresponding time constant?

2.18 What is the "half-life" for
   a) the thermometer of Problem 2.10,
   b) the RC circuit of Problem 2.14a,
   c) the RL circuit of Problem 2.15a?

2.19 If a first order system is subjected to a step function wherein the following time-magnitude data is determined, what is the value of the time constant for the following

$$t = 0, \qquad P_A = 10 \text{ units},$$

$$t = 5 \text{ sec}, \qquad P = 20 \text{ units},$$

$$t \to \infty, \qquad P \approx P_0 = 40 \text{ units}.$$

2.20 Arrange two open containers side by side. Maintain water at the boiling point in one container and ice water in the other. Using a mercury-in-glass thermometer, first permit the thermometer reading to come to equilibrium in the boiling water; then quickly withdraw it and plunge it into the ice water. By repeated manipulation and by use of a stopwatch, obtain data on the times required to reach various predetermined temperatures in the 32–212° range. Plot temperature versus time and investigate the characteristics of the system. Do they approximate those of a first-order system?

It is probable that deviations from single-order system response will be noted. Refer to Article 15.10c for a possible explanation.

Write a laboratory note on your findings.

2.21 Repeat Problem 2.20, but reverse the step, i.e., allow the thermometer to come to equilibrium in the ice water and then plunge it into the boiling water.

2.22 Repeat Problem 2.20, employing a bimetal-type thermometer instead of the mercury-in-glass type. A greater deviation from the simple first-order system should be noted in this case.

2.23 Repeat Problem 2.20, employing an unshielded thermocouple connected to a suitable recorder such as a stylus-type oscillograph.

2.24 Assemble an RC circuit (See Article 7.25a), and by means of an oscilloscope and camera, obtain a voltage-time trace for
  a) the "charging" process, and
  b) the "discharging" process.
     Do the traces adhere to simple first-order theory?

2.25 Repeat Problem 2.24, employing the RL circuit.

2.26 Derive an expression for the velocity of a sphere falling through a viscous fluid with gravity as the motivating force. Neglect buoyancy. *Hint:* We may write

$$w - c\dot{s} = (w/g)\ddot{s},$$

where

$$c\dot{s} = \text{viscous force opposing the motion},$$

$$w = \text{weight of the sphere}.$$

The solution will involve a time constant. What is its value? Write an expression for the terminal velocity.

2.27 Solve Problem 2.26, but include the effect of buoyancy.

2.28 Show that a phase lag greater than 25% of the exciting period is impossible for a harmonically excited first-order system.

2.29 Many dynamic systems of the second-order type are designed with a damping ratio in the range of 60–70% of critical. Why is this so?

2.30 A force transducer behaves like a simple second-order system. If the natural frequency of the transducer is 800 Hz and its damping is 40% of critical, what will be the inherent error in amplitude for an harmonic input signal of 400 Hz.? Calculate the phase lag.

2.31 A pressure transducer employs a diaphragm as a pressure-summing device. In application, the diaphragm and fluid behave as a second-order single-degree system. The static displacement is proportional to the applied force (pressure). Given that the natural frequency of the system is 1200 Hz and the total viscous damping is 60% of critical, determine the frequency range(s) over which the ratio of dynamic amplitude to static amplitude (inherent error) deviates from unity by an amount no greater than 5%.

2.32 Forcing functions $F(t)$ of many different forms may be applied to first and second-order systems. Two are discussed in Article 2.8 and 2.9. Others could include functions linearly dependent on time (the ramp function), exponential (e.g., parabolic, logarithmic), and many more complex forms.

For the ramp-type forcing function, $F(t) = Ct$,

$$\tau \dot{s} - s = Ct/k, \qquad (2.22)$$

where C is a constant.

A solution is

$$\frac{sk}{c\tau} = \frac{t}{\tau} - (1 - e^{-t/\tau}). \qquad (2.23)$$

For this case, plot the dimensionless output, $sk/c\tau$ versus t/τ, and show that the lag between the output and the ideal, "no-lag" output approaches unity as t/τ becomes large.

CHARACTERISTICS OF DYNAMIC SIGNALS

3.1 TYPES OF INPUT QUANTITIES

Mechanical quantities, in addition to their inherent defining characteristics, also have distinctive time-amplitude properties, which may be classified as follows:

    I. Static

    II. Dynamic

        A. Steady-state periodic

        B. Transient

Of course, a static, nonchanging magnitude is the most easily measured. If the system is terminated by some form of meter indicator, the meter pointer has no difficulty in eventually reaching a definite indication. The rapidly changing input presents a considerably different problem, however.

There are two general forms of dynamic input, steady-state periodic and transient. The steady-state periodic quantity is one whose magnitude has a definite repeating time cycle, whereas the time variation of a transient magnitude does not repeat. Sixty-cycle line voltage is an example of a steady-state periodic signal. So also are many mechanical vibrations, after a balance has been reached between a constant input exciting energy and energy dissipated by damping.

An example of a transient quantity would be an acceleration-time relation accompanying an isolated mechanical impact. The magnitude is temporary, sometimes being completed in a matter of milliseconds, with the portions of interest perhaps existing for only a few microseconds. Extremely high rates of change, or wavefronts, exist, 'placing severe demands on the measuring system.

Dynamic inputs require system components which are sufficiently fast-acting to faithfully follow the input. The term *response* is often used in this connection and will be discussed in a later section.

3.2 SIMPLE HARMONIC RELATIONS

A function is said to be simple harmonic in terms of a variable when its second derivative is proportional to the function but of opposite sign. More often than not, the independent variable is time t, although any two variables may be related harmonically.

One of the most common harmonic functions in mechanical engineering is one relating displacement and time. In electrical engineering many of the variable quantities in a-c circuitry are harmonic functions of time. The relation is quite basic to dynamic functions, and most quantities which are time functions may be expressed harmonically.

In its most elementary form, *simple harmonic motion* is defined by the relation

$$s = s_0 \sin \omega t, \tag{3.1}$$

in which

$s = $ instantaneous displacement from equilibrium, in.,*

$s_0 = $ amplitude, or maximum displacement from equilibrium, in.,

$\omega = $ circular frequency, rad/sec,

$t = $ any time interval measured from the instant when $s = 0$, sec.

Pendulum motions of small amplitude, a mass on a beam, a weight suspended by a rubber gum band—all vibrate with simple harmonic motion, or very nearly so.

By differentiation, the following relations may be derived from Eq. (3.1):

$$v = \frac{ds}{dt} = s_0 \omega \cos \omega t \tag{3.2}$$

and

$$v_0 = s_0 \omega. \tag{3.2a}$$

Also,

$$a = \frac{dv}{dt} = -s_0 \omega^2 \sin \omega t \tag{3.3}$$

$$= -s \omega^2. \tag{3.3a}$$

In addition,

$$a_0 = -s_0 \omega^2. \tag{3.3b}$$

In the above equations

$v = $ velocity, in./sec,

$v_0 = $ maximum velocity or velocity amplitude, in./sec,

$a = $ acceleration, in./sec^2,

$a_0 = $ maximum acceleration or acceleration amplitude, in./sec^2.

---

* The more common units are indicated. However, other units appropriate to the relations may be used provided they are consistent.

Equation (3.3a) satisfies our word description of simple harmonic motion expressed in the first paragraph of this article: the acceleration a is proportional to the displacement s, but is of opposite sign. The proportionality factor is ω^2.

3.3 THE SIGNIFICANCE OF CIRCULAR FREQUENCY

The idea of *circular frequency* ω, as used above, is very useful in studying cyclic relations. Even if the student is completely familiar with its use, the mechanical analogy in the form of the well-known *Scotch-yoke* mechanism may be a helpful adjunct to his thinking.

Figure 3.1(a) shows the elements of the Scotch yoke, consisting of a crank, OA, with a slider-block driving the yoke-piston combination. If we measure the piston displacement from its mid-stroke position, the displacement amplitude will be $\pm OA$. If the crank turns at ω radians per second, then the crank angle θ may be written as ωt. This, of course, is convenient because it introduces time t into the relationship, which is not directly apparent in the term θ. Piston displacement may now be written as

$$s = s_0 \sin \omega t.$$

[This is the same as Eq. (3.1).] One cycle takes place when the crank turns

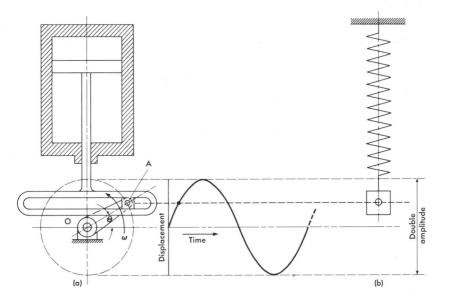

(a)   (b)

Fig. 3.1 (a) The Scotch-yoke mechanism, which provides a simple harmonic motion to the piston. (b) A simple spring-mass system which moves with simple harmonic motion.

through 2π radians, and if f is the frequency in Hertz*, then

$$\omega = 2\pi f. \tag{3.4}$$

The displacement equation shows that the yoke-piston combination moves with simple harmonic motion. As mentioned before, there are many other simple harmonic relationships in mechanical and electrical fields. A spring-mass system, as shown in Fig. 3.1(b), is an example. If its amplitude and natural frequency just happen to match the amplitude and frequency of the Scotch-yoke mechanism, then the mass and the piston may be made to move up and down in perfect synchronization.

Or to put it in another way, *for every simple harmonic relationship, an analogous Scotch-yoke mechanism may be devised or imagined.* The crank length OA will represent the vector amplitude, and the angular velocity of the crank, ω in radians per second, will correspond to the circular frequency of the harmonic relation. If the mass and piston have the same frequencies and simultaneously reach corresponding extremes of displacement, their motions are said to be *in phase*. When they both have the same frequency, but do not oscillate together, the time relation (lag or advance) between their motions may be expressed by an angle referred to as the *phase angle*, ϕ.

3.4 COMPLEX RELATIONS

Most complex dynamic-mechanical signals, steady-state or transient, whether they are time functions of pressure, displacement, strain, or something else, may be expressed as a combination of simple harmonic components. Each component will have its own amplitude and frequency and will be combined in various phase relations with the other components. A general mathematical statement of this may be written as follows [1]:

$$f(t) = \frac{A_0}{2} + \sum_{n=1}^{\infty} (A_n \cos n\omega t + B_n \sin n\omega t), \tag{3.5}$$

in which

A_0, A_n, and B_n = amplitude-determining constants called *harmonic coefficients*,

n = integers from 1 to ∞, called *harmonic orders*.

---

* Adoption of the frequency unit "Hertz," abbreviated Hz and carrying the units 1/sec, was announced by the National Bureau of Standards in 1964 [22]. This conforms with the International System of Units given official status in a resolution of the 11th General Conference on Weights and Measures meeting in October 1960 (see also, Appendix IV).

When n is unity, the corresponding sine and cosine terms are said to be *fundamental*. For $n = 2, 3, 4$, etc., the corresponding terms are referred to as 2nd, 3rd, 4th *harmonics*, and so on.

Equation (3.5) may be written in the two equivalent forms:

$$f(t) = \frac{A_0}{2} + \sum_{n=1}^{\infty} C_n \cos{(n\omega t - \phi_n)} \tag{3.5a}$$

or

$$f(t) = \frac{A_0}{2} + \sum_{n=1}^{\infty} C_n \sin{(n\omega t + \phi'_n)}, \tag{3.5b}$$

where the harmonic coefficients, C_n, are determined by the relation

$$C_n = \sqrt{A_n^2 + B_n^2}$$

and the phase relation ϕ_n and ϕ'_n are determined as follows:

$$\tan{\phi_n} = \left(\frac{B_n}{A_n}\right), \quad \text{and} \quad \tan{\phi'_n} = \left(\frac{A_n}{B_n}\right).$$

The *phase angles* ϕ_n and ϕ'_n provide necessary timewise relationships among the various harmonic components.

While Eq. (3.5) indicates that all harmonics may be present in defining the signal-time relation, actually such relations often include only a limited number of harmonics. In fact, all measuring systems have some upper- and some lower-frequency limit beyond which further harmonics will be attenuated. In other words, no measuring system can handle an infinite frequency range.

While it would be utterly impossible to catalog all the many possible harmonic combinations, nevertheless it may be useful to consider the effects of some of the variables such as relative amplitudes, harmonic orders n, and phase relations ϕ Therefore Figs 3.2 through 3.6 are presented for two component relations in each case showing the effect of one variable only on the overall waveform. Figure 3.2 shows the effect of relative amplitudes; 3.3 shows the effect of relative frequencies; 3.4 shows the effect of various phase relations; 3.5 shows the appearance of the waveform for two components having considerably different frequencies; and Fig. 3.6 shows the effect of two frequencies which are very nearly the same.

Example. As an example of a relation made up of harmonics, let us consider a relatively simple pressure-time function consisting of two harmonic terms:

$$P = 100 \sin{(80t)} + 50 \cos{\left(160t - \frac{\pi}{4}\right)}. \tag{3.6}$$

Inspection of the equation shows that the circular frequency of the fundamental has a value of 80 radians per second, or $80/2\pi = 12.7$ Hz.

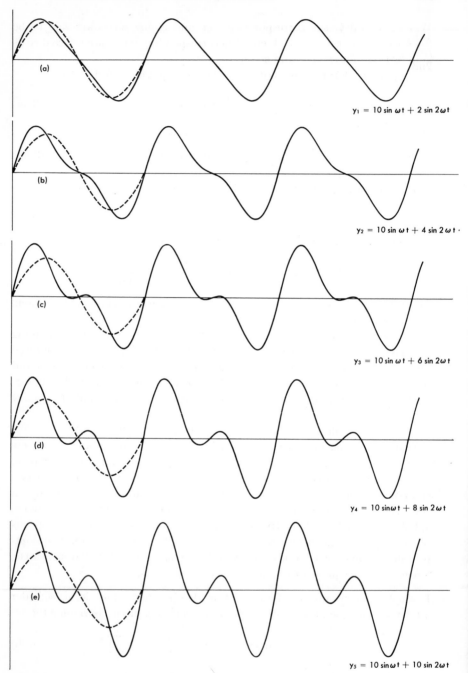

(a)

$y_1 = 10 \sin \omega t + 2 \sin 2\omega t$

(b)

$y_2 = 10 \sin \omega t + 4 \sin 2\omega t$ ·

(c)

$y_3 = 10 \sin \omega t + 6 \sin 2\omega t$

(d)

$y_4 = 10 \sin \omega t + 8 \sin 2\omega t$

(e)

$y_5 = 10 \sin \omega t + 10 \sin 2\omega t$

Fig. 3.2 Examples of two component waveforms with second harmonic component of various relative amplitudes.

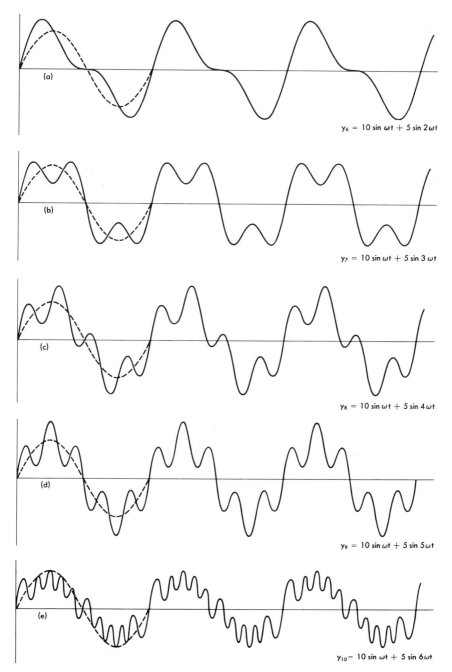

Fig. 3.3 Examples of two component waveforms with second term of various relative frequencies.

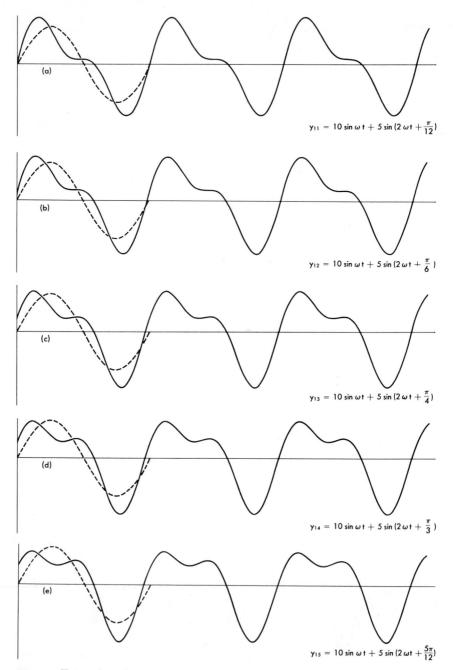

(a)

$$y_{11} = 10 \sin \omega t + 5 \sin (2 \omega t + \frac{\pi}{12})$$

(b)

$$y_{12} = 10 \sin \omega t + 5 \sin (2 \omega t + \frac{\pi}{6})$$

(c)

$$y_{13} = 10 \sin \omega t + 5 \sin (2 \omega t + \frac{\pi}{4})$$

(d)

$$y_{14} = 10 \sin \omega t + 5 \sin (2 \omega t + \frac{\pi}{3})$$

(e)

$$y_{15} = 10 \sin \omega t + 5 \sin (2 \omega t + \frac{5\pi}{12})$$

Fig. 3.4 Examples of two component waveforms with the second harmonic having various degrees of phase shift.

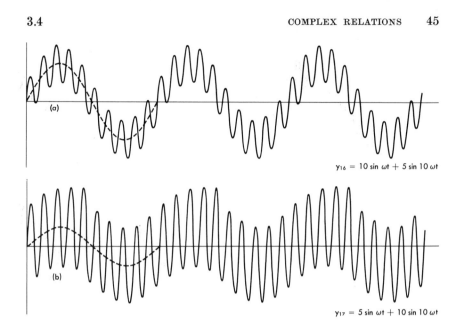

Fig. 3.5 Examples of waveforms with the two components having considerably different frequencies.

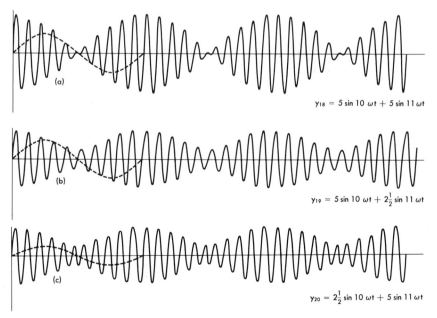

Fig. 3.6 Examples of waveforms with two components having frequencies that are very nearly the same.

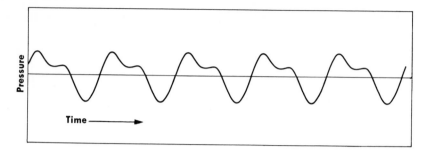

Fig. 3.7 Pressure-time relation, $P = 100 \sin (80) + 50 \cos (160 - \pi/4)$.

The period for the pressure variation is therefore $1/12.7 = 0.0788$ sec. The second term has a frequency twice that of the fundamental, as indicated by its circular frequency of 160 radians per second. It also lags the fundamental by one-eighth cycle, or $\pi/4$ radians. In addition, the equation indicates that the amplitude of the fundamental, which is 100, is twice that of the second harmonic which is 50 psi. A plot of the relation is shown in Fig. 3.7.

Example. As another example, suppose that an acceleration-time relation is expressed by the following equation:

$$a = 3800 \sin (2450t) + 1750 \cos \left(7350t - \frac{\pi}{3} \right)$$

$$+ 800 \sin (36{,}750t), \tag{3.7}$$

where

$$a = \text{acceleration, in./sec}^2,$$

$$t = \text{time, sec.}$$

The relation consists of three harmonic components having circular frequencies in the ratio 1 to 3 to 15. Hence the components may be referred to as the fundamental, the third harmonic, and the fifteenth harmonic. Corresponding frequencies in Hertz are 390, 1170, and 5850 Hz.

3.5 SPECIAL WAVEFORMS

There are a number of special waveforms whose equations may be written as infinite trigonometric series. Several of these are shown in Fig. 3.8. Table 3.1 lists the corresponding equations.

Both the square wave and the saw-tooth are useful in checking the response of dynamic measuring systems. In addition, the skewed saw-tooth form, Fig. 3.8(c), is of the form required for the voltage-time relation necessary for driving the horizontal sweep of a cathode-ray oscilloscope. All these

TABLE 3.1

EQUATIONS FOR SPECIAL PERIODIC WAVEFORMS SHOWN IN FIG. 3.8

Fig.	Equation*
3.8a	$y = \dfrac{4A_0}{\pi}\left(\sin \omega t + \tfrac{1}{3}\sin 3\omega t + \tfrac{1}{5}\sin 5\omega t + \cdots\right) = \dfrac{4A_0}{\pi}\sum_{n=1}^{\infty}\left[\dfrac{1}{2n-1}\sin(2n-1)\omega t\right]$
3.8b	$y = \dfrac{2A_0}{\pi}\left(\sin \omega t + \tfrac{1}{2}\sin 2\omega t + \tfrac{1}{3}\sin 3\omega t + \cdots\right) = \dfrac{2A_0}{\pi}\sum_{n=1}^{\infty}\left[\dfrac{1}{n}\sin(n\omega t)\right]$
3.8c	$y = \dfrac{2A_0}{\pi}\left(\sin \omega t - \tfrac{1}{2}\sin 2\omega t + \tfrac{1}{3}\sin 3\omega t - \tfrac{1}{4}\sin 4\omega t + \cdots\right) = \dfrac{2A_0}{\pi}\sum_{n=1}^{\infty}\left[\dfrac{(-1)^{n+1}}{n}\sin(n\omega t)\right]$
3.8d	$y = \dfrac{A_0}{2} - \dfrac{4A_0}{(\pi)^2}\left(\cos \omega t + \dfrac{1}{(3)^2}\cos 3\omega t + \dfrac{1}{(5)^2}\cos 5\omega t + \cdots\right) = \dfrac{A_0}{2} - \dfrac{4A_0}{\pi^2}\sum_{n=1}^{\infty}\left[\dfrac{1}{(2n-1)^2}\cos(2n-1)\omega t\right]$
3.8e	$y = \dfrac{8A_0}{(\pi)^2}\left(\sin \omega t - \dfrac{1}{(3)^2}\sin 3\omega t + \dfrac{1}{(5)^2}\sin 5\omega t - \cdots\right) = \dfrac{8A_0}{(\pi)^2}\sum_{n=1}^{\infty}\left[\dfrac{(-1)^{n+1}}{(2n-1)^2}\sin(2n-1)\omega t\right]$
3.8f	$y = -\dfrac{8A_0}{(\pi)^2}\left(\cos \omega t + \dfrac{1}{(3)^2}\cos 3\omega t + \dfrac{1}{(5)^2}\cos 5\omega t + \cdots\right) = -\dfrac{8A_0}{(\pi)^2}\sum_{n=1}^{\infty}\left[\dfrac{1}{(2n-1)^2}\cos(n\omega t)\right]$
3.8g	$y = \dfrac{2A_0}{\alpha(\pi-\alpha)}\left(\sin \alpha \sin \omega t + \dfrac{1}{(2)^2}\sin 2\alpha \sin 2\omega t + \dfrac{1}{(3)^2}\sin 3\alpha \sin 3\omega t + \cdots\right) = \dfrac{2A_0}{\alpha(\pi-\alpha)}\sum_{n=1}^{\infty}\left[\dfrac{1}{n^2}\sin(n\alpha)\sin(n\omega t)\right]$

*n as used in these equations does not necessarily represent the harmonic order.

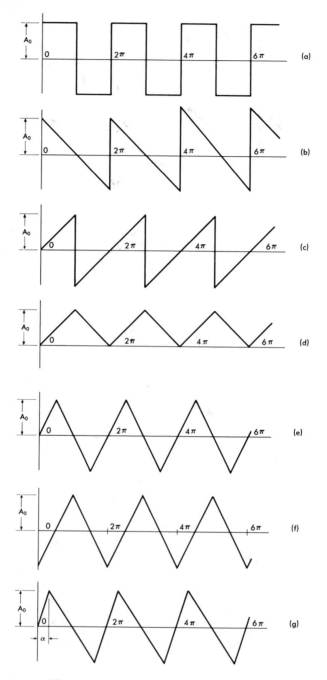

Fig. 3.8 Various special waveforms of harmonic nature.

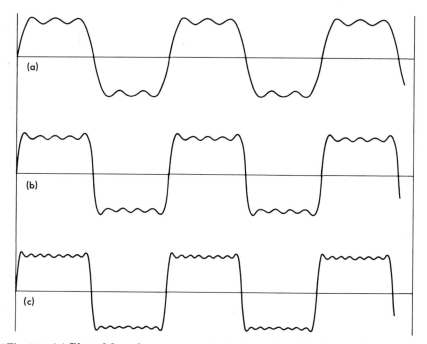

Fig. 3.9 (a) Plot of first three terms only (includes the fifth harmonic), from the square-wave relation. (b) Plot of the first five terms in the square-wave relation (includes the ninth harmonic). (c) Plot of the first eight terms (includes the fifteenth harmonic).

forms may be obtained as voltage-time relations from electronic signal generators (Article 4.5b).

In each case shown in Fig. 3.8 all the terms in the infinite series are necessary if the precise waveform indicated is to be obtained. Of course, with increasing harmonic order their effect on the whole becomes less and less.

As an example of this, consider the square wave shown in Fig. 3.8(a). The complete series includes all the terms indicated in the relation

$$y = \frac{4A_0}{\pi} \left(\sin \omega t + \tfrac{1}{3} \sin 3\omega t + \tfrac{1}{5} \sin 5\omega t + \cdots \right).$$

By plotting the first three terms only, which includes the fifth harmonic, the waveform shown in Fig. 3.9(a) is obtained. Figure 3.9(b) shows the result of plotting terms through and including the ninth harmonic, and Fig. 3.9(c) shows the form for the terms including the fifteenth harmonic. As more and more terms are added, the waveform approaches closer and closer to the square wave which results from the infinite series.

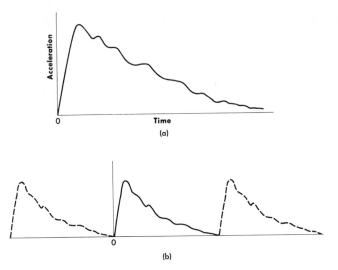

Fig. 3.10 (a) Acceleration-time relation resulting from shock test. (b) Considering the nonrepeating function as one real cycle of a periodic relation.

In the foregoing examples of special waveforms, various combinations of harmonic components were used. In each case the result was a periodic relation repeating indefinitely in every detail. Many mechanical inputs are not repetitive, as for example the acceleration-time relation (Fig. 3.10a) resulting from an impact test. Although such a relation is transient, it may be thought of as one cycle of a periodic relation in which all other cycles are fictitious (Fig 3.10b). On this basis nonperiodic functions may be analyzed in exactly the same manner as are periodic functions.

3.6 HARMONIC ANALYSIS*

In the previous article we have shown various forms plotted from given equations. In actual practice, however, our problem would be to obtain the important information from the recorded waveform. We would have the plot, from which we would like to obtain the harmonic components with their relative amplitudes, frequencies, and phase relations, rather than vice versa.

Harmonic analysis, which is the term applied to this process, is a subject of considerable extent, which most certainly cannot receive a thorough discussion here. However, it is possible to present a practical approach to the problem which will serve in most cases. (For detailed information on the subject, the reader is referred to references for this chapter.)

---

* See Appendix B for the theoretical basis.

If we assume that our experimental data can be made to fit relations as expressed in Eqs. (3.5), (3.5a), or (3.5b), our problem becomes that of determining appropriate values for the harmonic orders, coefficients, and phase angles. This may be done by what amounts to a graphical integration, most easily accomplished by numerical methods. It should be noted, however, that the procedure requires considerable time and patience, and previous experience is most helpful in recognizing short cuts reducing the work required. Although digital computer solutions may be employed, each student should, at least once, carry out a "long-hand" analysis for experience.

Often before beginning actual numerical analysis, we may determine certain facts by inspection. For instance, it may be observed that the positive values for the first half-cycle are repeated as negative values in the second half-cycle. When this situation occurs, it may be shown that only the odd harmonics are present and that the even harmonics may be ignored in the analysis [2, 3]. As will soon be apparent, such an observation considerably reduces the numerical work to be done. Other useful relationships are shown in Table 3.2. In addition, comparison of the unknown relation with typical plots such as shown in Figs. 3.2 through 3.6 will often indicate the general nature of the function. The presence of higher harmonics in significant amplitudes may often be determined by observation.

3.7 ANALYTICAL PROCEDURE

The analytical procedure may be outlined as follows:

1. Establish the fundamental cycle and assign the values 0 and 2π to its limits. The general form of the desired equation is then

$$f(\theta) = \frac{A_0}{2} + (A_1 \cos \theta + A_2 \cos 2\theta + A_3 \cos 3\theta + \cdots)$$

$$+ (B_1 \sin \theta + B_2 \sin 2\theta + B_3 \sin 3\theta + \cdots). \qquad (3.8)$$

2. Divide the fundamental cycle into m equal intervals, each of width $\Delta\theta$, and determine the corresponding ordinates. (*Note:* Do not include the ordinates for both ends of the interval being analyzed, because this would be a duplication.)

We should also note that although the computer may make determination of high-order harmonic coefficients feasible, we should not expect to be able to find meaningful coefficients without correspondingly well-defined data. For example, we should not expect to find useful coefficients for the 10th harmonic with only 12 data intervals per cycle. As a rough rule of thumb we may put down, as a limit, $m \geq 4n$, where m is the number of intervals per cycle and n is the order of coefficient desired.

TABLE 3.2

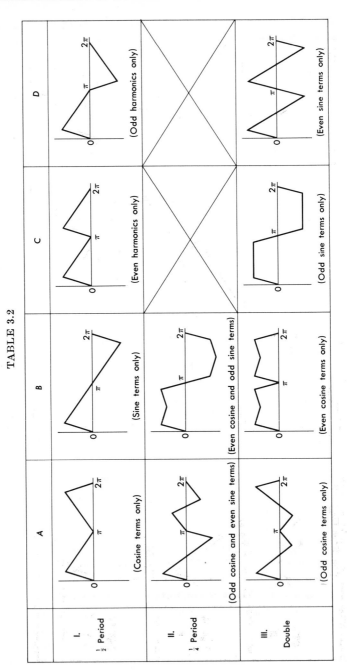

3. To determine a given coefficient, multiply each of the m ordinates determined in (2) by the corresponding numerical values of the desired trigonometric function. The average value of the resulting column of products will be one-half the coefficient being sought.· For example: To determine A_2, multiply each of the values of $f(\theta)$ as determined in step (2) by the corresponding values of cos 2θ. Add all the products together and divide by $m/2$ This will give the numerical value of A_2 Repeat this process for each of the values of A and B that are required.

4. Determine A_0, which is equal to twice the average of the values of $f(\theta)$.

Example. As an exercise, let us assume that we have just obtained the pressure-time trace of Fig. 3.7 by means of an appropriate pressure pickup and recording system. Let us also assume that we now wish to analyze the plot to determine the amplitudes, frequencies, and phase relations that are involved. Since in this example the equation is known, we will be able to easily check our result. As a first step we must determine the unit cycle as shown in Fig. 3.11. The zero pressure and time coordinates will result from our measurements; however, in general, zero time would not be a critical quantity. The pressure scale would be predetermined by system calibration, and the time scale would be established by recorder chart speed. In determining the period, the total time for a number of cycles (say 10 or 20) should be used, provided the frequency did not change with time. A check shows that for our example the period T is equal to 0.079 sec, hence $f =$ 12.75 Hz and $\omega = 80$ rad/sec.

Let us say that the information shown in Fig. 3.11 is the result of our preliminary inspection of the trace. We will now attempt to predict the important components in the function. Comparison of our plot with Figs. 3.2 through 3.6 indicates that we are probably working with a function in which higher harmonics are negligible. We will, therefore, begin our analysis

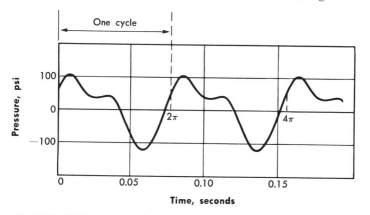

Fig. 3.11 Enlargement of pressure-time plot shown in Fig. 3.7.

TABLE 3.3

(1)	(2)	(3)	(4)	(5)	(6)	(7)	(8)	(9)	(10)
θ	$P = f(\theta)$	$\sin \theta$	$f(\theta) \sin \theta$	$\cos \theta$	$f(\theta) \cos \theta$	$\sin 2\theta$	$f(\theta) \sin 2\theta$	$\cos 2\theta$	$f(\theta) \cos 2\theta$
0	35	0.0	0.0	1.000	35.00	0.0	0.0	1.000	35.00
10	63	0.174	10.96	0.985	62.06	0.342	21.55	0.940	59.22
20	84	0.342	28.73	0.940	78.96	0.643	54.01	0.766	64.34
30	98	0.500	49.0	0.866	84.87	0.866	84.87	0.500	49.0
40	105	0.643	67.51	0.766	80.43	0.985	103.42	0.174	18.27
50	105	0.766	80.43	0.643	67.51	0.985	103.42	−0.174	−18.27
60	99	0.866	85.73	0.500	49.50	0.866	85.73	−0.500	−49.50
70	90	0.940	84.60	0.342	30.78	0.643	57.87	−0.766	−68.94
80	77	0.985	75.84	0.174	13.40	0.342	26.33	−0.940	−72.38
90	65	1.000	65.00	0.0	0.0	0.0	0.0	−1.000	−65.00
100	53	0.985	52.20	−0.174	−9.22	−0.342	−18.13	−0.940	−49.82
110	44	0.940	41.36	−0.342	−15.05	−0.643	−28.29	−0.766	−33.70
120	38	0.866	32.91	−0.500	−19.00	−0.866	−32.91	−0.500	−19.00
130	36	0.766	27.58	−0.643	−23.15	−0.985	−35.46	−0.174	−6.26
140	36	0.643	23.15	−0.766	−27.58	−0.985	−35.46	0.174	6.26
150	37	0.500	18.50	−0.866	−32.04	−0.866	−32.04	0.500	18.50
160	39	0.342	13.34	−0.940	−36.66	−0.643	−25.08	0.766	29.87
170	39	0.174	6.79	−0.985	−38.41	−0.342	−13.34	0.940	36.66

angle								
180	35	0.0	0.0	−1.000	−35.00	0.0	0.0	1.000	35.00
190	28	−0.174	−4.87	−0.985	−27.58	0.342	9.58	0.940	26.32
200	16	−0.342	−5.47	−0.940	−15.04	0.643	10.29	0.766	12.26
210	−2	−0.500	1.00	−0.866	1.73	0.866	−1.73	0.500	−1.00
220	−23	−0.643	14.79	−0.766	17.62	0.985	−22.66	0.174	−4.00
230	−48	−0.766	36.77	−0.643	30.86	0.985	−47.28	−0.174	8.35
240	−74	−0.866	64.08	−0.500	37.00	0.866	−64.08	−0.500	37.00
250	−98	−0.940	92.12	−0.342	33.52	0.643	−63.01	−0.766	75.07
260	−120	−0.985	118.20	−0.174	20.88	0.342	−41.04	−0.940	112.80
270	−135	−1.000	135.00	0.0	0.00	0.0	0.0	−1.000	135.00
280	−144	−0.985	141.84	0.174	−25.06	−0.342	49.25	−0.940	135.36
290	−144	−0.940	135.36	0.342	−49.25	−0.643	92.59	−0.766	110.30
300	−135	−0.866	116.91	0.500	−67.50	−0.866	116.91	−0.500	67.50
310	−118	−0.766	90.39	0.643	−75.87	−0.985	116.23	−0.174	20.53
320	−93	−0.643	59.80	0.766	−71.24	−0.985	91.60	0.174	−16.18
330	−63	−0.500	31.50	0.866	−54.56	−0.866	54.56	0.500	−31.50
340	−30	−0.342	10.26	0.940	−28.20	−0.643	19.29	0.766	−22.98
350	4	−0.174	−0.70	0.985	−3.94	−0.342	−1.37	0.940	3.76
	$\sum = -11$		$\sum = 1800.61$		$\sum = -10.23$		$\sum = 635.62$		$\sum = 637.84$

$$A_0 = \frac{-11}{18} = -0.6$$

$$A_1 = \frac{-10.23}{18} = -0.56$$

$$B_1 = \frac{1800.61}{18} = 100.03$$

$$A_2 = \frac{637.84}{18} = 35.4$$

$$B_2 = \frac{635.62}{18} = 35.4$$

with the assumption that second harmonics only occur in addition to the fundamental. If we are in error on this point, a comparison of our resulting equation with the data will so indicate.

Our next step will be to lift the data from the curve and put it into tabular form. This is done in columns (1) and (2) of Table 3.3. The remainder of the table consists of the numerical determination of the coefficients according to the rules set down in the preceding pages. The columns are calculated as indicated and their sums obtained. (See Appendix C for a table of harmonic sines and cosines.)

The values of $A_0/2$ and A_1 indicate that perhaps they should actually be equal to zero, and for the time being at least, we will assume that this is the case. Our equation then may be written as:

$$P = 100.03 \sin \theta + 35.4 \cos 2\theta + 35.4 \sin 2\theta \qquad (3.9)$$

If we convert the 2θ terms into the form of Eq. (3.5b) and substitute $\theta = \omega t = 80t$, we obtain

$$P = 100.03 \sin (80t) + 50 \cos \left(160t - \frac{\pi}{4}\right). \qquad (3.10)$$

Having used, for this example, a trace whose equation was known, we have no problem in checking our analysis. Direct comparison of the above relation with Eq. (3.6) indicates that we have obtained the correct answer. In the usual case, however, we would now use our resulting function to calculate values of pressure for various values of time t. These calculated values would then be compared with the corresponding values on the curve. Reasonable agreement or lack thereof would indicate the measure of our success.

Note also that in the usual case it would probably be more logical to select our time origin ($t = 0$) corresponding to zero pressure. In this case the final equation would contain phase angles other than zero for the fundamental and $\pi/4$ for the second harmonic. However, the same waveform would be defined.

Example. We will now apply the procedure we have established to a second example. Figure 3.12(a) shows a plot of strain versus time taken from a shaft connecting an intermittent drive with a feeding mechanism on an experimental machine. The time scale was obtained from the chart speed of the recorder, and the ordinate scale from a system calibration.

In analyzing such a record the first step is to convert the time scale into an angular relation such that the period is represented by 2π radians, or 360 degrees. Actually, of course, the only angular relation involved in the quantity represented is through the concept of circular frequency (Article 3.3). In our example, the period is found to be 0.031 sec, which means that the frequency of the fundamental is 32.3 Hz and $\omega = 206$ rad/sec. Figure 3.12(b) shows the plot to an enlarged scale.

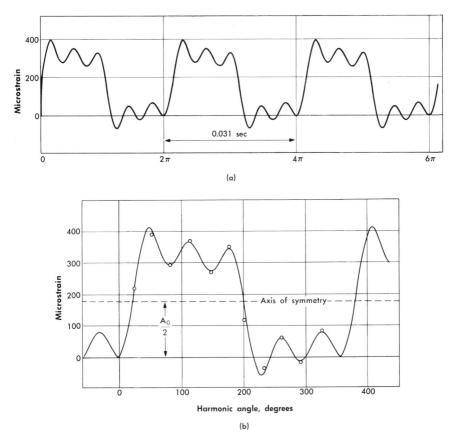

Fig. 3.12 (a) Strain-time record from an experimental machine. (b) One cycle of the strain-time relation shown in (a).

Several observations may be made before the numerical analysis actually begins. It will be noted that the trace appears to be symmetrical above and below a strain value of about 175 microstrain.* If this is true, it indicates two things. First, what may be considered a static strain exists, about which a completely reversing plus and minus dynamic strain amplitude oscillates. The value of this static strain component will correspond to our value of $A_0/2$ (Eq. 3.8). Secondly, if the trace is symmetrical about such an axis, only odd harmonics will be present (Case I-D, Table 3.2). This means that only the first half-cycle need be considered in the analysis, thereby materially reducing the work involved.

Let us assume that this is the case, as indeed it so appears, and further let us assume, at least for the time being, that harmonics above the fifth

_____

* See Article 11.1

TABLE 3.4

DATA FROM FIG. 3.12(b) REFERRED TO AVERAGE
VALUE OF ORDINATE

θ	ϵ = strain, Microstrain (as recorded)	$\epsilon' = \epsilon - \epsilon_{avg}$ Microstrain
0	0	−168
10	21	−147
20	103	−65
30	224	56
40	337	169
50	397	229
60	391	223
70	343	175
80	297	129
90	287	119
100	313	145
110	347	179
120	359	191
130	337	169
140	297	124
150	266	98
160	273	105
170	309	141
180	336	168
190	315	147
200	233	65
210	112	−56
220	−1	−169
230	−61	−229
240	−55	−223
250	−7	−175
260	39	−129
270	49	−119
280	23	−145
290	−11	−179
300	−23	−191
310	−1	−169
320	43	−125
330	70	−98
340	63	−105
350	27	−141
	$\sum \epsilon = 6048$	

$$\frac{A_0}{2} = \frac{6048}{36} = 167.8, \text{ say } 168, \quad \epsilon' = \epsilon - 168$$

may be ignored. Should higher harmonics be significant, the results of our analysis will so indicate.

Our first step will be to determine the value of $A_0/2$, and to accomplish this we will require data for the complete cycle. Table 3.4 lists the data taken directly from the plot and includes the calculation of $A_0/2$, which is determined as 168 microstrain.

We will now refer the plot and the data to a new axis of symmetry, Fig. 3.12(b). We do this in order to take advantage of the fact that even harmonics may be ignored. The resulting values appear in column 3 of Table 3.4 and column 2 of Table 3.5.

Based on these assumptions, the numerical analysis is carried out as shown in Table 3.5. Sine and cosine values for the fundamental, third, and fifth harmonics are multiplied by the values of $f(\theta)$, and the harmonic coefficients are calculated. The resulting equation is

$$\epsilon = 167.8 - 64.1 \cos 206t + 185.3 \sin 206t - 66.3 \cos 618t$$
$$+ 11.8 \sin 618t - 41.3 \cos 1030t - 54.2 \sin 1030t. \quad (3.11)$$

This relation may then be reduced by use of Eq. (3.5a) to read

$$\epsilon = 167.8 - 196 \cos (206t + 1.24) - 67.3 \cos (618t + 0.18)$$
$$- 68.1 \cos (1030t - 0.92). \quad (3.11a)$$

The correctness of our analysis may then be checked by comparing values of $f(\theta)$ as calculated from our equation against the original data. The plotted points shown in Fig. 3.12(b) are obtained from the above equation and indicate good agreement.

3.8 SYSTEM RESPONSE

Quite simply, *response* is the measure of a system's fidelity to purpose. It may be defined as an evaluation of the system's ability to faithfully transmit and present all the pertinent information included in the input signal and to exclude all else.

We would like to know if the output information truly represents the input. If the input information is in the form of a sine wave, a square wave or a saw-toothed wave, does the output appear as a sine wave, a square wave or a saw-toothed wave, as the case may be? Is each of the harmonic components in a complex wave treated equally, or are some attenuated, completely ignored, or perhaps shifted timewise relative to the others? These questions are answered by the response characteristics of the particular system.

Response may be considered under the following headings: (a) amplitude response, (b) frequency response, (c) phase response, and (d) delay or rise time.

TABLE 3.5

HARMONIC ANALYSIS OF STRAIN-TIME RELATION

(1) θ	(2) e' = f(θ)	(3) sin θ	(4) f(θ) sin θ	(5) sin 3θ	(6) f(θ) sin 3θ	(7) sin 5θ	(8) f(θ) sin 5θ	(9) cos θ	(10) f(θ) cos θ	(11) cos 3θ	(12) f(θ) cos 3θ	(13) cos 5θ	(14) f(θ) cos 5θ
0	-168	0.0	00.0	0.0	0.0	0.0	0.0	1.0	-168.0	1.0	-168.0	1.0	-168.0
10	-147	0.174	-21.6	0.500	-73.5	0.766	-112.6	0.985	-144.8	0.866	-127.3	0.643	-94.5
20	-65	0.342	-22.2	0.866	-56.3	0.985	-64.0	0.940	-61.1	0.500	-32.5	-0.174	-11.3
30	56	0.500	28.0	1.000	56.0	0.500	28.0	0.866	48.5	0.000	0.0	-0.866	-48.5
40	169	0.643	108.7	0.866	146.4	-0.342	-57.8	0.766	129.5	-0.500	-84.5	-0.940	-158.9
50	229	0.766	175.4	0.500	114.5	-0.940	-215.3	0.643	134.4	-0.866	-198.3	-0.342	-78.3
60	223	0.866	193.1	0.0	0.0	-0.866	-193.1	0.500	111.5	-1.000	-223.0	0.500	111.5
70	175	0.940	164.5	-0.500	-87.5	-0.174	-30.5	0.342	59.9	-0.866	-151.6	0.985	172.4
80	129	0.985	127.1	-0.866	-111.7	0.643	82.9	0.174	22.4	-0.500	-64.5	0.766	98.8
90	119	1.000	119.0	-1.000	-119.0	1.000	119.0	0.0	0.0	0.0	0.0	0.0	0.0
100	145	0.985	142.8	-0.866	-125.6	0.643	93.2	-0.174	-25.2	0.500	72.5	-0.766	-111.1
110	179	0.940	168.3	-0.500	-89.5	-0.174	-31.1	-0.342	-61.2	0.866	155.0	-0.985	-176.3
120	191	0.866	165.4	0.0	0.0	-0.866	-165.4	-0.500	-95.5	1.000	191.0	-0.500	-95.5
130	169	0.766	129.5	0.500	84.5	-0.940	-158.9	-0.643	-108.7	0.866	146.4	0.342	57.8
140	125	0.643	80.4	0.866	108.3	-0.342	-42.8	-0.766	-95.8	0.500	62.0	0.940	117.5
150	98	0.500	49.0	1.000	98.0	0.500	49.0	-0.866	-84.9	0.0	0.0	0.866	84.9
160	105	0.342	35.9	0.866	90.9	0.985	103.4	-0.940	-98.7	-0.500	-52.5	0.174	18.3
170	141	0.174	24.5	0.500	70.5	0.766	108.0	-0.985	-138.9	-0.866	-121.1	-0.643	-90.7
			1667.8		106.0		-488.0		-576.6		-596.4		-371.9

$B_1 = 1667.8 \times 2/18 = 185.3$

$B_3 = 106.0 \times 2/18 = 11.8$

$B_5 = -488.0 \times 2/18 = -54.2$

$A_1 = -576.6 \times 2/18 = -64.07$

$A_3 = -596.4 \times 2/18 = -66.3$

$A_5 = -371.9 \times 2/18 = -41.3$

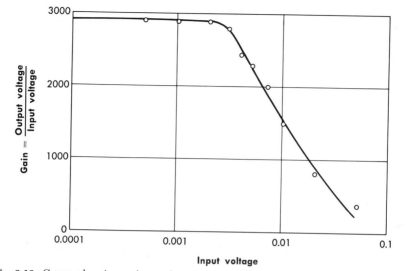

Fig. 3.13 Curve showing gain vs. input voltage for a three-stage amplifier used for strain measurement.

a) Amplitude response. Amplitude response is governed by the system's ability to treat all input amplitudes uniformly. If an input of 5 units is fed into a system and an output of 25 indicator divisions is obtained, it is usually expected that an input of 10 units will result in an output of 50 divisions. While this is the common situation, there are other special nonlinear responses that are sometimes required. But in any event, whatever the arrangement, whether it be linear, exponential, or some other amplitude function, discrepancy between design expectations in this respect and the actual performance results in poor amplitude response.

Of course, no system exists that is capable of faithfully responding over an unlimited range of amplitudes. Figure 3.13 shows the amplitude response of a three-stage voltage amplifier suitable for connecting a strain-gage bridge to an oscilloscope. The usable range of the amplifier is restricted to the horizontal portion of the curve. The plot shows that for inputs above about 0.002 volts the amplifier becomes overloaded and amplification ceases to be linear.

b) Frequency response. Good frequency response is obtained when a system treats all *frequencies* with equal faithfulness. If a 100-Hz sine wave having an input amplitude of 5 units is fed into a system, and a peak-to-peak output of $2\frac{1}{2}$ in. results on an oscilloscope screen, it would be expected that a 500-Hz sine wave input of 5 units amplitude would also result in a $2\frac{1}{2}$-in. peak-to-peak output. Changing the frequency of the input signal should not alter the system's output magnitude so long as the input amplitude remains unchanged.

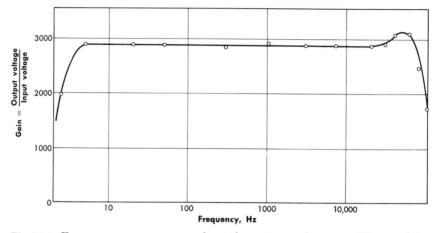

Fig. 3.14 Frequency-response curve for a three-stage voltage amplifier used for strain measurement.

Here again, there must be a limit to the range over which good frequency response may be expected. This is true for any dynamic system, regardless of quality. Figure 3.14 illustrates the frequency-response relations for the same voltage amplifier used for Fig. 3.13. Frequencies both below 5 Hz and above about 30,000 Hz are attenuated, and only strain inputs within these limits are amplified in the correct relative proportions.

Example. Let us consider an example which illustrates the significance of a system's frequency response when it is called upon to handle a complex input. Referring to Fig. 3.12(b), let us assume that this represents a true strain-time situation recorded by a *perfect* system, which we shall designate as A. Now suppose the identical signal is fed into system B, which has the frequency-response characteristics shown in Fig. 3.15.*

We found that the fundamental frequency of our strain signal was 32.3 Hz and that third and fifth harmonics existed. This would mean that our signal involves components having frequencies of 32.3, 96.9, and 161.5 Hz. Comparing the harmonic frequencies with the response curve for system B, we see that the fundamental and the third harmonic should be transmitted faithfully; however, the fifth harmonic will be seriously attenuated. In fact, it will be transmitted in only about 15% of its correct relative value. Figure 3.16 shows the plot of the modified or distorted output compared with the true values. The error is apparent.

---

* Figure 3.15 is typical of most commercially available stylus-type recorders. The example should not be interpreted, however, as casting discredit on this very useful type of equipment. Rather, the example represents a misuse of equipment if higher frequencies are required. In certain cases the mechanical filtering ability of such instruments may be used to a decided advantage.

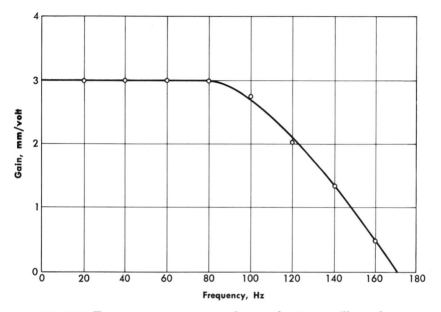

Fig. 3.15 Frequency-response curve for a stylus-type oscillograph.

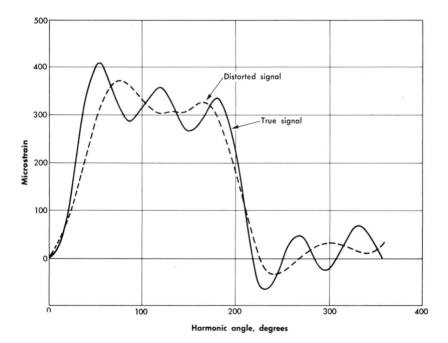

Fig. 3.16 Effect of poor frequency response on recording of strain-time relation shown in Fig. 3.12(b).

This illustrates how the higher-frequency components of a complex signal may be degraded by poor response. Poor frequency response does not always behave in this fashion, however. In some cases nonlinear amplifications result in the higher frequencies being *accentuated* out of proportion to the lower frequencies.

Poor frequency response is not limited to electronic components by any means. Inertia of mechanical elements attenuates output. The stylus of the recorder referred to above is almost entirely responsible for the drop in output above 100 Hz. Its mass simply cannot be driven at higher rates. On the other hand, a mechanical element operating near its resonant vibrational frequency will accentuate signal components of nearby frequencies.

c) Phase response. Amplitude and frequency responses are important for all types of input waveforms, simple or complex. *Phase response*, however, is of importance primarily for the complex wave only.

Time is required for the transmission of a signal through any measuring system. When a sine-wave voltage is amplified by a simple single-stage amplifier, the output trails the input, timewise, by approximately 180 degrees, or one-half cycle. For two stages, the shift is approximately 360 degrees, and so on. Actually the shift will not be exact multiples of half wavelengths and will differ depending on the equipment and also on the frequency. It is a shift of this type depending on frequency that is important in determining phase response.

For the single sine-wave input, any shift would normally be unimportant. The output viewed on an oscilloscope screen would show the true waveform, and the proper amplitude could be determined. The fact that the shape being viewed was actually formed a few microseconds or a few milliseconds after being generated would be of no consequence.

However, consider the complex wave made up of numerous harmonics. Suppose that the time lag for each component is different, each one being delayed a different amount. The harmonic components would then emerge from the system in a phase relation different from when they entered. The whole waveform and amplitudes would be altered.

Let us use our previous example again, this time to illustrate the effect of phase shift. Suppose once more that the relation defined by Eq. (3.11a) and shown in Fig. 3.12(b) represents the true strain-time function that we are attempting to measure. Suppose further that our system, in transmitting the information, causes the third harmonic to be retarded 15 degrees with respect to its correct relation to the fundamental, and that it also causes the fifth harmonic to lag by 0.87 radians. The resulting output could then be expressed as:

$$\epsilon = 167.8 - 196 \cos{(206t + 1.24)} - 67.3 \cos{(618t - 0.09)}$$
$$(- 68.1) \cos{(1030t - 1.79)}.$$

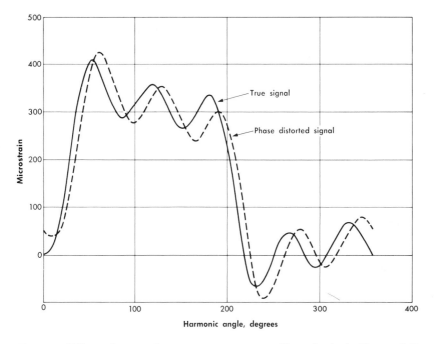

Fig. 3.17 **Fig. 3.17** Effect of poor phase response on recording of strain-time relation shown in Fig. 3.12(b).

Figure 3.17 illustrates the result. Obviously the system has distorted the truth and an incorrect result is obtained.

d) Delay or rise time. Finally, a fourth type of response, which is actually another form of frequency response, is *delay*, or *rise time*. When a *stepped* or *instantaneous* input is applied to a system, the output may lag as shown in Fig. 3.18. The time delay, Δt, after the step is applied but before proper output magnitude is reached is known as delay, or rise time. It is a measure of the system's ability to handle transients.

3.9 USE OF SPECIAL WAVEFORMS

System response is often checked *qualitatively* by using input signals of known harmonic content and observing the output function versus time. Inputs that are often used are square and saw-tooth waves (see Fig. 3.8). The saw-tooth wave possibly has a slight theoretical advantage for this purpose in that all the integral harmonics are present, whereas in the square wave only odd harmonic terms are present. The square wave, however, is the most popular form, undoubtedly because it is more easily obtained not only in electrical form but also pneumatically, hydraulically, and even approximately

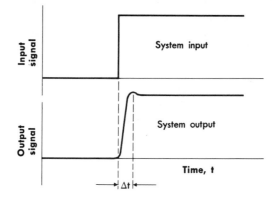

Fig. 3.18   Response of a typical system to a pulse-type input; Δt is the rise time.

as temperature functions. Of course, in most measurement applications it is hoped that if a square wave is fed into the system, a true square waveform will be indicated at the output.

3.10 CLOSING REMARKS

The purpose of this chapter has been to increase the student's awareness of the harmonic nature of dynamic signal information and to give him a basis for understanding the response characteristics of a measuring system. The approach has been deliberately kept on an elementary level, and the student should realize that the problem lends itself to advanced mathematical treatment beyond the intended scope of this book. For those students interested in pursuing the subject further, references are listed at the end of the book.

SUGGESTED READINGS

Keast, D. N., *Measurements in Mechanical Dynamics*. New York: McGraw-Hill Book Co., 1967.

Manley, R. G., *Waveform Analysis*. New York: John Wiley & Sons, Inc., 1945.

PROBLEMS

3.1 A simple linkage, familiar to all, is shown in Fig. 3.19.
   a) Write an exact equation for the displacement of the piston in terms of time t and constant values of R, L, and ω.

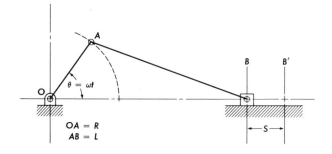

Figure 3.19

b) Show that the displacement may be approximated by
$$s \approx R - R \cos \omega t + (R^2/2L) \sin^2 \omega t + (R^4/8L) \sin^4 \omega t.$$

c) Show that the piston velocity may be approximated by
$$v \approx R\omega \sin \omega t + (R^2/2L)\omega \sin 2\omega t;$$

or if the fourth harmonic is included,
$$v \approx R\omega \sin \omega t + R^2/2L[1 + R^2/4L^2]\omega \sin 2\omega t - (R^4\omega/16L^3) \sin 4\omega t.$$

d) Determine expressions for the piston acceleration based on each of the two expressions in part (c).

(*Note:* This is a simple problem in the kinematics of machinery illustrating one source of higher-order displacement, velocity, and acceleration harmonics. Refer to books on mechanism or mechanics of machinery.)

3.2 A pressure-time relation is expressed by the equation
$$p = 110 + 40 \sin 5t + 30 \cos 5t + 10 \sin 10t - 15 \cos 10t.$$

a) What is the static component of pressure?
b) What harmonics are present?
c) What is the fundamental period in seconds?
d) Write the equation in terms of cosines only.
e) Write the equation in terms of sines only.

3.3 A very simple mechanical-vibration amplitude-time relationship is expressed by the following equation:
$$s = 0.01 \sin (314t) + 0.005 \sin (628t),$$

where

$$s = \text{displacement, in., and}$$

$$t = \text{time, sec.}$$

a) What is the fundamental frequency in Hertz?
b) What is the fundamental period?
c) Write an expression for the readout if the measuring system attenuates the second harmonic by 40%.
d) Write an expression for the readout if the measuring system introduces a 5 millisecond delay in the second harmonic.

e) Write an expression for the readout if both response characteristics expressed in (c) and (d) occur simultaneously.

3.4 Why is it especially important that a dynamic measuring system have "good" frequency and phase response when the input has a complex form? Is phase response a problem when the input signal has a simple harmonic form?

3.5 A pressure measuring system employs an oscillographic readout having frequency response characteristics illustrated in Fig. 3.16 and phase characteristics as given by the accompanying table. Write an expression for the output of the system, when the "true" input is given as

$$p = 2.5 + (7.8) \cos (377t) + (3.6) \cos (754t) + (2.1) \cos (1131t)$$

Frequency, Hz.	Phase lag, degrees
0	0
30	0
60	8
90	20
120	36
150	56
180	77
210	108
240	150

3.6 Write the equation for a distorted square wave in which the components are amplitude attenuated by the factors $C(2n - 1)$, and a phase lag exists expressed by the relation $2k(n - 1)\phi$, where C and k are coefficients, and n is the harmonic order (for the fundamental, $n = 1$). Write the equation in series form, including the first four terms.

3.7 The tabulated data fit the equation

$$y = 5 + 4 \sin \theta + 3 \sin (3\theta).$$

As an exercise, assume that the equation is unknown.

a) Plot the data.

b) "Guess" that the equation will contain fundamental and third harmonic sine terms.

c) By harmonic analysis, determine the coefficients.

d) Check several points.

θ	y	θ	y
0	5.00	195	1.84
15	8.16	210	0.00
30	10.00	225	0.05
45	9.95		
		240	1.54
60	8.46	255	3.26
75	6.74	270	4.00
90	6.00		
		285	3.26
105	6.74	300	1.54
120	8.46	315	0.05
135	9.95		
		330	0.00
150	10.00	345	1.84
165	8.16	360	5.00
180	5.00		

3.8 Figure 3.20 illustrates an oscillographic record showing the displacement-time relationship for the motion of a cam follower. The stroke is 10 in. and the period is 5 sec.

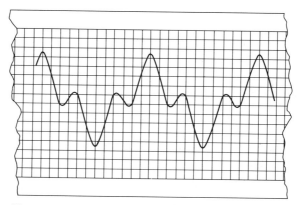

Fig. 3.20 Oscillograph record of cam-follower displacement-time relation.

a) Assuming that the motion contains the fundamental plus a third harmonic, determine, by harmonic analysis, the displacement-time equation.

b) Check a sufficient number of points to satisfy yourself with the accuracy of the result.

3.9 Figure 3.21 shows a pressure-time trace as it appears on a CRO screen. System calibration gives the following: Each small square vertically equals 0.5 psig, and each small square horizontally equals 0.10 sec. Use pressure and time zeros as indicated.

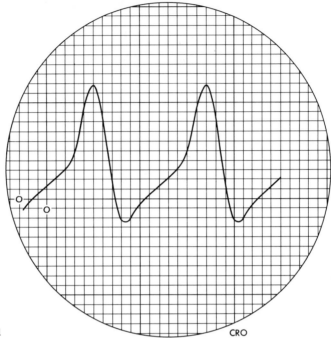

Figure 3.21   CRO

a) Determine the fundamental frequency, circular frequency, and period.
b) By harmonic analysis, determine an equation to fit the trace.
c) Check the equation against the trace.

3.10 Arrange an instrument-type amplifier, a signal generator, ACVM and CRO, as shown in Fig. 3.22.

a) For a given generator frequency, increase the signal input until distortion is detected in the output waveform displayed on the CRO screen. (Note that distortion is often more readily detected when a Lissajous pattern is displayed. See also Article 9.5). Determine the limiting input voltages for a range of frequencies.

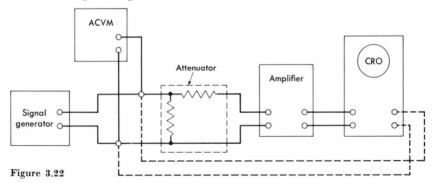

Figure 3.22

b) While keeping the input amplitude at a constant value well below the distortion level, determine the output voltages displayed on the CRO. Plot the voltage gain versus frequency, thereby obtaining a frequency-response curve for the amplifier.

c) By sampling the generator output and applying it to the horizontal input terminals of the CRO (connections are shown as dotted lines), determine the phase lag by using Lissajous diagrams. (*Note:* Beware of phase shift through any attenuator that may be employed, and also in the CRO).

3.11 A square-wave or stepped pulse is often used to obtain qualitative information regarding performance characteristics of a system. Such an input is not only used for electrical systems, but also for pneumatic, hydraulic, and thermodynamic systems for which inputs may be in the form of such quantities as pressure, temperature, etc. Briefly discuss the reasons for such testing and what information may be obtained.

The saw-tooth waveform, referred to as "white noise," is often substituted for the square wave as a test input. What advantage may be claimed for its use?

CHAPTER 4

STANDARDS OF MEASUREMENT

4.1 INTRODUCTION

Regardless of the measurement methods used, some basis for the comparison, a standard unit, must be employed. Also, there must be general agreement as to the exact value of such standard units. In addition, for international trade to function properly, the standards for the various countries must be convertible on some basis mutually agreed to. The relation between the centimeter and the inch must be definitely established, as must the relations among all the other basic units. Certain standards fall naturally into place; the average length of the day is one. Even in this case, however, the starting point or the zero must be defined.

4.2 LEGAL STATUS OF STANDARDS OF MEASUREMENT

In regard to the legal status of measurement standards in the United States, the following will be of interest. First of all, who or what body shall have authority to control such matters? Quoting from Article 1, Section 8, Paragraph 5, of the United States Constitution: "The congress shall have power to . . . fix the standard of weights and measures." Although Congress was given the power, considerable time elapsed before anything was done about it. In 1866, as found in the Revised Statutes of the United States, Sec. 3569: "It shall be lawful throughout the United States of America to employ the weights and measures of the metric system . . . ." This simply makes it clear that the metric system *may* be used. In addition, this act established the following relation for conversion:

$$\text{One meter} = 39.37 \text{ in. (exactly).}$$

An international convention held in Paris in 1875 resulted in an agreement signed for the United States by the U.S. Ambassador to France. The following is quoted therefrom: "The high contracting parties engage to establish and maintain, at their common expense, a scientific and permanent international bureau of weights and measures, the location of which shall be Paris" [1, 15]. While this established a central bureau of standards, which was set up at Sèvres, a suburb of Paris, it of course did not bind the United States to make use of or adopt such standards.

On April 5, 1893, in the absence of further Congressional action, Superintendent Mendenhall of the Coast and Geodetic Survey issued the following order [1, 2]:

The Office of Weights and Measures with the approval of the Secretary of the Treasury, will in the future regard the international prototype meter and the kilogram as *fundamental standards*, and the customary units, the yard and pound, will be derived therefrom in accordance with the Act of July 28, 1866.

The Mendenhall Order turned out to be a very important action. First, it recognized the meter and the kilogram as being fundamental units on which all other units of length and mass should be based. Second, it ties the metric and English systems of length and mass together in a definite relationship, thereby making possible international exchange on a more exact basis.

In response to requests from scientific and industrial sources, and to a great degree influenced by the establishment of like institutions in Great Britain and Germany,* Congress, on March 3, 1901, passed an act providing that "The office of Standard Weights and Measures shall hereafter be known as 'The National Bureau of Standards' " [16]. Expanded functions of the new Bureau were set forth and included development of standards, research basic to standards, and the calibration of standards and devices. The NBS was formally established in July, 1910, and its functions were considerably expanded by an amendment passed in 1950.

Commercial standards are largely regulated by state laws, and to maintain uniformity, regular meetings (National Conferences on Weights and Measures) are held by officials of NBS and officers of state governments. Essentially all state standards of weights and measures are in accordance with the Conference's standards and codes. International uniformity is maintained through regularly scheduled meetings (held at about six-year intervals), termed the *General Conference on Weights and Measures* and attended by representatives from most of the civilized countries of the world. In addition, numerous interim meetings are held to consider solutions to more specific problems for later action by the General Conference.

---

* The National Physical Laboratory. Teddington, Middlesex, and Physikalisch-Technische Reichsanstalt, Braunschweig.

4.3 THE STANDARD OF LENGTH

Originally, the meter was intended to be one ten-millionth of the earth's quadrant. Also, it was planned that the unit of mass, the kilogram, should be equal to the mass of one one-thousandth of one cubic meter of water at maximum density. Until recently the magnitude of the meter was defined as the length of the International Prototype Meter, the distance between two finely scribed lines on a platinum-iridium bar when subject to certain specified conditions. However, on October 14, 1960, the Eleventh General Conference on Weights and Measures adopted a new definition of the meter as 1,650,763.73 wavelengths, in vacuum, of the radiation corresponding to the transition between the levels $2p_{10}$ and $5d_5$ of the krypton atom. The National Bureau of Standards of the United States has adopted this standard and the inch now becomes 41,929.399 wavelengths of the krypton light [18]. (See also Article 10.10.)

The relation between the meter and the inch as specified by the Mendenhall Order (1 m = 39.37 in.) results in

$$1 \text{ in.} = 2.54000508 \text{ cm (approximately).}$$

However, the convenient relation

$$1 \text{ in.} = 2.54 \text{ cm (exactly)}$$

has been used in industry and engineering throughout the world. This latter relation has never been recognized officially by the Congress of the United States although it is used by the American Standards Association as the basis for length conversions. The difference between these two "standards" may be written as

$$\frac{2.54000508}{2.54} - 1 = 0.000002,$$

or 0.0002%. This is about $\frac{1}{8}$ inch per mile.

A better idea of the significance of the difference may be gained by considering the following situations. The work of the United States Coast and Geodetic Survey is based on the 39.37 in./m relation and a coordinate system with origin located in Kansas [4]. Changing the metric relation from 39.37 in./m (exactly) to 2.54 cm/in. (exactly) would cause discrepancies of almost 16 ft at a distance of 1500 mi. One can only imagine the confusion over property lines if such a change were made. On the other hand, gage blocks, which are the manufacturing secondary standards, are established on the basis of 2.54 cm/in. If they were measured on the basis of 39.37 in./m, errors of 2 μin./in. would be found. This is in the same order of magnitude as the *tolerance* of the better-quality blocks.

All this would present a very real problem were it not for the fact that geodetic distances and small mechanical displacements need seldom be compared. Hence it is quite probable that the two nearly equal equivalents (sic) will be in use for many years to come.

4.4 THE STANDARD OF MASS

The *kilogram* is defined by the mass of the International Prototype Kilogram, a platinum-iridium mass kept at the International Bureau of Weights and Measures near Paris. Secondary standards of known relative masses are maintained by each of the primary industrial countries of the world. In the United States, the basic unit of mass is the "United States National Prototype Kilogram No. 20," carefully maintained by the National Bureau of Standards.

The Mendenhall Order of 1893 included the following relationship:

$$1 \text{ lb avoirdupois } = 453.592\ 427\ 7 \text{ g.}$$

For 58 years, following the founding of the National Bureau of Standards, the above relation applied. However, on July 1, 1959, the equivalent was altered to

$$1 \text{ lb avoirdupois } = 453.592\ 37 \text{ g.}$$

This change was brought about by the desire to unify the equivalencies employed by Australia, Canada, New Zealand, South Africa, the United Kingdom, and the United States. This is the relation in use today (1969).

4.5 TIME AND FREQUENCY STANDARDS

A paradox of scientific measurement is that we measure time without having a very good idea of the nature of the subject we are dealing with [19]. This has been very ably described by G. M. Clemence [5] of the United States Naval Observatory as follows:

It is not possible in a scientific journal (or anywhere for that matter) to say much about the nature of time itself. That subject belongs to philosophy rather than to science. It may be permissible, however, to recall a few propositions, not contradicted by experience, which may be thought to be facts.

Time seems to pass more quickly as we grow older. I used to think time was like a suitcase that seemed to grow smaller because there were more and more things to put in it, but later I found that even persons who had nothing to occupy their time experienced the same illusion.

As a rule time passes very quickly in our dreams, and here we have the ability, as we have not when we are awake, to move backward and forward in time, seemingly without effort, and perhaps even against our will.

Time moves, or we move through it, in one direction only. Our senses receive no impression of the future, and we commonly think that we live only in the present. In strictness, however, we know nothing of the present, but only of the past. We are able to speak of the present only because the lag between events in the immediate vicinity and our appreciation of them is so small as to be negligible for most practical purposes. When we regard events outside the earth, the truth is apparent. We observe the moon, not as it is now, but as it was somewhat more than a second ago. Light from the sun requires eight minutes to reach us; from Pluto, five hours. If we could see events on the earth reflected in mirrors placed on suitable stars at different distances, we could observe the history of the United States, and even of mankind itself, taking place before our very eyes.

What we, as scientists, know about time itself is very little indeed. We can say much more about the measurement of it.

(For additional philosophical discussion of time, see Reference [17].)

The engineer is not very often interested in the precise *time of day*. His interest is usually in terms of duration of some event or phenomenon. He requires a time base for his observations, but is not particularly interested in knowing the precise instant in history at which an event occurs. This may not always be the case, however, because it is sometimes easier to determine the relative occurrence of events by establishing a time *fix* than by determining the time interval by intraobservation. Even here, of course, the *fix* need only be mutually agreed to and need have no time relation to extraneous events.

Until 1956 the second was defined as 1/86,400 of the average period of revolution of the earth on its axis. Although this seems to be a relatively simple and straightforward definition, problems remained. There is a gradual slowing of the earth's rotation (about 0.001 sec/century) [5], and, in addition, the rotation is irregular.

Therefore, in 1956 an improved standard was agreed upon; the second was defined as 1/31,556,925.9747 of the time required by the earth to orbit the sun in the year 1900. This is termed the *ephemeris second*. Although the unit is defined with a high degree of exactness, implementation of the definition is dependent on astonomical observation which is incapable of realizing the implied precision.

In the 1950's, atomic research led to the observation that oscillations associated with certain atomic transitions may be measured with great repeatability. One, the hyperfine transition of the cesium atom, was related to the ephemeris second with an estimated accuracy of two parts in 10^9. On October 13, 1967, in Paris, the 13th General Conference on Weights and Measures officially adopted the unit of time of the International System of Units as the second, defined in the following terms: "The second is the duration of 9,192,631,770 periods of the radiation corresponding to the transition between the two hyperfine levels of the fundamental state of the atom of cesium-133" [20].

An atomic beam apparatus [28], commonly termed an atomic "clock," is employed to produce the frequency of transition. Heated metallic cesium is caused to emit a beam of atoms which is separated into two beams of differing energies depending on the alignment of nuclei and electrons. When an oscillating electromagnetic field is applied having a frequency characteristic of the particular transition, the frequency agreement is detected and appropriately indicated. The frequency of the "master" oscillator may then be employed as a standard with which the outputs of other oscillators may be compared.

a) *Frequency standards.* The cesium "clock" is a basic frequency standard. Pendulums, tuning forks, electronic oscillators, etc., may be used as secondary standards. *Frequency* is the number of recurrences of a phenomenon or series of events during a given time interval, and the reciprocal of frequency is *period*. A frequency standard *chops* time into discrete bits which may be used as time standards and, through comparative means, for timing events. The actual source of such a frequency may be mechanical or electrical, or in fact, pneumatic, hydraulic, thermal, etc. In certain cases mechanical frequency sources are used because of their long-time stability. Or the mechanical source, such as a pendulum or tuning fork, may be combined with the electrical, the mechanical being used to control the electrical. Or a strictly electronic source may be used. Of course, a very familiar electrical frequency which may be used as a secondary standard is the ordinary 60-Hz line voltage.

Probably the most basic example of a mechanical frequency standard is the *seconds* pendulum, cut to have a frequency or rate of 1 Hz. (See Reference [21] for a very interesting summary of the application of pendulums for the accurate measurement of time.)

Another well-known frequency standard is the balance wheel, such as that used in a watch. This is a form of torsional pendulum depending on mass-elastic relations for its characteristics, and which in general may be constructed in a wide number of forms.

A third mechanical frequency standard is the tuning fork. As with the torsional pendulum, the detail construction of the tuning fork may be in a large number of different forms (see Reference [14], Chapter 2).

An example of a mechano-electrical frequency source is the piezoelectric crystal (Article 6.17), which provides means for maintaining very precise frequencies in the range of 4 kilocycles per second to 100 megacycles per second [6]. Such crystals possess the ability to convert mechanical energy into electrical energy, or vice versa. Materials exhibiting this characteristic include quartz, barium titanate, and various crystalline salts. When a small plate or bar of such materials is mechanically strained, a voltage develops across its faces; conversely, when a voltage is applied to the faces, a mechanical strain results. If the voltage is an alternating voltage, the plate or bar may be made to vibrate, and because of its mechanical mass-elastic characteristics, such a member will have a natural frequency of vibration. This fundamental frequency is often used as the basis for a very stable control of electronic oscillator output [7]. The tuning fork is also used for this purpose.

b) Electronic oscillators. Electronic oscillators are sources of periodic voltage variation of either fixed or variable frequency. The rotating a-c generator is a form of nonelectronic oscillator whose primary purpose is to provide a source of power rather than voltage. However, 60-Hz line voltage is often quite useful as a frequency source. For the purposes of mechanical measurement, it is the voltage output from the oscillator that is of primary value.

In general, oscillators are used for a wide variety of purposes; for example, as energy sources for circuitry measurement, as audio sources for electronic musical instruments, for radio and TV signal propagation, for supersonic testing, etc. Of course they can also be employed as frequency standards which, by suitable comparative means, may be used for timing and phase measurement. This is discussed in detail in Chapter 9.

Electronic oscillators may be classified as follows:

I. Fixed-frequency oscillators
   A. Simple electronic
   B. Tuning-fork controlled
   C. Crystal controlled

II. Variable-frequency oscillators
   A. Sine wave
      1. Audio frequency (0 to 20,000 Hz)
      2. Supersonic (20,000 to 50,000 Hz, roughly)
      3. Radio frequency (50,000 to 10,000,000,000 Hz)
   B. Nonsine wave
      1. Square wave
      2. Saw-tooth wave
      3. Random noise

c) Fixed-frequency oscillators. Fixed-frequency oscillators are of primary value in mechanical measurements for the purpose of calibrating and recording standard timing signals. For measurement purposes, frequencies above 20,000 Hz are seldom useful except in the field of supersonic nondestructive testing.

One commercial vacuum-tube tuning-fork oscillator* has a specified frequency of 1000 ± 0.5 Hz at 77°F, and a temperature coefficient of -0.08 Hz/°F. Such an oscillator is quite adequate as a frequency or timing standard for the majority of mechanical measurement problems. The stability of this unit is intermediate between that of the simple electronic-type oscillator and that of the temperature-compensated (or controlled),

---

* Type 723C, General Radio Company, Cambridge, Mass.

crystal-stabilized oscillator. Although simple electronic oscillators are seldom of fixed frequency, accuracies of $1\frac{1}{2}$ to 3% of specified frequency may be expected with warm-up drifts of 1 to 2%.

Precise frequency and time standards are available to all through reception of transmissions from the National Bureau of Standards radio stations [23]. Stations WWV, WWVB, and WWVL are located at Fort Collins, Colorado [24]. Station WWVH is in Hawaii. WWV and WWVH are classified as *high-frequency* (HF) stations, while WWVB transmits at 60 kHz and is classified as *low frequency*, and the frequency of WWVL is 20 kHz, which is termed *very low frequency* (VLF) [25]. Low and very low frequencies have the advantage of providing more stable reception at a distance because of reduced variations in the transmission paths peculiar to those frequencies. Eventually NBS expects the two lower-frequency stations to have sufficient power to provide a worldwide frequency and time service.

Although WWV information programming changes from time to time, the following services, as quoted from NBS Letter Circular LC 1023, are typical:

The National Bureau of Standards' Radio Stations WWV (in operation since 1923) and WWVH (since 1949) broadcast six widely used technical services: 1. Standard Radio Frequencies, 2. Standard Audio Frequencies, 3. Standard Time Intervals, 4. Standard Musical Pitch, 5. Time Signals, 6. Radio Propagation Forecasts.

The WWV-WWVH broadcasts are a convenient means of transferring the national standard of frequency and time interval and making it readily available throughout the United States and over much of the world. [The broadcast program is shown schematically in Fig. 4.1.]

Station WWV broadcasts on standard radio frequencies of 2.5, 5, 10, 15, 20, and 25 MHz. The broadcasts are continuous, night and day, except WWV is off the air for approximately 4 minutes each hour. The silent period commences at 45 minutes, plus 0 to 15 seconds, after each hour.

The accuracy of each of the radio frequencies as transmitted is better than 1 part in 100,000,000.

Two standard audio frequencies, 440 Hz and 600 Hz, are broadcast on each radio carrier frequency. The audio frequencies are given alternately starting with 600 Hz on the hour for three minutes, interrupted two minutes, followed by 440 Hz for three minutes and interrupted two minutes. Each 10-minute period is the same except for transmitter interruptions mentioned above.

The two standard audio frequencies are useful for accurate measurement or calibration of instruments operating in the audio or ultrasonic regions of the frequency spectrum. The frequencies broadcast were chosen because 440 Hz is the standard musical pitch and 600 Hz has the maximum number of integral multiples and sub-multiples; also, 600 Hz is conveniently used with the standard power-frequency 60 Hz.

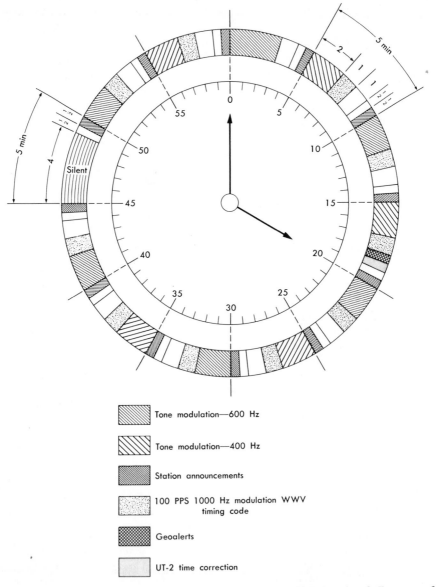

Fig. 4.1 Hourly broadcast schedule of Station WWV (National Bureau of Standards).

The accuracies of the audio frequencies as transmitted is better than one part in 100,000,000. Changes in the transmitting medium (Doppler effect, etc.) result at times in fluctuations in the audio frequencies as received.

Seconds pulses at intervals of precisely one second are given as double side-band amplitude modulation on each radio carrier frequency. The pulse duration is 0.005 second. At WWV each pulse consists of five cycles of a 1000 Hz frequency.

The seconds pulses provide a useful standard time interval for quick and accurate measurement or calibration of time and frequency standards and timing devices. Intervals of one minute are marked by omitting the pulse at the beginning of the last second of every minute and by commencing each minute with two pulses spaced by 0.1 second. The two-minute, three-minute and five-minute intervals are synchronized with the seconds pulses and are marked by the beginning or ending of the periods when the audio frequencies are off.

The time interval as broadcast from WWV is accurate to 1 part in 10^8 plus or minus 1 microsecond. Received pulses have random phase shifts or jitter because of changes in the propagating medium. The magnitudes of these changes range from practically zero for the direct or ground wave to about 1000 microseconds when received via a changing ionosphere.

Most *short-wave* radios are capable of receiving one of the WWV radio frequencies.* The audio-frequency signals are then available as voltages at the loudspeaker terminals. In this manner any laboratory has frequency standards available direct from the nation's legally authorized standardizing agency.

Calibration procedures and other applications are discussed in Chapter 9.

d) *Variable-frequency oscillators.* VFO (variable-frequency oscillators), used for the purposes of mechanical measurement, normally produce sine-wave outputs covering a frequency range from about 20 to 20,000 Hz. Since this is usually thought of as approximating the audible range, such devices are also called audio oscillators or audio-signal generators.

Figure 4.2 shows a typical audio-frequency oscillator which produces a sine-wave output over a frequency range of 1 to 100,000 Hz. Frequency is adjusted through the main dial and a stepped multiplier. Outputs to 1 watt at $24\frac{1}{2}$ volts may be obtained through adjustment of the amplitude control.

---

* In addition to the United States, many other nations also broadcast timing signals of various types. Among them are, Argentina, South Africa, England, New Zealand, USSR, Japan, Italy, Belgium, and Canada. The Canadian signals are broadcast by the Dominion Observatory Radio Station CHU, Ottawa, on the following frequencies: 3.330, 7.335, and 14.670 MHz [8].

Fig. 4.2 General-purpose sinusoidal
oscillator having a frequency range
from 1 to 100,000.   (Courtesy
Hewlett-Packard Company, Palo
Alto, California.)

Calibration of a general-purpose oscillator of this type may be checked against
frequency standards such as 60 Hz line frequency, a fork-controlled frequency
standard, or the National Bureau of Standards broadcasts. An instrument
of this type is very useful as a timing source and as a frequency comparison
for many mechanical measurements.

e) *Complex-wave oscillators.* Sine-wave oscillator outputs may be shaped
to provide a variety of waveforms for special-purpose applications. Square,
saw-tooth, pulsed, and random forms are some of the more common shapes
obtainable (see Fig. 3.8). Undoubtedly the nonsine waveform most used
for test purposes is the square wave. Electronically, such a waveform is
obtained by chopping off most of the sine wave, leaving only a small portion
of the base, giving a good approximation of the square wave. The necessary
wave-shaping circuitry may be incorporated in a general-purpose oscillator,

making possible the desired waveform simply by throwing a switch. More often, however, special-purpose oscillators are used.

4.6 TEMPERATURE STANDARDS

In 1927 the national laboratories of the United States, Great Britain, and Germany proposed a temperature standard that became known as the International Temperature Scale of 1927 (ITS-27). This standard, adopted by 31 nations, conformed as closely as possible to the thermodynamic scale proposed by Lord Kelvin in 1854. It was based on six fixed-temperature points dependent upon physical properties of certain materials, including the ice and steam points of water. Revisions have been made by succeeding conferences, notably in 1948 and 1968. Currently, the International Practical Temperature Scale of 1968 (IPTS-68), adopted by the International Committee on Weights and Measures and authorized by the Thirteenth General Conference, is in effect [9].

The basic unit of temperature, the Kelvin (K), is defined as the fraction 1/273.16 of the thermodynamic temperature of the triple point of water; the temperature at which the solid, liquid, and vapor phases of water exist in equilibrium. Degrees Celsius (°C) is defined by the relationship

$$t = T - 273.15$$

where t and T are *degrees Celsius* and *Kelvins*, respectively.

In reality, two temperature scales are defined, a *thermodynamic* and a *practical* scale. The latter is the more common basis for measurement. The thermodynamic scale is put in terms of such domains as magnetic, ultrasonic, gas, and optical principles, whereas the practical scale is established in terms of temperature-related physical properties such as thermal expansion and thermoelectrical variations. Both systems are anchored to fixed reference points, with the difference lying in the methods of interpolation [10]. Engineering measurement of temperatures is primarily concerned with practical applications, and hence the following discussion is in terms of the International Practical Temperature Scale (IPTS-68).

The fixed points are temperatures corresponding to thermal states of materials and are highly reproducible. Zero Celsius is the temperature established by equilibrium between pure ice and air-saturated pure water at normal atmospheric pressure. It has been found, however, that a more precise datum, independent of ambient pressure and possible contaminants, utilizes the so-called *triple point* of water. This is the temperature at which the solid, liquid, and vapor phases of water exist in equilibrium. The value of 0.0100°C is assigned to this temperature. Simple, easily used apparatus are available for reproducing this fixed temperature point [26].

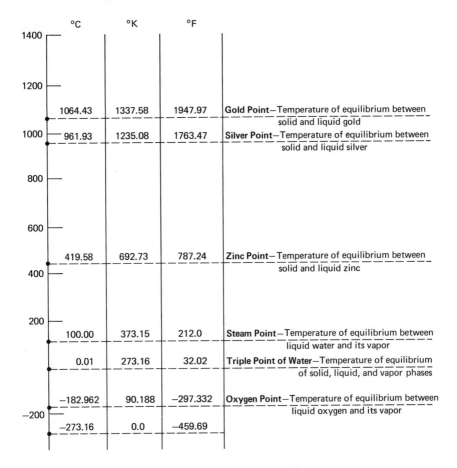

Fig. 4.3 Some fixed points established by IPTS-68. Others, in Kelvins, are: triple point of hydrogen, 13.81; boiling point of hydrogen, (25/76 atmos.), 17.042; boiling point of hydrogen, 20.28; boiling point of neon, 27.102; triple point of oxygen, 54.361.

Intermediate temperatures are standardized through specification of a standard platinum resistance thermometer or a platinum and platinum-10% rhodium thermocouple.

For the range from the ice point to the freezing point of antimony, the intermediate temperatures are defined by the equation

$$R_T = R_0(1 + AT + BT^2), \tag{4.2}$$

Figure 4.3 illustrates the scope of the scale established by the fixed points prescribed by the IPTS-68 (six are shown on the diagram and the remaining five are identified in the caption). In addition to the points used to establish the scale, numerous other secondary fixed points are agreed to, some of which are listed in Table 15.8.

Although the fixed points spot-determine the scale, there remains the matter of establishing reliable procedures for determining intermediate values. A reproducible interpolating or transfer method (or methods) must be prescribed to establish the continuous scale. No single defining means has been found; the IPTS-68 divides the temperature span into four "ranges," the lowest range being further divided into four parts with different interpolating relationships for each segment. The division is as follows:

Range 1.   Pt. 1 – From t.p.* hydrogen to b.p. hydrogen
              Pt. 2 – From b.p. hydrogen to t.p. oxygen
              Pt. 3 – From t.p. oxygen to b.p. oxygen
              Pt. 4 – From b.p. oxygen to f.p. water

Range 2.   From f.p. water to f.p. antimony

Range 3.   From f.p. antimony to f.p. gold

Range 4.   Above f.p. gold.

Standard resistance-type thermometers (see Article 15.5b) are used as the interpolating devices for Ranges 1 and 2. For Range 3 a platinum-10% rhodium and platinum thermocouple is prescribed (see Article 15.6). For Range 4, temperatures are based on Planck's law, referred to the gold point.

The application of relationships for making interpolations for Ranges 1 and 2 is not simple, and space will not permit inclusion of the details. Briefly, however, for Ranges 1 and 2, polynomial equations with as many as twenty terms may be required, along with the use of up to five fixed points; computer reduction programs are dictated. For further information the reader is referred to reference [9].

The following relationship is specified for Range 3:

$$E = a + bt + ct^2,$$

where

E = emf from the thermocouple referred to the ice point,

t = any temperature within the limits of Range 3, °C,

---

* t.p., triple point; b.p., boiling point; f.p., freezing point.

a, b, and c = constants determined by measuring the value of E with one junction at the ice point and the other successively at the silver and gold points, plus a further determination at 630.74°C as measured with a platinum resistance thermometer.

Above the gold point (Range 4), temperatures are determined by optical means which evaluate the radiant energy from the unknown temperature source and compare it with that from a source at the gold point. The following relationship is prescribed:

$$\frac{J_T}{J_{Au}} = \frac{\exp\left[\dfrac{C_2}{\lambda T_{Au}}\right] - 1}{\exp\left[\dfrac{C_2}{\lambda T}\right] - 1},$$

where

J_T, J_{Au} = radiant energy emitted per unit time, per unit area, and per unit wavelength at wavelength, λ, at temperature T, and gold point temperature T_{Au}, respectively,

C_2 = 0.014388 meter-Kelvin,

λ = wavelength.

Methods for establishing any temperature in the range from 13.81K (−259.34°C) to above the gold point are therefore defined. In application to mechanical development problems the standardized resistance thermometer, the standardized thermocouple or the standardized pyrometer would be used as secondary devices for calibration of working instruments. (See Chapter 4.)

4.7 ELECTRICAL UNITS

Absolute electrical units are fundamentally derivable from the mechanical units of length, mass, and time and an assumed value for the permeability of space, μ_v [11]. Hence with mechanical units standardized, electrical values are also established. There remains, however, the problem of laboratory usage. How shall a laboratory provide a basis for calibration? This requires some form of reproducible source, preferably one having a magnitude as close as possible to the absolute unit.

Prior to 1948, electrical standards were based on the "International" Ohm, Ampere, and Volt, adopted in 1893. The International Ohm was defined as "the resistance of a column of mercury of uniform cross section, having a length of 106.300 cm and a mass of 14.4521 gm when the temperature is 0°C." The International Ampere was defined as "the unvarying current which, when passed through a solution of silver nitrate in water in accordance with standard specifications, deposits silver at the rate of 0.001118 grams per second." The International Volt was defined so that a Clark cell at 15°C had an emf of 1.434 volts.

A primary trouble with this system lay in the fact that as measuring methods inevitably improved, discrepancies between the absolute and the International systems were bound to appear. It soon became apparent that the units as defined above were not consistent with Ohm's law. This was corrected in 1908 by allowing the volt to become a derived unit based on the 1893 definitions of the ampere and the ohm. At that time, however, it was agreed that the *practical* units should be based directly on the derivable absolute units. It was decided that steps should be taken to establish such a standard.

The basis for establishing absolute electrical units lies in theoretically derived relations [12] based on definitions and two experimental procedures. By the first procedure an inductor of accurately determined physical dimensions is constructed and the inductance *calculated*. The reactance of the inductor is then compared with a resistance standard whose resistance is thus determined in dimensional terms. The second procedure consists of accurately determining the force or torque exerted between two coils, in the form of what is known as a *current balance*, when carrying a current [13]. The current is also passed through a resistance, standardized by the first procedure. The potential drop across the resistor is then compared with the output of a *standard-type* cell. On the basis of the data obtained from the current balance, the absolute volt is calculated, thereby standardizing the cell, which in addition to being a standard *type* now becomes a true standard.

From the volt and ohm determined by these experiments, the other electrical units such as the ampere, the henry, and the farad may be derived in absolute terms. Accuracies of a few parts in ten million can be obtained [4].

It was not until 1935, however, that the International Committee on Weights and Measures finally decided that the conversion should be made, setting January 1, 1940, as the date for change-over. The war intervened, and the change to the absolute system was finally made on January 1, 1948. At that time it was found necessary to equate the absolute system to the old International System through use of the following relations:

1 International Ohm   = 1.00049 absolute ohms,

1 International Volt   = 1.000330 absolute volts,

1 International Ampere = 0.99835 absolute amperes.

Secondary standards in the form of.*standard cells* for voltage and *standard resistors* for electrical resistance remain as practical laboratory calibration sources.

SUGGESTED READINGS

Application Note 52, "Frequency and Time Standards," Palo Alto, Calif.: Hewlett Packard Co., 1965.

Cochrane, R. C., *Measures for Progress, A History of the National Bureau of Standards.* U.S. Dept. of Commerce, 1966.

National Bureau of Standards, Tech. Note No. 262, "Accuracy in Measurements and Calibrations," 1965, U.S. Dept. of Commerce, 1965.

National Bureau of Standards, *Precision Measurement and Calibration.* Handbook 77, U.S. Dept. of Commerce, 1961. Vol. I, *Electricity and Electronics*, Vol. II, *Heat and Mechanics*, Vol. III, *Optics, Metrology and Radiation.*

PROBLEMS

4.1 In 1790, Secretary of State Thomas Jefferson proposed that a standard unit of length be established equal to the length of a seconds pendulum [27]. Write a short report analyzing the merits and problems inherent in such a unit.

4.2 Figure 4.4 shows a "constant moment" beam employed as a secondary strain standard. By use of such an apparatus, the accuracy of strain-measuring systems* may be evaluated simply by comparing strain readouts with computed strains. Design a standard such as shown in the figure, i.e., after careful consideration, select a material and assign values to dimensions a, b, w, and t; then analyze the design for accuracy. Within what limits do you expect to be able to produce a "standard strain"?

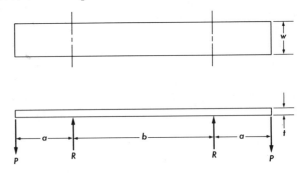

---

* See Chapter 11 for discussion of strain measurement.

4.3 With reference to the apparatus you designed in Problem 4.2:

a) What range or span of strain should the apparatus have?

b) List factors that must be considered in selecting the beam material. If metallurgical experts are available, make use of them. Specify a material.

c) How accurately can the beam be made? Assign dimensional tolerances. After it is made, how accurately can the dimensions be checked? Which dimensions are of greatest importance? The value of a on one side may be slightly different than on the other side. What significance would this have?

d) How accurately can the applied loads be determined? If dead weights are to be employed, how certain will you be of their magnitudes? In your present circumstance, how would you go about checking the accuracies of any available dead weights? If you were given the job of obtaining "accurate" weights, what would be your procedure? Investigate this matter. The values of P may be slightly different on each side. How nearly equal can they be made? How important is this? Analyze.

e) What value of Young's modulus and Poisson's ratio will you employ in your design calculations? (Poisson's ratio may be important if transverse strains are to be considered.) You may wish to determine E and ν experimentally. What facilities are available to you to carry out these measurements? With the available facilities, how accurately can you expect to determine E and ν? How accurately can the specimens be made, and how accurately can dimensions, loads, and strains be measured? In summary, how accurately do you think you can determine E and ν for the material you are using? After careful consideration of all the factors, provide numbers with supporting discussion and findings for the accuracy with which you think values of E and ν can be determined.

f) Evaluate the effects of changing ambient temperatures on the strain standard. What problems would be presented if the apparatus were used at extreme temperatures (elevated or cryogenic)?

g) The assumption of "constant moment" between the supports ignores the distributed weight of the beam. Analyze this source of error.

h) There will be certain frictional effects at the support and load points. Analyze them.

i) Bonded-type gages may be applied to the standard for evaluation. They may be used, removed, and others applied. If the surface is mechanically cleaned, how important may the dimensional changes become? Estimate, or better still, experimentally determine such dimensional changes and evaluate the problem. Is it one to worry about? Can local values of E and ν be changed by mechanical cleaning?

j) Most of the error sources mentioned above are inherent in the apparatus. Carefully consider any procedural problems that may occur in *using* the apparatus.

k) Can you think of any other factors that may affect the accuracy of the standard? If so, analyze them. What about supporting structure? What about changing barometric pressure? Corrosion? Wear?

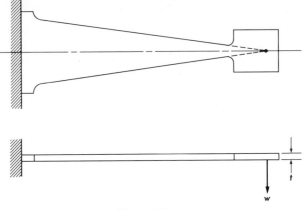

Figure 4.5

4.4 Figure 4.5 illustrates an alternative "constant-strain" design. Following the assignment for Problems 4.2 and 4.3, design a strain standard of this type.

4.5 A bending-type strain standard may be designed wherein a controlled deflection may be employed rather than a controlled load, as specified in Problems 4.2 and 4.4. Compare the two approaches to the problem. Submit your findings in the form of a report.

CHAPTER 5

STATISTICAL TREATMENT OF DATA

5.1 INTRODUCTION

When the value of a physical quantity is obtained by experimental methods, only through sheer luck will the measured magnitude coincide exactly with the true value—and even should this extremely unlikely event occur, it would go unrecognized by the experimenter. The magnitude of the true value will be approached, but in the strictest sense, never evaluated. The difference between the true value (unknown) and the result (best experimentally determined value) is the *error*.

Although the error may not be known, it is still possible through application of experience, good judgment, and theoretical methods to obtain estimates of true magnitudes and associated uncertainties. The following articles establish important nomenclature and outline some of the procedures applicable to data treatment. It must be understood, however, that only a brief summary can be attempted here; the reader is referred to the references at the end of the chapter for greater detail.

5.2 ERROR CLASSIFICATION

Errors may be classified in a variety of ways, the following being useful for our purposes [5]:

I. Systematic or fixed errors
   A. Calibration errors
   B. Human errors
   C. Experimental errors
   D. Errors of technique
   E. Loading error

II. Random or accidental errors
   A. Errors of judgment
   B. Variation of conditions
   C. Definition

III. Illegitimate errors
  A. Blunders or mistakes
  B. Computational errors
  C. Chaotic errors

Systematic errors are of constant or similar form, often resulting from improper conditions or procedures that are consistent in action. Human errors may well be systematic, as in the case of an individual's tendency to consistently read high or to "jump the gun" when synchronized readings are to be taken, and the like.

The equipment itself may introduce built-in errors resulting from incorrect design, fabrication, or maintenance. Such errors may be caused by what have been termed *false elements* [1], such as incorrect scale graduations, defective gearing, or linkages of wrong proportions. Certain electronic-amplifier faults would also fall under this heading. Errors of these types are consistent with regard to both sign and magnitude, and because of their consistency may sometimes be corrected by calibration. This is not always the case, however, for if the input is of complex form, distortion caused by poor system response would not ordinarily be correctable by calibration.

Loading error is of special significance. It results from the influence exerted by the act of measurement on the physical system being tested. It is basic that *the measurement process inevitably alters the characteristics of both the source of the measured quantity and the measuring system itself, from which it must follow that there will always be some difference between the measured indication and the corresponding to-be-measured quantity.*

This is an important fact which must always be borne in mind. The process must take some energy from the input source; therefore the source will be changed by the process. Placing a probe in an air stream changes the flow pattern. A bonded strain gage mounted on a foil test subject will alter the stiffness. Or the mass of a vibrometer placed on a small beam will alter the characteristics of the quantity being measured. These are obvious examples, but similar conditions exist in all measuring systems, including even optical methods [2].

Random errors are distinguished by their lack of consistency. An observer may not be consistent in his method of estimating readings, or the process involved may include certain uncontrolled, or poorly controlled, variables causing changing conditions. In the instrumentation itself, such errors may result from what have been termed *disturbed elements* [1], usually coming from some outside influence such as vibration or temperature variations (vibration- and temperature-measuring systems excepted). Errors of these types are normally of limited time duration or are inherent to specific environment.

Variation of conditions would also include the effects of unconstrained elements caused by such things as backlash and friction. These are design problems involving materials and dimensional tolerances. One means of detecting and often correcting this type of error is to determine results while quantity is first increased and then decreased in magnitude. This is sometimes called the *method of symmetry.*

As their name implies, *illegitimate errors* should not exist. They include outright mistakes which may be eliminated through exercise of care and repetition of the measurement.

Chaotic errors include random disturbances introducing errors of sufficient magnitude to hide the test information. Extreme vibration, mechanical shock of the equipment, or pickup of extraneous electrical noise may be of sufficient magnitude to make testing meaningless. In such cases the test should be stopped until the disturbing elements can be eliminated.

5.3 NOMENCLATURE

Before proceeding further it is desirable that we define certain terms applied to the statistical treatment of data.

a) *Data.* Elemental bits of information obtained by experimental means, herein assumed to be in numerical form.

b) *Population* (*also termed "universe"*). A collection of data, either finite or infinite in number. The term carries the connotation of completeness.

c) *Sample.* A portion of a population. A set of values, experimentally obtained, represents a sample of a theoretically possible larger population; that m values are obtained does not preclude the possibility of $m + n$ values.

d) *Multisample test.* Repeated measurement of a given quantity, using altered test conditions, such as different observers and/or different instrumentation. Merely taking repeated readings with the same procedure and equipment does not provide multisample results.

e) *Single-sample test.* A single reading *or* succession of readings taken under identical conditions excepting for time.

The exactness of a measurement may only be determined by application of statistics to multiple tests, using as many means and procedures and experimenters as practicable, assuming of course that all methods and equipment are appropriate to the test. When multiple tests, in this sense, are not made, the results are single-sample.

f) *True or actual value, V_a.* Actual magnitude of an input signal to a measuring system. Evaluation of this quantity may be approximated, but in the strictest sense never truly determined.

g) *Indicated value, V_i.* The magnitude indication directly supplied by the measuring system. This is the supply of raw or directly recorded data.

h) Correction. The revision applied to the indicated value, which it is assumed improves the worth of the result. Such revision may be in the form of either an additive factor or a multiplier or both.

i) Result, V_r. Obtained by making all known corrections to the indicated value;

$$V_r = A V_i + B, \tag{5.1}$$

where A and B are multiplying and additive corrections, respectively.

j) Discrepancy. Difference between two indicated values or results determined from a supposedly fixed true value.

k) Error. The *actual* difference between the true value and the result;

$$\text{Error} = V_r - V_a. \tag{5.2}$$

It is seen that the value of the error is never really known.

l) Accuracy. The *maximum* amount by which the result differs from the true value—measurement with small systematic error;

$$\text{Accuracy} = \text{maximum error} = V_{r(\text{max or min})} - V_a. \tag{5.3}$$

In many instances accuracy is expressed as a percentage, based either on the *actual scale reading* or on what is termed *full-scale reading.* The latter is the maximum value of the particular range being employed:

$$\text{Percent accuracy based on reading} = \frac{V_{r(\text{max or min})} - V_a}{V_a} \times 100, \tag{5.3a}$$

$$\text{Percent accuracy based on full scale} = \frac{V_{r(\text{max or min})} - V_a}{V_{fs}} \times 100, \tag{5.3b}$$

where V_{fs} is the maximum reading the measuring system is capable of for the particular setting or scale being employed. It is most common to find accuracies expressed in terms of full scale.

It should be pointed out that the percent accuracies indicated by Eqs. (5.3a) and (5.3b) are normally used to express equipment accuracies, and when so used do not include procedural and personnel performance. The values do not, therefore, express the true total performance of the measuring system.

m) Precision. The degree of agreement between repeated results—measurement with small random error. To cite an extreme example [4]: If all clocks in a jewelry store are set at 8:20 but are not running, the indicated values show precision but are accurate only twice in 24 hours.

Precise data have small dispersion (spread or scatter), but may be far from the true value.

n) *Definition*. An evaluation of the consistency of the measured quantity is its definition. For example, the width of a desktop may not be exactly defined. The edges are not perfect, nor are the sides exactly straight or parallel. For these reasons, even if our measuring system were perfect we could not always expect to obtain the same results. This variation may be closely associated with uncertainty. However, it is only one component of the latter, involving the uncertainty of definition of the measured quantity itself.

o) *Uncertainty*. Possible error or what one thinks may be the range of error. This differs from accuracy, for although accuracy may not actually be known, it is a definite concrete number for a given situation. Uncertainty, on the other hand, is the region in which one believes (or guesses) the error to be. This relation may be described in terms of an analogous idea, a limit dimension. A machined dimension may be specified on a drawing through use of tolerances. When the drawing is made, the dimension of a *particular* part is unknown; however, the range is definitely known. After a part is produced, it then has a definite dimension, deviating a definite amount from nominal. It may be said that this deviation expresses the dimension's *accuracy*, whereas the tolerance range expresses the *uncertainty*.

p) *Propagation of uncertainty*. The manner in which the uncertainties affect the result.

Certain errors may add directly, whereas others may not. For example, if a micrometer having an inherent error in thread pitch is used to measure two dimensions of a rectangular area, the area calculation will involve errors in both width and thickness and the total error will depend on the interaction of the two errors. On the other hand, if pressure is measured with the simple tire gage shown in Fig. 2.1, errors may be introduced by incorrect piston size and by incorrect spring constant. In the latter case it is clear that the errors could possibly be compensating, whereas this would be impossible in the first example. The former are termed *dependent* errors and the latter *independent* errors. A study of propagation of errors must include consideration of the interrelationship of the various types of error.

A relationship often employed for summing *equally weighted, independent* errors is

$$e_i = \sqrt{(e_1)^2 + (e_2)^2 + \cdots + (e_n)^2}, \tag{5.4}$$

where

$$e_i = \text{overall independent error,}$$

$$e_1, e_n = \text{independent errors.}$$

In addition, what are termed *correlated* errors may be present when a systematic relation exists between the two. This would occur, for example, when temperature and pressure measurements are both required and a poorly temperature-compensated pressure gage is employed.

q) Mean (or average). Summation of results divided by the number of results. This is considered the "best" approximation of the true value.

r) Median. Middle value of a series of data—as many values fall to one side of the median as to the other side.

s) Mode. The most frequently occurring result. More than one mode is possible.

t) Range. The difference between the largest and the smallest result.

u) Dispersion (or scatter). The manner in which the results lie about the mean—a measure of reliability.

v) Deviation (also called the "residual"). The difference between a single result and the mean of many results.

w) Mean deviation (also called "probable error"). The sum of the absolute deviations divided by their number. There is equal probability that the error will be either larger than or smaller than this value.

x) Standard deviation (also called "mean square error"). The square root of the mean of the squares of the deviations. This quantity represents a common measure of the preciseness of a sample of data. [*Note:* A "better" value is obtained when the summation of the square is divided by $(N - 1)$ rather than by (N)].

y) Percent standard deviation. The ratio of the standard deviation to the mean, expressed as a percentage.

z) Variance. The square of the standard deviation.

5.4 TREATMENT OF MULTISAMPLE DATA

When a number of multisample observations are experimentally obtained, discrepancies occur displayed by the dispersion or scatter of the data about some average result. The distribution may be presented in the form of a *histogram** (Fig. 5.2), in which the height of each bar represents the frequency of occurrence of the indicated value—, in this case, pressure. The worthiness of data of this sort may be evaluated through application of certain statistical rules. First of all, there is a variety of possible theoretical frequency distributions; e.g., linear, binomial, Poisson, Gaussian, etc. [3].

---

* In the example, each bar of the histogram represents a single value of pressure. In general, each bar is made to represent a range of values. Selection of different ranges results in different histograms. One criterion is known as Sturges' rule [9]:

$$k = 1 + 3.3 \log_{10} N,$$

where k is the number of class intervals (bars) and N is the number of items of data. In general, the calculated value of k will be nonintegral but may arbitrarily be adjusted to the nearest round figure.

The latter two distributions are variations of the binomial. Certain sources of data usually fit one kind of distribution more consistently than they do another kind. For example, the frequency of the possible sums of two thrown dice behaves in accordance with linear distribution. The binomial distribution adapts to situations involving two possible results, such as two sides of a coin. The Poisson distribution is of particular importance to the nuclear physicist. It is a fact, however, that many experimentally determined sets of data with random error fit the Gaussian distribution (also known as the normal error distribution) better than any other. For this reason it is widely employed as the assumed distribution for evaluating experimentally determined data. The remainder of this discussion will therefore be concerned only with the Gaussian function.

When data abide by Gaussian or normal distribution rules, plus and minus errors are equally likely, and small errors are more probable than large errors; but there is no real limit to the magnitude of large errors. The equation for such a distribution, assuming an infinite population, may be written as follows:

$$y_x = \frac{h}{\sqrt{\pi}} e^{-h^2(x-m)^2}, \tag{5.5}$$

where

$x = $ the magnitude of a result,

$y_x = $ the frequency of occurrence of a given result, or the probability of its occurrence,

$m = $ the mean value of the population,

$h = $ a constant depending on the data distribution.

Figure 5.1 is a plot of this function for two different values of h. It is seen that the distribution represented by the dashed curve indicates a greater number of small errors and a correspondingly smaller number of large errors. Data represented by the dashed curve are said to be more precise than those represented by the solid curve.

It can be shown [5] that if a large number of measurements are all taken with equal care, the most probable value of the measurement is the arithmetic mean, or

$$\bar{x} = \frac{x_1 + x_2 + \cdots + x_n}{n} = \frac{\sum x_i}{n}. \tag{5.6}$$

By definition the deviation, d, is the difference $x - \bar{x}$, or

$$d_1 = x_1 - \bar{x}, \qquad d_2 = x_2 - \bar{x}, \qquad d_n = x_n - \bar{x}.$$

It can be shown that the arithmetic mean $\bar{x}$ is such that the sum of the squares of the deviations is a minimum. Conversely, if the sum of

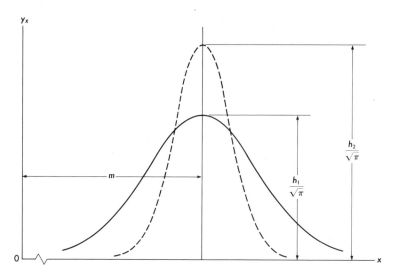

Fig. 5.1 Normal distribution curve. More *precise* data are represented by the dashed curve than by the solid curve.

the squares of the deviation is a minimum, the most probable value of the measured quantity is $\bar{x}$.

Mean and standard deviations. The following equation defines the *mean deviation:*

$$\text{mean deviation} = d_m = \frac{|d_1| + |d_2| + \cdots + |d_n|}{n}. \tag{5.7}$$

Note that the numerical values of the deviations are added without regard to algebraic sign.

Standard or mean-square deviation is defined by the relation

$$\sigma = \sqrt{\frac{(d_1)^2 + (d_2)^2 + \cdots + (d_n)^2}{n-1}}. \tag{5.8}$$

As stated previously, the actual value by which a result is in error is never known. There are, however, various ways of estimating error. Mean deviation (also known as *probable error*) and standard deviation are two. From the definition of mean deviation, it is seen that for a given population, there is an equal probability that a randomly selected value will differ from the mean by an amount greater than (or less than) the average deviation. In this connection the term "probable error" is applied to the mean deviation. This quantity is not, however, the most popularly used error estimate.

TABLE 5.1

SUMMARY OF ERROR ESTIMATES BASED ON NORMAL DISTRIBUTION

Name of error	Symbol	Value in terms of σ	Percent certainty	Probability that a single value will be greater
Probable error (also mean deviation)	E_p	0.6745σ	50	1 in 2
Standard deviation	σ	σ	68.3	1 in 3 (approx.)
90% error	E_{90}	1.6449σ	90	1 in 10
Two sigma error	2σ	2σ	95	1 in 20
Three sigma error	3σ	3σ	99.7	1 in 370
Maximum* error	E_{max}	3.29σ	99.9 +	1 in 1000

* Some regard the 95% error as "maximum." In any case, of course, a practical maximum is being considered. The actual maximum is theoretically infinite.

Standard deviation, σ, has an analogous mechanical counterpart in the radius of gyration, as applied to the moment of inertia. If the area under the normal distribution curve is considered a two-dimensional body rotating about the mean, then the value of σ corresponds to its radius of gyration. Standard deviation is particularly useful in further statistical treatment of data, and is therefore a very commonly used quantity for expressing error estimate.

For normal distribution, the probability of obtaining a deviation greater than 2σ is about 5%; greater than 3σ, about $\frac{1}{3}$%; and greater than 4σ, approximately 0.006%. One criterion for discarding suspected data is the 3σ value; any data having a deviation greater than 3σ are arbitrarily thrown out. Table 5.1 compares several of these error estimates.

5.5 EXAMPLE OF THE TREATMENT OF MULTISAMPLE DATA

Let us employ an example to illustrate an application of the statistical treatment of test data. Columns 1 and 2 in Table 5.2 represent values from a fictitious test in which a pressure is measured 100 times with variations in apparatus and procedures. We will assume that all the known corrections have been applied and that column 1 lists the results. It will be noted that the maximum discrepancy is 8 psi, as determined from the greatest differences in results. Column 2 lists the frequency of each value, and Fig. 5.2 shows the distribution. Columns 3 and 4 are the deviations and squares thereof, respectively.

The average of the readings is calculated as 400.78 psia, the mean deviation is 1.028, and the standard deviation is 1.389. The latter figure

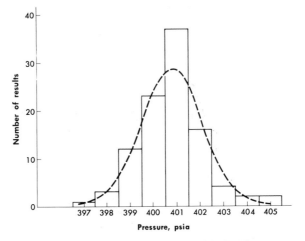

Fig. 5.2 Distribution of data listed in Table 5.2.

TABLE 5.2

Pressure, P, psia	Number of results, n	Deviation, d	d^2
397	1	−3.78	14.288
398	3	−2.78	7.728
399	12	−1.78	3.168
400	23	−0.78	0.608
401	37	0.22	0.048
402	16	1.22	1.488
403	4	2.22	4.928
404	2	3.22	10.368
405	2	4.22	17.808
40,078	100	102.84	191.12

Arithmetical average =

$P_0 = 40,078/100 = 400.78$ psia,

d_m = mean deviation = $102.84/100 = 1.028$ psia,

σ = standard deviation = $\sqrt{191.12/99} = 1.389$ psia,

means that should we arbitrarily select any single result, there is a 68.3% chance that it will fall within the region $(400.78 + 1.389) = 402.17$ and $(400.78 − 1.389) = 399.39$ psia, provided (and this is an important proviso) that the data are truly distributed in accordance with Eq. (5.5). In this connection, it should be clear that the various numerical values relating such quantities as mean deviation and standard deviation (Table 5.1) assume a true normal distribution. Although experimental data may approximate normal distribution, they seldom, if ever, fit the bell-shaped curve exactly. Goodness of fit, therefore, becomes a legitimate question, and various methods have been devised to evaluate this factor. The reader is referred to references at the end of this chapter for further discussion of this matter.

The question may be asked, "Is it not possible to obtain more than one sample of a given population? If this is so, should we not expect that each sample would yield a slightly different average or mean? Hence should it not be possible to treat the resulting means in a statistical manner?" This is indeed possible, and it can be shown that the standard deviation of the resulting means may be expressed by the relationship

$$\sigma_m = \frac{\sigma}{\sqrt{n}}, \tag{5.9}$$

where σ_m is termed the *standard deviation of the mean* and is a measure of the probable distribution of the means that would be obtained by taking many samples of a given population.

For our example above,

$$\sigma_m = \frac{1.389}{\sqrt{100}} = 0.14 \text{ psia.}$$

If, as a final result of our example, we should write

$$P = 400.78 \pm 0.14 \text{ psia,}$$

this should be interpreted as meaning that the final arithmetic average (400.78 psia) has a 68% chance of falling within the interval $400.78 - 0.14 = 400.64$ and $400.78 + 0.14 = 400.92$ psia. Perhaps we would wish to increase the probability range by using a $2\sigma_m$ tolerance. In this case we would assume a 95% chance that the correct mean would fall within the interval 400.78 ± 0.28 psia.

If multiple samples of a given population are taken, then the resulting standard deviation should be susceptible to similar treatment, and a standard deviation of the standard deviation, σ_σ, should have meaning. It may be shown that

$$\sigma_\sigma = \frac{\sigma}{\sqrt{2n}} = \frac{\sigma_m}{\sqrt{2}}. \tag{5.10}$$

Returning to our example, we have

$$\sigma_\sigma = \frac{1.389}{\sqrt{2 \times 100}} = 0.099 \text{ psia.}$$

This result means that there is a 68% chance that σ will fall within the interval 1.389 ± 0.099 psia, giving us a measure of confidence in this value.

Inspection of Eqs. (5.8), (5.9), and (5.10) shows that in each case, n, the number of experimental data points, enters the relationships under the radical. It is seen therefore that results are improved rather slowly as additional data are accumulated.

It must be completely clear that the statistical methods discussed above apply only to random error; they do not account for systematic errors. In truth, therefore, it is more a treatment of precision than it is of error. Systematic error is a much more insidious factor than is random error and must be minimized through careful attention to procedures, calibration and measuring-system design, selection, and application.

It is seen that the final expression for the result of our test must be based on an exercise of judgment on the part of the experimenter, but that statistical methods can be employed as a powerful assistance.

5.6 TREATMENT OF SINGLE-SAMPLE DATA

When the available data are single-sample, methods discussed in the preceding paragraphs are not directly applicable, and it becomes necessary to modify our approach. Because of the single-sample nature of our data, we cannot directly observe their scatter by plotting a frequency distribution curve. We are in the position, however, to indicate what we *think* its nature might be, and the more experienced we are in the nature of the test, the more accurate should be our prediction. By applying the methods of probability and statistics, Kline and McClintock [10] have proposed an analysis employing *uncertainty distribution* rather than frequency distribution. They define uncertainty distribution as the error distribution the experimenter believes would exist if the situation would permit multisampling. They also make a good case for treating single-sample systematic errors in the same manner as single-sample random errors.

After considering various possible approaches, Kline and McClintock suggest that a single-sample result be expressed in terms of a mean value and an uncertainty interval, based on stated odds. This result may be written as follows:

$$V_a = V_m \pm w \qquad (b \text{ to } 1),$$

where

V_m = the value if only one reading is available, or the arithmetic mean of several readings,

w = the uncertainty interval,

b = odds, or the chance that the true value lies within the stated range, based on the *opinion* of the experimenter (for who has better knowledge of the equipment and conditions?).

As an example, an experimentally determined pressure may be indicated as

$$P = 27.8 \pm 0.3 \text{ psia} \qquad (20 \text{ to } 1).$$

This in effect states that the experimenter is willing to wager 20 to 1 that the true value of pressure lies between 27.5 and 28.1 psia.

This approach is of particular value in setting up an experiment, especially if the test is expensive in terms of manpower, time, and equipment. It provides a basis for establishing *predetermined* estimates of the reliability of results through a study of propagation of uncertainties [11]. By this means judgment as to the value of the test may be made before the test is run.

SUGGESTED READINGS

Beers, Y., *Introduction to the Theory of Error.* Reading, Mass.: Addison-Wesley Publishing Co., 1957.

Natrella, M. G., *Experimental Statistics.* National Bureau of Standards Handbook 91, United States Department of Commerce, 1963.

Neville, A. M., and J. B. Kennedy, *Basic Statistical Methods for Engineers and Scientists.* Scranton, Pa.: International Textbook Co., 1964.

Young, H. D., *Statistical Treatment of Experimental Data.* New York: McGraw-Hill Book Co., 1962.

PROBLEMS

5.1 A car travels from A to B at an average speed of 30 mph, and from B to A at an average speed of 60 mph. What is the average speed for the entire trip?

5.2 If the car in Problem 5.1 traveled from point A to point B at an average speed of 30 mph, at what speed must it travel from B to A to maintain an overall average of 60 mph?

5.3 Three teachers reported the mean grades in their classes as follows:

Teacher	Mean grade	Number of students
1	81	28
2	78	23
3	86	17

What is the mean grade for all the classes taken together?

5.4 Given the following two sets of data:

A. 16, 12, 10, 10, 11, 9, 7, 22, 19, 14

B. 13, 15, 7, 11, 12, 22, 12, 13, 13, 12

a) For each set separately, find: (1) range, (2) mean, (3) variance, (4) standard deviation, (5) standard deviation of the mean, (6) standard deviation of the standard deviation.

b) Express the meanings of (4), (5), and (6) in words.

c) Do part (a) for the total set of data, A and B combined.

5.5 The accompanying table lists a "sample" of experimental data along with certain calculations that have been made from it.

Frequency, n	Value, x_i	$x_i - \bar{x}$	$(x_i - \bar{x})^2$
1	3	−4.062	16.500
2	4	−3.062	9.376
3	5	−2.062	4.252
6	6	−1.062	1.128
7	7	−0.062	0.004
6	8	0.938	0.880
4	9	1.938	3.756
2	10	2.938	8.632
1	11	3.938	15.509
$\sum n = 32$	$\sum nx_i = 226$		$107.881 = \sum n(x_1 - \bar{x})^2$

$$\bar{x} = \frac{226}{32} = 7.062$$

On the basis of the tabulated data:

a) Calculate the mean deviation.

b) Calculate the standard deivation.

c) Calculate the percent standard deviation.

d) Calculate the standard deviation of the mean, σ_m. What is the meaning of this quantity?

e) Calculate the standard deviation of the standard deviation σ_σ.

f) It is sometimes said that any data falling outside the three-sigma limits should be discarded. On this basis, should any of the above data be dropped?

g) What value would you give as the best estimate of the true value?

5.6 A mechanical comparator with a resolution of 0.001 in. was used to measure the diameters of a sample of 300 glass marbles selected at random from a population of 2500 marbles. (The marbles were employed in a special chemical absorption column.) The distribution of diameters was as follows (the first number represents the frequency of occurrence, and the second represents the diameter in inches):

1	0.590	3	0.614	8	0.629	3	0.644
1	0.595	10	0.615	8	0.630	1	0.646
1	0.596	7	0.616	7	0.631	2	0.647
2	0.597	8	0.617	11	0.632	1	0.649
2	0.602	8	0.618	10	0.633	2	0.650

2	0.603	12	0.619	8	0.634	3	0.651
3	0.604	10	0.620	6	0.635	1	0.654
2	0.606	3	0.621	6	0.636	2	0.655
5	0.607	8	0.622	2	0.637	1	0.657
1	0.608	10	0.623	11	0.638	1	0.659
4	0.609	13	0.624	2	0.639		
3	0.610	13	0.625	11	0.640		
3	0.611	16	0.626	5	0.641		
5	0.612	13	0.627	2	0.642		
5	0.613	4	0.628	8	0.643		

a) Plot a histogram. Use Sturges' rule for determining the number of class intervals. (Try other numbers of class intervals and note the very considerable changes in histogram shapes.)

b) Determine: (1) the mean, (2) the mean deviation, (3) standard deviation, (4) percent standard deviation, (5) standard deviation of the mean, (6) standard deviation of the standard deviation.

c) Write each diameter on a separate card or slip of paper. Place the cards in a container and make a random drawing of 30 cards. Using the 30 items of data, carry out parts (a) and (b) above. How well do the results compare?

CHAPTER 6

BASIC DETECTOR-TRANSDUCER ELEMENTS

6.1 PRELIMINARY COMMENTS

The first contact that a measuring system has with the quantity to be measured is through the input sample accepted by the detecting element of the first stage (see Article 2.3). This act is usually accompanied by the input's being immediately transduced into an analogous form.

The media handled in this process is *information*. The detector senses the information input, and the transducer puts it into a more convenient form. Simpson [1] uses the term *transfer efficiency* as a basis for comparing first-stage devices. The term may be defined by the relation

$$\text{Transfer efficiency} = \frac{I_{\text{out}}}{I_{\text{in}}}, \qquad (6.1)$$

where

$I_{\text{out}} = $ information delivered by the pickup,

$I_{\text{in}} = $ information received by the pickup.

He states that the transfer efficiency may not be more than unity, because the pickup cannot generate information, but can only receive and process it. Obviously it is desirable to have as high a transfer efficiency as possible.

6.2 LOADING OF THE SIGNAL SOURCE

Energy will always be taken from the signal source by the measuring system, which means that the information source must always be changed by the act of measurement. This is a measurement axiom. This effect may be referred to as *loading*. The smaller the load placed on the signal source by the measuring system, the better.

Of course, the problem of loading occurs not only in the first stage, but throughout the entire chain of elements. While the first-stage detector-transducer loads the input source, the second stage loads the first stage, and finally the third stage loads the second stage. In fact, the loading problem may be carried right down to the basic elements themselves.

106

In measuring systems made up primarily of electrical elements, the loading of the signal source is almost exclusively a function of the detector. Intermediate modifying devices and output indicators or recorders receive most of the energy necessary for their functioning from sources *other* than the signal source. A measure of the quality of the first stage, therefore, is its ability to provide a usable output without draining an undue amount of energy from the signal.

6.3 THE SECONDARY TRANSDUCER

As an example of a detector in its simplest form, consider the ordinary dial indicator shown in Fig. 6.1. In this case the spindle acts as a detector through its contact with the signal source. However, it does no more than furnish this information to the stage-two linkage. It simply performs the function of detection, and nothing else.

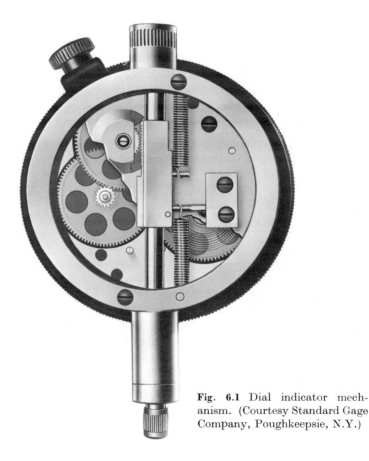

Fig. 6.1 Dial indicator mechanism. (Courtesy Standard Gage Company, Poughkeepsie, N.Y.)

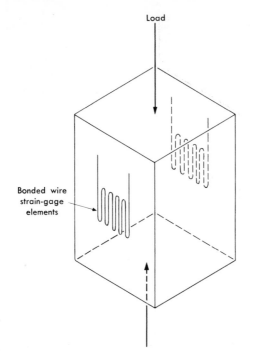

Load

Bonded wire
strain-gage
elements

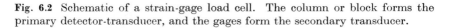

Fig. 6.2 Schematic of a strain-gage load cell. The column or block forms the primary detector-transducer, and the gages form the secondary transducer.

On the other hand, let us analyze a simplified compressive-type force-measuring *load cell* consisting of a short column or strut, with strain gages attached thereto (Fig. 6.2). Let us assume that a force deflects or strains the block, which thereby senses the input. It converts the load into a deflection analogous to the weight or force. The unit deflection is, in turn, transformed into an electrical output by the strain gages acting as *secondary transducers*.

As another example, consider the familiar Bourdon tube, which acts as a primary detector-transducer sensing pressure, but whose output in the form of displacement is used directly to drive the linkage chain of the stage-two elements. We find that in this case a secondary change in signal form is not necessary.

On the other hand, the displacement from the Bourdon tube might be used to move the core of a differential transformer (see Article 6.14) from which a voltage output would be obtained. In this case the transduction would be through the following sequence: pressure to displacement to voltage. The differential transformer would perform the function of a secondary transducer.

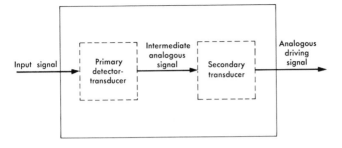

Fig. 6.3 Block diagram of a first-stage device with primary and secondary transducers.

6.4 CLASSIFICATION OF FIRST-STAGE DEVICES

It appears, therefore, that the stage-one instrumentation may be of varying basic complexity, depending on the number of operations performed. This leads to a classification of first-stage devices as follows:

CLASS I. First-stage element used as detector only.

CLASS II. First-stage elements used as detector and single transducer.

CLASS III. First-stage elements used as detector with two transducer stages.

A generalized first stage may therefore be shown schematically as in Fig. 6.3.

Stage-one instrumentation may be very simple, consisting of no more than a mechanical spindle or contacting member used to transmit the quantity to be measured to a secondary transducer. Or, it may consist of a much more complex assembly of elements. In any event the primary detector-transducer is an *integral* assembly whose function is (1) to selectively sense the quantity of interest, and (2) to process the sensed information into a form acceptable to stage-two operations. It does not present an output in immediately usable form.

More often than not the initial operation performed by the first-stage device is to transduce the input quantity into an analogous displacement. Without attempting to formulate a completely comprehensive list, let us consider Table 6.1 as representing the general area of the primary detector-transducer in mechanical measurements.

Several general observations may be made before we discuss some of the specific entries in Table 6.1. Note that in by far the greatest number of cases, displacement is the *output quantity* of the mechanical or hydropneumatic devices, while it is the *input* to many of the electrical types. This results in a very workable combination, with the mechanical device serving as the *primary detector-transducer*, the electrical device serving as the *secondary transducer*, and with displacement serving as the intermediate signal form.

TABLE 6.1

TYPICAL PRIMARY DETECTOR-TRANSDUCER ELEMENTS AND THE
OPERATION THEY PERFORM

Type	Operation
I. Mechanical	
A. Contacting spindle, pin, or finger	Displacement to displacement
B. Elastic member	
1. Proving ring	Force to displacement
2. Bourdon tube	Pressure to displacement
3. Bellows	Pressure to displacement
4. Diaphragm	Pressure to displacement
5. Spring	Force to displacement
C. Mass	
1. Seismic mass	Forcing function to displacement
2. Pendulum scale	Force to displacement
3. Manometer	Pressure to displacement
D. Thermal	
1. Thermocouple	Temperature to electric current
2. Bimaterial (includes mercury in glass)	Temperature to displacement
3. Temp-stik	Temperature to phase
E. Hydropneumatic	
1. Static	
a) Float	Fluid level to displacement
b) Hydrometer	Specific gravity to displacement
2. Dynamic	
a) Orifice	Velocity to pressure
b) Venturi	Velocity to pressure
c) Pitot tube	Velocity to pressure
d) Vanes	Velocity to force
e) Turbines	Linear to angular velocity
II. Electrical	
A. Resistive	Displacement to resistance change
1. Contacting	Displacement to resistance change
2. Variable-length conductor	Displacement to resistance change
3. Variable area of conductor	
4. Variable dimensions of conductor	Strain to resistance change
5. Variable resistivity of conductor	Air velocity to resistance change
	Temperature to resistance change

TABLE 6.1 (*Continued*)

Type	Operation
B. Inductive	
1. Variable coil dimensions	Displacement to inductance change
2. Variable air gap	Displacement to inductance change
3. Changing core material	Displacement to inductance change
4. Changing coil positions	Displacement to inductance change
5. Changing core positions	Displacement to inductance change
6. Moving coil	Velocity to inductance change
7. Moving permanent magnet	Velocity to inductance change
8. Moving core	Velocity to inductance change
C. Capacitive	
1. Changing air gap	Displacement to capacitance change
2. Changing plate areas	Displacement to capacitance change
3. Changing dielectric	Displacement to capacitance change
D. Electronic	Displacement to current
E. Piezoelectric	Displacement to voltage
F. Photoelectric	Light intensity to voltage
G. Streaming potential	Flow to voltage

No attempt will be made to discuss all the many combinations of detector-transducer elements possible in Table 6.1. Many are self-evident to the mechanical engineer; however, others warrant a few words of explanation.

6.5 MECHANICAL MEMBERS AS PRIMARY DETECTORS

Most mechanical primary detecting means are well enough known so as to require little more than simple enumeration at this point.

Elastic members are used to change force into displacement. There are many forms from which to choose, but basically they fall into one or a combination of the three following categories: (1) direct tension or compression, (2) bending, and (3) torsion.

Proving rings (Fig. 12.7) have long been used as secondary force standards for calibrating testing and weighing machines. Micrometer arrangements or dial gages are often used for measuring the deflections; however, electrical transducer elements may also be used. For capacities higher than proving rings can accommodate, direct tension or compression members are often used. Strain gages may be employed as secondary transducers to measure the deflections (see Article 12.4c).

Torque meters usually, although not always, make use of elastic torsion members (Article 12.8). The member twists in proportion to the applied torque, and the deformation is used as a measure of torque. Here again, secondary transducer elements are employed to provide a usable output.

Most pressure-measuring devices use elastic members of one type or another (Chapter 13). The actions of the diaphragm, the bellows, or the Bourdon tube are all based on elastic deformations brought about by the force resulting from pressure summation. Indeed, the manometer may also be considered a form of elastic device whose deflection is proportional to force.

In the case of pressure-measuring devices, as with force measurement, the mechanical displacements are usually so small (except for the manometer) that secondary transducers must be used in conjunction with stage-two devices to provide interpretable outputs.

Another basic mechanical detector-transducer element is provided by the inertia of a concentrated mass (Chapter 16). In the form of an accelerometer (Fig. 16.4) this serves to measure the characteristics of dynamic motion, i.e., displacement, velocity, acceleration, and frequency, through application of Newton's laws of motion.

The pendulum, or any simple mechanically vibrating member, may serve as a *time* or frequency transducer, chopping the passage of time into discrete bits. The balance wheel and escapement in a watch is an example.

Temperature detection (Chapter 15) may be based on differential expansion of two different materials, such as in the case of the simple thermometer, as one example, or the bimetal strip thermometer as another. Or, temperature may be determined from thermoelectric properties of materials or combinations of materials. The resistance thermometer and the thermocouple are examples of the latter.

Of course, the simplest types of hydropneumatic devices are based on the ordinary float idea, or that of the hydrometer. The simple float is used primarily as a liquid-level detector and makes no allowance for change in density of the supporting liquid, it being assumed that the float is always immersed to the same depth. On the other hand, the hydrometer uses the depth of immersion as a means for detecting variations in specific gravity of the supporting liquid. Both these examples, of course, apply to static conditions.

Orifices and venturis (see Article 14.3) provide flow information in the form of pressure change as a result of energy transformation. Because of the change in area of the passageway, the necessary increase in velocity results at the expense of pressure. Therefore, when pressure change is measured, a measure of velocity is obtained.

Fluid velocity may also be measured by reference to aero- or hydrodynamic principles. The pitot tube is used to determine pressures resulting from total flow-rate rather than change of rate. Vanes in the form of airfoils or turbine wheels may also be used to sense fluid flow. Vanes may be elastically supported such that their displacement becomes a measure of flow, or the angular velocity of the turbine wheel can be used.

Most of these mechanical transducer principles are discussed in considerable detail in later chapters and therefore are only mentioned in passing at this point.

6.6 ELECTRIC TRANSDUCER ELEMENTS

Almost without exception, the *electrical* stage-one element serves as a secondary transducer. In most cases it converts mechanical displacement to voltage. The quantity of interest is first detected and transduced to displacement by some form of mechanical element; then the electric element serves as the secondary transducer, transforming the analogous displacement into an analog voltage or current. Actually the basic electrical change may be resistive, inductive, capacitive, etc., from which the voltage or current change results.

The two most commonly used principles of operation employed for this purpose are variable resistance and variable inductance. In addition, there are a number of other modes of operation, such as variable capacitance, piezoelectric, photoelectric, etc. These are of more limited use and are usually employed for special-purpose applications.

6.7 ADVANTAGES OF ELECTRICAL-SYSTEM ELEMENTS

In addition to the inherent compatibility of mechano-electrical transducer principles, electrical elements used in systems for mechanical measurements have several other very important advantages, which may be itemized as follows:

1. Amplification or attenuation may be easily obtained.
2. Mass-inertia effects are minimized.
3. The effects of friction are minimized.
4. An output with sufficient power for control may be provided.
5. Remote indication or recording is feasible.
6. The transducers are usually susceptible to miniaturization.

6.8 VARIABLE-RESISTANCE TRANSDUCER ELEMENTS

Resistance of an electrical conductor varies according to the following relation:

$$R = \frac{\rho L}{A}, \tag{6.2}$$

where

R = resistance, ohms,

L = length of conductor, cm,

A = cross-sectional area of conductor, cm²,

ρ = resistivity of material, ohms-cm.

Probably the simplest mechanical-to-electrical transducer is the ordinary *switch*. It is a yes-no, conducting-nonconducting device which can be used to operate an indicator. Here a lamp is fully as useful for readout as a meter, since only two values of quantitative information can be obtained. In its simplest form, the switch may be used as a limiting device operated by direct mechanical contact (as for limiting the travel of machine-tool carriages) or it may be used as a position indicator. When actuated by a diaphragm or bellows, it becomes a pressure-limit indicator, or if controlled by a bimetal strip, it is a temperature-limit indicator. It may also be combined with a proving ring to serve as either an overload warning device or as a device actually limiting load-carrying, such as a safety device for a crane.

6.9   SLIDING-CONTACT DEVICES

Sliding-contact resistive transducers convert a mechanical displacement input into an electrical output, either voltage or current. This is accomplished by changing the effective length, L, of the conductor in Eq. (6.2). Some form of electrical-resistance element is used, with which a contactor or brush maintains electrical contact as it moves. In its simplest form, the device may consist of a stretched resistance wire and slider, as in Fig. 6.4. The effective resistance existing between either end of the wire and the brush thereby becomes a measure of the mechanical displacement. Devices of this type have been used for sensing relatively large displacements [2].

More commonly, the resistance element is formed by wrapping a resistance wire around a form or *card*. The turns are spaced to prevent shorting,

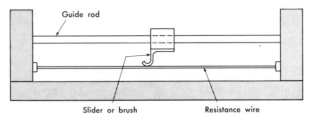

Guide rod

Slider or brush   Resistance wire

Fig. 6.4 Variable resistance consisting of a wire and movable contactor or brush. This is often referred to as a slide wire.

and the brush slides across the turns from one turn to the next. In actual practice, the arrangement may be either wound for a rectilinear movement or the resistance element may be formed into an arc and angular movement used, as shown in Fig. 6.5.

These devices are commonly termed *resistance potentiometers,** or simply *pots*. Variations of the basic angular or rotary form are the multiturn, the low-torque, and various nonlinear types [3]. Multiturn potentiometers are available with various numbers of turns, up to 40. Also see Article 7.16.

a) *Potentiometer resolution.* Resistance variation available from a sliding contact moving over a wire-wound resistance element is not a continuous function of contact movement. The smallest increment into which the whole may be divided determines the *resolution*. In the case of resistance winding, the limiting resolution equals the reciprocal of the number of turns. If 1200 turns of wire are used and the winding is linear, the resolution will be 1/1200, or 0.09083%. The meaning of this quantity is apparent: no matter how refined the remainder of the system may be, it will be impossible to divide, or resolve, the input into parts smaller than 1/1200 of the total potentiometer range.

b) *Potentiometer linearity.* When used as a measurement transducer, a linear potentiometer is normally required. Use of the term *linear* assumes

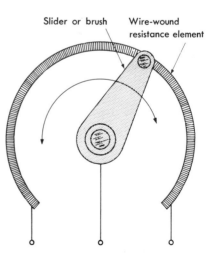

Slider or brush   Wire-wound
                    resistance element

Fig. 6.5 Angular-motion variable resistance, or potentiometer.

---

* Unfortunately, another entirely different device is also called a potentiometer. It is the voltage-measuring instrument wherein a standard reference voltage is adjusted to counterbalance the unknown voltage (see Article 7.17). The two devices are different and must not be confused.

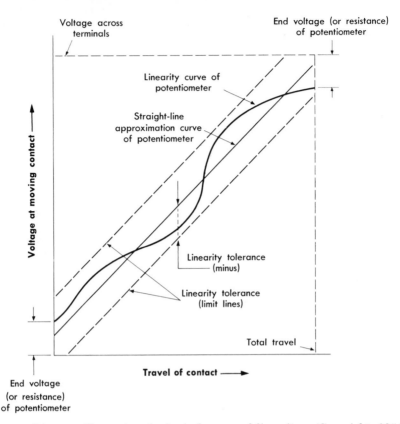

Fig. 6.6 Diagram illustrating the basis for normal linearity. (Copyright, 1954, Beckman Instruments, Inc., Fullerton, California. Reprinted by permission of Helipot Division.)

that the resistance measured between one of the ends of the element and the contactor is a direct linear function of contactor position relative to that end. Linearity, however, is never completely achieved, and the deviation from the ideal is termed linearity tolerance. Figure 6.6 illustrates the basis for *normal* linearity tolerance.

6.10 THE WIRE RESISTANCE STRAIN GAGE

The wire resistance strain gage is a very interesting device, and experiment has shown that each term in Eq. (6.2) is simultaneously affected by the input strain. The wire is cemented to the surface of the strained member, and as the wire elongates with application of strain (assuming a tensile strain) its diameter reduces and a longer length of smaller wire results. However, simply accounting for these dimensional changes does not explain the gage action; there is also a change in resistivity with strain.

This device is of sufficient importance in the field of mechanical measurements to warrant a fuller discussion than can be given at this point. Chapter 11 is devoted entirely to the theory and use of strain gages.

6.11 THERMISTORS [4]

Thermistors are thermally sensitive variable resistors made of ceramic-like semiconducting materials. Oxides of manganese, nickel, and cobalt are used in formulations having resistivity values of 100 to 450,000 ohms-cm.

There are two basic applications for these devices: (1) as temperature-detecting elements used to sense temperature changes for the purpose of measurement or control, and (2) as electric-power-sensing devices wherein the thermistor temperature and hence resistance is a function of the power being dissipated by the device. The second application is particularly useful for measuring radio-frequency power.

Further discussion of thermistors will be found in Article 15.5c.

6.12 THE THERMOCOUPLE

While two dissimilar metals are in contact, an electromotive force exists whose magnitude is a function of several factors, including *temperature.* Junctions of this sort, used to measure temperature, are called *thermocouples.* Often the junction is formed by twisting and welding together two wires.

Because of its small size, its reliability, and its relatively large range of usefulness, the thermocouple is a very important primary element. Further discussion of its application will be reserved for Chapter 15 (see especially Article 15.6).

6.13 VARIABLE-INDUCTANCE TRANSDUCER ELEMENTS

A classification of inductive transducers, based on the fundamental principles employed, is as follows:

I. Variable self-inductance
   A. Single-coil (simple variable permeance)
   B. Two-coil (or single-coil with center tap) connected for inductance ratio

II. Variable mutual inductance
   A. Simple two-coil
   B. Three-coil (using series opposition)

III. Variable reluctance
   A. Moving iron
   B. Moving coil
   C. Moving magnet

Inductive reactance, which is the evaluation of the inductive effect, may be expressed by the relation

$$X_L = 2\pi f L,$$ (6.3)

where

X_L = inductive reactance, ohms,

f = frequency of applied voltage, Hz,

L = inductance, henrys.

Inductance, L, is influenced by a number of factors, including the number of turns in the coil, the coil size, and especially the permeability of the flux path. Some coils are wound with only air as the core material. These are usually used at relatively high frequencies; however, they will occasionally be found in transducer circuitry. Often some form of magnetic material will be used in the flux path, commonly in conjunction with one or more air gaps.

An expression which may be used to estimate the inductance of an air-core coil is as follows [5]:

$$L = \frac{0.2a^2N^2 \times 10^{-6}}{3a + 9b + 10c},$$ (6.3a)

where

L = inductance, henrys,

a = mean diameter of winding, in.,

b = length of winding, in.,

c = radial depth of winding, in.,

N = total number of turns in winding.

When the flux path includes both a magnetic material (usually iron) and an air gap or gaps, the inductance may be estimated by use of the following relation [6]:

$$L = \frac{3.19N^2 \times 10^{-8}}{(h_i/\mu a_i) + (h_a/a_a)},$$ (6.3b)

in which

h_i = length of iron circuit, in.,

h_a = length of air gaps, in.,

a_i = cross-sectional area of iron, in.^{2},

a_a = cross-sectional area of air gap, in.^{2},

μ = permeability of the magnetic material at maximum flux density.

In many instances the permeability of the magnetic material is sufficiently high so that only the air gaps need be considered. In such cases,

Eq. (6.3b) reduces to

$$L = 3.19N^2 \frac{a_a}{h_a} \times 10^{-8}. \qquad (6.3c)$$

The total impedance of a coil may be expressed by the relation

$$Z = \sqrt{X_L^2 + R^2}, \qquad (6.4)$$

in which R is the d-c resistance of the coil. The higher the inductance of a coil relative to its d-c resistance, the higher is said to be its quality, which is designated by the symbol Q. In most cases high Q is desired.

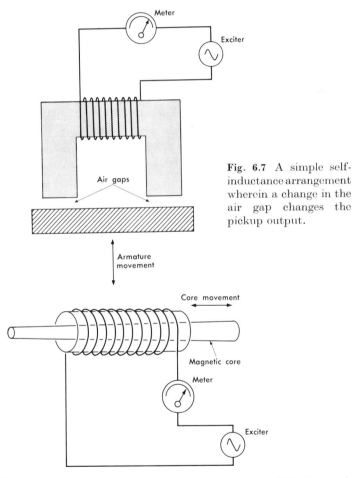

Fig. 6.7 A simple self-inductance arrangement wherein a change in the air gap changes the pickup output.

Fig. 6.8 A self-inductance arrangement wherein the permeance of the flux path is determined primarily by the amount of magnetic material centered in the coil. As the core is moved, the inductance is changed, thereby changing the output.

Inductive transducers may be based on any of the variables indicated in the above equations, and most have been tried at one time or another. The following are representative.

a) *Simple inductance types.* When a simple single coil is used as a transducer element, the mechanical input usually changes the permeance of the flux path generated by the coil, thereby changing its inductance. The change in inductance is then measured by suitable circuitry, indicating the value of the input. The flux path may be changed by a change in air gap; however, a change in either the amount or type of core material may also be employed (Figs. 6.7 and 6.8).

Figure 6.9 illustrates a form of *two-coil* self-inductance. (This may also be thought of as a single coil with center tap.) Movement of the core or armature alters the relative inductance of the two coils. Devices of this type are usually incorporated in some form of inductive bridge circuit (see Article 7.19) in which variation in the inductance ratio between the two coils provides the output. An application of a two-coil self-inductance used as a secondary transducer for pressure measurement is described in Article 13.7b.

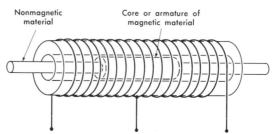

Fig. 6.9 Two-coil (or center-tapped single-coil) inductance-ratio transducer.

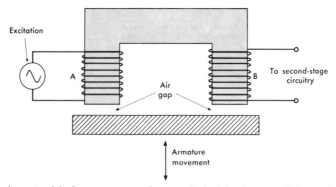

Fig. 6.10 A mutual-inductance transducer. Coil A is the energizing coil, and B is the pickup coil. As the armature is moved, thereby altering the air gap, the output from coil B is changed, and this may be used as a measure of armature displacement.

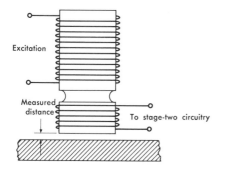

Excitation

Measured distance

To stage-two circuitry

Fig. 6.11 Two-coil inductive pickup for an electronic micrometer.

b) Two-coil mutual-inductance arrangements. Mutual-inductance arrangements using two coils are shown in Figs. 6.10 and 6.11. Figure 6.10 illustrates the manner in which these devices function. The flux from a power coil is coupled to a pickup coil which supplies the output. Input information, in the form of armature displacement, changes the coupling between the coils. In the arrangement shown, the air gaps between the core and the armature govern the degree of coupling. In other arrangements the coupling may be varied by changing the relative positions of either the coils or armature, linearly or angularly (see Fig. 10.8 for an application).

Figure 6.11 shows the detector portion of an *electronic micrometer* [7]. Inductive coupling between the coils, which depends on the permeance of the magnetic-flux path, is changed by the relative proximity of a permeable material. A variation of this has been used in a transducer for measuring small inside diameters [8]. In this case the coupling is varied by relative movement between the two coils.

6.14 THE DIFFERENTIAL TRANSFORMER [10]

One of the most useful of the variable-inductance transducers is the differential transformer, which provides an a-c voltage output proportional to the displacement of a core passing through the windings. It is a mutual-inductance device making use of three coils arranged generally as shown in Figs. 6.12(a) and (b).

The center coil is energized from an external a-c power source, and the two end coils, connected together in phase opposition, are used as pickup coils. Output amplitude and phase depend on the relative coupling between the two pickup coils and the power coil. Relative coupling is, in turn, dependent upon the position of the core. Theoretically, there should be a core position for which the voltage induced in each of the pickup coils will be of the same magnitude, and the resulting output should be zero. As will be seen later, this condition is difficult to attain perfectly.

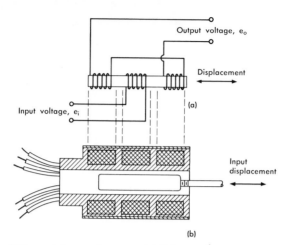

Fig. 6.12 The differential transformer. (a) Schematic arrangement. (b) Section through typical transformer.

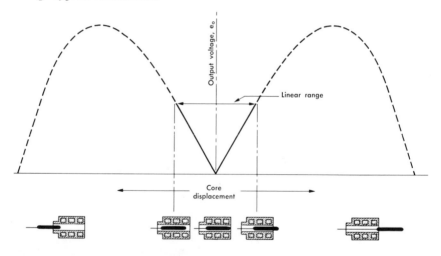

Fig. 6.13 Typical differential transformer characteristics.

Typical differential-transformer characteristics are illustrated in Fig. 6.13, which shows output versus core movement. Within limits, on either side of the null position, core displacement results in a proportional output. In general, the linear range is primarily dependent on the length of the secondary coils. While the output voltage magnitudes are ideally the same for equal core displacements on either side of null balance, the phase relation existing between power source and output changes 180° through null. It is therefore possible, through phase determination or by use of a phase-sensitive

TABLE 6.2
CHARACTERISTICS OF TYPICAL DIFFERENTIAL TRANSFORMERS
Courtesy Schaevitz Engineering Company, Camden, New Jersey.

Type number	Linear range, in.	Transformer size Diameter, in.	Bore, in.	Length, in.	Core size Diameter, in.	Length, in.	Sensitivity mv/0.001 in./V input for 0.5 megohm load — Excitation frequency 60 Hz	400 Hz	2000 Hz	5000 Hz	10,000 Hz	20,000 Hz
HR Series												
050 HR	0.050–0–0.050	$1\frac{3}{16}$	$\frac{3}{8}$	$1\frac{1}{8}$	0.25	0.80	0.73	3.00	3.65	3.64	3.75	
200 HR	0.200–0–0.200	$1\frac{3}{16}$	$\frac{3}{8}$	$2\frac{1}{2}$	0.25	1.90	1.40	2.50	2.44	2.24	2.38	
400 HR	0.400–0–0.400	$1\frac{3}{16}$	$\frac{3}{8}$	$4\frac{23}{64}$	0.25	3.15	0.76	1.01	0.99	0.40	0.61	
1000 HR	1.000–0–1.000	$1\frac{3}{16}$	$\frac{3}{8}$	$6\frac{5}{8}$	0.25	4.25	0.09	0.25	0.33	0.34	0.35	
5000 HR	5.000–0–5.000	$1\frac{3}{16}$	$\frac{3}{8}$	$17\frac{7}{8}$	0.25	6.00	0.04	0.13	0.15	0.16	0.16	
10,000HR	10.000–0–10.000	$1\frac{3}{16}$	$\frac{3}{8}$	$30\frac{27}{32}$	0.25	9.50	0.02	0.07	0.09	0.10	0.11	
HR(operating temperatures from −65°F to +300°F)												
005MS-L	0.005–0–0.005	$\frac{3}{8}$	$\frac{1}{8}$	$1\frac{7}{32}$	0.108	0.180	0.40		1.88			4.55
005MS-LT	0.005–0–0.005	$\frac{3}{8}$	$\frac{1}{8}$	$1\frac{17}{32}$	0.108	0.180	0.24		1.28			4.34
020MS-L	0.020–0–0.020	$\frac{7}{16}$	$\frac{1}{8}$	$1\frac{19}{32}$	0.108	0.233	0.84		3.54			6.20
020MS-LT	0.020–0–0.020	$\frac{7}{16}$	$\frac{1}{8}$	$1\frac{19}{32}$	0.108	0.233	0.47		2.10			4.10
040MS-L	0.040–0–0.040	$\frac{7}{16}$	$\frac{1}{8}$	$\frac{3}{4}$	0.108	0.475 to	1.06		3.00			3.34
040MS-LT	0.040–0–0.040	$\frac{7}{16}$	$\frac{1}{8}$	$\frac{3}{4}$	0.108	0.425 *	0.75		2.42			3.17
100MS-L	0.100–0–0.100	$\frac{7}{16}$	$\frac{1}{8}$	$5\frac{9}{64}$	0.108	0.620	1.24		3.00			2.89
100MS-LT	0.100–0–0.100	$\frac{7}{16}$	$\frac{1}{8}$	$5\frac{9}{64}$	0.108	0.600 *	1.24		3.00			2.89

MS-L (operating temperatures from −40°F to +200°F
MS-LT (operating temperatures from −40°F to +450°F
*Depends on chosen frequency

circuit arrangement, to distinguish between outputs resulting from displacements on each side of null.

Table 6.2 lists information on typical commercially available linear-variable differential transformers (LVDT).

Input power. Input voltage is limited by the current-carrying ability of the primary coil. In most applications, LVDT sensitivities are great enough so that very conservative ratings can be applied. Many commonly used commercial transformers are made to operate on 60 Hz at 6.3 V. Most of the 60-Hz differential transformers draw less than one watt of excitation power.

Higher frequencies provide increased sensitivities. However, in order to maintain linearity, design differences, primarily core length, may be required for different frequencies; and, in general, a given LVDT is designed for a specific input frequency.

Exciting frequency, sometimes referred to as *carrier* frequency, limits the dynamic response of a transformer. The desired information is super-imposed on the exciting frequency, and a minimum ratio of 10 to 1 between carrier and signal frequencies is usually considered to be the limit. For ratios less than 10 to 1, signal definition tends to become lost. This, therefore, has an important bearing on the selection of an operating frequency.

Transformer sensitivity is usually stated in terms of *millivolts output per volt input per 0.001-inch core displacement*. It is directly proportional to exciting voltage and, as indicated above, also increases with frequency. Of course, the output also depends on LVDT design, and in general, the sensitivity will increase with increased number of turns on the coils. There is a limit, however, determined by the solenoid effect on the core. In many applications this effect must be minimized; hence design of the general-purpose LVDT is the result of compromise [9].

Solenoid or axial force exerted by the core is zero when the core is centered and increases linearly with displacement. Increasing the excitation frequency reduces this force. Typically, an LVDT having a linear range of ±0.03 in. exerts an axial force of about 0.080 gm at 60 Hz and about 0.008 gm at 1000 Hz, for a driving voltage of 7 V RMS.

Input harmonics and anything distorting the symmetry of the flux pattern, such as adjacent metallic materials, magnetic or nonmagnetic, may hinder obtaining a sharp null balance. In many cases this is not serious. But, especially when utmost sensitivity is desired and the transformer output is amplified, it may be necessary to use external circuitry to improve the null balance condition. Satisfactory results are usually obtained by an arrangement as shown in Fig. 6.14. A voltage-dividing potentiometer of sufficiently high resistance to avoid appreciably loading the transformer secondary is placed across the output; 20,000 ohms or higher may be used. The potentiometer is then adjusted to obtain best null, the adjustment only being made once for a given setup. Capacitance may be added by trial, as

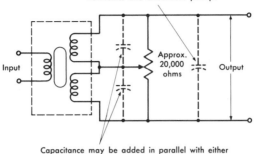

Capacitance may be inserted to eliminate some
harmonics and extraneous pickup.

Capacitance may be added in parallel with either
secondary to further reduce minimum balance voltage.
Especially applicable for high frequencies.

Fig. 6.14 Arrangement for improving the sharpness of null balance. (Courtesy Schaevitz Engineering Company, Camden, N.J.)

shown, to further reduce the null voltage. More elaborate arrangements are sometimes required to realize the maximum inherent potential of the device.

The LVDT offers several distinct advantages over many competitive transducers. First, serving as a primary detector-transducer, it converts mechanical displacement into a proportional electrical voltage. As we have found, this is a fundamental conversion. In contrast, the electrical strain gage requires the assistance of some form of elastic member. In addition, the LVDT cannot be overloaded mechanically, since the core is completely separable from the remainder of the device. It is also relatively insensitive to high or low temperatures or to temperature changes, and it provides comparatively high output, often usable without intermediate amplification. It is reusable and of reasonable cost.

Probably its greatest disadvantages lie in the area of dynamic measurement. Its core is of appreciable mass, particularly compared with the mass of the bonded strain gage. And the exciting frequency of the carrier may also be a limiting factor, particularly if the readily available 60-Hz source is used. In addition, the advantage of simple circuitry is lost if the direction from null must be indicated.

6.15 VARIABLE-RELUCTANCE TRANSDUCERS

In transducer practice, use of the term *variable reluctance* usually assumes some form of inductance device incorporating a *permanent magnet*. In most cases these devices are limited to dynamic application, either steady state or transient, where the flux lines supplied by the magnet are cut by the turns of the coil. Some means for providing relative motion is incorporated into the device.

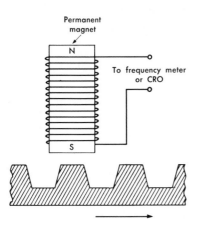

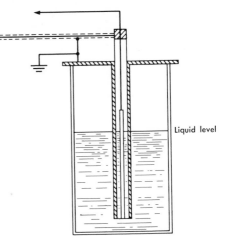

Fig. 6.15 A simple variable-reluctance pickup.

Fig. 6.16 Capacitance pickup for determining level of liquid hydrogen.

In its simplest form, the variable-reluctance device consists simply of a coil wound on a permanent magnet core (Fig. 6.15). Any variation of the permeance of the magnetic circuit causes a change in the flux. As the flux field expands or collapses, a voltage is developed in the coil. Practical applications of this arrangement are discussed in Articles 9.7 and 14.8a.

While the above arrangement depends on changing permeance, other devices based on variable reluctance depend on relative movement between the flux field and the coil. Notable examples are the two vibration pickups discussed in Article 16.8. In the first case, the coil serves both as pickup and seismic mass, and in the second case the magnet is used as the seismic mass.

6.16 CAPACITIVE TRANSDUCERS

An equation for calculating capacitance is [11]

$$C = \frac{0.244KA(N-1)}{d},$$

where

C = capacitance, picofarads,

K = dielectric constant (= 1 for air),

A = area of one side of one plate, in^2,

N = number of plates,

d = separation of plate surfaces, in.

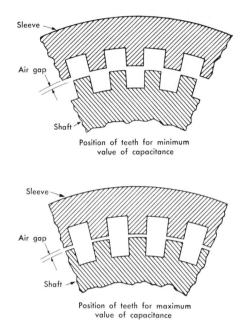

Position of teeth for minimum
value of capacitance

Position of teeth for maximum
value of capacitance

Fig. 6.17 Section showing relative arrangement of teeth in capacitance torque meter.

All the terms represented in the above equation, except possibly the number of plates, have been used in transducer applications. The following are examples of each.

Changing dielectric constant. Figure 6.16 shows a device developed for the measurement of level in a container of liquid hydrogen [12]. The capacitance between the central rod and the surrounding tube varies with changing dielectric constant brought about by changing liquid level. The device readily detects liquid level even though the difference in dielectric constant between the liquid and vapor states may be as low as 0.05.

Changing area. Capacitance change depending on changing effective area has been used for the secondary transducing element of a torquemeter [13]. The device uses a sleeve with teeth or serrations cut axially, and a matching internal member or shaft with similar axially cut teeth. Figure 6.17 illustrates the arrangement. A clearance is provided between the tips of the teeth, as shown. Torque carried by an elastic member causes a shift in the relative positions of the teeth, thereby changing the effective area. The resulting capacitance change is calibrated in terms of torque.

Tube for circulating
cooling water

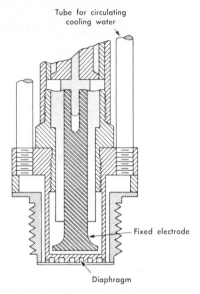

Fixed electrode

Diaphragm

Fig. 6.18 Section through capacitor-type pressure pickup.

Changing distance. Changing distance between plates of a capacitance is undoubtedly the more commonly employed method for using capacitance in a pickup.

Figure 6.18 illustrates a capacitor-type pressure transducer wherein the capacitance between the diaphragm to which the pressure is applied and the electrode foot is used as a measure of the diaphragm's relative position [14, 15, 16]. Flexing of the diaphragm under pressure changes the distance between itself and the electrode.

6.17 THE PIEZOELECTRIC EFFECT

Certain materials possess the ability to generate an electrical potential when subjected to mechanical strain, or conversely, to change dimensions when subjected to voltage (Fig. 6.19). This is known as the *piezoelectric** *effect.* Pierre and Jacques Curie are credited with its discovery in 1880. Notable among these materials are quartz, Rochelle salt (potassium sodium tartarate), properly polarized barium titanate, ammonium dihydrogen phosphate, and even ordinary sugar [17].

Of all the materials which exhibit the effect, there is no one possessing all the desirable properties, such as stability, high output, insensitivity to

---

* The prefix *piezo* is derived from the Greek *piezein*, meaning *to press* or *to squeeze.*

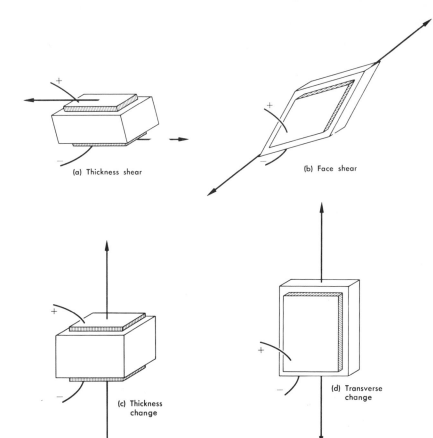

Fig. 6.19 Basic deformations of piezoelectric plates.

temperature extremes and humidity, and the ability to be formed into most any desired shape. Quartz is undoubtedly the most stable; however, its output is quite low. On the other hand, Rochelle salt provides the highest output of any of the materials, but requires protection from moisture in the air and cannot be used above about 115°F.

Because of its stability, quartz is quite commonly used for stabilizing electronic oscillators (Articles 4.5c and 7.10). It is ground to the shape of a rectangular or square plate and firmly held between two electrodes contacting its faces. The thickness of the plate is ground to the dimension that provides a mechanical resonant frequency corresponding to the desired electrical frequency. The crystal is incorporated in an appropriate electronic circuit whose frequency it controls.

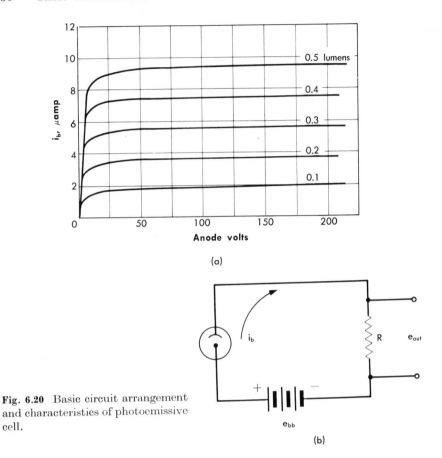

Fig. 6.20 Basic circuit arrangement and characteristics of photoemissive cell.

Barium titanate, rather than being a single crystal, as are many piezoelectric materials, is polycrystalline and hence may be formed into a variety of sizes and shapes. The piezoelectric effect is not present until the element is subjected to a polarizing treatment. Exact polarizing procedure varies with the manufacturer. However, the following procedure has been used [18]. The element is heated to a temperature above its Curie point of 120°C, and a high d-c potential is applied across the faces of the element. The magnitude of this voltage depends on the thickness and is in the order of 10,000 V/cm. The element is then cooled with the voltage applied, which results in an element that exhibits the piezoelectric effect.

The use of piezoelectric transducer elements is primarily limited to dynamic measurements. The potential developed by application of strain is not held under static conditions. Hence the elements are primarily used in the measurement of such quantities as surface roughness and in accelerometers and vibration pickups (see Articles 10.18, 10.19, and 16.9).

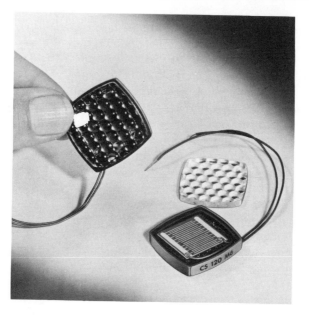

Fig. 6.21 Cadmium sulphide photocell.   (Courtesy: International Rectifier Corporation, El Segundo, California.)

Ultrasonic generator elements also use barium titanate. Such elements are used in industrial cleaning apparatus and also in underwater detection systems known as *Sonar*.

6.18 PHOTOELECTRIC TRANSDUCERS [19, 32]

Light-sensitive detectors, photosensors, or photocells may be categorized in several different ways. They may be divided into two types, (a) electronic and (b) solid state, or they may be grouped under the classifications: (a) photoconductive, (b) photogenerative (photovoltaic), and (c) photoemissive.

Electronic types consist of a cathode-anode combination within an evacuated glass or quartz envelope. In the proper circuit (commonly requiring a d-c source of 100 to 200 V), light impingement on the cathode frees electrons to flow, thereby providing a small current. This also comprises the *photoemissive* class. Figure 6.20 illustrates the general characteristics of this type.

Semiconductor photosensors employ the electron-hole principle exhibited by certain materials: the number of electron-hole pairs depends on the intensity of the incident light and the surface area exposed. This general type of sensor may be had in many different forms. Figure 6.21 illustrates only one of many possible configurations.

Photoconductive cells consist of a thin film of material such as selenium, several of the metallic sulphides, or germanium, coated between electrodes on a glass plate. The cell behaves as a light-controlled variable resistor whose resistance is reduced when exposed to a light source. In conjunction with resistance-sensitive circuitry, an output may be obtained that is proportional to the intensity of the light source.

The *photogenerative* type consists of a sandwich of unlike materials such as an iron base covered with a thin layer of iron selenide. When it is exposed to light, a voltage is developed across its section. An important application of this type is the so-called "sun battery" employed in a number of space vehicles.

More sophisticated forms of solid-state photosensors include photodiodes, phototransistors, and photoswitches [19, 31, 32, 33].

Photosensors may be made selectively sensitive not only to light in the visible spectrum but also to light in the infrared and ultraviolet ranges as well. Heat-seeking, infrared sensors are commonly of the photoconductor type [34].

The response of photosensors to sudden variations in light intensity is not instantaneous. It is determined both by the cell itself and by related circuitry. Rise and fall times, as determined by the cell type, may range from a fraction of a millisecond to several thousand milliseconds [19, 31, 32].

Photocell applications in mechanical measurements would include simple counting, where the interruption of a beam of light could be employed (Fig. 9.14). It has also been used for strain gages [20], dew-point controls [21], and edge and tension controls [22].

6.19 ELECTRONIC TRANSDUCER ELEMENT

Figure 6.22 shows schematically a very interesting transducer element [23]. Essentially, it is an electronic tube in which certain of the elements are movable. In this case, the plates are mounted on an arm which extends through a flexible diaphragm in the end of the tube. In this manner a mechanical movement applied to the external end of the rod is transferred to the plates within the tube, thereby changing the tube's characteristics.

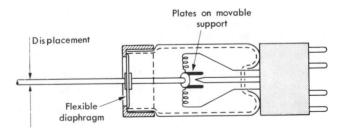

Fig. 6.22 Electronic transducer element.

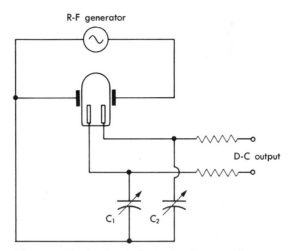

Fig. 6.23 Basic circuit for ionization transducer.

A commercially available tube with moving anode is the RCA 5734. These transducer elements have been used for numerous purposes, for example as surface-roughness indicators [24], accelerometers [25], and pressure and force transducers [26].

6.20 IONIZATION TRANSDUCER

An interesting transducer element is based on observations of Kurt S. Lion [27]. It is an ionization transducer [28] consisting of a glass envelope with two internal electrodes and filled with a gas or gases under reduced pressure.

A usable circuit shown in Fig. 6.23 makes use of a radio-frequency power source which ionizes the gas in the transducer by the field from the two external electrodes. A space charge is created, furnishing a d-c output signal, which is a function of the configuration and of the potential of the electrodes. A variation in either of the capacitances C_1 and C_2 of the circuit shown in the figure tends to change the balance of the electric field and thus to produce an output.

Of course, in this form the ionization tube really is a part of the circuitry, while the capacitors serve as the secondary transducers. However, if, using the same circuitry arrangement, we move the tube relative to the external electrodes, a d-c output proportional to the movement results. Hence, the tube and electrode combination may also serve as a transducer.

Capacitance changes of 10^{-3} pf or motion of 10^{-6} in. are readily measured.

Somewhat different arrangements may be used for large displacements. A ring electrode is used which surrounds the tube. Motions of 1 to 20 in. have been measured with a linearity better than 1%.

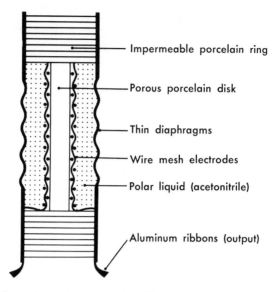

Impermeable porcelain ring

Porous porcelain disk

Thin diaphragms

Wire mesh electrodes

Polar liquid (acetonitrile)

Aluminum ribbons (output)

Fig. 6.24 Section through an electrokinetic cell. (Courtesy Consolidate Electro-dynamics Corporation, Pasadena, California.)

Pickups for mechanical measurements have been developed for pressures, displacements, acceleration, and humidity.

6.21 ELECTROKINETIC TRANSDUCERS

The electrokinetic phenomenon [29, 30], also referred to as *streaming potential*, occurs when a polar liquid such as water, methanol, or acetonitrile is forced through a porous disk. When the liquid flows through the pores, a voltage is generated that is in phase with and directly proportional to the differential pressure across the faces of the disk. When the direction of flow is reversed, the polarity of the electrical signal is reversed.

To measure a static differential pressure with the electrokinetic pickup, an unlimited supply of liquid would be required on the upstream side. Since this is impractical, a finite amount of liquid is constrained within the cell. This means that the device is inherently suitable for measuring dynamic rather than steady pressure values.

A typical cell, such as that shown in Fig. 6.24, consists of a porcelain disk glazed into the center of an impermeable porcelain ring. Diaphragms tightly sealed on each side of the ring retain the polar liquid, which fills the space between the diaphragms. A wire-mesh electrode is mounted on each side of the porous disk, with electrical connections brought out via aluminum strips. The cell assembly is then clamped within a suitable housing.

The electrokinetic principle is particularly applicable to the measurement of small dynamic displacements, pressures, and accelerations.

SUGGESTED READINGS

Bube, R. H., *Photoconductivity of Solids.* New York: John Wiley & Sons, Inc., 1960.

Doeblin, E. O., *Measurement Systems: Application and Design.* New York: McGraw-Hill Book Co., 1966.

Lion, K. S., *Instrumentation in Scientific Research.* New York: McGraw-Hill Book Co., 1959.

Minnar, E. J., *ISA Transducer Compendium.* Pittsburgh, Pa.: Instrument Society of America, 1963.

PROBLEMS

6.1 Why are electrical secondary transducers usually more popular than mechanical, hydraulic, or pneumatic types? When are the latter types superior to the electrical types?

6.2 List advantages and disadvantages of the voltage-dividing potentiometer as a displacement-to-voltage transducer.

6.3 List the factors influencing the output of a linear variable differential transformer (LVDT).

6.4 Write a general expression for the sensitivity (henrys/in.) of the device shown in Fig. 6.7. Plot sensitivity versus h over the range $0.010 < h < 0.100$ for gap dimensions $\frac{3}{8} \times \frac{3}{8}$ in. and $N = 1200$.

6.5 Calculate the maximum capacitance of a torque-meter of the type shown in Fig. 6.17 if the device has 50 pairs of teeth, a uniform gap of 0.010 in., and each tooth face has dimensions $\frac{1}{2} \times 10$ in. Calculate the sensitivity in picofarads/degree of rotation.

6.6 Obtain an inexpensive crystal-type phonograph pickup designed for replaceable phonograph needles. The least expensive will be quite satisfactory. Place a small overhanging mass in the needle chuck, connect the pickup output to an oscilloscope, and use it to investigate various sources of mechanical vibration. [*Note:* This device will not be adequate to determine vibrational amplitudes. Its response will display too many resonances. It is useful, however, as a "frequency" pickup.]

CHAPTER 7

INTERMEDIATE MODIFYING SYSTEMS

7.1 INTRODUCTION

After a mechanical quantity is detected and possibly transduced, it is usually necessary that the stage-one output be further modified before it is in satisfactory form for driving an indicator or recorder. It is the purpose of this chapter to outline some of the methods used in this intermediate step.

Measurement of dynamic mechanical quantities places severe requirements on the elements of the intermediate modifying stage. Large amplifications are often desired, coupled with good transient response, both of which are difficult to obtain by mechanical, hydraulic, or pneumatic methods. As a result, electrical or electronic elements are usually required.

7.2 MECHANICAL SYSTEMS: INHERENT PROBLEMS

An input signal is often converted by the detector-transducer to a mechanical displacement. It is then commonly fed to a secondary transducer which converts it into a form, often electrical, that is more easily processed by the intermediate stage. In some cases, however, such a displacement is fed to mechanical intermediate elements in the form of linkages, gearing, cams, etc. Such mechanical elements present design problems of considerable magnitude, particularly if dynamic inputs are to be handled.

To illustrate some of the problems involved, let us consider a well-known, strictly mechanical, dynamic measuring instrument, namely the piston and spring engine-pressure indicator, such as is shown in Fig. 7.1. The primary detector-transducer is the piston-cylinder and spring combination. These serve in a manner similar to the way the primary elements in the tire gage (Article 2.3) serve. The tire gage, however, is intended for static-pressure measurement, whereas the engine indicator is used for rapidly varying pressures.

After the pressure in the engine cylinder has been detected and transduced to a displacement, it must be further modified before it is useful for recording. Full-stroke piston-rod movement in this type of indicator is

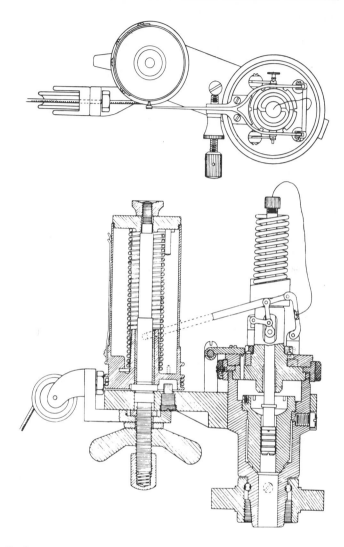

Fig. 7.1 Spring and piston engine-pressure indicator.   (Courtesy Bacharach Industrial Instrument Company, Pittsburgh, Pa.)

about 0.3 in., a magnitude so small that the rod's movement is not immediately suitable for accurate output. Some form of amplification is required. For this purpose a pantograph linkage driven by the piston rod is used to magnify the rod movement mechanically. For the instrument shown in the figure, the amplification, or ratio of stylus movement to piston-rod movement, is six. This ratio may also be referred to as *gain*.

    Mechanisms of this kind present quite difficult kinematic and kinetic design problems, some of which will be discussed in the following paragraphs.

7.3 KINEMATIC LINEARITY

Any linkage like that shown in Fig. 7.1 must be designed so that, as a mechanical amplifier, it provides the same amplification or *gain* over its entire range of output. It must be linear, otherwise poor amplitude response will result. In addition, this application must provide a straight-line stylus movement or the recorded trace will be distorted along the time axis. This would introduce what could be interpreted as poor phase response. Initially, of course, *proper kinematic layout* is required to ensure linearity and straight-line stylus movement. Also, control of *tolerances* on link dimensions and fixed pivot-point locations is important.

7.4 MECHANICAL AMPLIFICATION

Inertial loading, elastic deformation, frictional loading, and backlash all present problems when mechanical amplification is used. In line with our discussion in Article 5.2, errors resulting from the first two factors could be classified as *systematic*, whereas those due to the last two would be *random*.

Mechanical amplification may be expressed in terms of mechanical advantage as follows:

$$\text{Gain} = \text{mechanical advantage}$$

$$= \frac{\text{output displacement}}{\text{input displacement}}$$

$$= \frac{\text{output velocity}}{\text{input velocity}}. \tag{7.1}$$

In this respect, we usually think of the quantity as being greater than unity, although, of course, that would not be necessary.

7.5 REFLECTED FRICTIONAL AMPLIFICATION

There will be no change in energy or power brought about by mechanical amplification, other than losses due to friction. *This is an important limitation of mechanical intermediate devices.* The only amplifications (or attenuations) that can take place are those of force and displacement, which will be affected in inverse relation. A small frictional force in the mechanism will reflect back to the input as a magnified load, and this magnification will be equal to the gain between the two points. Using our earlier example, we find that for a mechanical amplification of, say, six (electrical amplifications may be several thousands) and a stylus friction of one ounce at the output, a force of six ounces would be required at the input. The well-known mechanical work relation holds in this case. We shall refer to this effect as *reflected frictional amplification.* Numerically it will be equal to the gain, as defined above, between the frictional source and the point of input of energy to the mechanism.

Total frictional drag reflected to the input (the piston in our example) may be expressed as follows:

$$F_{fr} = \sum A F_f, \tag{7.2}$$

in which

F_{fr} = the total reflected frictional force at the input of the system, lb,

F_f = actual frictional force at its source, lb,

A = mechanical gain or amplification.

7.6 REFLECTED INERTIAL AMPLIFICATION

*Inertial forces** cause problems similar to those caused by frictional forces. That is, their effects may be considered as reflected back to the input in proportion to the mechanical amplification existing between the source of the force and the input point. There is no loss of energy, however, since it is merely stored temporarily to be regained later in the cycle. We may therefore use the term *reflected inertial amplification* in the same sense that we use *reflected frictional amplification*, although the sources of inertial forces are distributed, whereas the source of frictional force is concentrated. We may write

$$F_{ir} = \sum A \, \Delta F_i, \tag{7.3}$$

in which

F_{ir} = total reflected inertial force to be overcome at the system input, lb,

ΔF_i = inertial force increment at any point in the system, lb,

A = mechanical amplification or gain.

Another basic difference between reflected frictional and inertial effects is that friction is primarily a *static* quantity, whereas inertia is strictly *dynamic*. An inertial force results from acceleration, and in general accelerations vary directly with the square of velocity. Hence the inertial effects become *increasingly important* as speeds increase and the duration of the signal interval reduces.

Adding the reflected forces, we have

$$F_r = F_{fr} + F_{ir}, \tag{7.4}$$

where F_r is the total reflected force, in pounds.

---

* Discussion of procedures for determining inertial forces in mechanisms is beyond the scope of this book. Description of graphical methods usually employed may be found in any good text on dynamics of machinery. References [1] and [2] for this chapter are cited for this purpose.

The influence of forces reflected in this manner will differ with different applications. In our example, however, the effect would be the same as though the spring calibration were in error, since a portion of the pressure force would be counteracted by the reflected force rather than by the spring. This effect would not be constant, but would vary throughout the pressure cycle.

7.7 AMPLIFICATION OF BACKLASH AND OF ELASTIC DEFORMATION

Elastic deformation is the *give*, or *spring*, in the parts, resulting directly from transmitted and inertial loads that are being carried. Backlash results from a temporary nonconstraint in a linkage, caused by clearances required in mechanical fits where relative motion occurs. Backlash and elastic deformation both cause a *lost motion* at the output equal to the actual backlash or deformation multiplied by the amplification between the source and the output. We may therefore refer to these effects as *backlash amplification* and *elastic amplification*, respectively.

Elastic deformation is brought about only by loads or forces carried by the links. In our example, such forces would come from the writing load imposed at the stylus, from frictional loads, and, most important, from inertial loads caused by the dynamic nature of the input.

In considering the effects of mechanical clearances or backlash and elastic deformation, it is convenient to think in terms of displacement losses being projected ahead to the output rather than being reflected back to the input.

If we let

$y_b =$ backlash or any mechanical clearance resulting in loss of motion in the kinematic train, in.,

$y_{bp} =$ total projected displacement loss resulting from backlash and mechanical clearances, in.,

$A =$ amplification or gain between source of loss and output,

then

$$y_{bp} = \sum A y_b. \qquad (7.5)$$

Sometimes springs are used to minimize backlash by applying constant loads to pin connections, thereby always taking up the slack in the same direction. Gear backlash may also be reduced by use of antibacklash gears. In this arrangement a gear is made of two halves divided along a plane of diameters, and the two halves are displaced sufficiently by spring load to absorb the backlash. Spring loading, however, can only be used at the expense of increased friction, and the loss must be balanced against the gain to determine the advantage, if any.

In a manner similar to the situation regarding forces, backlash losses result from point sources, while elastic deformations are distributed throughout the kinematic train. Hence

$$y_{ep} = \sum A \, \Delta y_e,$$ (7.6)

in which

y_{ep} = total projected displacement loss at the output, caused by elastic deformation, in.,

Δy_e = elastic deformation increment at any point in the system, in.,

A = amplification as defined above.

Total projected displacement loss, y_p, may be expressed as follows:

$$y_p = y_{bp} + y_e \, .$$ (7.7)

Applied to our example, this effect, as in the case of reflected forces, results in output indications that are too small, just as though the spring were actually stiffer than calibrated.

We may summarize all the effects just discussed by saying that to one degree or another they all degrade an instrument's response. Inertial loading is, of course, one of the primary factors limiting the use of strictly mechanical instruments for dynamic measurement. In the case of the engine indicator, severe detonation or other unusual engine operation has, on occasion, actually produced inertial forces great enough to bend the stylus arm into a U-shape even though it is made of a heat-treated high-carbon steel.

The engine indicator shown in Fig. 7.1 is actually a very unusual example of a strictly mechanical dynamic measuring system. Modifications of this instrument produce reliable results from four-cycle engines running as fast as 4000 rpm. When all the problems involved are considered, it is realized that a truly remarkable job of instrument engineering has gone into its design.

7.8 TOLERANCE PROBLEMS

Still another problem particularly inherent in mechanical instrumentation is that of dimensional tolerance. Let us illustrate this by another example.

The well-known polar planimeter, Fig. 7.2, may be used to measure small areas by tracing the area boundary with the planimeter stylus (A) while point (B) remains as a fixed pivot point. If exactly one circuit of the boundary is made, the indicator wheel (C) will provide the information necessary to determine the area traced. In so doing, the wheel actually indicates the net distance traveled by the wheel arm (D), in the direction normal to itself, but ignores the component along the arm. If point (B)

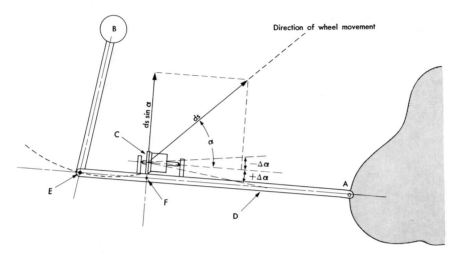

Fig. 7.2 Polar planimeter with misaligned wheel axis.

is located *outside the area being traced*,* the planimeter will behave according to the following relation [3]:

$$A = L\int dP = (\pi D)LN, \qquad (7.8)$$

where

A = area traced by stylus,

L = effective length of wheel arm,

P = distance traveled by wheel arm in direction normal to itself $= (\pi D)N$,

D = diameter of wheel,

N = turns of wheel.

Planimeter accuracy is dependent upon adherence to proper dimensional values at the time of manufacture. Tolerances will, of course, be assigned to the important dimensions. Dimensions having greatest effect on the accuracy will be: (1) wheel diameter, (2) length of wheel arm, and (3) parallelism between wheel axis and effective arm centerline. Let us investigate the effect of variations from correct wheel diameter on planimeter accuracy. We may

---

* When point (B) is located *within* the measured area, the planimeter reading will be in error by an amount equal to the area of the "zero circle." The zero-circle correction may be experimentally determined by preparing a known area, measuring it [with point (B) located within its boundary], and calculating the discrepancy.

rewrite Eq. (7.8) to read

$$A = \frac{LD\beta}{2},$$ (7.9)

where β is the angle turned by the wheel, in radians.

For any area A, the error introduced by a variation ΔD from the correct wheel diameter D will be recorded as an unwanted increment in wheel rotation, $\Delta\beta$.

Solving Eq. (7.9) for β yields

$$\beta = \frac{2A}{LD},$$ (7.9a)

and we may write

$$\beta + \Delta\beta = \frac{2A}{L(D + \Delta D)}.$$ (7.10)

Dividing Eq. (7.10) by (7.9a) gives

$$\frac{\beta + \Delta\beta}{\beta} = \frac{D}{D + \Delta D}$$ (7.11)

or

$$\frac{\Delta\beta}{\beta} = -\frac{\Delta D}{(D + \Delta D)}.$$ (7.11a)

Hence, the percentage of error resulting from variation from correct wheel diameter may be expressed by the relation

$$E_w = \left(\frac{-\Delta D}{D + \Delta D}\right) \times 100.$$ (7.12)

The negative sign indicates that a positive tolerance will result in a reading smaller than the correct one.

Inspection of Eq. (7.9a) indicates that the arm length enters the relation in the same manner as the wheel diameter. Therefore we may write

$$E_L = \left(\frac{-\Delta L}{L + \Delta L}\right) \times 100,$$ (7.13)

where E_L is the percentage of error introduced by variation in arm length.

The effect of angular misalignment between wheel axle and wheel arm is much more involved than the preceding, and only an indication of the resulting error can be given here.

Referring to Fig. 7.2, let ds be an increment of displacement along the path of wheel contact, and let α be the angle between the wheel path and the wheel axis. For increment ds the wheel will record

$$ds \sin \alpha,$$

and the wheel rotation will be

$$d\beta = \frac{2}{D} \sin \alpha \, ds \qquad \text{or} \qquad \beta = \frac{2}{D} \int_s \sin \alpha \, ds.$$

By introducing a small angular misalignment of the wheel axle, $\pm \Delta\alpha$, we find that our equation becomes

$$d(\beta + \Delta\beta) = d\beta + d(\Delta\beta)$$

$$= \frac{2}{D} \sin (\alpha \pm \Delta\alpha) \, ds,$$

in which $\Delta\beta$ is the error in wheel rotation introduced by the misalignment. By substitution,

$$d(\Delta\beta) = \frac{2}{D} [\sin (\alpha \pm \Delta\alpha) - \sin \alpha] \, ds.$$

However,

$$\sin (\alpha \pm \Delta\alpha) = (\sin \alpha \cos \Delta\alpha \pm \cos \alpha \sin \Delta\alpha).$$

And because $\Delta\alpha$ is quite small,

$$\cos \Delta\alpha \to 1 \qquad \text{and} \qquad \sin \Delta\alpha \to \Delta\alpha.$$

Therefore we may write

$$d(\Delta\beta) = \frac{2}{D} (\pm \Delta\alpha \cos \alpha) \, ds,$$

$$\Delta\beta = \frac{2}{D} \int_s (\pm \Delta\alpha \cos \alpha) \, ds.$$

The expression for error then becomes

$$E_\alpha = \frac{\Delta\beta}{\beta} = \frac{\int_s (\pm \Delta\alpha \cos \alpha) \, ds}{\int_s \sin \alpha \, ds}. \tag{7.14}$$

As can be seen, evaluation of expression (7.14) is dependent on the shape of area traced and does not yield a solution depending on $\Delta\alpha$ alone. The relation is further complicated by the fact that ds is measured along the path traced by the wheel rather than the area boundary.

The foregoing discussion was presented to outline the problem of dimensional tolerance as applied to a simple instrument. Of course, the problem has many facets which cannot be covered by one or two examples. For instance, many instruments depend on the elastic properties of materials, particularly the modulus of elasticity, for their proper functioning.

Spring scale. As another example, suppose that a spring scale such as the one shown in Fig. 7.3 is to be designed. Let us assume that the scale capacity

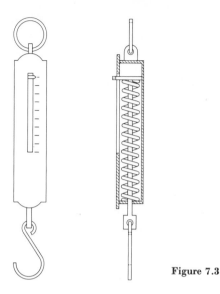

Figure 7.3

is to be 10 lb and that a maximum deflection of 4 in. is to be used. The spring modulus, M, therefore would be $\frac{10}{4}$, or $2\frac{1}{2}$ lb/in.

Conventional coil-spring design relations are as follows [4]:

$$\sigma = \frac{8KFD_m}{\pi D_w^3} \tag{7.15}$$

and

$$y = \frac{8FD_m^3 N}{E_s D_w^4}, \tag{7.16}$$

in which

σ = design stress, psi,

K = the Wahl factor = $\dfrac{4C - 1}{4C - 4} + \dfrac{0.615}{C}$,

C = spring index = D_m/D_w,

F = load on spring, lb,

D_m = mean coil diameter, in.,

D_w = wire diameter, in.,

y = deflection, in.,

N = number of active coils,

E_s = torsional modulus of elasticity, psi.

If we assume that $E_s = 11.5 \times 10^6$ psi and $D_m = 1$ in., we find that #14 W&M gage wire (0.080-in. diameter) results in a stress of 55,000 psi, which is satisfactory for carbon steel wire in an application such as this if mechanical stops are provided to prevent overloading. Using Eq. (7.16), we also find that 23.6 coils are required to provide the necessary spring modulus of $2\frac{1}{2}$ lb/in.

These values, however, are all nominal, and we must account for possible dimensional variations. If we assume that our scale is not intended as a laboratory instrument, the following tolerances, based on economic considerations should be reasonable:

$$D_w = 0.080 + 0.0000 \quad \text{and} \quad -0.0010 \text{ inch,}$$

$$D_m = 1.000 \pm 0.020 \text{ inch,}$$

$$N = 23.6 \pm 0.2 \text{ turns,}$$

$$E_s = 11.5 \times 10^6 \pm 0.5 \times 10^6 \text{ psi.}$$

On this basis,

$$M_{\max} = \frac{E_{s\max} D^4_{w\max}}{8 D^3_{m\min} N_{\min}} = 2.80 \text{ lb/in.}$$

$$M_{\min} = \frac{E_{s\min} D^4_{w\min}}{8 D^3_{m\max} N_{\max}} = 2.22 \text{ lb/in.}$$

These values represent errors of $+12\%$ and -11.2% relative to the desired modulus value of 2.5 lb/in.

When inexpensive measuring devices such as this are being manufactured in quantity, the calibration problem is sometimes handled simply by the selection of one of several different scale plates or dials, each having slightly different scales. For our example, three plates might be used, one for a range on each side of optimum, one for those devices calibrating high, and the other for those calibrating low. In this case optimum maximum deflection is to be 4 in. By providing three different indicator plates, A, B, and C, having lengths of 4.32, 4.00, and 3.68 in. respectively, the maximum possible error may be reduced from 12% to 4% by a simple production check and plate selection at the time of assembly.

7.9 TEMPERATURE PROBLEMS

An ideal measuring system will react to the design signal only and ignore all else. Of course, this is an ideal that is never completely fulfilled. One of the more insidious extraneous stimuli adversely affecting instrument operation is temperature. It is insidious in that it is almost impossible to maintain a constant temperature environment for the general-purpose measuring system. The usual solution is to accept the temperature variation and to devise methods to compensate for it.

Temperature variations cause dimensional changes and changes in physical properties, both elastic and electrical, resulting in deviations referred to as *zero shift* and *scale error* [5, 6]. Zero shift, as the name implies, results in a change in the no-input reading. Factors other than temperature may cause zero shift; however, temperature is probably the most common cause. In most applications the zero indication on the output scale would be made to correspond to the no-input condition. For example, the indicator or the spring scales referred to earlier should be set at zero pounds when there is no weight in the pan. If the temperature changes after the scale has been set to zero, there may be a differential dimensional change between spring and scale, altering the no-load reading. This change would be referred to as *zero shift*. Zero shift is primarily a function of linear dimensional change caused by expansion or contraction with changing temperature.

Dimensional changes are expressed in terms of the coefficient of expansion by the following familiar relations:

$$\alpha = \frac{1}{\Delta T} \frac{\Delta L}{L_0} \qquad (7.17)$$

and

$$L_1 = L_0(1 + \alpha \, \Delta T), \qquad (7.18)$$

in which

$\alpha =$ coefficient of thermal expansion, in./in. $\cdot$ °F,

$\Delta L/L_0 =$ unit change in length, in./in.,

$\Delta T =$ change in temperature $= (T_1 - T_0)$, °F,

$L_0 =$ dimension in inches, at reference temperature, T_0,

$L_1 =$ dimension in inches, at any other temperature, T_1.

In addition to causing zero shift, temperature changes usually affect scale calibration when resilient load-carrying members are involved. The coil and wire diameters of our spring would be altered with temperature change, and so also would the modulus of elasticity of the spring material. These variations would cause a changed spring constant, hence changed load-deflection calibration, resulting in what is referred to as *scale error*.

The thermoelastic coefficient is defined by the relations

$$C = \frac{1}{\Delta T} \frac{\Delta E}{E_0} \qquad (7.19)$$

and

$$E_1 = E_0(1 + C \, \Delta T), \qquad (7.20)$$

in which

$C =$ coefficient for tensile modulus of elasticity, psi/psi $\cdot$ °F,

$\Delta E/E_0 =$ unit change in tensile modulus of elasticity,

$E_0 =$ tensile modulus of elasticity, psi, at T_0,

$E_1 =$ tensile modulus of elasticity, psi, at any other temperature, T_1.

Similarly, the coefficient for torsional modulus may be written

$$m = \frac{1}{\Delta T} \frac{\Delta E_s}{E_{s_0}} \qquad (7.21)$$

and

$$E_{s_1} = E_{s_0}(1 + m\,\Delta T), \qquad (7.22)$$

where

m = coefficient for torsional modulus of elasticity, psi/psi · °F,

$\Delta E_s/E_{s_0}$ = unit change in torsional modulus of elasticity,

E_{s_0} = torsional modulus of elasticity, psi, at reference temperature, T_0,

E_{s_1} = torsional modulus of elasticity, psi, at any other temperature, T_1.

Representative values of these quantities are given in Table 7.1.

The manner in which temperature changes in elastic properties affect instrument performance may be demonstrated by the following example. Assume that a restoring element in an instrument is essentially a single-leaf cantilever spring of rectangular section, for which the deflection equation at reference temperature T_0 is

$$K_0 = \frac{F}{y} = \frac{3E_0 I_0}{L_0^3} = \frac{E_0 w_0 t_0^3}{4L_0^3}, \qquad (7.23)$$

in which

K_0 = deflection constant, lb/in.,

I_0 = moment of inertia, in^4 = $\frac{1}{12}w_0 t_0^3$,

w_0 = width of section at reference temperature, in.,

t_0 = thickness of section at reference temperature, in.,

L_0 = length of beam at reference temperature, in.

A second equation may be written for any other temperature, T_1, as follows:

$$K_1 = \frac{[E_0(1 + C\,\Delta T)][w_0(1 + \alpha\,\Delta T)][t_0(1 + \alpha\,\Delta T)]^3}{4[L_0(1 + \alpha\,\Delta T)]^3}. \qquad (7.24)$$

$$\begin{array}{l}\text{Percent error in} \\ \text{deflection scale}\end{array} = \left(\frac{K_0 - K_1}{K_0}\right) \times 1000$$

$$= [1 - (1 + C\,\Delta T)(1 + \alpha\,\Delta T)] \times 100$$

which we may simplify, by expanding and discarding the second-order term, to read

$$\text{Percent scale error} = -(C + \alpha)\,\Delta T \times 100. \qquad (7.25)$$

TABLE 7.1

TEMPERATURE CHARACTERISTICS OF INSTRUMENT SPRING MATERIALS*

Material	Composition, %		Tensile mod. of elasticity, E, psi	Torsional mod. of elasticity, E_s, psi	Coeff. of linear expansion, α, in./in.°F	Coeff. for tensile mod. of elasticity, C, psi/psi°F	Coeff. for torsional mod. of elasticity, m, psi/psi°F
Flat spring steel	C Mn	0.65–0.80 0.50–0.90	30,000,000		6.6×10^{-6}	-133×10^{-6}	-133×10^{-6}
High-carbon wire	C Mn	0.85–0.95 0.25–0.60	30,000,000	11,500,000	6.4×10^{-6}	-117×10^{-6}	-111×10^{-6}
Oil-tempered wire ASTM A229-41	C Mn	0.60–0.70 0.60–0.90	29,000,000	11,500,000	6.6×10^{-6}	-139×10^{-6}	-156×10^{-6}
SAE 6150	C Mn Chrome V	0.45–0.55 0.50–0.80 0.80–1.10 0.15–0.18	30,000,000	11,500,000	6.8×10^{-6}	-145×10^{-6}	-145×10^{-6}
Stainless Type 302	Chrome Ni C Mn Si	17 to 20 7 to 10 0.08–0.15 2 max 0.30–0.75	28,000,000	10,000,000	9.3×10^{-6}	-239×10^{-6}	-244×10^{-6}
Spring brass	Cu Zn	64 to 72 remainder	15,000,000	5,500,000	11.2×10^{-6}	-217×10^{-6}	-217×10^{-6}
Phosphor Bronze	Cu Sn	91–93 7–9	15,000,000	6,250,000	9.9×10^{-6}	-200×10^{-6}	-222×10^{-6}
Invar	Ni Fe	36 remainder	21,400,000	8,100,000	0.61×10^{-6}	$+26.7 \times 10^{-6}$	
Isoelastic	Ni Cr Mn Fe	36 8 0.05 remainder	26,000,000	9,200,000	4.0×10^{-6}	-20×10^{-6} to $+7.3 \times 10^{-6}$	

* Reprinted from the April, 1955, issue of *Control Engineering*, by permission of McGraw-Hill Publishing Company, Inc., N.Y. See also Ref. 6.

If our spring is made of spring brass,

Percent scale error/°F $= -(-217 + 11.2) \times 10^{-6} \times 100 = 0.0205\%$.

Hence a temperature change of $+50°F$ would result in a scale error of about $+1\%$. (This means that the reading is too high; our spring is too flexible, and a given load deflects the spring more than it should.)

It is interesting to note that for our example the scale error is a function of material or materials. It should be clear that we are speaking of the load-deflection relation for resilient members in this connection, and that this would not include members whose duty it is simply to transmit motion, such as the linkage in a Bourdon-tube pressure gage.

Although not a mechanical quantity, another item affected by temperature change is electrical resistance. The basic resistance equation may be written in the form

$$R = \rho \frac{L}{A}, \tag{7.26}$$

where

$R =$ electrical resistance, ohms,

$\rho =$ resistivity, ohms · cm,

$L =$ length of conductor, cm,

$A =$ cross-sectional area of conductor, cm^2.

As temperature changes, a change in the resistance of an electrical conductor will be noted. This will be caused by two different factors: dimensional changes due to expansion or contraction, and changes in the current-opposing properties of the material itself. For an unconstrained conductor, the latter is much more significant than the former, causing more than 99% of the total change for copper [7]. Therefore, in most cases it is not very important whether the dimensional effect is accounted for or not. If dimensional changes caused by temperature are ignored, change in resistivity with temperature may be expressed as follows:

$$b = \frac{1}{\Delta T} \frac{\Delta \rho}{\rho_0} \tag{7.27}$$

or

$$\rho_1 = \rho_0(1 + b\,\Delta T), \tag{7.28}$$

in which

$b =$ temperature coefficient of resistivity,

ohms · cm/ohms · cm · °F,

$\Delta T =$ temperature change, °F,

$\Delta\rho/\rho_0 =$ unit change in resistivity,

$\rho_0 =$ resistivity at reference temperature, T_0, ohms,

$\rho_1 =$ resistivity at any temperature, T_1, ohms.

TABLE 7.2

RESISTIVITY AND TEMPERATURE COEFFICIENTS OF RESISTIVITY FOR
SELECTED MATERIALS

Material	Composition (for alloys)	Resistivity at 20°C (68°F), ohm · cm × 10^6	Coefficient of resistivity, b, per °F
Aluminum	—	2.82	0.002 17
Constantan (advance)	60% Cu, 40% Ni	47–49	0.000 006
Copper (annealed)	—	1.724	0.002 18
Iron	—	44	0.000 006
Isoelastic	36% Ni, 8% Cr 4% Mn, Si, & Mo Fe remainder	48	0.000 26
Manganin	9–18% Mn $1\frac{1}{2}$–4% Ni Cu remainder	44	0.000 006
Monel	33% Cu, 67% Ni	42	0.001 1
Nichrome	75% Ni, 12% Fe 11% Cr, 2% Mn	100	0.000 22
Nickel	—	6.84	0.003 55
Silver	—	1.51	0.002 25

Accounting for temperature-dimensional changes, the equation would read

$$\rho_1 = \frac{R_0 A_0}{L_0}(1 + b\,\Delta T)(1 + \alpha\,\Delta T)$$

$$= \rho_0(1 + b\,\Delta T)(1 + \alpha\,\Delta T). \tag{7.29}$$

Table 7.2 lists values of the coefficients of resistivity for selected materials.

7.10 METHODS FOR LIMITING TEMPERATURE ERRORS

Three approaches to a solution of the temperature problem in instrumentation are as follows: (1) *minimization* through careful selection of materials and operating temperature ranges, (2) *compensation* through balancing of inversely reacting elements or effects, and (3) *elimination* through temperature control. Although each situation is a problem unto itself, thereby making specific recommendations difficult, a few general remarks with regard to these possibilities may be made.

Minimization. As pointed out earlier, temperature errors may be caused by thermal expansion in the case of simple motion-transmitting elements, by thermal expansion and modulus change in the case of calibrated resilient transducer elements, and by thermal expansion and resistivity change in the case of electrical resistance transducers. All these effects may be minimized by selecting materials with low-temperature coefficients in each of the respective categories. Of course, minimum temperature coefficients are not always combined with other desirable features such as high strength, low cost, corrosion resistance, etc.

Compensation. This approach may take a number of different forms, depending on the basic characteristics of the system. If a mechanical system is being used, a form of compensation making use of a composite construction may be employed. If the system is electrical, compensation is usually possible in the electrical circuitry.

An example of composite construction may be had by considering the balance wheel in a watch or clock. As the temperature rises, the modulus of the spring material reduces and, in addition, the moment of inertia of the wheel (if of simple form) increases because of thermal expansion, both of which cause the watch to *slow down*. By incorporating a bimetal element of appropriate characteristics in the rim of the wheel, the moment of inertia can be caused to decrease with temperature enough to compensate both for expansion of the wheel spokes and change in spring modulus. (See also Article 10.6 for a discussion of temperature effect on linear measuring devices.)

Electrical circuitry may employ various means for compensating temperature effects [21]. The thermistor, discussed in some detail in Articles 6.11 and 15.5c, is quite useful for this purpose. Most circuit elements possess the characteristic of increasing d-c resistance with rising temperature. The thermistor has an opposite temperature-resistance property, along with reasonably good stability, both of which make it ideal for simple temperature-resistance compensation.

Resistance-type strain gages are particularly susceptible to temperature variations. The actual situation is quite complex, involving thermal-expansion characteristics of both the base material and all the gage materials (support, cement, and grid), temperature-resistivity properties of the grid material, all combined with the fact that heat is dissipated by the grid since it is a resistance device. Temperature compensation is very nicely handled, however, by pitting the temperature-effect output from like gages against one another, while subjecting them differentially to strain. This is accomplished by use of a resistance bridge circuit arrangement, which is used extensively in strain gage work (see Article 11.12). In addition, through careful selection of grid materials, the so-called "self-compensating" gages have been developed. (Also see Article 12.4d.)

Elimination. The third method, that of eliminating the temperature problem altogether by temperature control, really requires no discussion. However, an interesting application of this method, carried to an extreme degree, is the temperature control applied to the oscillators at the National Bureau of Standards Radio Station WWV. The oscillators are installed in an oven consisting of concentric cubical chambers, each with temperature-controlled heaters, all placed in a room approximately 25 feet below ground level. Temperature control to better than 0.001°C is obtained [8].

7.11 REASONS FOR THE POPULARITY OF ELECTRICAL METHODS

Our discussion of the generalized measuring system, to this point, should have made it clear that inertia and friction are rather severe limiting factors in the use of strictly mechanical systems. In addition, required rigidity results in relatively heavy and bulky apparatus. Electrical devices, on the other hand, particularly, for the intermediate modifying stage, are not burdened to as great an extent by these factors.

There is still another reason, however, why electrical apparatus is used for mechanical measurement—one that is perhaps as important as any of the others. The reason is that electrical equipment may be used for *power amplification*. Additional power may be fed into the system to provide an increased power output beyond that of the input. This is accomplished through the use of power amplifiers, which have no important mechanical counterpart. This is of particular value where recording procedures employ stylus-type recorders, mirror galvanometers, or magnetic-tape methods.

It is true that hydraulic and pneumatic systems may be set up to increase signal power; however, their use is limited to relatively slow-acting control applications, primarily in the fields of chemical processing and electric-power generation. As in the case of mechanical systems, friction and inertia severely limit transient response of the type required for dynamic measurement.

7.12 ELECTRICAL INTERMEDIATE MODIFYING DEVICES

As has been pointed out previously, many mechanical inputs are transduced to analogous electrical signals by the stage-one detector-transducer. Such transduced information must usually be modified before becoming useful for driving the final indicator or recorder. Such modification more often than not consists of an amplification of voltage, power, or both. If an indicator is to be used as the terminating device, voltage amplification is usually sufficient. If a recorder is to be driven, however, power amplification is also normally required.

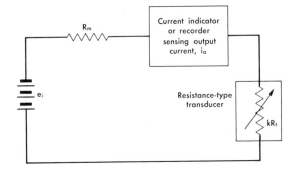

Fig. 7.4 Simple current-sensitive circuit.

7.13 INPUT CIRCUITRY

Electrical detector-transducers are of two general types: (1) *passive*, those requiring an auxiliary source of energy, and (2) *active*, those which are self-powering. The simple bonded wire strain gage is an example of the former, whereas the piezoelectric accelerometer is an example of the latter.

While it may be possible to employ the active, or self-powering, detector-transducer directly with a minimum of circuitry, the passive type, in general, requires special arrangements to introduce the auxiliary energy. The particular arrangement required will depend on the operating principle involved. For example, resistive-type pickups may be powered by either an a-c or d-c source, whereas capacitive and inductive types, with an exception or two, require an a-c source.

Although far from being all-inclusive, the following list classifies the most common forms of input circuits used in transducer work: (1) simple current-sensitive circuits, (2) ballast circuits, (3) voltage-dividing circuits, (4) voltage-balancing potentiometer circuits, (5) bridge circuits, and (6) resonant circuits. These will be discussed in the following articles.

7.14 THE SIMPLE CURRENT-SENSITIVE CIRCUIT

Figure 7.4 illustrates a simple current-sensitive circuit in which the transducer may employ any one of the various forms of variable-resistance elements. We will let the transducer resistance be kR_t, where R_t represents the maximum value of transducer resistance and k represents a percentage factor which may vary between 0.0 and 1.0 (0 and 100%), depending on the input-signal magnitude. Should the transducer element be in the form of a sliding contact resistor, the value of k could vary through the complete range of 0 to 100%. On the other hand, if R_t represents, say, a thermistor, then k would fall within some limiting range not including 0.0%. We will let R_m represent the remaining circuit resistance.

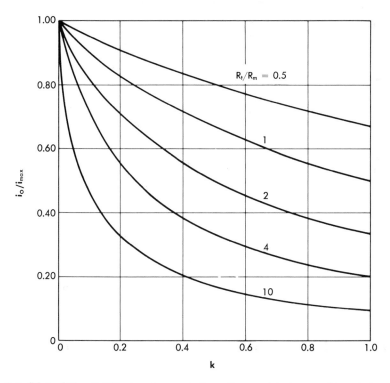

Fig. 7.5 Plot of Eq. (7.31), showing variation of current in terms of input signal k for a simple current-sensitive circuit.

If i_0 is the current flowing through the circuit and hence the current indicated by the readout device, we have, using Ohm's law,

$$i_0 = \frac{e_i}{kR_t + R_m} \cdot \tag{7.30}$$

This may be rewritten as

$$\frac{i_0}{i_{\max}} = \frac{i_0 R_m}{e_i} = \frac{1}{1 + \left(\dfrac{R_t}{R_m}\right)k} \cdot \tag{7.31}$$

Note that maximum current flows when $k = 0$, at which time the current is e_i/R_m.

Figure 7.5 shows plots of Eq. (7.31) for various values of resistance ratio. The abscissa is a measure of *signal input* and the ordinate a measure of *output*. First of all, it is observed that the input/output relation is nonlinear, which of course would generally be undesirable. In addition, the higher the

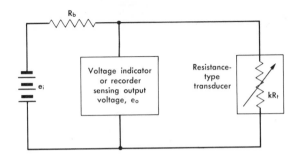

Fig. 7.6   Schematic of a ballast circuit.

relative value of transducer resistance R_t to R_m, the greater will be the output variation or sensitivity.   It will also be noted that the output is a function of i_{max}, which in turn is dependent on e_i.   This means that careful control of the driving voltage is necessary if calibration is to be maintained.

7.15 THE BALLAST CIRCUIT

Now let us look at a variation of the current-sensitive circuit, often referred to as the *ballast circuit*, shown in Fig. 7.6.   Instead of a current-sensitive indicator or recorder through which the total current flows, we shall use a voltage-sensitive device (some form of voltmeter), placed across the transducer.   The *ballast resistor*, R_b, is inserted in much the same manner as R_m was used in the previous circuit.   It will be observed that in this case, were it not for R_b, the indicator would show no change with variation in R_t; it would always indicate full source voltage.   So some value of resistance R_b is necessary for the proper functioning of the circuit.

Two different situations may exist, depending on the relative impedance of the meter.   First, the meter may be of high impedance, such as would be the case if some form of vacuum-tube voltmeter (Article 8.3) were used, in which case any current flow through the meter may be neglected.   Second, the meter may be of low impedance, and consideration of such current flow is required.

Assuming a high-impedance meter, we have by Ohm's law

$$i = \frac{e_i}{R_b + kR_t} .$$
(7.32)

Then, if $e_0 =$ voltage across kR_t (which is indicated or recorded by the readout device),

$$e_0 = i(kR_t) = \frac{e_i kR_t}{R_b + kR_t} .$$
(7.33)

This may be written as

$$\frac{e_0}{e_i} = \frac{kR_t/R_b}{1 + (kR_t/R_b)} . \tag{7.34}$$

For a given circuit, e_0/e_i is a measure of the output, and kR_t/R_b is a measure of the input.

Defining η as the sensitivity, or the ratio of change in output to change in input,

$$\eta = \frac{de_0}{dk} = \frac{e_i R_b R_t}{(R_b + kR_t)^2} . \tag{7.35}$$

We may change R_b by inserting different values of resistance. If this is done, the sensitivity should be altered, which would mean that there may be some optimum value of R_b so far as sensitivity is concerned. By differentiation with respect to R_b, we should be able to determine this value:

$$\frac{d\eta}{dR_b} = \frac{e_i R_t(kR_t - R_b)}{(R_b + kR_t)^3} . \tag{7.36}$$

The derivative will be zero under two conditions: (1) for $R_b = \infty$, which results in minimum sensitivity, and (2) for $R_b = kR_t$, for which maximum sensitivity is obtained.

The second relation indicates that for full-range usefulness, the value R_b must be based on compromise because R_b, a constant, cannot always

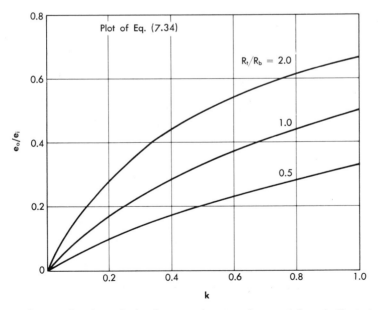

Fig. 7.7 Curves showing relation between input and output for a ballast circuit.

have the value of kR_t, a variable. However, R_b may be selected to give maximum sensitivity for a certain point in the range by setting its value to correspond to that of kR_t.

This circuit is sometimes used for dynamic applications of resistance-type strain gages [9, 10]. In this case the change in resistance is quite small compared with the total gage resistance, and the above relations indicate that a ballast resistance equal to gage resistance is optimum.

Figure 7.7 shows the relation between input and output for a circuit of this type as given by Eq. (7.34).

It will be noted that the same disadvantages apply to this circuit as to the current-sensitive circuit discussed previously, namely: (1) a percentage variation in the supply voltage, e_i, results in a greater change in output than does a similar percentage change in k, hence very careful voltage regulation must be employed, and (2) the relation between output and input is not linear.

7.16 THE VOLTAGE-DIVIDING POTENTIOMETER CIRCUIT

Figure 7.8 shows a very useful circuit arrangement for sliding-contact resistance transducer elements. This is known as the *voltage-dividing potentiometer circuit*. It will be noted that the voltage source is connected, not to the slider as it would be in the ballast circuit, but across the complete resistance element. The terminating or readout device is connected to sense the voltage drop across the portion of resistance element R_p as determined by k.

Two different situations may occur with this arrangement, depending on the relative impedance of the resistance element and the indicator-recorder. If the terminating instrument is of sufficiently high relative impedance, no appreciable current will flow through it, and it may be considered a simple "*pressure*-measuring" device. The circuit then becomes a true voltage divider, and the indicated output voltage e_0 may be determined

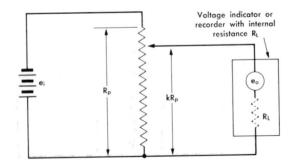

Fig. 7.8 Simple voltage-dividing potentiometer circuit.

from the following relation.

$$e_0 = k e_i \qquad (7.37)$$

or

$$k = \frac{e_0}{e_i}. \qquad (7.37a)$$

a) Loading error. On the other hand, if the readout device draws appreciable current, a *loading error* (see Article 6.2) will result. This may be analyzed as follows. Referring to Fig. 7.8, we find that the total resistance *seen* by the source of e_i will be

$$R = R_p(1 - k) + \frac{k R_p R_L}{k R_p + R_L}$$

and

$$i = \frac{e_i}{R} = \frac{e_i(k R_p + R_L)}{k R_p^2(1 - k) + R_p R_L}.$$

The output voltage will then be

$$e_0 = e_i - i R_p(1 - k)$$

or

$$\frac{e_0}{e_i} = \frac{k}{1 + (R_p/R_L)k - (R_p/R_L)k^2}. \qquad (7.38)$$

If we assume the simpler relation given by Eq. (7.37) to hold, an error will be introduced according to the following relation:

$$
\begin{aligned}
\text{Error} &= e_i\left[k - \frac{k}{k(1 - k)(R_p/R_L) + 1} \right] \\
&= e_i\left[\frac{k^2(1 - k)}{k(1 - k) + (R_L/R_p)} \right]. \qquad (7.39)
\end{aligned}
$$

Based on *full-scale output* (see Article 5.31), this relation may be written as

$$\text{Percent error} = \left[\frac{k^2(1 - k)}{k(1 - k) + (R_L/R_p)} \right] \times 100. \qquad (7.40)$$

Except for the endpoints ($k = 0.0$ or 1.0), where the error is zero, the error will always be on the negative side, i.e., the actual value of voltage will be lower than would be the case if the system performed as a linear voltage divider. Figure 7.9 shows a plot of the variation in error with slider position for various ratios of load to potentiometer resistance. Obviously, the higher the value of load resistance compared with potentiometer resistance, the lower will be the error.

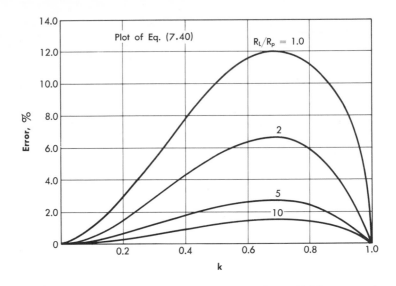

Fig. 7.9 Curves showing error caused by loading a voltage-dividing potentiometer circuit.

Fig. 7.10 Methods for improving linearity of potentiometer circuits when low impedance indicating devices are used. Resistors R_e are termed end resistors.

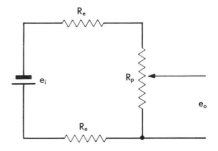

b) Use of end resistors. It will be observed that the nonlinearity in the relation between the potentiometer output and the input displacement k may be reduced if only a portion of the available potentiometer range is used. For example, a 1000-ohm potentiometer may be selected, but the input limited to only a 500-ohm portion of the total range. This would reduce the potentiometer resolution and would be generally impractical; however, it would result in a reduction in the deviation from linearity. A similar result may be obtained through use of what are known as *end* resistors (Fig. 7.10). Either an upper or a lower-end resistor, or both, may be used [11, 12]. When this method is employed, it is often possible to compensate for reduced potentiometer output caused by the increased resistance by increasing the voltage input e_i by a proportional amount.

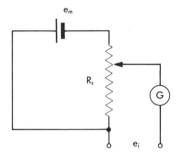

Fig. 7.11 Schematic showing principle of operation of voltage balancing potentiometer circuit.

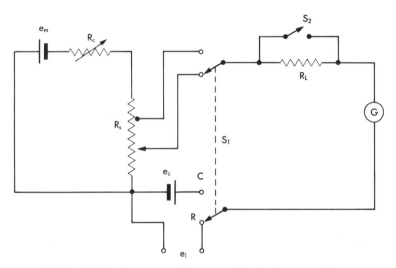

Fig. 7.12 Voltage-balancing potentiometer circuit incorporating means of standardization.

7.17   THE VOLTAGE-BALANCING POTENTIOMETER CIRCUIT

Figure 7.11 illustrates a simplified form of a circuit that has been used for years, primarily for measuring thermocouple output. Basically the circuit measures small electrical potentials by *comparison*. A known portion of voltage e_m is balanced against the unknown voltage e_i through use of a variable resistor R_s. A galvanometer, G, is employed to determine balance. Readout, or output indication, is obtained from the position of the slider relative to a calibrated scale. Of course, this simplified circuit would be impractical for making careful measurement because no provision is made to compensate for variation of battery voltage with use or age.

Figure 7.12 illustrates a more useful form of voltage-balancing circuit. Here provision is made for calibrating or standardizing the circuit, thereby

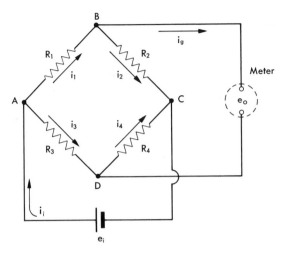

Fig. 7.13 Simple Wheatstone bridge circuit.

providing adjustment for variation in voltage e_m. This is accomplished through use of a standard cell, e_c. When switch S_1 is thrown to position C, a predetermined voltage is introduced in place of the unknown. Resistance R_c is then adjusted for balance, as indicated by the galvanometer G. After the circuit has been standardized by this means, the precalibrated slide wire becomes usable. Resistance R_L is employed to protect the galvanometer when large unbalance exists. As soon as approximate balance is achieved, switch S_2 is closed for more sensitive adjustment.

A general-purpose instrument of this kind would include additional means for adjusting R_s in the form of shunt and series resistances which could be switched into the circuit, thereby increasing the overall scale of the instrument (see Fig. 15.13).

In operation, the instrument is first standardized by means of the standard cell. The unknown voltage is then measured by balancing it against voltage e_m through use of R_s.

7.18 RESISTANCE BRIDGES

Use of some form of bridge circuit is the most common method for connecting passive transducers to associated equipment in making up a measuring system. Of all the possible configurations, the Wheatstone resistance bridge [14] devised by S. H. Christie in 1833 [13] is undoubtedly used to the greatest extent. Figure 7.13 shows a d-c Wheatstone bridge consisting of four resistance *arms* with a source of energy (battery) and a detector (meter). Measurement may be accomplished either by *balancing* the bridge by making known adjustments in one or more bridge arms until the voltage across the

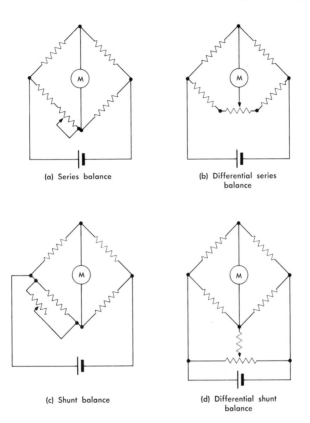

(a) Series balance

(b) Differential series
balance

(c) Shunt balance

(d) Differential shunt
balance

Fig. 7.14 Methods used to balance d-c resistance bridges.

meter is zero, or by determining the magnitude of *unbalance* from the meter reading. Typical resistance transducers using a circuit of this kind may include resistance thermometers, thermistors, or resistance-type strain gages.

Using Fig. 7.13, we may analyze the requirements for balance as follows: for balance, no current may flow through the meter, hence $i_g = 0$. If this is true, we also know that $i_1 = i_2$ and $i_3 = i_4$. In addition, the potential across the meter must be zero, or $i_1 R_1 = i_3 R_3$ and $i_2 R_2 = i_4 R_4$. By eliminating i_1 and i_3 from the above relations, we obtain the condition for balance, namely,

$$\frac{R_1}{R_2} = \frac{R_3}{R_4} \tag{7.41}$$

or

$$\frac{R_1}{R_3} = \frac{R_2}{R_4}. \tag{7.41a}$$

From the above two equations we may formulate a statement which should assist us in remembering the necessary balance relation. *For the Wheatstone resistance bridge to balance, the ratio of resistances of any two adjacent arms must equal the ratio of resistances of the remaining two arms, taken in the same sense.* (*Note:* "Taken in the same sense" means that if the first resistance ratio is formed from two adjacent resistances reading from left to right, the balancing ratio must also be formed by reading from left to right, etc.)

When used with a resistance-type transducer, any one or several of the bridge arms may consist of transducer resistance elements with one or more of the remaining arms variable so that adjustment may be made to maintain balance. Certain of the arms may be fixed resistances. Such a circuit may be used either as a *null-balance* bridge or as a *deflection* bridge.

As an example of the *null balance*, let R_1 represent the resistance element in a sliding-contact variable resistance which may be actuated by a Bourdon-tube pressure element. We may also assume that R_2 is variable and is used to manually maintain bridge balance. Initially R_2 would be adjusted to obtain a zero galvanometer reading. When pressure changes take place, the pressure pickup would cause R_1 to change value. The calibrated adjustment required in R_2 to maintain balance could then be used to determine pressure change. It is obvious that the null-balance method is usable only for static or very slowly changing inputs.

Various arrangements may be used for accomplishing bridge balance. Figure 7.14 shows four possibilities. One of the important factors in determining the kind to use is bridge sensitivity. If large resistance changes are to be accommodated, large resistance adjustments must be provided and one of the series arrangements would be indicated. This could well be the type to use for sliding-contact variable-resistance transducers or for thermistors.

When small resistance changes are to take place, as when resistance strain gages are used, then the shunt balance would be employed. In order to provide for a range of resistances, a bridge employing both series and shunt balances might be used.

When the *deflection* bridge is used, bridge unbalance, as indicated by the meter reading, is used as the measure of input. When this system is employed, provision is usually made for initial zero balance through adjustment of one of the resistance arms. For static inputs, an ordinary meter or galvanometer may be used; for dynamic outputs, however, the signal may be displayed by a cathode-ray oscilloscope (Article 8.6) or recorded by a stylus type or a light-beam oscillograph (Article 8.9).

When a deflection bridge is used, the bridge output may be connected to a high- or a low-impedance device. If the bridge output were connected directly to a simple D'Arsonval meter or galvanometer, the output circuit would be of low impedance, and appreciable current would flow through the indicator. In most cases where amplification is used, the bridge output would be fed to the grid of the amplifier input tube, and the circuit would be of high impedance ($i_g \rightarrow 0$). This would be the case if an oscilloscope or vacuum-tube voltmeter were used. In the first case, the bridge could be referred to as being *current-sensitive*, whereas, in the second case, the bridge would be *voltage-sensitive*.

a) The voltage-sensitive Wheatstone bridge. Let us consider the simplest case first, in which the bridge output is connected directly to a high-impedance device, say an oscilloscope. Referring to Fig. 7.13, we find that

$$e_0 = i_{ABC}R_1 - i_{ADC}R_3,$$

and making use of relations developed in the derivation for null-balance condition and Ohm's law, we may write

$$
\begin{aligned}
e_0 &= e_i\left(\frac{R_1}{R_1 + R_2} - \frac{R_3}{R_3 + R_4}\right) \\
&= e_i\left(\frac{R_1 R_4 - R_2 R_3}{(R_1 + R_2)(R_3 + R_4)}\right).
\end{aligned}
\tag{7.42}
$$

We will now assume that resistance R_1 changes by an amount ΔR_1, or

$$
\begin{aligned}
\frac{e_0 + \Delta e_0}{e_i} &= \left[\frac{(R_1 + \Delta R_1)(R_4) - R_2 R_3}{(R_1 + \Delta R_1 + R_2)(R_3 + R_4)}\right] \\
&= \left\{\frac{1 + (\Delta R_1/R_1) - (R_2 R_3/R_1 R_4)}{[1 + (\Delta R_1/R_1) + (R_2/R_1)][1 + (R_3/R_4)]}\right\}.
\end{aligned}
\tag{7.43}
$$

The relation may be simplified by assuming all resistances to be initially equal (in which case $e_0 = 0$). Then

$$\frac{\Delta e_0}{e_i} = \frac{\Delta R_1/R}{4 + 2(\Delta R_1/R)}.
\tag{7.44}$$

Figure 7.15, plotted from Eq. (7.44), shows the relation for the output of a voltage-sensitive deflection bridge whose resistance arms are initially equal. Inspection of the curve indicates that this type of resistance bridge is inherently nonlinear. In many cases, however, the actual resistance change is so small that the arrangement may be assumed linear. This applies to most resistance strain-gage circuits.

b) The current-sensitive Wheatstone bridge. When the deflection-bridge output is connected to a low-impedance device such as a galvanometer, appreciable current flows and the galvanometer resistance must be considered in the bridge equation. Galvanometer current may be expressed by the

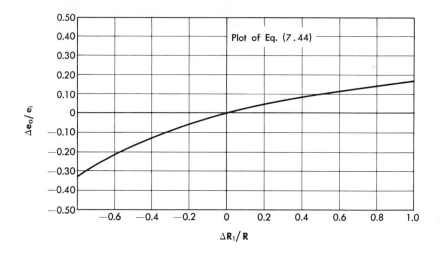

Fig. 7.15 Output from a voltage-sensitive deflection bridge whose resistance arms are initially equal.

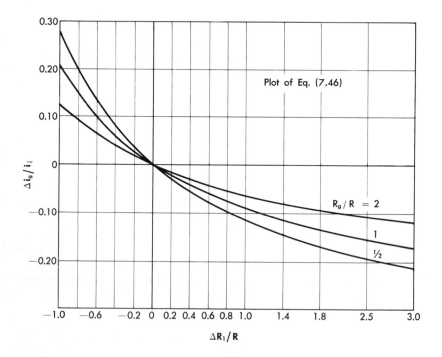

Fig. 7.16 Output from a current-sensitive deflection bridge whose resistance arms are initially equal, plotted for different relative galvanometer resistances.

following relation [15]:

$$i_g = \frac{i_i(R_2R_3 - R_1R_4)}{R_g(R_1 + R_2 + R_3 + R_4) + (R_2 + R_4)(R_1 + R_3)} , \tag{7.45}$$

in which

i_g = galvanometer current,

i_i = input current,

R_g = galvanometer resistance, ohms.

The remaining symbols are as defined in Fig. 7.13.

If we assume that an initial bridge balance is upset by an incremental change in resistance ΔR_1 in arm R_1 and all arms are of equal initial resistance R, we may write

$$\frac{\Delta i_g}{i_i} = \frac{-\Delta R_1/R}{4[1 + (R_g/R)] + [2 + (R_g/R)](\Delta R_1/R)} . \tag{7.46}$$

Figure 7.16 shows Eq. (7.46) plotted for various values of R_g/R.

c) *The constant-current bridge* To this point our discussion of bridge circuits has assumed a constant-voltage energizing source (a battery for example). As the bridge resistance is changed the total current through the bridge will, therefore, also change. In certain instances (see Article 11.14b), use of a *constant-current* bridge* may be desirable [24, 25]. Such a circuit is usually obtained through the application of a commercially available *current-regulated* d-c power supply†, whereby the total current flow through the bridge, i_i (Fig. 7.13), is maintained at a constant value. It should be noted that such a bridge may still be either voltage-sensitive or current-sensitive, depending on the relative impedance of the readout device.

Relationships for the voltage-sensitive *constant-current* bridge may be developed as follows. Referring to Fig. 7.13, we may write,

$$i_i = \frac{e_i}{R_1 + R_2} + \frac{e_i}{R_3 + R_4} , \tag{7.47}$$

or

$$e_i = i_i\left[\frac{(R_1 + R_2)(R_3 + R_4)}{R_1 + R_2 + R_3 + R_4}\right].$$

---

* The term "Wheatstone," as applied to bridge circuits, is commonly limited to the *constant-voltage resistance bridge*. We shall abide by this convention and avoid referring to the constant-current bridge as a Wheatstone bridge.

† Constant current is obtained by employing the voltage drop across a series resistor in the supply-output line to provide a regulating feedback voltage.

Substituting in Eq. (7.42), we have

$$e_o = i_i \left[\frac{R_1 R_4 - R_2 R_3}{R_1 + R_2 + R_3 + R_4} \right]. \tag{7.48}$$

This is the basic equation for the voltage-sensitive constant-current bridge, provided that i_i is maintained at a constant value. If the resistance of one arm, say R_1, is changed by an amount ΔR, then

$$e_o + \Delta e_o = i_i \left[\frac{(R_1 + \Delta R)(R_4) - R_2 R_3}{(R_1 + \Delta R) + R_2 + R_3 + R_4} \right]$$

and

$$\Delta e_o = i_i \left[\frac{(R_1 + \Delta R)R_4 - R_2 R_3}{(R_1 + \Delta R) + R_2 + R_3 + R_4} - \frac{R_1 R_4 - R_2 R_3}{R_1 + R_2 + R_3 + R_4} \right]. \tag{7.49}$$

For equal initial resistances ($R_1 = R_2 = R_3 = R_4 = R$),

$$\Delta e_o = i_i \left[\frac{\Delta R}{4 + \Delta R/R} \right]. \tag{7.50}$$

There is an improved linearity as a result of the constant-current bridge, which is apparent from a comparison of Eqs. (7.44) and (7.50). This coupled with the extreme sensitivity of the semiconductor strain gage is a major reason for the interest in the constant-current bridge. (See Articles 11.11a and 11.14b.)

d) *The a-c resistance bridge.* Resistance bridges powered by a-c sources may also be used. An additional problem, however, is the necessity for providing reactance balance. In spite of the fact that the Wheatstone bridge, strictly speaking, is a resistance bridge, it is impossible to completely eliminate stray capacitances and inductances resulting from such factors as closely placed lead wires in cables to and from the transducer, and wiring and component placement in associated equipment. In any system of reasonable sensitivity, such unintentional reactive components must be accounted for before satisfactory bridge balance can be accomplished.

Reactive balance can usually be accomplished by introducing an additional balance adjustment in the circuit. Figure 7.17 shows how this may be provided. Balance is accomplished by alternately adjusting the resistance and reactive balance controls, each time reducing bridge output, until proper balance is finally achieved.

e) *Adjusting bridge sensitivity.* There are several reasons for desiring to adjust bridge sensitivity. (1) Such adjustment may be used to attenuate inputs which are larger than desired. (2) It may be used to provide conven-

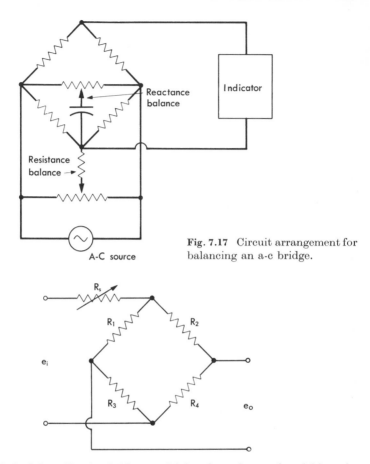

Fig. 7.17 Circuit arrangement for balancing an a-c bridge.

Fig. 7.18 Method for adjusting bridge sensitivity through use of variable series resistance, R_s.

ient relation between system calibration and the scale of the readout instrument. (3) It may be used to provide adjustment for adapting individual transducer characteristics to precalibrated systems. (This method is used to insert the gage factor for resistance strain gages in some commercial circuits.) (4) It provides a means for controlling certain extraneous inputs such as temperature effects, etc. (see Article 12.4d).

A very simple method of adjusting bridge output is to insert a variable series resistor in one or both of the input leads, as shown in Fig. 7.18. If we assume equal initial resistance R in all bridge arms, the resistance seen by the voltage source will also be R. If a series resistance is inserted, as shown, then thinking in terms of a voltage-dividing circuit, we see that the input to the

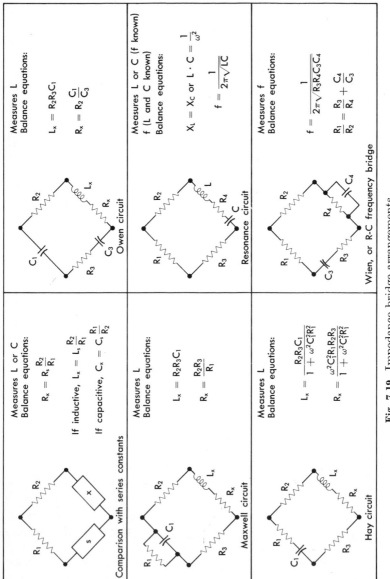

Measures L or C
Balance equations:

$$R_x = R_s \frac{R_2}{R_1}$$

If inductive, $L_x = L_s \frac{R_2}{R_1}$

If capacitive, $C_x = C_s \frac{R_1}{R_2}$

Comparison with series constants

Measures L
Balance equations:

$$L_x = R_2 R_3 C_1$$

$$R_x = \frac{R_2 R_3}{R_1}$$

Maxwell circuit

Measures L
Balance equations:

$$L_x = \frac{R_2 R_3 C_1}{1 + \omega^2 C_1^2 R_1^2}$$

$$R_x = \frac{\omega^2 C_1^2 R_1 R_2 R_3}{1 + \omega^2 C_1^2 R_1^2}$$

Hay circuit

Measures L
Balance equations:

$$L_x = R_2 R_3 C_1$$

$$R_x = R_2 \frac{C_1}{C_3}$$

Owen circuit

Measures L or C (f known)
f (L and C known)
Balance equations:

$$X_L = X_C \quad \text{or} \quad L \cdot C = \frac{1}{\omega^2}$$

$$f = \frac{1}{2\pi\sqrt{LC}}$$

Resonance circuit

Measures f
Balance equations:

$$f = \frac{1}{2\pi\sqrt{R_3 R_4 C_3 C_4}}$$

$$\frac{R_1}{R_2} = \frac{R_3}{R_4} + \frac{C_4}{C_3}$$

Wien, or R-C frequency bridge

Fig. 7.19 Impedance bridge arrangements.

bridge will be reduced by the factor

$$n = \frac{R}{R + R_s} = \frac{1}{1 + (R_s/R)} \, . \tag{7.51}$$

We call n the *bridge factor*. The bridge output will be reduced by a proportional amount, which makes this method very useful for controlling bridge sensitivity.

7.19 REACTANCE OR IMPEDANCE BRIDGES

Reactance or impedance bridge configurations are of the same general form as the Wheatstone bridge, except that reactance elements (capacitors and inductances) are involved in one or more of the arms. Because such elements are inherently frequency-sensitive, impedance bridges are a-c excited. Obviously the multitude of variations that are possible preclude more than a very general discussion in a work of this nature; thus the reader is referred to more specialized works for detailed coverage [16].

Figure 7.19 shows several of the more common a-c bridges, along with the type of element usually measured and the balance requirements.

7.20 RESONANT CIRCUITS

Capacitance-inductance combinations (Fig. 7.20) present varying impedance, depending on their relative values and the frequency of the applied voltage. When connected in parallel, as in Fig. 7.20(a), the inductance offers small opposition to current flow at low frequencies, while the capacitive reactance is low at high frequencies. At some intermediate frequency, the opposition to current flow, or impedance, of the combination is a maximum [Fig. 7.20(b)]. A similar but opposite variation in impedance is presented by the series-connected combination.

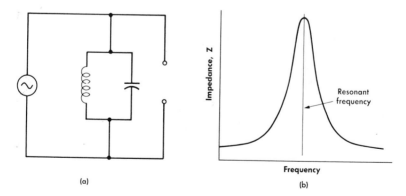

Fig. 7.20 Parallel L-C circuit with curve showing frequency-impedance characteristics.

The frequency corresponding to maximum effect, known as the *resonant frequency*, may be determined by the relation

$$f = \frac{1}{2\pi\sqrt{LC}}, \qquad (7.52)$$

in which

$$f = \text{frequency, Hz,}$$
$$L = \text{inductance, henrys,}$$
$$C = \text{capacitance, farads.}$$

It is evident that should, say, a capacitive transducer element be used, it could be employed in combination with an inductive element to form a resonant combination. Variation in capacitance caused by variation in an input signal (e.g., mechanical pressure) would then alter the resonant frequency, which could then be used as a measure of input.

a) Undesirable resonant conditions. On occasion, resonant conditions occur, introducing spurious outputs. Most circuits are susceptible because they employ some combination of inductance and capacitance and most are called upon to handle dynamic signal inputs. In certain cases the capacitance and inductance may be not more than the stray values existing between the circuit components, including the wiring. This means that resonant conditions are possible which can result in nonlinearities at certain input or exciting frequencies.

Normally such situations are avoided in the design of commercial equipment insofar as possible. However, the instrument designer is not always in a position to predict the exact manner in which general-purpose components may be assembled or the exact nature of the input signal fed to the equipment. As a result, it is quite possible to unintentionally set up arrangements of circuit elements combined with frequency conditions which result in undesirable resonant conditions.

7.21 ELECTRONIC AMPLIFICATION OR GAIN

Electronic amplification ratio, or *gain*, may be defined in two different ways. It may be defined simply as a ratio of the output to input voltages, currents, or powers; that is,

$$\text{Voltage gain} = \frac{\text{voltage output}}{\text{voltage input}},$$

$$\text{Current gain} = \frac{\text{current output}}{\text{current input}},$$

$$\text{Power gain} = \frac{\text{power output}}{\text{power input}}.$$

Another way to specify amplification is to use *decibels*. A decibel is one-tenth of a *bel*, and is based on a ratio of powers as follows:

$$\text{Decibels} = \text{db} = 10 \log_{10}\left(\frac{P_2}{P_1}\right), \tag{7.53}$$

where P_2 = output power, in watts, and P_1 = input power, in watts.

The average human ear can just detect a loudness change from an audio-amplifier when a power ratio change of one decibel is made. It has also been observed that this is approximately true regardless of power level.

Solving Eq. (7.53) for the ratio P_2/P_1 corresponding to one decibel yields a ratio of 1.26. In other words, for the average human ear to just detect an increase in sound output from an amplifier (feeding some form of earphone or loudspeaker), approximately 26% increase in power is required.

Electrical power may be written either as e^2/Z or as $i^2 Z$, where e is voltage, i is current, and Z is impedance. Substituting either of these forms in Eq. (7.53), and canceling impedance values, yields

$$\text{db} = 10 \log_{10}\left(\frac{e_2}{e_1}\right)^2 = 20 \log_{10}\left(\frac{e_2}{e_1}\right) \tag{7.54}$$

or

$$\text{db} = 10 \log_{10}\left(\frac{i_2}{i_1}\right)^2 = 20 \log_{10}\left(\frac{i_2}{i_1}\right). \tag{7.54a}$$

It is important to observe, however, that in dropping the Z's it is assumed that the impedances are equal. Decibels are basically measures of *power* gain rather than voltage or current gain.

Amplification calculations based on the decibel have a decided advantage over simple ratios because combined effects are determined simply by addition or subtraction. The amplification of a system in decibels is equal to *the sum* of the individual amplifications in decibels.

7.22 ELECTRONIC AMPLIFIERS

It is not the purpose of this article, or of the book, to be concerned with electronics or electronic theory beyond the barest minimum required to make intelligent use of such equipment for the purposes of mechanical measurement. The following discussion, therefore, will be brief and will be directed primarily to applications rather than to specific theory of operation.

Some form of amplification is almost always employed in circuitry intended for mechanical measurement. Traditionally, the term "electronic," as opposed to the word "electrical," assumes that in some part of the circuit electrons are caused to flow through space in the absence of a physical conductor, thus assuming the use of vacuum tubes. With the advent of *solid-state* devices, diodes, transistors, and the like, the word electronics

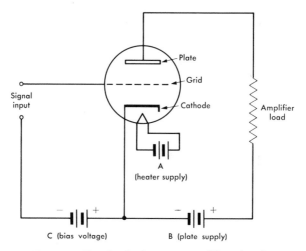

Fig. 7.21 Simple single-stage amplifier circuit.

has taken on broader meaning. Throughout the remainder of the book it will be understood that, unless more specific reference is made, the word "electronics" is being used in its broadest sense.

7.23 VACUUM-TUBE AMPLIFIERS

Electronic amplification by vacuum tubes is based on the fact that electrons emitted from a heated cathode (Fig. 7.21) are attracted to a positively charged plate, thereby causing current to flow in the plate circuit. The flow of electrons may be controlled by a third element, called a *grid*, interposed between the cathode and plate, provided it is properly negatively charged relative to the cathode. The negative voltage on the grid is referred to as *bias*. Variations in the charge on the grid supplied by the input signal may therefore be used to control the current flow in the plate circuit, including the amplifier load. In the figure, C supplies the necessary bias, B provides the plate supply, and A heats the filament and hence the cathode. Actually, of course, in the usual practical amplifier, the various voltages are obtained from a common supply through use of voltage dividers or dropping resistors.

This illustrates, in simplest form, a single stage of amplification. Often tubes with more elements are used, and usually greater amplification is desired than can be provided by one stage. Stages are then connected together so that the load for the input stage is a second stage, and so on.

a) Voltage and power amplifiers. Amplifiers of different types may be designed, depending primarily on the characteristics of the tubes used. In many measurement problems, amplitude is all that is required, and therefore a *voltage* amplifier may be employed. In this case, some form of indicator such

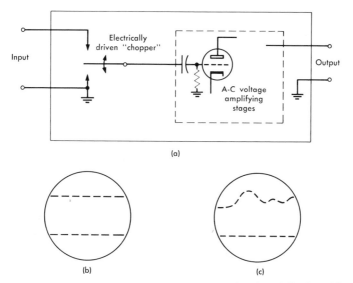

Fig. 7.22　(a) Schematic of a chopper amplifier. (b) d-c signal displayed by a CRO attached to a chopper amplifier. (c) Dynamic signal displayed by a CRO attached to a chopper amplifier.

as a cathode-ray oscilloscope or a vacuum-tube voltmeter might be used as the terminating device. On the other hand, if a recorder or some form of controller must be driven, then a *power* amplifier would be necessary to boost the energy available to drive the terminating device.

b) a-c and d-c amplifiers. Amplifiers may also be classified according to the character of the input signal which they will accept. Reliable amplifiers that will amplify d-c inputs are difficult to design. A constant d-c amplification or gain is difficult to maintain, as is a drift-free zero. As a result, a-c amplifiers are used wherever possible. As far as application is concerned, the difference between the two may be summed up by the statement that the a-c amplifier will accept only varying inputs, whereas the d-c amplifier will amplify constant* as well as varying inputs. A d-c voltage amplifier therefore will amplify such things as the voltage from a battery as well as a varying voltage, while the a-c amplifier will ignore a d-c voltage and amplify only the varying components.

c) Chopper amplifiers. A simple a-c amplifier may be used to amplify a d-c input through use of an additional circuit component known as a *chopper*. In its simplest form, the chopper is merely an electrically driven on-off switch, Fig. 7.22(a). Such devices may be driven at 60, 100, or 400 Hz, or at

---

* The word *constant* should not be confused with *steady state*. A battery voltage is constant, while the usual 110-V house current is steady state.

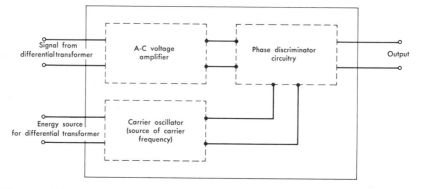

Fig. 7.23 Block diagram of a carrier amplifier for use with a differential transformer.

some other speed. When a d-c signal input is connected, the amplifier receives a chopped or square-wave voltage which it amplifies without trouble. More sophisticated chopped amplifiers employ electronic switching.

Figures 7.22(b) and (c) indicate the appearance of the amplifier output displayed on a CRO screen for a constant and a dynamic input, respectively. In each case the lower trace is present, resulting from the shorted input, providing a zero reference line. Of course, a problem is introduced when the input signal is varying at a frequency approaching that of the chopping frequency. In fact, the 10 to 1 ratio mentioned in Article 6.14 is usually considered the limit for chopper amplifiers also, i.e., the input signal frequency should be limited to $\frac{1}{10}$ of the chopping frequency.

d) Carrier amplifiers. Input signal information is often *carried* on an a-c frequency, in which case the input signal is said to modulate the carrier frequency. Certain special-purpose amplifiers incorporate the carrier source as a part of the amplifier. These arrangements are referred to as *carrier* amplifiers. Simple carrier systems provide an a-c output whose amplitude is blind to sign of the input signal. For example, the amplitude from an a-c powered strain-gage bridge does not in itself indicate strain direction from null. However, the phase relation between bridge power source and bridge output does depend on direction of bridge unbalance. This information may be used to determine sign. Phase-sensitive arrangements employed to accomplish this are referred to as phase-discriminator or mixer-demodulator circuitry. Figure 7.23 shows a block diagram of a carrier amplifier system for a differential transformer.

e) Tuned amplifiers. Amplifiers called upon to amplify a predetermined frequency, such as found in a carrier system, are sometimes *tuned*. This consists of incorporating a tuned filter (see Article 7.25c) that ideally allows only the predetermined frequency to pass, rejecting all others. Actually

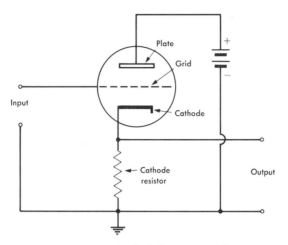

Fig. 7.24 Cathode-follower amplifier.

the sharpness of such tuning is limited, and a band of frequencies is allowed to pass. The advantage gained by using a tuned amplifier is that extraneous signals at other frequencies may be attenuated.

f) *The cathode-follower amplifier.* While Fig. 7.21 shows the most common simple circuit used for electronic amplification with vacuum tubes, actually the triode-type tube can be used in six different amplifier configurations [18], depending on the particular tube elements used for input and output. Normally the signal is introduced through the grid, as shown, and the plate is the output element. Of the remaining possible arrangements, the *cathode follower* is undoubtedly the second most useful in instrumentation circuitry. When this arrangement is used, the signal input is accomplished by means of the grid, as before, but the output is taken from across the cathode resistor (Fig. 7.24). The voltage gain obtained by this means is always less than unity. However, the primary advantage of the circuit is that it transforms a high-impedance source at the grid to a low-impedance output.

This circuit is often used as the output stage of an oscillator or a voltage amplifier and is quite commonly used for coupling high-impedance transducers, such as the piezoelectric type, to low-impedance intermediate or terminating devices.

g) *The charge amplifier.* Many electrical-type transducers have low to moderately high output impedances (several ohms to perhaps 500,000 ohms). No particular impedance-matching problem is normally encountered when they are connected to the grid circuit of a vacuum-tube amplifier. Piezoelectric devices, on the other hand, commonly have output impedances of many megohms. Although cathode-follower amplifiers may be employed to advantage, another problem often remains, namely, the effects of unavoidable capacitance in attached cables, connectors, etc., which may seriously attenuate the transducer output. Special cables are sometimes supplied with the cable-transducer combination calibrated and used as a unit.

An important solution to this problem (which also permits more reliable calibration techniques) is to employ a *charge amplifier* instead of the voltage amplifier or cathode follower. Through proper application of negative feedback, the charge amplifier senses charge, or quantity of electrons at its input, rather than voltage. The output of a charge amplifier is expressed by [23]

$$e_o = \frac{Q}{C_f} \times \frac{G_o C_c}{C_t + C_f + (1 + G_o)C_c}, \qquad (7.55)$$

where

e_o = output voltage (readout), volts,

Q = charge developed by transducer, coulombs,

G = open circuit gain of amplifier,

C_t = capacitance of transducer,

C_c = capacitance of connecting cable,

C_f = capacitance of amplifier feedback circuit.

If C_t and C_f are small in comparison to $(1 + G_o)C_c$ (as may be easily arranged), and G_o is made large in comparison to unity (it is usually 1000 or more), then Eq. (7.55) reduces to

$$e_o = Q/C_f. \qquad (7.56)$$

Therefore the output voltage is directly proportional to the charge supplied by the transducer and is independent of connecting capacitances, e.g., satisfactory use of cables as long as one mile is claimed by some manufacturers.

7.24 SOLID-STATE AMPLIFIERS [22]

"Transistorized" measurement devices are quite common. The transistor can perform most of the functions of the vacuum tube and can do so without the heated filament, high voltages, and shock-sensitive elements inherent in the construction and operation of the vacuum tube. It is not surprising, then, that transistorized measurement apparatus such as amplifiers and oscillators are gradually taking the places of their vacuum-tube counterparts.

The basic transistor is a three-element device (Fig. 7.25) and in this respect is similar to the triode vacuum tube. It is also adaptable to various circuit arrangements as shown in Fig. 7.25. Although comparisons are not directly possible, inspection of the impedance and gain relationships listed

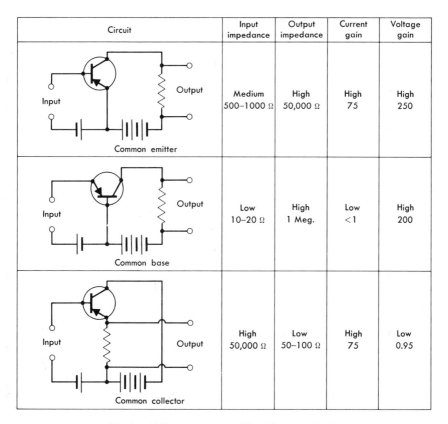

Circuit	Input impedance	Output impedance	Current gain	Voltage gain
Common emitter	Medium 500–1000 Ω	High 50,000 Ω	High 75	High 250
Common base	Low 10–20 Ω	High 1 Meg.	Low <1	High 200
Common collector	High 50,000 Ω	Low 50–100 Ω	High 75	Low 0.95

Fig. 7.25 Transistor amplifier characteristics.

in the figure indicate rough equivalencies between the common emitter, common base, common collector-transistor arrangements, and the voltage- and cathode-follower vacuum-tube amplifiers, respectively.

Several advantages accrue from the use of transistors in measurement apparatus. The transistor, coupled with module-type integrated circuits, permits considerable weight saving and reduced size. In addition, the inherent characteristics of the transistor permit replacement of relatively high-voltage plate supplies required by vacuum tubes with much lower voltages, often conveniently supplied by batteries. The latter permits easy field use. Vibration and shock occasionally make vacuum-tube "micro-phonics" a problem; this is essentially eliminated in transistorized equipment. On the debit side, transistors are more sensitive than are tubes to misuse through voltage extremes or incorrect polarities. Proper performance is also more temperature dependent.

Undoubtedly, however, the use of transistors coupled with integrated circuitry will grow very rapidly in the area of mechanical measurements.

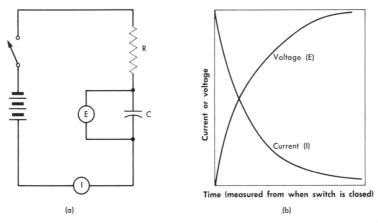

(a)   (b)

Fig. 7.26 Circuit illustrating the voltage-time and current-time relations for a resistance and capacitance connected in series.

7.25 SPECIAL CIRCUITRY

a) Time constant. If a capacitance and a resistance are connected in series with a battery as shown in Fig. 7.26(a), the time required to charge the condenser will be a function of the values of both R and C. From the instant the switch is closed, the current-time and voltage-time relations will be as shown in Fig. 7.26(b).

The product of the capacitance C, in farads, and the resistance R, in ohms, is referred to as the *time constant* for the circuit. This product is equal to the time, in seconds, required for the *voltage* across the condenser to reach a value of $1 - (1/e)$, or about 63% of the value of the battery voltage. (e is the base of the natural logarithms.) (See Article 2.8a.)

A similar relation may be had for an inductive-resistive circuit. In this case the time constant is equal to L/R, where L is in henrys and R is in ohms.

Relatively simple circuits involving combinations of resistance, capacitance, and inductance are used for filtering and for differentiation or integration.

b) Differentiation and integration. Differentiation is obtained when the output is proportional to the time rate of change of the input. Figure 7.27(a) shows a simple R-C differentiation circuit for which we will assume a capacitor reactance that is small compared with the resistance R. Actual resistance and capacitance values, however, must be selected so that the time constant for the circuit is small compared with the period of the input signal. The capacitor reactance depends on frequency, and the output appearing across the resistor will be a function of the rate of change of the input signal, hence its derivative. As a result, a differentiating action is obtained.

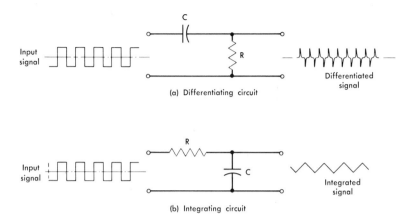

(a) Differentiating circuit

(b) Integrating circuit

Fig. 7.27 Differentiation and integration circuits and their treatment of a square-wave input.

By using a somewhat similar circuit, but with component values which provide a time constant that is long compared with the signal frequency, an integrator is obtained. Such a circuit is shown in Fig. 7.27(b). When a long time constant is provided, the capacitor charges slowly and the output becomes a function of the time summation of the input, or its time integral.

Figure 7.27 illustrates typical differentiator and integrator treatments of a square-wave input. Similar action may be obtained by means of circuits incorporating inductance.

c) Filtering. Attenuation of unwanted signal components may also be accomplished by *R-C-L* arrangements in the form of filter circuits. Filtering action may be of the *high-pass, low-pass,* or *band-pass* type. A high-pass filter permits the higher frequencies to pass while attenuating or blocking the lower frequencies. The low-pass filter performs the opposite function, while the band-pass filter allows only a controlled range of frequencies to pass, attenuating frequencies on either side. Another kind of filter, known as the *band-elimination* filter, attenuates frequency components falling within a predetermined range or band of frequencies.

High-pass and low-pass filters are evaluated in terms of *cut-off* frequency and rates of attenuation. The theoretical frequency at which discriminatory action commences is referred to as the *cut-off frequency* and is determined by filter configuration and component values. Of course, the attenuation that is provided is not sudden in the sense that all frequencies on one side of the cutoff are passed while all on the other side are eliminated. Sharpness of attenuation is determined by the filter configuration and is usually evaluated in terms of *db per octave.* (One octave corresponds to a doubling or halving of the frequency.)

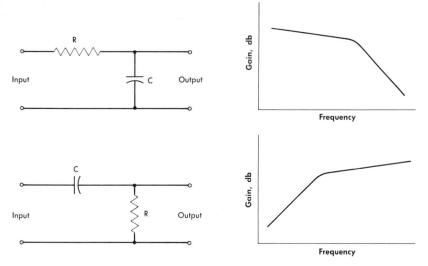

Fig. 7.28 Simple R-C filter arrangements and curves illustrating output characteristics.

Filters may be of many different configurations, and the problems and possibilities can only be briefly outlined here. In general, resistance elements introduce energy losses and are used only where power losses are not important. Simple R-C circuits as shown in Fig. 7.28 may be used for filtering when proper component values are employed. More efficient filtering is accomplished with L-C circuit arrangements as shown in Fig. 7.29.

An additional problem in filter selection and design is that of providing proper matches between the filter and preceding and following system components. The general requirement is that the impedances be matched.

Variable filters may be had through use of adjustable filter elements. For rf (radio-frequncy) signals, variable capacitances may be employed. Most mechanical signals, however, are in the audio-frequency range, and for these frequencies relatively large capacitors are normally required which cannot easily be made variable.   Therefore, adjustable inductances are employed in variable filters covering the audio range. Commercial variable-range filters of the high-pass, low-pass, and band-pass types are available.

Filters discussed to this point are known as "passive-network" filters, i.e., the only source of energy is that of the signal itself. Filters incorporating electronic amplifiers and matching circuitry are also used [19, 20]. In such cases the same filtering principles are employed; however, the problem of signal losses may be solved by amplification. Such filters, sometimes referred to as *electronic* filters, are usually inserted in a measuring system at a point where the signal is at a low energy level. R-C circuitry is normally used, thereby facilitating the design of the variable type. Such filters are often

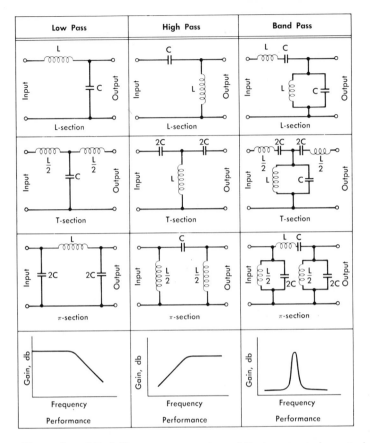

Fig. 7.29 Examples of L-C filter arrangements and their output characteristics.

employed in vibration and sound analyzers for the purpose of separating the various harmonic components. (See Article 17.4d.)

d) *Component coupling methods*. When electrical circuit elements are connected, it is often necessary to give special attention to the coupling methods used.   In certain cases transducer-amplifier, amplifier-recorder, or other component combinations are inherently incompatible, making direct coupling impossible or, at best, causing nonoptimum operation. Coupling problems include obtaining proper impedance match and maintaining proper circuit requirements such as damping.   Problems of this nature are usually brought about by desire for maximum energy transfer and optimum fidelity of response.

The importance of exact impedance match, however, varies considerably from application to application.   For example, the input impedances of most cathode-ray oscilloscopes and vacuum-tube voltmeters are relatively

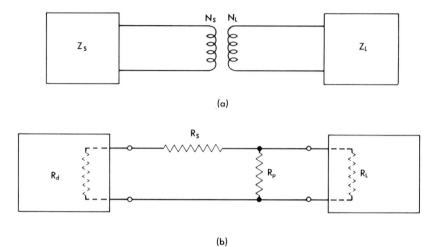

Fig. 7.30 (a) Impedance matching by means of a transformer. (b) Impedance matching by means of a resistance pad.

high, yet satisfactory operation may be obtained from directly connected low-impedance transducers. In this case, voltage is the measured quantity and power transfer is incidental. In most cases, driving a high-impedance circuit component with a low-impedance source presents far fewer problems than does the reverse situation.

In addition to allowing power transfer, proper coupling may be important in determining dynamic response. An example of this is presented by the amplifier-oscillograph combination. Oscillograph response depends on proper damping, which, in turn, is affected by the electrical circuitry. (Oscillograph damping requirements are discussed in Articles 8.11 and 8.11c.)

There are three special methods commonly employed for coupling, depending on the circuit elements used. The three methods make use of (1) matching transformers, (2) cathode-follower amplifiers, and (3) coupling networks.

An example of transformer coupling is the use of electronic power amplifiers to drive vibration exciters such as discussed in Article 16.17a. The problem of connecting the amplifier to the shaker head is similar to that of connecting a speech amplifier to a loudspeaker. In both cases, the plate output impedance of the driving power tubes in the amplifier is normally much higher than the load impedance to which they must be connected.

In this case, transformer coupling is usually employed as illustrated in Fig. 7.30(a). Matching requirements may be expressed by the relation

$$\frac{N_s}{N_L} = \sqrt{\frac{Z_s}{Z_L}}, \tag{7.55}$$

in which

$$Z_s = \text{source impedance,}$$
$$Z_L = \text{load impedance,}$$
$$N_s/N_L = \text{turns ratio of transformer.}$$

General-purpose devices, such as simple voltage amplifiers and oscillators, often incorporate a final cathode-follower stage to supply a low-impedance output. Cathode followers are also used immediately after a high-impedance pickup, such as a piezoelectric accelerometer. By reducing the impedance source, losses in the connecting lines and the possibility of extraneous signal pickup are minimized (Article 16.9).

Proper coupling may also be accomplished through use of matching resistance pads. Figure 7.30(b) illustrates what is probably the simplest form.

If we assume that the driver output and load impedances are resistive, then the requirements for proper matching may be put in simple form as follows: The driving device, which may be a voltage amplifier, *looks* into the resistance network and *sees* the resistance R_s in series with the paralleled combination of R_L and R_p. Hence, for proper matching,

$$R_d = R_s + \frac{R_p R_L}{R_p + R_L}, \tag{7.56}$$

in which

$$R_d = \text{output impedance of the driver, ohms,}$$
$$R_L = \text{load resistance, ohms,}$$
$$R_p = \text{paralleling resistance, ohms,}$$
$$R_s = \text{series resistance, ohms.}$$

The driven device *sees* two parallel resistances, made up of R_p and the series-connected resistances R_s and R_d. Hence, for matching,

$$R_L = \frac{R_p(R_s + R_d)}{R_p + (R_s + R_d)}. \tag{7.56a}$$

Solving for R_s and R_p, we have

$$R_s = [R_d(R_d - R_L)]^{1/2} \tag{7.57}$$

and

$$R_p = \left[R_L \left(\frac{R_d R_L}{R_d - R_L} \right) \right]^{1/2}. \tag{7.57a}$$

Now if R_d and R_L are known, values of R_s and R_p may be determined to satisfy the matching requirements by use of Eqs. (7.57) and (7.57a). It must be realized, however, that in using resistive elements there will be an unavoidable loss in signal energy. Such losses are often referred to as *insertion losses*. In general, however, by providing proper match, the network will provide optimum gain.

7.26 TELEMETRY

Telemetry, which is the technique of measuring from a distance, is a very important part of the intermediate measurement stage in systems used for missile and aircraft flight testing. Applications such as these require radio links which permit the use of readout devices located on the ground. Telemetry is also employed in industrial, medical, and transportation applications. Sometimes the term is applied to certain hydraulic and pneumatic links; however, in the discussion that follows we shall be primarily interested in the transmission of data by radio link.

Telemetering provides several distinct advantages over recording of data at the source. The weight of telemetering equipment located at the data source is usually less than that of recording equipment of comparable capacity. In addition, as many channels as necessary may be individually and continuously monitored during the test without requiring the direct attention of the pilot or operator. When safe limits are exceeded, they may be immediately recognized and corrective measures taken. When equipment is operated in untested performance where there is some possibility of destructive failure, telemetered data ensures a complete record up to the final moment. This is especially important in missile testing when the test item may not be recoverable. Another advantage of certain telemetered systems is that the practical recording time is not limited, which may be the case when paper, film, or tape must be carried with the data source.

Of course, there are also certain disadvantages in the use of telemetering systems. The complexity and therefore the expense of the system is unavoidably increased. In addition, the required extra processing of the data provides greater chance for error and, in particular, greater opportunity for the introduction of extraneous noise.

Basically the measuring system differs from the conventional only in the addition of the telemetering link. Standard detector-transducer elements and recording systems may be employed, along with conventional amplification and calibration equipment. Assuming the use of a radio link, it is of course necessary that the input be transduced to electrical form for telemetering.

Various telemetering systems are employed for flight testing, only one of which will be described (briefly) below. The most common systems employ what are termed subcarrier oscillators (SCO), whose frequencies are controlled by the input signals through appropriate transducer elements. A range of *audio*-frequency channels are employed, with the frequency of

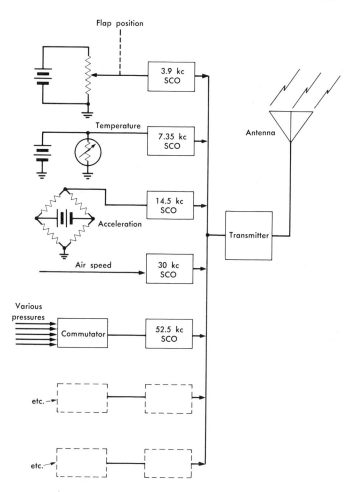

Fig. 7.31(a) Block diagram of a telemetering transmitting system.

each SCO modulated by the magnitude of the input signal. The outputs from the various subcarrier oscillators are then mixed and fed to a phase-modulated transmitter (Fig. 7.31(a)) which relays the combined information to the remote receiving station (Fig. 7.31(b)).

Additional information may be transmitted by a single SCO channel through use of commutation. A commutator repeatedly connects and disconnects a sequence of signal inputs to the input of a subcarried oscillator. As many as 60 functions may be fed to one SCO. Of course, only slowly changing functions may be handled in this manner because of the relatively short-pulse time bases. In order to decommutate the information, it is necessary to provide a synchronizing pulse for identification.

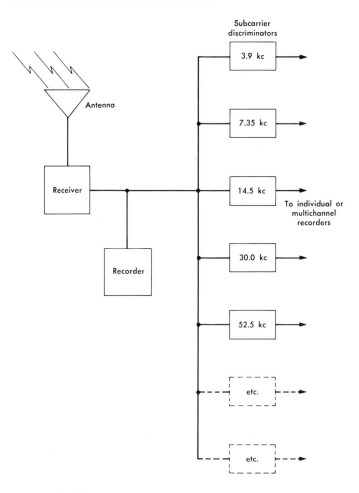

Fig. 7.31(b)   Block diagram of a telemetering receiving system.

At the receiving end, the various subcarrier frequencies are separated through use of filtering or discriminating circuitry, and the information from the individual channels is recorded by any of the conventional methods. To guard against loss of important test data, it is often the practice to also record the directly received signal on tape.

Calibration may be provided by any of the electrical methods appropriate to the transducer. This operation may be either time controlled, initiated by the operator or pilot, or controlled from the recording installation through additional radio links.

Although the emphasis in this article has been placed on telemetry as applied to missile and flight testing, the methods are applicable to any situation where positive physical continuity throughout the test system is difficult to maintain. In some instances the problem may be concerned with danger to life or equipment. This would be the case in certain nuclear test facilities, although direct-wire connections would usually be feasible. In other cases, the mechanical situation may make direct connections difficult, as would be the case if direct contact is to be made to rotating machinery. A solution to a problem of this type through use of telemetry is referred to in Article 11.17.

In this chapter we have discussed system elements under the general heading, "Intermediate Modifying Devices." The devices discussed have included those elements required to alter detector-transducer outputs into forms acceptable to the terminating devices, whose duty it is to present the information in intelligible form. The next chapter considers the special requirements of this third stage of the measurement system.

SUGGESTED READINGS

Laws, F. A., *Electrical Measurements*, 2nd. Ed. New York: McGraw-Hill Book Co., 1938.

Michels, W. C., *Electrical Measurements and Their Applications*. Princeton, N.J.: D. Van Nostrand Co., 1957.

Stout, M. B., *Basic Electrical Measurements*. Englewood Cliffs, N.J.: Prentice-Hall, Inc., 1950.

PROBLEMS

7.1 Describe precisely and concisely the two following concepts, and indicate how they affect mechanical measurements:
a) reflected frictional amplification, and
b) reflected inertial amplification.

7.2 Referring to the sketch of the polar planimeter, Fig. 7.2, show that the area of the "zero circle" is equal to

$$(\pi) [(BE)^2 - (EF)^2 + (FA)^2].$$

7.3 A ballast circuit is used with a resistance-type transducer (say, a thermistor), which has a nominal resistance of 100 ohms.
a) What should be the value of the ballast resistor, R_b, for maximum sensitivity at the nominal transducer resistance?
b) If the resistor recommended in part (a) is used along with an input of 10 V, what output voltage would be had if the transducer resistance remains unchanged from nominal?

c) If the conditions of part (b) remain, except that the transducer resistance is reduced by 5%, what output voltage would be indicated?

d) What are the sensitivities corresponding to parts (b) and (c) above? What units?

7.4 Equations (7.34) and (7.36) are derived on the basis of a high-impedance indicator. Analyze the circuit, assuming that the indicator resistance, R_m, is comparable in magnitude to R_t.

7.5 A simple voltage-dividing potentiometer circuit is to be used as the secondary transducer in a pressure pickup.

a) If the output load is resistive and is 5 times the potentiometer resistance, what loading error, based on full scale, will be inherent at 60% of full-scale potentiometer position?

b) What will be the above error based on reading?

c) What errors will occur at 0% and 100% positions?

7.6 Equation (7.40) expresses the loading error for a voltage-dividing potentiometer circuit *based on full scale*. Show that the following equation yields the error based on reading:

$$\text{Percent error} = k(1 - k)(R_b/R_L) \times 100.$$

[*Hint:* Note that Eq. (7.38) yields the reading, and Eq. (7.39) yields the error.]

7.7 A 5000-ohm voltage-dividing potentiometer in a conventional circuit (Fig. 7.8) feeds an 8000-ohm load. Sketch a curve showing the percent error *based on reading*, plotted against the potentiometer input. Calculate enough points to provide a reasonably definitive curve.

7.8 Employing methods similar to those used in Article 7.16, analyze the voltage-divider circuit including end resistors (Fig. 7.10).

7.9 A resistance bridge has the configuration illustrated in Fig. 7.13, in which $R_1 = 120.4$ ohms, $R_2 = 119.7$ ohms and $R_3 = 119.0$.

a) What resistance must R_4 have for resistance balance?

b) If R_4 has a value of 121.2 ohms, and if the input voltage is 12 V dc, what will be the meter reading in volts, assuming a voltage-sensitive bridge?

7.10 A resistance-type bridge circuit is made up of the following resistances: $R_1 = 9725$ ohms, $R_2 = 8820$ ohms, $R_3 = 8550$ ohms, and $R_4 = 9875$ ohms.

a) If the bridge is of the voltage-sensitive type and if the input voltage is 24 V, what should be the meter reading?

b) If R_4 is variable, what value should it have for null balance?

7.11 A simple equal-arm voltage-sensitive resistance bridge is initially in balance (null). Three of the arms consist of ordinary resistors, while the fourth is a thermistor. Each arm has a nominal resistance of 10,000 ohms, and the bridge is energized with a 6 V d-c source.

a) If temperature change causes a +5% change in the thermistor resistance, what output voltage will be indicated from the bridge?

b) If a −5% change in resistance is caused, what will be the bridge output?

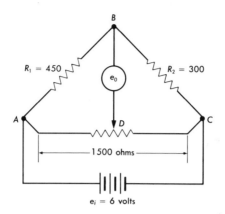

$R_1 = 450$ e_0 $R_2 = 300$

A D C

1500 ohms

$e_i = 6$ volts

Figure 7.32

7.12 Figure 7.14c shows a bridge employing shunt balance. Abiding by the symbolism of the "standard" bridge, Fig. 7.13, $R_1 = 420$ ohms, $R_2 = 360$ ohms, $R_3 = 480$ ohms, and $R_4 = 300$ ohms. What value of shunting resistance will be necessary, under these conditions, for null?

7.13 Figure 7.32 shows a simple d-c resistance bridge. Resistance ADC is provided by a 1500-ohm linear 10-turn potentiometer. The dial reads "zero" when slider D is nearest to A, and it reads 1000 when it is nearest to C.

a) What should be the dial reading corresponding to null?

b) If the bridge is voltage-sensitive, what will be the output when the dial reads 350?

7.14 Describe what is meant by:

a) a carrier amplifier,

b) a cathode-follower amplifier,

c) a high-pass filter, and

d) a power gain of 6 db.

7.15 Assume that the resistances, R, of the four arms of a constant-current bridge are initially equal. Assume further that the resistances of any two opposite arms change by an amount ΔR and that the resistances of the two remaining arms change by the amount $-\Delta R$. Under these conditions, show that

$$\frac{\Delta e_o}{i_i R} = \frac{\Delta R}{R}.$$

See Article 11.14(b) for application.

CHAPTER 8

TERMINATING DEVICES AND METHODS

8.1 INTRODUCTION

Final usefulness of any measuring system depends on its ability to present the measured output in a form which is comprehensible to the human operator or the controlling device. The primary function of the terminating device is to accept the analogous driving signal presented to it and to either provide the information in a form for immediate reading or record it for later interpretation.

For direct human interpretation, except for simple yes-or-no indication, the terminating device presents the readout in one of two forms: (1) a relative displacement, or (2) in digital form. Examples of the first are: a pointer moving over a scale, a scale moving past an index, a light beam and scale, and a liquid column and scale. Examples of digital output are: odometer indicators in automobile speedometers, electronic decade counters, rotating drum mechanical counters, etc.

Examples of exceptions to the two forms above are: any form of yes-or-no limiting-type indicator, such as the red oil-pressure lights in some automobiles, pilot lamps on equipment, and—an unusual kind—litmus paper, which also provides a crude measure of magnitude in addition to a yes-or-no answer. Perhaps the reader can think of other examples.

By far the most common form of indication is relative displacement.

As has been stated on numerous occasions in previous chapters, measurement of *dynamic* mechanical quantities practically presupposes use of electrical equipment for stages one and two. In many cases the electronic components used consist of rather elaborate systems within themselves. This is true, for example, of the cathode-ray oscilloscope. Sweep circuitry is involved, providing a time basis for the measurement. In addition, the input is carried through further stages of amplification before final presentation. The primary purpose of the complete system, however, is to present the input analogous signal in a form acceptable for interpretation. Such a self-contained system will therefore be classified as an integral part of the terminating device itself.

For the most part, dynamic mechanical measurement requires some form of voltage-sensitive terminating device. Rapidly changing inputs preclude strictly mechanical, hydraulic-pneumatic, and optical systems, either because of their extremely poor response characteristics or because the output cannot be interpreted. Therefore the major portion of this chapter will be concerned with electrical indicators and recorders.

8.2 METER INDICATORS

Pointer-and-scale meters are useful for static and steady-state dynamic indications. They are not generally usable for transient measurement. This is due to the relatively high inertia of the meter movement. In addition, assuming the meter could follow a transient input, it would be impossible for the human eye to follow the pointer.

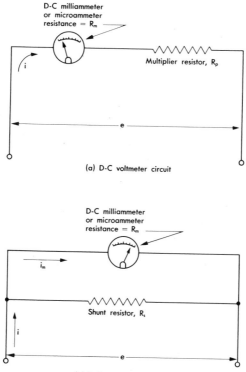

Fig. 8.1   (a) D-C voltmeter circuit, and (b) d-c current-meter circuit.

Meter indicators may be classified as follows: (1) simple D'Arsonval types for measurement of current or voltage, (2) ohmmeters and volt-ohm milliammeters (VOM, or multimeters), and (3) vacuum-tube voltmeters (VTVM).

In the majority of cases, the simple D'Arsonval meter movement is employed as the final indicating device. However, moving-iron meters may be used for measuring a-c current. In the more versatile types, such as the volt-ohm milliammeter or multimeter, internal shunts or multipliers are provided with switching arrangements for increasing the usefulness of the instrument.

Basically, the D'Arsonval movement is current-sensitive; hence regardless of the application, whether it be as a current meter or as a voltmeter, current must flow. Naturally in most applications, the smaller the current flow, the lower will be the *loading* on the circuit being measured. The meter movement itself possesses internal resistance varying from a few ohms for the less sensitive milliammeter to roughly 2000 ohms for the more sensitive micro-ammeter. Actual meter range, however, is primarily governed by associated range resistors.

Figure 8.1 shows schematically the basic d-c voltmeter and d-c current-meter circuits. Either multiplier or shunt resistors are used in conjunction with the same basic meter movement. To minimize circuit loading, it is desirable that total *voltmeter* resistance be much greater than the resistance of the circuit under test. For the same reason, the *current-meter* resistance should be as low as possible. In both cases, meter movements providing large deflections for given current flow through the meter are required for high sensitivity.

a) Voltmeter sensitivity. Voltmeter resistance is determined primarily by the series multiplier resistance. High multiplier resistance means that the current available to actuate the meter movement is low and that a sensitive basic movement is required. Because sensitivity may differ from meter to meter even though the meters may be of the same range, it is insufficient to rate voltmeters simply by stating total resistance. Rating is commonly stated in terms of *ohms per volt. This value may be thought of as the total voltmeter resistance that a given movement must possess in order for the application of one volt to provide full-scale deflection.* This value combines both resistance and movement sensitivity, and the higher the value, the lower will be the loading effect for a given meter indication.

Simple pocket multimeters usually employ a meter of 1 ma and 1000 ohms/volt rating, whereas more expensive multimeters may use movements with a rating of 50 μa and 20,000 ohms/volt.

The value of the series multiplying resistor, R_p, may be determined from the relation (see Fig. 8.1a)

$$R_p = \frac{e}{i} - R_m.$$

Fig. 8.2 Simpson Model 260 volt ohm milliammeter (VOM). (Courtesy Simpson Electric Company, Chicago, Illinois.)

b) The current meter. Since current meters are connected in series with the test circuit, the voltage drop across the meter must be kept as low as possible. This means that the combination of meter and shunt must have as low a combined resistance as practical. Referring to Fig. 8.1(b), we may write the relation, based on equal voltage drops across meter and shunt,

$$R_s = \frac{i_m R_m}{i - i_m}.$$

c) a-c meters. Provision for measuring a-c voltages is made through use of a rectifier in conjunction with a d-c meter movement. Meters of this type are usually calibrated to read in terms of the root-mean-square (rms) value of a sine-wave input. For this reason the careful interpretation of rectifier-type a-c meter indications is necessary when other than sine-wave inputs are measured.

d) The ac-dc multimeter. Figure 8.2 shows a typical ac-dc multimeter, often referred to as a volt-ohm milliammeter (VOM). A meter of this sort uses switching arrangements for connecting multiplier and shunt resistors and rectifier circuits. In addition, the meter is arranged to measure resistance, using an internal battery as a source of energy. By switching to the ohm-meter function and connecting the leads to the unknown resistance, one can determine from the meter movement the current flowing through the resistor. The current flow indication is calibrated in terms of resistance, thereby providing direct means of measurement.

8.3 THE VACUUM-TUBE VOLTMETER

As the name indicates, the vacuum-tube voltmeter (VTVM) is a voltage-measuring instrument incorporating one or more electronic tubes in its circuitry. The tubes are used for amplifying the input and for rectification. The a-c or d-c input to the meter is applied through appropriate multiplier resistors to the control grid of the tube, and the resulting change in plate current is connected to the meter movement, causing a proportional meter deflection. If the input is in a-c form, rectification immediately precedes the meter.

Because grid circuit resistance is usually quite high, the input resistance of a vacuum-tube voltmeter is normally much higher than that of the simple meter. This, for the most part, is a decided advantage in favor of the vacuum-tube voltmeter because it minimizes loading of the signal source. It is not, however, an unmixed blessing, inasmuch as the instrument is thereby

Fig. 8.3 Model 300D Ballantine vacuum-tube voltmeter. (Courtesy Ballantine Laboratories, Inc., Boonton, N.J.)

capable of sensing very low energy voltages. For this reason, it is susceptible to miscellaneous electrostatic voltages or *noises*, the most troublesome being an extraneous 60-cps hum radiated from the power lines. It may even indicate voltages from radio-frequency sources such as radio stations located in the vicinity. Unwanted pickup of this sort can only be combated by shielding, which is often a nuisance to use.

In addition to the inherent high input impedance, a second advantage enjoyed by the VTVM is the possibility of amplification. This permits use of relatively rugged meter movements without sacrificing sensitivity.

Figure 8.3 shows a laboratory quality a-c VTVM having ranges from 0.01 to 100 V rms full scale. The input impedance for this instrument is 500,000 ohms.

8.4 MECHANICAL COUNTERS

Mechanical counting devices are usually of the decade drum type used in the conventional automobile odometer. Such counters may either be mechanically driven by direct coupling to a rotating input shaft, or they may be actuated by oscillating or reciprocating motions introduced through appropriate linkage-and rachet mechanisms. Mechanical counters of this sort are also available actuated by electrical impulses through the use of solenoids.

Electrically-actuated counters may be energized by a simple switch or relay, by photocells (with amplification), or by any source of pulse able to supply a power of several watts. One high-speed counter of this type is capable of making 1000 counts/min. This counter requires a 5-watt power source and will respond to a *make* pulse as short as 0.024 sec or a *break* of 0.036 sec duration.

8.5 THE ELECTRONIC COUNTER

Of course, there are many mechanical measurement problems requiring counting speeds greater than a mechanical counter can accommodate. Such problems may involve measurement of vibration frequencies, use of a turbine-type flow meter [Article 14.8(a)], tachometry, etc. It is in the high-speed counting area that the electronic counter enters the picture.

Figure 8.4 shows an example of a multipurpose electronic counter, which is actually much more than a simple counting device. In addition to performing direct counts, this instrument is also capable of measuring events per unit time (EPUT), time interval or period, average period, frequency, or the ratio of two different frequencies. Its capacities are 1,000,000 counts at rates up to 2 MHz, and time intervals from 10 μ-sec to 10^7 sec.

The basic unit in an electronic counter is the counter decade. As each pulse is received, a progressive circuit change occurs, which is indicated by

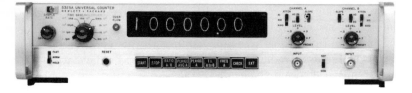

Fig. 8.4 Hewlett-Packard Model 5325A Electronic Counter. (Courtesy: Hewlett-Packard Co., Palo Alto, Calif.)

the counter tube. After ten such changes, the decade produces an output pulse which may be used to drive the following decade. A series of decades may be used in this manner to record units, tens, hundreds, etc.

Counters of this type have high-impedance inputs and require little energy for pulsing. The input impedance of the counter shown in Fig. 8.4 is one megohm.

a) Time-interval meter. If a crystal-controlled time base and an electronic switch or *gate* are incorporated into the system, the time interval during which an event occurs may be determined by a counter. All that is necessary is that a pulse be provided at the beginning of the event and that a second pulse occur at the end. These pulses are used to start counting and to stop counting the cycles of the time-base oscillator.

Figure 8.5 illustrates schematically how the system operates. The counter actually counts the cycles from the built-in time base, over a period determined by the external event. The typical counter shown in Fig. 8.4 is provided with the time bases of 1 MHz to 0.1 Hz in decade steps.

b) Events-per-unit-of-time (EPUT) meter. By rearranging the basic elements in the universal timer somewhat, an EPUT meter (Fig. 8.6) may be obtained. In this case the time base is used to control the electronic gate, and the number of pulses from the external source is counted for a preset time interval. This change-over is accomplished by simple switching on the front panel. The counted input pulses may be at a fixed rate, such as from a steady-state vibration, or at an erratic rate, such as would be obtained from a Geiger-counter tube (Article 18.5b).

c) Variations from the basic universal counter. There are many variations from the basic counter system used for special-purpose applications. For mechanical measurements, a high counting rate may not be required. When such is the case, two or three electronic decades are sometimes used, which feed a simple, solenoid-operated mechanical counter. Other variations use the basic counter as a control unit by providing a control output pulse when a preset total count is reached. A unit of this type may be used for such applications as automatic packaging, coil winding, sorting, etc. Digital

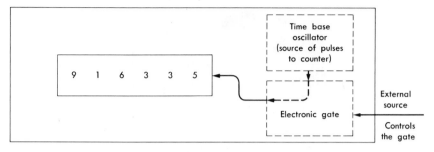

Fig. 8.5 Schematic of a time-interval meter. If the time base were set to 10 kHz, the indicated time interval would be 91.6335 sec.

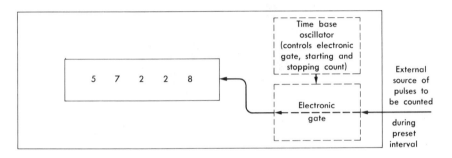

Fig. 8.6 Schematic of events-per-unit-time (EPUT) meter. If the gate time were set at 10 sec, the indicated frequency would be 5722.8 Hz.

print-out recorders are also available, providing automatic count recording. Further application of electronic counters is discussed in Chapter 9.

d) Counter error. Counting accuracy may involve time-base error, trigger error, and a "± 1 count ambiguity." Time-base error is concerned with any deviation of the time-base oscillator frequency from intended frequency. This may result from a lack of both short- and long-term frequency stability. For most mechanical measurements, this error may be negligible (e.g., after proper warm-up a drift or aging rate of ± 3 parts in 10^9 per day may be attained). Trigger error is concerned with the preciseness with which the gating action is known or controlled. The uncertainty of this may be reduced by period averaging.

Finally, a ± 1 count error is inherent in electronic counting. This exists because of the normal lack of synchronization between the gating and the measured pulses (whether from internal or external sources). It results from the possibility that the gate closing (or opening) may occur and barely miss the count of a passing cycle, but still, in fact, include (or exclude) the greater part of that particular cycle's period.

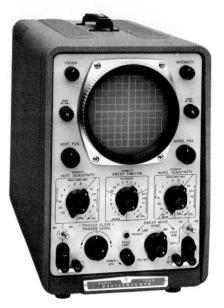

Fig. 8.7 Panel arrangement of Model 130B Hewlett Packard cathode-ray oscilloscope. (Courtesy Hewlett Packard Co., Palo Alto, Calif.)

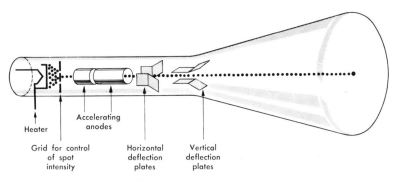

Heater

Accelerating anodes

Grid for control of spot intensity

Horizontal deflection plates

Vertical deflection plates

Fig. 8.8 Elements of a cathode-ray tube.

8.6 THE CATHODE-RAY OSCILLOSCOPE (CRO)

Probably the most versatile readout device used for mechanical measurements is the cathode-ray oscilloscope (CRO). This is a voltage-sensitive instrument, much the same as the VTVM, but with an inertialess (at mechanical frequencies) beam of electrons substituted for the meter pointer and a fluorescent screen replacing the meter scale. Figure 8.7 shows a typical general-purpose CRO.

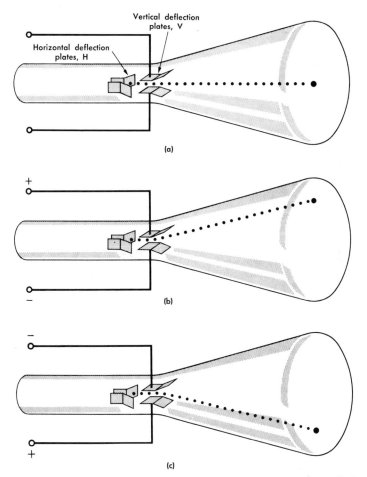

Fig. 8.9 Electrostatic deflection principle. (a) With no voltage applied to the vertical deflection plates, the electron beam is not deflected. (b) When a positive voltage is applied to the upper plate, the electron beam is deflected upward. (c) When the polarity is reversed and positive voltage is applied to the lower plate, the beam is deflected downward.

The heart of the instrument is the cathode-ray tube, shown schematically in Fig. 8.8. A stream of electrons emitted from the cathode is focused sharply on the fluorescent screen which glows at the point of impingement forming a bright spot of light. Deflection plates control the direction of the electron stream and hence the position of the bright spot on the screen. If an electrical potential is applied across the plates, the effect is to bend the pencil of electrons, as shown in Fig. 8.9. With the use of two sets of deflection

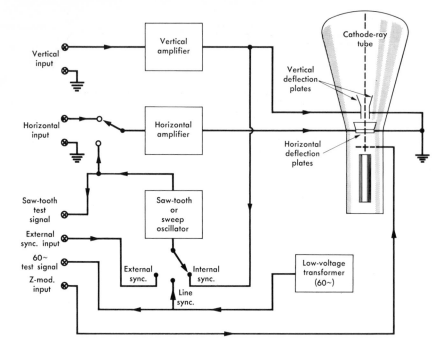

Fig. 8.10   Block diagram of a typical general-purpose oscilloscope.

plates arranged to bend the electron stream both vertically and horizontally, an instantaneous relation between two separate deflection voltages may be obtained.

Figure 8.10 is a block diagram of a typical general-purpose cathode-ray oscilloscope. The nature of the CRO is such that it may appear with many different variations in the form of special controls and input and test terminals. The diagram shown is not for any particular commercial instrument. Certain oscilloscopes will have features not shown here, and others may not employ certain ones that are shown.

a) *Oscilloscope amplifiers.* The sensitivity of the typical electrostatic cathode-ray *tube* is relatively low, varying from about 0.004 to 0.06 in. deflection per d-c volt, or from about 16 to 250 V/in. of deflection. This means that in order to be widely useful for measurement work, the CRO should provide means for signal amplification before the signal is applied to the deflection plates. All general-purpose oscilloscopes provide such amplification. Most are equipped for both d-c and a-c amplification on both the vertical and horizontal plates (Article 7.23b).

Some means for varying gain is provided in order to control the amplitude of the trace on the screen. This is often accomplished through use of fixed gain amplifiers, preceded by variable attenuators.

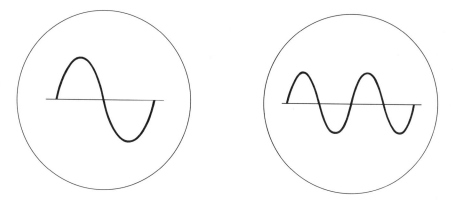

Fig. 8.11 Trace obtained when 60 Hz line voltage is applied to the vertical deflection plates and (a) the horizontal sweep is adjusted to 60 Hz (b) 30 Hz saw-tooth voltage is applied to the horizontal deflection plates.

b) Saw-tooth oscillator or time-base generator. Except for special-purpose applications, the usual cathode-ray oscilloscope is equipped with an integral *saw-tooth* or *sweep* oscillator. This is a variable-frequency oscillator which produces an output voltage-time relation in the form shown in Fig. 3.8c. (See also Articles 3.4e and 3.5.) Ideally, the voltage increases uniformly with time until a maximum is reached, at which point it collapses almost instantaneously.

When the output from the saw-tooth oscillator is applied to the horizontal deflection plates of the cathode-ray tube, the bright spot of light will traverse the screen face at a uniform velocity. As the voltage reaches a maximum and collapses to zero, the spot is whipped back across the screen to its starting point, from which it repeats the cycle. The length of the path will then be a measure of the period of the oscillator frequency (called sweep frequency) in seconds, and each point along the path will represent a proportional time interval measured from the beginning of the trace. (By convention, increasing time is measured to the right.) In this manner a very useful *time base* is obtained along the x-axis of the tube face.

As a simple example, let us suppose that ordinary 60-Hz line voltage is applied to the y-deflection plates of the tube and the output from a variable-frequency sweep oscillator is applied to the x-deflection plates. (Usually the saw-tooth oscillator is within the case of the CRO and a knob is simply set to "Internal Sweep.") With the two voltages applied, the frequency of the saw-tooth oscillator would be adjusted by means of the sweep range and sweep vernier controls on the control panel. If the saw-tooth frequency is adjusted *exactly* to 60 Hz, then one complete cycle of the vertical input waveform will appear stationary on the screen, as shown in Fig. 8.11(a). If the sweep frequency is slightly greater or slightly less than 60 Hz, then the waveform will appear to creep backward or forward across the screen. *The reciprocal of the time in seconds required for the waveform to creep exactly one complete wavelength on the screen will be the discrepancy in cycles per second between the sweep frequency and the input frequency.* In certain cases this relationship may be employed in making precise measurement of frequency or period.

If the sweep frequency is changed to *exactly* 30 Hz, then two complete cycles of the input signal will appear and remain stationary on the screen, as shown in Fig. 8.11(b).

On occasion the saw-tooth waveform is desired for test applications in a manner similar to the use of square waves for checking amplifier response (see Article 3.9). Because the saw-tooth generator is required in the CRO, but is seldom otherwise available in a laboratory, many oscilloscopes provide a terminal from which the built-in oscillator may be tapped. This is shown in Fig. 8.10.

c) Synchronization. In the example just referred to, one cycle of the 60-Hz waveform will appear stationary on the screen only if the sweep frequency is exactly 60 Hz. Frequencies from all types of electronic generators tend to shift or drift with time. This is caused by a change in component characteristics brought about by temperature changes due to the warm-up of the instrument. Therefore, to hold a pattern on the screen without creep, it would be necessary for one to continuously monitor the trace, making adjustments in sweep frequency as required.

When a steady-state signal is applied to the vertical terminals, however, it is possible to lock the sweep oscillator frequency to that of the input frequency, provided the sweep frequency is *first* adjusted to approximately the input frequency or some multiple thereof. This is controlled through use of "Sync. Selector" and "Sync. Amplitude" controls.

In our example, we would wish to use the vertical input as our synchronizing signal source, so we would set the synchronization selector to "Internal" (see Fig. 8.10). Voltage pulses from the input signal would then be applied to the sweep oscillator and would be used to control the oscillator frequency over a small range. If the frequency is initially adjusted to some integral multiple of the input frequency, the sweep oscillator would then lock in step with the input signal and the trace would be held stationary on the screen. Since excessive synchronization voltage tends to distort the trace, only the amount required for synchronization must be applied.

Electrical engineers are often concerned with measurements at 60 Hz; hence oscilloscopes are usually equipped to provide a direct synchronizing signal from the power line. This setting on the synchronization selector is often simply marked "Line."

Finally, it is often desirable to synchronize a CRO trace from an external source closely associated with the input signal. As an example, suppose some form of electrical pressure pickup is being used for measuring the cylinder pressures in a reciprocating-type air compressor. Although the pressure signal from the pickup may be steady state, making internal synchronization a possibility, changing load, or erratic valve action, or the like, may make this signal an undesirable source for synchronization. An external circuit may be used in a case of this sort. A simple make-and-break contactor could be attached to the compressor shaft, and a voltage pulse could be provided for synchronization through use of a simple dry cell. Such a circuit would be connected between the external synchronization input and the ground, and the Sync. Selector would be set at "External." In this case the horizontal sweep would be set at "Driven" rather than "Recur," and sweeps would take place only when initiated by the external contactor.

This arrangement is also useful when a *single sweep* only is desired. When this is the case, the synchronizing contactor or switch may be simply hand-operated, or it may be incorporated in the test cycle. As an example, a photocell circuit could be arranged so that a beam of light intercepted by a projectile or the like would provide the initiating pulse in synchronization with the test signal of interest.

When the driven sweep is initiated as outlined above, through use of an external source of triggering, the sweep occurs once for each synchronizing pulse. The sweep rate in this case is still controlled by the sweep range and sweep vernier. Of course, the sweep cannot be pulsed at a rate greater than that provided by the sweep-control settings. That is, the electron beam must have returned from the previous excursion before it can be triggered again.

d) Intensity or Z-modulation. The fluorescent trace produced by the electron beam may be brightened or darkened by applying a positive or a negative voltage component, respectively, to the grid of the cathode-ray tube. Actually, this is what is done in a television receiver tube to produce the light and dark picture areas. Most oscilloscopes make provision for applying a brightness-modulating voltage from an external source, either through a terminal on the front panel or through a connection on the back of the instrument. This is known as *intensity* or *Z-modulation.* (Z is used in the sense that, along with the x and y trace deflections, intensity variation provides the third coordinate.)

If on a normal input trace, say from a pressure pickup, an alternating Z-modulation is superimposed, the trace becomes a dashed line providing timing calibration as well as the usual y-input information. (See Article 9.4c for further discussion of this application.)

e) External horizontal input. Of course, it is not necessary to use the sweep oscillator for the horizontal input. Input terminals are provided for connecting other sources of voltage. This permits a comparison of voltages, frequencies, and phase relations. (See Article 9.5.)

8.7 ELECTRONIC SWITCH

An electronic switch is a useful accessory to the cathode-ray oscilloscope. This device provides a means whereby signals from two sources may be viewed simultaneously on the screen of an ordinary CRO. The device alternately *switches* from input A to input B at a rate determined by an internal oscillator, and means is thereby provided for direct comparison of two different inputs (see Article 9.4b).

8.8 CRO RECORDING TECHNIQUES

Direct observation of an oscilloscope trace often provides sufficient information. In other cases, however, particularly when transient conditions are being studied, some form of recording is mandatory. This normally dictates the use of photographic methods. Various forms of photographic equipment may be employed, but the most satisfactory are special-purpose cameras which may be attached directly to the oscilloscope bezel. Several types are available, including those employing ordinary photographic film, the Polaroid Land "picture-in-ten-seconds" camera, and moving-film cameras.

Only very simple photographic techniques are required in using the first two types. When the trace is from a steady-state source, the sweep may be synchronized to hold the trace stationary on the screen. It is then only necessary to make an appropriate exposure to capture the record.

When a transient input is to be recorded, single sweep, along with "time" or "bulb" shutter setting on the camera, may be employed. The camera shutter is opened, the sweep initiated either internally or externally, and the trace recorded.

Moving-film camera. One of the difficulties inherent in the use of recurring sweep in oscillography is the fact that uninterrupted records cannot be made. A single trace is limited to the diameter of the cathode-ray screen, and the beam of electrons must continually stop, retrace, and begin again.

A solution to this problem is to substitute motion of the recording film for that of the electron beam. To accomplish this a camera employing continuously moving film is used, and the internal horizontal sweep is turned off. The electron beam, therefore, simply whips up and down along the vertical center line of the screen while the recording film moves past it. A section of a typical record is shown in Fig. 8.12.

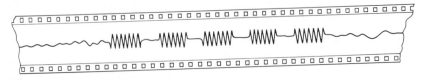

Fig. 8.12 Example of CRO record taken with moving-film camera.

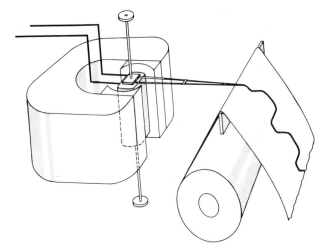

Fig. 8.13 Essential parts of a stylus oscillograph. (Courtesy Sanborn Div., Hewlett-Packard, Waltham, Mass.)

8.9 OSCILLOGRAPHS

An oscillograph differs basically from an oscilloscope in that the former is a writing instrument while the latter is primarily intended for viewing. In both cases an electrical input is converted into a mechanical displacement. In the case of the CRO, the displacement is displayed as a trace on a fluorescent screen, whereas the oscillograph displacement is recorded directly on paper. Another basic difference between the two instruments is that the oscillograph is inherently a low-impedance device, while the oscilloscope input is of high impedance.

Oscillographs are available in two broad classifications, the direct-writing type which employs some form of stylus that directly contacts a moving paper strip, and the mirror type which employs a light beam for writing on photographic film or paper. Various forms of stylus types may be had, depending on whether the recording is accomplished through use of ink, or by a heated stylus on treated paper, or by means of a stylus and pressure-sensitive paper.

Figure 8.13 illustrates the essentials of a stylus instrument consisting of a current-sensitive movement and a paper drive mechanism. As the stylus is deflected by the signal input, the paper is moved under it at a fixed rate, recording the time function of the input. Figure 8.14 shows the essentials of the light-beam type, again consisting of a current-sensitive movement or galvanometer, a paper transport mechanism, plus an optical system for transmitting galvanometer rotation to a displacement on the photographic paper for recording.

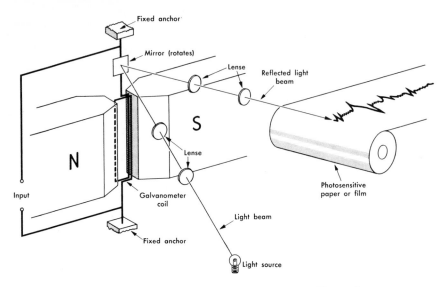

Fig. 8.14 Essential parts of a light-beam oscillograph.

The heart of the oscillograph is the galvanometer, which is basically a D'Arsonval movement. As we shall see later, the response of the galvanometer to rapid changes in input is determined by several things, including the mass moment of inertia of the coil. Sensitivity is determined partly by the number of turns in the coil; however, at the same time, it is desirable to keep the mass moment of inertia to a minimum. An actual design is therefore the result of compromise. In most cases a true coil of wire is used, such as shown in Figs. 8.13 and 8.14.

8.10 COMMERCIALLY AVAILABLE OSCILLOGRAPH SYSTEMS

Many different forms of oscillographs are available commercially, of which the two discussed below are representative. Stylus oscillographs providing one, two, four, eight, or more recording channels may be had, while the light-beam type is generally supplied with as few as four or as many as 50 separate channels.

A typical multichannel oscillographic recording system of the stylus type is shown in Fig. 8.15. This system employs eight heated stylus-type galvanometers, shown in the center section of the rack. The upper section houses plug-in amplifiers, selected on the basis of application, while the lower section holds the system's power supply. Flat response is claimed for this system to 100 Hz, down 3 db at 120 Hz.

Figure 8.16 illustrates a typical light-beam-type oscillograph. Incorporated within the housing are the paper drive, galvanometers, and optical system. The model shown provides for recording 12 channels of information, using 12 galvanometers of the kind shown in Fig. 8.17.

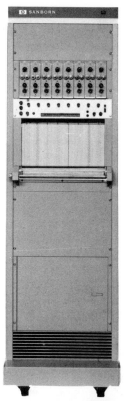

Fig. 8.15 Hewlett-Packard Model 7708B, eight channel stylus-type oscillograph. (Courtesy: Hewlett-Packard Co., Waltham Division, Waltham, Mass.)

Fig. 8.17 Mirror-type galvanometer used in the oscillograph shown in Fig. 8.16. (Courtesy: Honeywell Inc., Honeywell Test Instruments Division, Denver, Colo.)

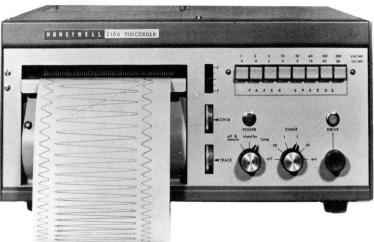

Fig. 8.16 Honeywell Model 2106 Visicorder light-beam type oscillograph capable of simultaneously recording 12 channels of data. Other models have as many as 36 channels. (Courtesy: Honeywell Inc., Honeywell Test Instruments Division, Denver, Colo.)

8.11 GALVANOMETER THEORY

Fundamentally, the galvanometer suspension consists of a mass (the coil and mirror or stylus), whose motion is constrained by a spring, and built-in damping either due to the reversed electromagnetic effect alone or combined with damping from enclosed viscous fluid. The system is driven from an external source, which may be static but which is usually some form of harmonic input. Therefore the motion of the galvanometer suspension must abide by the theory of a damped forced torsional vibration.

Assuming the driving torque to be a simple harmonic function of time, and the damping to be purely viscous, we may write the familiar relation

$$I \frac{d^2\Theta}{dt^2} + \zeta \frac{d\Theta}{dt} + \lambda\Theta = T_0 \cos \Omega t, \qquad (8.1)$$

in which

I = mass moment of inertia, lb·in·sec²,

ζ = coefficient of viscous damping, in·lb·sec/rad,

λ = torsional spring constant, in·lb/rad,

T_0 = amplitude of the driving torque, in·lb,

Θ = angular displacement of suspension, rad,

t = time, sec,

Ω = circular frequency of driving torque, rad/sec.

If we assume that the deflections are small, we may write

$$T_0 = k i_0,$$

where

i_0 = amplitude of electrical current driving the galvanometer, amperes,

k = proportionality constant, in·lb/amp.

When we make this substitution, Eq. (8.1) becomes

$$I \frac{d^2\Theta}{dt^2} + \zeta \frac{d\Theta}{dt} + \lambda\Theta = k i_0 \cos \Omega t. \qquad (8.1a)$$

Comparison of this equation with Eq. (2.18) permits us to write relations comparable to Eqs. (2.20) and (2.21). For the galvanometer these would be

$$\frac{\Theta_0 \lambda}{k i_0} = \frac{\Theta_0}{\Theta_s} = \frac{1}{\sqrt{[1 - (\Omega/\omega)^2]^2 + 2[(\zeta/\zeta_c)(\Omega/\omega)]^2}} \qquad (8.2)$$

and

$$\phi = \tan^{-1} \left\{ \frac{2[(\zeta/\zeta_c)(\Omega/\omega)]}{1 - (\Omega/\omega)^2} \right\}, \qquad (8.3)$$

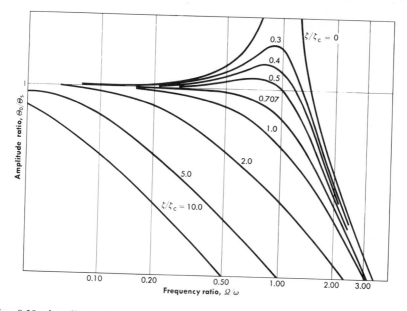

Fig. 8.18 Amplitude-frequency characteristics of a galvanometer movement.

where ζ_c is critical damping. Figure 7.18 is a plot of Eq. (8.2) for various values of damping ratio.

The ratio $\Theta\lambda/ki_0$ or Θ_0/Θ_s is a measure of galvanometer dynamic response, which, of course, should be a constant over the range of frequencies to which the galvanometer will be subjected. Inspection of Fig. 8.18 forces the conclusion that the ratio is reasonably constant only for a limited frequency range and for certain damping ratios. The figure indicates that if a galvanometer-type indicator is to be employed, ideal conditions for a given instrument occur only at one or two frequencies. If the device is to be used for general dynamic measurement applications, rather definite damping must be used, and an upper limit must be adhered to with respect to input frequencies. It appears that although there is no ideal, a damping ratio in the neighborhood of 65 to 75% must be employed. This will provide a value for the ratio very nearly unity over a range of frequency ratios from zero to about 40%, or perhaps somewhat higher at the expense of increased error.

Inspection of Fig. 2.13 (which applies to this case) indicates that damping ratios in the order of 65 to 75% of critical provide an approximately linear variation of phase shift with frequency. This is also desirable if proper time relationship is to be maintained between the harmonic components of a complex input (see Article 16.6b).

We may therefore say that any galvanometer intended for general dynamic measurement purposes should be provided with damping in the order of 70% of critical. Furthermore, such a galvanometer should not be expected to yield satisfactory information if used at frequencies above about 40 to 60% of its own natural frequency.*

Of course, certain assumptions were made in developing the foregoing relations. Those of greatest importance were: (1) that the driving torque was directly proportional to the input current, (2) that the input current was simple harmonic, and (3) that the only damping present was of the viscous type.

In the first place, the driving torque will be constant only as long as the coil turns are working in a constant magnetic field and are moving in a fixed direction relative to it. Although the galvanometer air gaps are arranged to meet this requirement, large excursions result in a certain amount of nonlinearity. In normal use, with a well-designed galvanometer, this is not a serious matter, however.

Dynamic input from a mechanical source is not usually simple harmonic, but is a complex waveform. It would seem, however, that if the complex input were made up of harmonics, all of whose frequencies are below the frequency limit of the instrument, then the waveform would be faithfully reproduced. This is actually the case, and if higher-frequency components are present, it is characteristic of the galvanometer with normal damping to attenuate such components. Usually such higher-frequency harmonics are of small amplitude; this means that although such harmonics will not be faithfully recorded, the instrument's failure will not seriously impair that part of the whole that is recorded. (Discussion of a very similar problem is included in Article 16.6, to which the reader is referred.)

Nonlinear damping, due primarily to ordinary friction and spring hysteresis, is always present to some extent in any oscillatory system. In the case of the stylus-type writing galvanometer, this is obviously so, even though the friction between the stylus and the paper is minimized. Hence the foregoing analysis does not apply quite as completely to this type as it does to the light-beam type. In general, however, both behave very nearly as elementary theory predicts.

a) Extending the range of acceptable galvanometer frequency response. The nature of the stylus-type galvanometer requires that the moving elements, the coil and writing arm, have appreciable mass moment of inertia as compared with the light beam of the mirror type. This in turn results in a relatively low natural frequency if the suspension stiffness is to be made low enough to provide practical sensitivity. However, the usable range is extended beyond the undamped natural frequency of the galvanometer

---

* The term "natural frequency" refers to the undamped natural frequency of the instrument, not to the resonant frequency of the instrument as it exists with built-in damping.

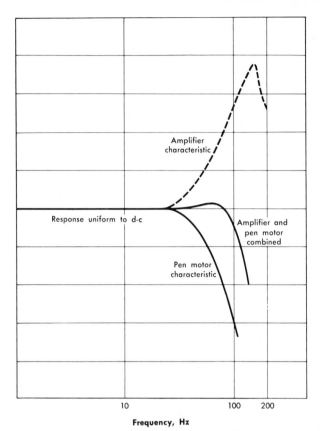

Fig. 8.19 Compensation of pen motor and amplifier characteristics.

itself through use of an amplifier with compensating characteristics [1]. This is illustrated in Fig. 8.19. The response of the galvanometer alone starts to drop at about 25 Hz. However, when it is combined with a matching amplifier, satisfactory response is extended to 80 or 100 Hz, depending on the acceptable error.

b) Galvanometer sensitivity. Galvanometer sensitivity is defined as the ratio of beam or stylus deflection to current, and may be given in inches per milliamp or inches per microamp. The deflection is measured at the recording end of either the stylus or light beam, the length of which must be considered in making comparisons. Other factors governing sensitivity are density of magnetic flux, number of turns on coil, coil shape, and stiffness of suspension. If all other factors are maintained, the *current sensitivity of a galvanometer is inversely proportional to the square of its natural frequency* [see also Article 2.5 and Eq. (2.3a)]. As a result, the sensitivity drops rapidly as the frequency range is increased. In general, the sensitivity of stylus-type galvanometers

TABLE 8.1

GALVANOMETER CHARACTERISTICS FOR HONEYWELL SERIES
SUBMINIATURE TYPE*

Type number	Nominal undamped natural frequency, Hz	Flat ($\pm 5\%$) frequency response, Hz	Required external damping resistance, ohms	Sensitivity (11.8" optical arm)	
				Current ($\pm 5\%$), μa/in.	Voltage, mv/in.
M24-350	24	0–15	350	1.50	0.203
M40-120A	40	0–24	120	8.00	0.240
M100-120A	100	0–60	120	10.0	0.522
M200-120	200	0–120	120	25.5	1.58
M400-120	400	0–240	120	77.0	9.63
M600–350	600	0–540	350	130.0	41.6
M1000	1000	0–600		2000.63	103.0
M5000	5000	0–3000		20000.0	640.0
M8000	8000	0–4800		40000.0	1400.0

*Courtesy of Honeywell Inc., Honeywell Test
Instrument Division, Denver, Colo.

is much lower than that of the light-beam variety. Typical stylus-type galvanometer sensitivity is 0.04 in./ma and an upper frequency limit (at reduced amplitude) of about 200 Hz. Representative characteristics of the light-beam types are given in Table 8.1.

c) *Methods of damping galvanometers.* There are two commonly used methods for damping galvanometers. They are magnetic damping and magnetic damping coupled with fluid damping.

When the galvanometer coil moves in its magnetic field, a back emf is established which produces a current whose effect is to oppose or damp the coil movement. This damping current is superimposed on the original input driving current; however, it may be considered as being independent from the latter. Damping produced in this manner is a velocity function because it depends on the rate at which the lines of flux are cut, hence it satisfies the manner in which we introduced damping in Eq. (8.1).

If the total input circuit resistance, including the source as well as the galvanometer coil, is quite high, little damping current will flow and the system will be underdamped. On the other hand, a low input circuit resistance will cause overdamping. It is seen, therefore, that proper matching of circuit components is quite important from the stand-point of obtaining the damping which we have already shown to be necessary.

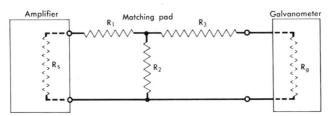

Fig. 8.20　Matching a galvanometer to an amplifier by means of a resistance pad.

In certain cases, magnetic damping is not sufficient to meet optimum requirements, and fluid damping is also used. This is done by filling the galvanometer case with a fluid of required viscosity. The viscous forces applied to the instrument's parts as they move in the fluid supply added damping. It must be realized, however, that magnetic damping is still present and the requirement for proper external circuit resistance is still present.

Matching resistance pads (see Article 7.25d) are normally used for coupling a galvanometer to an amplifier. Figure 8.20 shows an arrangement employing a pad. Here R_s and R_g represent the source and galvanometer resistances, respectively, and R_1, R_2, R_3 are the pad resistances. Optimum transfer conditions will occur when the amplifier "sees" the ideal load and the galvanometer "sees" the proper source. Amplifier load is normally equal to, or nearly equal to, the source resistance, R_s. However, the manufacturer may sometimes specify a somewhat different value, R_L. The resistance seen by the galvanometer must be that which provides proper damping as discussed above. This value, R_d, is supplied by the manufacturer of the galvanometer.

Referring to Fig. 8.20, we may write the following relations:

$$R_L = R_1 + \frac{R_2(R_g + R_3)}{R_2 + R_3 + R_g} \tag{8.4}$$

and

$$R_d = R_3 + \frac{R_2(R_1 + R_s)}{R_1 + R_2 + R_s}. \tag{8.5}$$

It is also desired that the galvanometer provide a specified trace for a given input current. (It is up to the amplifier to supply the necessary current, proportional to the transducer output.) Galvanometer sensitivity may be expressed by the relation

$$s = \frac{Ki_0}{D}, \tag{8.6}$$

in which

$$s = \text{sensitivity, ma/in.,}$$
$$K = \text{a constant,}$$
$$i_0 = \text{current, ma,}$$
$$D = \text{trace deflection, in.}$$

Using elementary circuit analysis, we may also write [2]

$$K = \frac{sD}{i_0} = \frac{R_2}{R_2 + R_3 + R_g} \,. \tag{8.7}$$

Pad resistances R_1, R_2, and R_3, must be selected to satisfy Eqs. (8.4), (8.5), and (8.7).

SUGGESTED READINGS

Bartholomew, D., *Electrical Measurements and Instrumentation*. Boston: Allyn and Bacon, 1963.

Cerni, R. H., and L. E. Foster, *Instrumentation for Engineering Measurement*. New York: John Wiley & Sons, Inc., 1962.

Doeblin, E. O., *Measurement Systems: Application and Design*. New York: McGraw-Hill Book Co., 1966.

Fribance, A. E., *Industrial Instrumentation Fundamentals*. New York: McGraw-Hill Book Co., 1962.

Partridge, G. R., *Principles of Electronic Instruments*. Englewood Cliffs, N.J.: Prentice-Hall, Inc., 1958.

Ruiter, J. H., Jr., *Modern Oscilloscopes and Their Uses*, Rev. Ed. New York: Rinehart Books, Inc., 1955.

PROBLEMS

8.1 A d-c millimeter (0.001 amperes, full-scale) has an internal resistance, R_m, equal to 140 ohms. What value of multiplier resistance will be required if the meter is to be used as a voltmeter with a 0 to 2 V d-c range?

8.2 If the meter movement in Problem 8.1 is to be used as an ammeter with a 1-amp range, what shunting resistance should be employed?

8.3 A certain battery has an open circuit voltage of 1.5 V dc and an internal resistance of 210 ohms. Show that the voltage, which would be indicated by the voltmeter of Problem 8.1, should be 1.357 V. What voltage would be indicated by a vacuum-tube voltmeter? Why should the two voltmeters yield different results?

8.4 Pursuant to Problem 8.3 above, obtain a new and a worn-out flashlight battery. Measure their voltages first with a 1000 ohm-per-volt meter and second with a VTVM. Explain the differences in readings.

8.5 Electronic counters are susceptible to a ± 1 count ambiguity when used for measuring frequency or period. How can this effect be minimized?

8.6 Obtain the instruction manual for an oscilloscope that is available in the laboratory. Abstract the operating instructions, and prepare a concise set of notes covering its manipulation.

8.7 Most oscilloscopes provide selectable a-c or d-c input amplification. Investigate the reasons for this.

8.8 High-frequency response and high sensitivity are conflicting specifications for the oscillograph galvanometer. Why is this so?

8.9 Why are oscillograph galvanometers usually damped?

8.10 A strain-gage pressure pickup is used for measuring the pressure-time relationship in the cylinder of an internal combustion engine. The output is amplified and applied to the vertical plates of an oscilloscope for photographic recording. Describe arrangements that might be employed for synchronizing the scope sweep with engine speed

a) using internal synchronization, and

b) using external synchronization.

c) Suppose that erratic engine operation is being experienced and that the "single-trace" mode of sweep is desired. Describe methods of triggering that might be employed.

PART 2

APPLIED MECHANICAL MEASUREMENT

CHAPTER 9

DETERMINATION OF COUNT, EVENTS PER UNIT OF TIME, AND TIME INTERVAL

9.1 INTRODUCTION

To be able to count items or events is basic to engineering. Items or events to be counted may be pounds of steam, cycles of displacement, number of lightning flashes, or anything divisible into discrete units. Also, time is often introduced, and the number of items or events per unit of time must be measured. The abbreviation EPUT may be used for events per unit of time. The expressions "EPUT" and "frequency" usually have slightly different connotations. Frequency is thought of as being the events per unit of time for phenomena under steady-state conditions, such as mechanical vibrations, a.c voltage, or current, etc. EPUT, however, is not dependent on a steady rate, and the term includes the counting of events which take place intermittently or sporadically. An example of this would be the counting of any of the various particles radiated from a radioactive source.

Time interval is often desired, and this becomes *period* if it is the duration of a cycle of a periodic event. Or the time interval desired may be that which occurs between events in an erratic phenomenon, or perhaps the duration of a "one-shot" event such as an impulsive pressure or force.

Of course, the problem of counting or timing emerges primarily when the events are too rapid to determine by direct observation, or the time intervals are of very short duration, or unusual accuracy is desired. In general, counting and timing-measurement problems may be classified as follows: (1) *Basic counting*, either to determine a total or to indicate the attainment of a predetermined count. (2) *Number of events or items per unit of time (EPUT)* independent of rate of occurrence. (3) *Frequency*, or the number of cycles of uniformly recurring events per unit of time. (4) *Time interval* between two predetermined conditions or events. (5) *Phase relation*, or percentage of period between predetermined recurring conditions or events.

9.2 USE OF COUNTERS

General-purpose counting equipment, including the various forms of me-chanical, electrical, and electronic counters, were discussed in Chapter 8. In addition, general laboratory equipment such as oscilloscopes and oscillo-graphs, used in conjunction with frequency standards, may also be employed in various EPUT and time-interval measuring systems, limited only by the ingenuity of the user.

The use of simple mechanical counters or electrically energized mechanical counters requires no particular technique, and further discussion at this point should be unnecessary.

a) Electronic counters. Electronic counters used either as basic counting devices or as EPUT meters require that the counted input be converted to simple voltage pulses, a count being recorded for each pulse. It should be clear that input functions used to trigger the counter need not be analogous to any quantity other than the count; hence even a simple switch may be employed, actuated by the function to be counted. In addition, photocells,

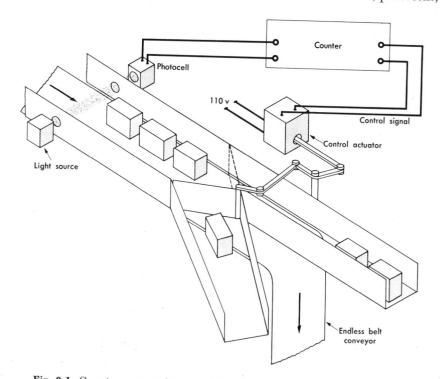

Fig. 9.1 Counter arranged to provide a control of a predetermined count.

variable resistance, inductance, or capacitance devices, Geiger tubes, and the like may be employed. Simple amplifiers may be employed, if necessary to raise the voltage level to that required by the counter, and because most electronic counters have a high-impedance input, no particular power requirement is imposed. Signal inputs may include almost any mechanical quantity, such as displacement, velocity, acceleration, strain, pressure, load, etc., as long as distinct cycles or pulses of the input are provided. The starting or stopping of the counting cycle may be controlled by direct manual-switch operation on the panel or by remote switching. One must not overlook, however, the ± 1 count ambiguity referred to in Article 8.5d.

A variation of the simple electronic counter is the *count-control* instrument. Provision is made for setting a predetermined count, and when the count is reached, the instrument supplies an electrical output that may be used as a control signal. Figure 9.1 shows how such a device could be used to prepare predetermined batches or lots for packaging.

b) EPUT meters. EPUT meters combine the simple electronic counter and an internal time base with means for limiting the counting process to preset time intervals. This permits direct measurement of frequency and is quite useful for accurate determinations of rotational speeds (see Article 9.7). The instrument is not limited, however, to an input varying at a regular rate; intermittent or sporadic events per unit of time may also be counted. Other applications include its use as a readout device for frequency-sensitive pickups such as resonant wire pressure pickups and turbine-type flow meters (see Articles 13.7c and 14.8a).

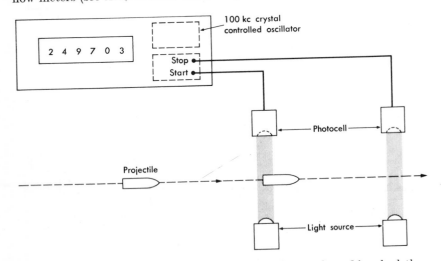

Fig. 9.2 Time-interval meter arranged to count the number of hundred thousandths of a second required for the projectile to traverse the known distance between photocells.

c) Time-interval meter. By modifying the arrangement of circuitry of an electronic counter, one can obtain a *time-interval meter.* In this case input pulses are employed to start and stop the counting process, and the pulses from an internal oscillator make up the counted information. In this manner the time interval taking place between starting and stopping may be determined, provided the frequency of the internal oscillator is known.

Figure 9.2 illustrates a simple application of the time-interval meter. Photocells are arranged so that the interruption of the beams of light provide pulses, first to start the counting process and, second, to stop it. The counter records the number of cycles from the oscillator, which has an accurately known stable output. In the example shown, the count would represent the number of one-hundred thousandths of a second required for the projectile to traverse the distance between the light beams.

9.3 THE STROBOSCOPE

The term *stroboscope* is derived from two Greek words meaning "whirling" and "to watch." Early stroboscopes employed some form of whirling disk arranged somewhat as shown in Fig. 9.3. During the interval when openings in the disk and in the stationary mask coincided, the operator could catch a fleeting glimpse of anything behind the disk. A moving object behind the window of the stroboscope could actually be made to appear stationary if the movement of the object during the interval of a single observation were small. If the object moved with a repeating or cyclic motion, the speed of the disk could be adjusted so that the object would be occupying the same position each time the windows came into coincidence. By this means, rotating or cyclically moving objects, such as rotating pulleys, gears, or reciprocating valve springs, could be made to appear motionless.

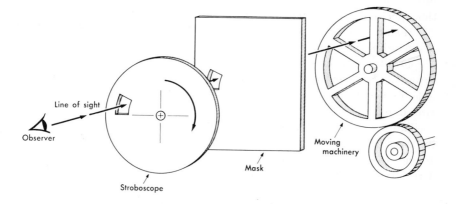

Fig. 9.3 Essential parts of early disk-type stroboscope.

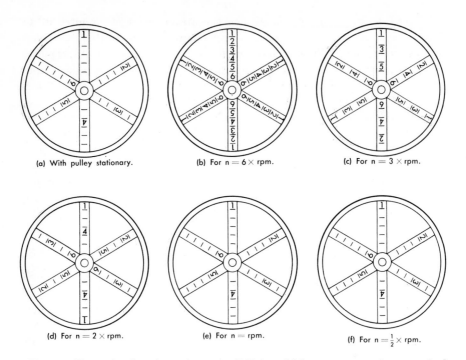

(a) With pulley stationary.   (b) For n = 6 × rpm.   (c) For n = 3 × rpm.

(d) For n = 2 × rpm.   (e) For n = rpm.   (f) For n = ½ × rpm.

Fig. 9.4 Example of various phase possibilities which may occur when a spoked wheel is observed with the light from a stroboscope. The symbol n represents the number of stroboscope flashes per minute.

If the disk were made to rotate with a period slightly less than that of the cyclically moving object, the object would have progressed somewhat farther in its cycle for each successive window alignment. In this manner the object could be studied at *slow motion* through every phase of its cycle.

Modern stroboscopes operate on a slightly different principle. Instead of a whirling disk arranged to give intermittent viewing of a moving object, a controllable flashing-light source is used. One such device is called a Strobotac.* Repeated short-duration (10 to 40 μsec) light flashes of adjustable frequency are supplied by the lamp. The frequency of flashing is controlled by an internal oscillator; to "stop" a rotating, reciprocating, or oscillating object, the frequency of flashing is varied by means of the knob on the side of the instrument until the flashes are synchronized with the frequency of motion. The range of frequencies for these instruments is about 1 to 2400 Hz. One of the disadvantages of this device is that is cannot be

---

* A registered trademark of General Radio Company, Cambridge, Mass.

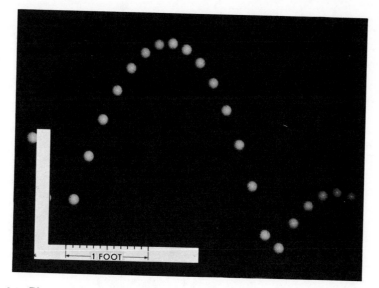

Fig. 9.5 Photo obtained by "open-shutter" camera technique and using a General Radio Company Strobotac (TM) set at a flash rate of 20 Hz. (Courtesy: General Radio Company, West Concord, Mass.)

used where the ambient lighting is above a certain value; for it to be most effective, the surrounding lighting must be subdued.

In certain instances it may be difficult to determine proper setting of the instrument for synchronization with the fundamental frequency rather than with some sub- or multifrequency. For example, if the light is directed onto a rotating pulley (Fig. 9.4) having six spokes, it is possible to obtain *apparent* synchronization at frequencies $N/6$ times the fundamental, where N represents any positive integral number. To explain this further, let us suppose a light flash occurs when spoke number one is in the vertical position, then again when spoke number two is vertical, and so on. Each spoke would be "stopped" in turn, in the vertical position, meaning that the stroboscope setting would be six times the correct value. The difficulty lies in determining that the setting is incorrect, for observation of the pulley indicates that it is stationary. Other possibilities would be to stop every other spoke, every third spoke, to catch spoke number one on every other revolution, and so on.

If the spokes are numbered or lettered as shown in Fig. 9.4, false indications can be avoided by comparing any assumed correct setting with submultiples.

Stroboscopic lighting can also be employed to study nonrepeating action (see Fig. 9.5).

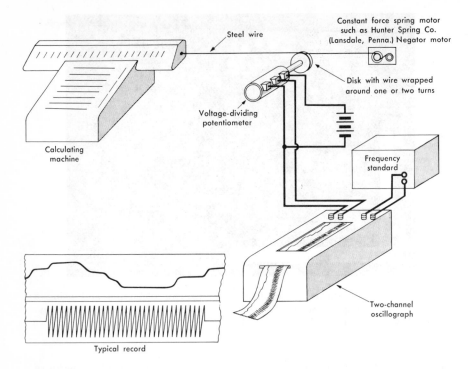

Fig. 9.6 Arrangement whereby the motion of a calculating-machine carriage may be studied.

9.4 DIRECT APPLICATION OF FREQUENCY STANDARDS BY COMPARATIVE METHODS

Probably the simplest and most basic method for measuring frequencies and short time intervals is to make a direct comparison of the unknown with a frequency standard. The problem lies in selecting a usable method for making the comparison.

When multichannel recording equipment is available, the solution is easily obtained. The input or inputs to be measured are simply recorded in terms of time in separate channels. In many cases the speed of the paper is known accurately enough to be used as the time reference, and the necessary time information is obtained automatically. In other cases it may be desirable to simultaneously record the output from a stable oscillator whose frequency is known, and to use this record as the measure of time. Some oscillographs provide a time base as a built-in part of the instrument.

Figure 9.6 shows a simple arrangement used to obtain a record of the motion of a calculating-machine carriage. An accurate time base was recorded on the second channel, using the output from a frequency standard oscillator.

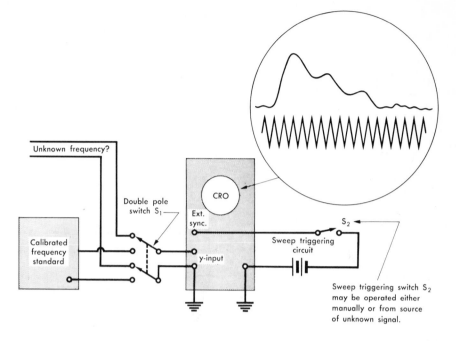

Fig. 9.7 Arrangement of equipment for frequency or time-interval determination through use of successive sweeps.

A greater challenge is presented to the engineer equipped only with the simple test equipment found in many laboratories, including the basic item, the general-purpose oscilloscope. The following several examples are presented to illustrate methods for accurately determining frequencies, short time intervals, and phase relations. These by no means exhaust the many possibilities, and the examples given will undoubtedly suggest other equally good arrangements or modifications.

a) Time calibration by substitution and comparison. Equipment: (1) cathode-ray oscilloscope, with provision for single-sweep triggering; (2) calibrated frequency standard capable of producing a frequency several times that expected from the unknown; (3) a means of recording the oscilloscope trace (the Polaroid Land camera would be convenient).

Method. Known and unknown signals may be introduced in succession to an oscilloscope, and calibration may be made through delayed comparison. Figure 9.7 shows an arrangement which may be used for this purpose.

With the camera in place, a record is made, first for the unknown signal and then for the known. If the camera position can be shifted slightly, the timing trace can be displaced on the film so that the two traces are not

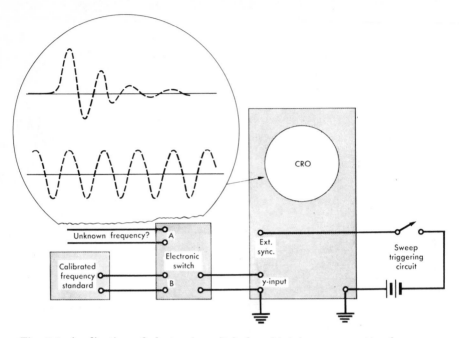

Fig. 9.8 Application of electronic switch for obtaining comparative-frequency or time-interval information.

superimposed. Single sweeps should be employed, either through use of an external synchronization circuit as shown, or perhaps by use of internal synchronization for the unknown signal and external synchronization for the calibrating signal. In many cases the synchronization may be obtained through use of a switch, S_2, actuated by some movement occurring in the system originating the signal source. For example, a lug attached to the side of a rotating gear could be used to close a spring-loaded microswitch. Exact synchronization could be provided by arranging the switch mounting so that the switch could be moved forward or backward, thereby controlling the relation between sweep and cycle. If sound is involved, use of a microphone for sweep triggering may prove feasible.

If we assume that the CRO sweep settings have remained unchanged between the two exposures, it is a simple matter to determine either an unknown frequency or a time interval by direct comparison.

b) Time calibration by use of an electronic switch. Equipment: same as for the previous example, except for the addition of an electronic switch (Article 8.7).

Method. The equipment is assembled as shown in Fig. 9.8. By proper adjustment of the oscilloscope, electronic switch, and frequency-standard

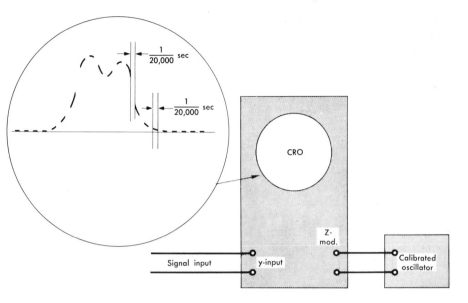

Fig. 9.9 Arrangement for use of Z-modulation for time-interval or frequency determination.

controls, a pattern may be obtained as shown. This may be recorded by approximately the same procedure outlined in the previous example. In this case, however, only a single photographic exposure will be required. In addition, only external synchronization can be considered. If internal synchronization is attempted, the transients introduced by the electronic switch will cause recurring sweep, resulting in a confused record.

Another limitation in the use of the method lies in the maximum switching rate of the electronic switch. This should be at least *ten times* that of either of the input frequencies. The maximum rate for typical switches is about 500 Hz, which would therefore limit the input frequency to about 50 Hz.

When this method is used, the possibility of changing sweep-rate settings between recording of signal and calibration traces is eliminated.

The unknown frequency or time interval may be easily determined by direct comparison of traces.

c) Frequency determination by Z-modulation (primarily for transient inputs). *Equipment:* (1) cathode-ray oscilloscope, with provision for Z-modulated input (see Article 8.6d); (2) calibrated frequency standard, preferably with square-wave output; (3) an oscilloscope camera for recording.

Method. Oscilloscope trace intensity may be increased or decreased by application of a voltage to the grid of the cathode-ray tube. This provides a very convenient method for supplying a timing calibration. For example, voltage from a calibrated oscillator may be applied to the Z-modulation terminals, as shown in Fig. 9.9. If the oscillator is set, say, at 10,000 Hz, the CRO intensity and the oscillator voltage output may be adjusted so that blanking occurs at intervals of one every 1/10,000 sec. Hence the time required for the trace, or any portion of it, may be determined by counting the markers.

Square-wave oscillators are preferred for this purpose because the sharp voltage changes provide corresponding intensity changes, thus supplying good definition to the blanking. Sweep triggering may be accomplished by any of the previously described methods, using either internal or external synchronization.

9.5 USE OF LISSAJOUS DIAGRAMS FOR DETERMINATION OF FREQUENCY AND PHASE RELATION

Equipment: (1) cathode-ray oscilloscope; (2) calibrated variable-frequency standard.

Procedure. Lissajous (Liss-a-ju) diagrams; first studied by Nathaniel Bowditch [1], and their interpretation, form a very basic approach to determining relative characteristics of two different frequency sources, primarily their frequency and phase relations.

Suppose two 60-Hz sinusoidal voltages from different sources are connected to a cathode-ray oscilloscope, one to the vertical and the other to the horizontal plates. Any of the following several patterns may result.

a) In-phase relations. If the two voltages are in phase, then as the x-voltage increases, so also does the y-voltage. The x-voltage will deflect the beam along the horizontal axis, and the y-voltage will deflect it in the vertical direction. The resulting trace, then, will be a line diagonally placed across the face of the tube, as shown in Fig. 9.10(a). The angle that the line makes

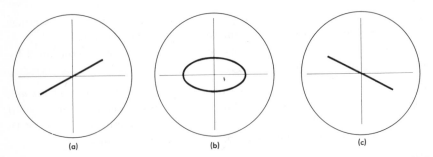

    (a)   (b)   (c)

Fig. 9.10 (a) In-phase Lissajous diagram. (b) Lissajous diagram for sinusoidal inputs $\pm 90°$ out of phase. (c) Lissajous diagram for inputs $180°$ out of phase.

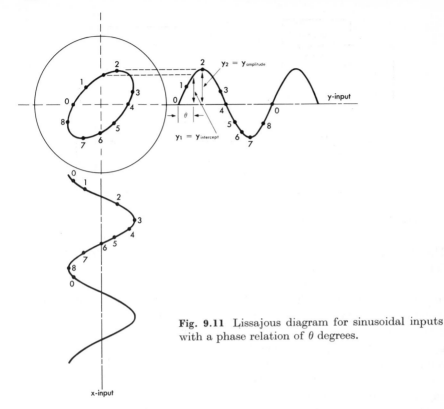

Fig. 9.11 Lissajous diagram for sinusoidal inputs with a phase relation of θ degrees.

with the horizontal will depend on the relative voltage magnitudes and the oscilloscope gain settings.

b) 90° phase relations. Suppose the two 60-Hz sinusoidal voltages are 90° out of phase. Then as one voltage passes through zero, the other will be at a maximum, and vice versa. The resulting trace will be that shown in Fig. 9.10(b). In general, it will be an ellipse with axes placed horizontal and vertical.

c) 180° phase relations. Figure 9.10(c) shows the pattern that results when the two voltages are 180° out of phase.

d) Other forms of Lissajous diagrams. Intermediate forms are ellipses with axes inclined to the horizontal. A study of Fig. 9.11 shows that when the horizontal input is at mid-sweep, the vertical precedes it by θ degrees, corresponding to a vertical input of y_1.

From the sine-wave plot of the curve we see that

$$\sin \theta = \frac{y_1}{y_2} = \frac{y\text{-intercept}}{y\text{-amplitude}} .$$

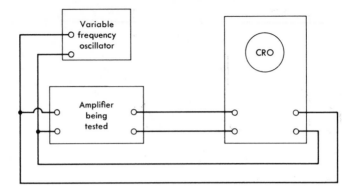

Fig. 9.12 Arrangement for measuring the phase shift in an amplifier.

Therefore, by determining the values of y_1 and y_2 from the ellipse, we may determine the phase relation between the two inputs.

An example of the application of this method would be in the determination of phase shift through an amplifier. A sampling of the amplifier input signal would be applied to the x-input terminals of a CRO, and the amplifier output would be connected to the y-input terminals as shown in Fig. 9.12. By scanning the frequency range for which the amplifier is intended, one could detect any shift in phase relation. Of course, it would be necessary to know that no shift occurs with frequency in the oscilloscope circuitry or in any of the circuitry external to the amplifier.

It should be obvious by this time how Lissajous diagrams may be used to determine frequencies. Suppose an unknown frequency source with voltage output is connected to the y-input terminals of an oscilloscope, and that the output of a variable-frequency oscillator is connected to the x-input terminals. In general, the two frequencies would be different. However, by adjusting the oscillator frequency, we may obtain equal frequency diagrams such as are shown in Figs. 9.10 and 9.11. When some form of ellipse results, proof would be had that the oscillator and unknown frequencies are equal. With one known, so also would be the other.

Fortunately the method is not limited to equal frequencies. Figure 9.13 shows Lissajous diagrams for several other frequency ratios. By studying these figures, it will be seen that a basic relation may be written as follows:

$$\frac{\text{Vertical input frequency}}{\text{Horizontal input frequency}} = \frac{\text{number of vertical maxima on Lissajous diagram}}{\text{number of horizontal maxima on Lissajous diagram}}.$$

It will also be realized that for the diagram to remain fixed on the screen,

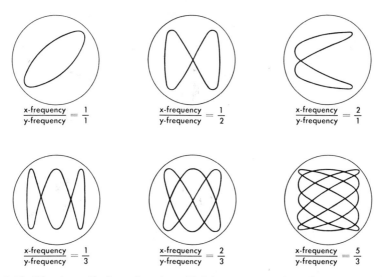

Fig. 9.13 Lissajous displays for sinusoidal inputs at various frequency ratios.

either the two input frequencies must each be fixed or they must be changing at proportional rates. In addition, it can be seen that the symmetry of the diagram will depend on the phase relation between the two inputs.

If the frequencies are reasonably fixed, ratios as high as ten to one may be determined without undue difficulty.

9.6 CALIBRATION OF FREQUENCY SOURCES

These methods suggest means whereby a variable-frequency source, such as an oscillator or signal generator, may be calibrated. By use of a fixed-frequency source, such at the 60-Hz line voltage (through a small step-down transformer), any variable-frequency source may be calibrated for a number of points. Using the ten-to-one relation mentioned above, 60-Hz line voltage may be used to spot calibrate from 6 to 600 Hz. A 1000-Hz tuning-fork standard could provide another range, from 100 to 10,000 Hz, or the National Bureau of Standards radio station WWV (Article 4.5c) provides another fixed-frequency source. It alternately broadcasts audio signals at 440 and 600 Hz, which would provide calibration points between 44 and 6000 Hz. If two or more variable-frequency sources are available, they may be used to extrapolate calibration points to higher frequencies.

It should also be pointed out that pure sine-wave inputs are not always necessary. Figure 9.14 shows a simple arrangement for determining the speed of a fan. In this case the resulting "one-to-one" Lissajous diagram takes a shape approximating a distorted parallelogram rather than an ellipse.

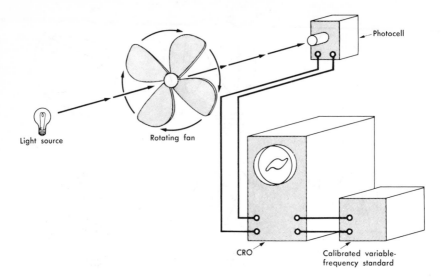

Fig. 9.14 A method for determining fan speed employing a photoelectric pickup and a frequency standard.

9.7 MEASUREMENT OF ANGULAR MOTION

Probably the most common example of direct counting and EPUT determination is the measurement of angular motion. Many different devices have been used for this purpose, most of which would fall under one of the headings in the following classification:

   I. Mechanical
      A. Direct counters
      B. Centrifugal speed indicators

  II. Electrical
      A. Generators (a-c and d-c)
      B. Reluctance-type proximity pickups

 III. Optical
      A. Stroboscope
      B. Photocells

Mechanical counters may be of the *direct-counting* digital type (Article 8.4), or may be counters with gear reducer. In the latter case, angular motion available at a shaft end is reduced by a worm and gear, and the output is indicated by rotating scales. In both examples, rpm is measured by simply counting the revolutions for a length of time as measured with a stopwatch, and calculating the turns per minute from the resulting data.

Modifications incorporate the timing mechanism in the counter. The timer is employed to actuate an internal clutch that controls the time interval during which the count is made.

Centrifugal rpm indicators employ the familiar *flyball*-governor principles which balance centrifugal force against a mechanical spring. An appropriate mechanism transmits the resulting displacement to a pointer, which indicates the speed on a calibrated scale. The term *tachometer* is often applied to an instrument of this sort, or to any *direct-indicating*, angular-speed measuring device.

Electrical tachometers usually employ a small permanent magnet-type d-c or a-c generator connected to a simple voltmeter. The d-c generator requires some form of commutation, which presents the problem of brush maintenance. On the other hand, the a-c generator requires an instrument rectifier if a simple d-c meter is to be used for indication. Of course, the advantage of the electrical kind over the mechanical is that the former provides continuous indication which may be displayed or recorded remotely.

A variable-reluctance pickup (discussed in Article 6.15) may be employed for angular-speed indication. If the pickup is placed near the teeth of a rotating gear, for example, extremely accurate speed measurements may be made, employing either an electronic counter, a frequency-sensitive indicator (frequency meter), or, if the speeds are constant, Lissajous techniques (Article 9.5). Photocells may also be used to provide voltage pulses originated by the interruption of a light beam from rotation or movement of a machine member. These pulses may be treated in the same manner.

Finally, any of the various stroboscopic methods discussed in Article 9.3 may also be used for speed measurements.

SUGGESTED READINGS

ASME Performance Test Code, PTC 19.13-1961, *Measurement of Rotary Speed.*
ASME Performance Test Code 19.12-1958, *Measurement of Time.*

PROBLEMS

9.1 Using an electronic counter, monitor the local power-line frequency. During the time period employed, what variations were noted?

9.2 Describe a method that might be used for measuring the frequency of a tuning fork, minimizing "loading."

9.3 Suppose that you are to calibrate a sinusoidal signal generator over the range 10 to 600 Hz. Describe a simple procedure employing power-line frequency as the "standard."

9.4 Employing a radio receiver and a CRO, compare the local power-line frequency with the 600-Hz audio tone received from WWV. (See Article 4.5c.)

9.5 A double-throw electrically actuated relay operates from a $1\frac{1}{2}$-V d-c source. Suggest a means for measuring the time of travel of the armature.

9.6 Obtain a common mousetrap at the hardware store. Measure the time required for the "bale" to swing from the "armed" position to the "closed" position.

By measurement, determine the torque-displacement characteristics of the spring. Calculate the theoretical "time of action." Also calculate the theoretical impacting velocity. How would you measure the impacting velocity?

CHAPTER 10

DISPLACEMENT AND
DIMENSIONAL MEASUREMENT

10.1 INTRODUCTION

Determination of linear displacement is one of the most fundamental of all measurements. The displacement may determine the extent of a physical part, or it may establish the extent of a movement. It is characterized by determining a magnitude of space. In *unit* form it may be either a measure of strain or a measure of angular displacement.

Probably to a greater extent than any other quantity, displacement lends itself to the simplest process of measurement, *direct comparison.* Certainly the most common form of displacement measurement is by direct comparison with a secondary standard. Measurements to least counts in the order of one-hundredth of an inch may be accomplished without undue difficulty with the use of nothing more than a standard in the form of a steel rule. For greater sensitivities than this, measuring systems of varying degrees of complexity are required.

For purposes of discussion, various measuring devices may be classified as in Table 10.1.

Most of the devices listed in the first two categories of Table 10.1 are undoubtedly quite familiar to the average reader. *Telescoping gages* and *expandable ball gages* are merely transfer devices for carrying a dimension to the measuring apparatus, a micrometer, etc. A dial indicator is shown in Fig. 6.1, and measuring microscopes are reviewed in Article 10.15.

With a few exceptions, these systems are used for measuring static physical dimensions. Before we discuss any of them in detail, let us consider the following measurement problem.

10.2 A PROBLEM IN DIMENSIONAL MEASUREMENT

Suppose a hole is to be bored to the dimensions shown in Fig. 10.1, and that the part is to be produced in quantity. Such a dimension would probably be checked with some form of plug gage, illustrated in Fig. 10.2. One end

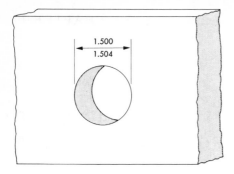

Fig. 10.1 Typical dimensioning of an internal diameter.

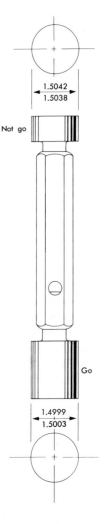

Fig. 10.2 A plug gage consisting of two gages assembled with a handle.

of the plug gage would be the *Go* end, and the other would be the *Not Go* end. If the *Go* end of the gage fits the hole, this would indicate that the hole had been bored large enough; if the *Not Go* end could not be inserted, that fact would indicate that the hole had not been bored too large.

Now, the plug gage itself would have to be manufactured, and no doubt drawings of it would be made. A rule of thumb is to dimension the plug gage with tolerances in the order of 10% of the tolerance of the part to be measured. If this rule is followed, the ends of the plug gage may be dimensioned as shown in Fig. 10.2.

It will be noted that the gage tolerance of 0.0004 in. (10% of the part tolerance, 0.004 in.) is applied symmetrically to the *Not go* end, corresponding to the upper limiting dimension of 1.504 in. On the other hand, the gage tolerance as applied to the *Go* end penalizes the machinist somewhat because, in effect, an extra ten-thousandth of an inch is taken away from what he might consider as being rightly his. This is often done to increase the life of the gage by letting the gage wear toward the specified limit. Ideally, the *Go* end will be inserted every time the gage is used and hence will wear, while the *Not go* end will never be inserted and will therefore experience no wear.

Provision has now been made for satisfactorily gaging the bored hole, provided the gages themselves are accurately made. How will we know if the

TABLE 10.1

CLASSIFICATION OF DISPLACEMENT MEASURING DEVICES

LOW-RESOLUTION DEVICES (TO 1/100 IN.)
1. Steel rule used directly or with assistance of
   a) Calipers
   b) Dividers
   c) Surface gage
2. Thickness gages

MEDIUM-RESOLUTION DEVICES (TO 1/10,000 IN.)
1. Micrometers (in various forms, such as ordinary, inside, depth, screw thread, etc.) used directly or with assistance of
   a) Telescoping gages
   b) Expandable ball gages
2. Vernier instruments (various forms, such as outside, inside, depth, height, etc.)
3. Specific-purpose gages (variously named, such as plug, ring, snap, taper, etc.)
4. Dial indicators
5. Measuring microscopes

HIGH-RESOLUTION DEVICES (TO A FEW MICROINCHES)
Gage blocks used directly or with assistance of some form of comparator, such as
a) Mechanical comparators
b) Electronic comparators
c) Pneumatic comparators
d) Optical flats and monochromatic light sources

SUPER-RESOLUTION DEVICES (TO FRACTIONS OF MICROINCHES)
Various forms of interferometers used with special light sources.

---

gages are within tolerance? We can only find out by measuring them. This leads directly to the *gage block* listed under "High-resolution devices" in Table 10-1. A gage-block set is the basic "company" standard for any small (0.01 to 10 in.) dimension.

By use of one of the comparison methods (to be described in a later section), the plug gage would be checked dimensionally. But, of course, we must not overlook the fact that to be useful, the gage blocks themselves must be measured, and so on ad infinitum, at least back to the basic length standard (Article 4.3).

An example may be employed to illustrate the extreme importance of measurement standards. Suppose that the 1.500-in hole described above is in a part to be used by an automobile manufacturer. Very probably the gage would be made by some other company, one that specializes in making gages. Both the gage maker and the automobile manufacturer would undoubtedly "standardize" their measurements by using gage blocks. It is clear that unless the different gage-block sets are accurately derived from the same basic standard, the dimension specified by the automobile manufacturer will not be reproduced by the gage maker.

10.3 GAGE BLOCKS

Gage-block sets are industry's dimensional standards. They are the *known* quantities used for calibration of dimensional measuring devices, setting special-purpose gages, and for direct use with accessories as gaging devices. They are simply small blocks of steel having parallel faces and dimensions accurate within the tolerances specified by their class. Blocks are normally available in the following classes:

Class B	"Working" blocks	Tolerance of ± 8 μin.,*
Class A	"Reference" blocks	Tolerance of ± 4 μin.,
Class AA	"Master" blocks	Tolerance of ± 2 μin. for all blocks up to 1 in. and ± 2 μin./$in.$ for larger blocks.

Gage blocks are supplied in sets, with those sets having the largest number of blocks being the more versatile. The ratio of costs for the different classes, based on prices quoted by one company, is approximately 1 to $1\frac{1}{4}$ to $4\frac{1}{2}$ for Class B, A, and AA blocks, respectively. Figure 10.3 shows a set made up of 81 blocks (plus two wear blocks) having dimensions as follows:

9 blocks with 0.0001-in. increments from 0.1001 to 0.1009 inclusive,

49 blocks with 0.001-in. increments from 0.101 to 0.149 inclusive,

19 blocks with 0.050-in. increments from 0.050 to 0.950 inclusive,

4 blocks with 1-in. increments from 1 to 4 inclusive,

2 tungsten-steel wear blocks, each 0.050 in. thick.

Blocks are made of steel that has been given a stabilizing heat treatment to minimize dimensional change with age. This consists of alternate heating and cooling until the metal is substantially without "built-in" strain. They are hardened to about 65 Rockwell C.

Distribution of sizes within a set is carefully worked out beforehand, and for the set in Fig. 10.3 accurate combinations are possible in steps of one ten-thousandth in over 120,000 dimensional variations.

---

* For very small displacements or tolerances, the mechanical engineer usually uses the microinch (1×10^{-6} in.). Physicists commonly use the micron (1×10^{-6} m) or the angstrom (1×10^{-7} mm). Equivalents are 1 micron = 39.37 μin., and 1 angstrom = 0.003937 μin.

Fig. 10.3 A set of 81 gage blocks. (Courtesy The DoAll Company, Des Plaines, Illinois.)

10.4 ASSEMBLING GAGE-BLOCK STACKS

Blocks may be assembled by *wringing* two or more together to make up a given dimension. Suppose that a dimensional standard of 3.7183 inches is desired. The procedure for arriving at a suitable combination might be determined by successive subtraction as indicated immediately below:

		Blocks used
Desired dimension = 3.7183		
Ten-thousandths place = 0.1003		0.1003
Remainder = 3.618		
Thousandths place = 0.108		0.108
Remainder = 3.51		
Hundredths place = 0.11		0.11
Remainder = 3.4		
Two wear blocks = 0.1 (0.05 each)		0.1
Remainder = 3.3		
Tenths place = 0.3		0.3
Remainder = 3.0		
Units place = 3.0		3.0
Remainder = 0.0 Check · · ·		3.7183

The resulting stack of blocks would appear as shown in Fig. 10.4.

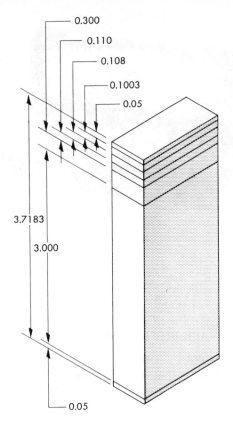

Fig. 10.4 Gage blocks assembled to a dimension of 3.7183 in.

Blocks are not stacked by simply resting them one on top of another. They must be wrung together in such a way as to eliminate all but the thinnest oil film between them. This oil film, incidentally, is an integral part of the block itself; it cannot be completely eliminated, since it was present even at manufacture. The thickness of the oil film is always in the order of 0.2 microinch [1].

A recommended procedure for assembling gage blocks is as follows [1]. Clean the surfaces with alcohol and absorbent cotton, then apply a thin film of petroleum jelly with an otherwise clean cloth. Rub the surface with clean surgical cotton until no trace of film can be observed. Bring opposite corners of the two blocks lightly together; then, applying moderate pressure, wipe the two blocks together, sweeping ahead any foreign particles that may be still remaining.

Properly wrung blocks markedly resist separation because the adhesion between the surfaces is about 30 times that due to atmospheric pressure.

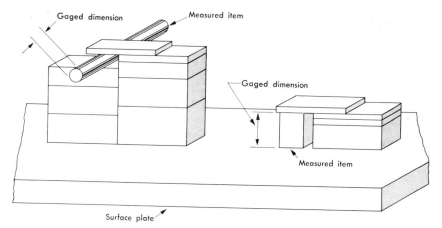

Fig. 10.5 Two methods for using gage blocks for *direct* comparison.

Unless the assembled blocks exhibit this characteristic, they have not been properly combined. The resulting assembly of blocks may be used for *direct* comparison in various ways. Two simple ways are shown in Fig. 10.5.

10.5 SURFACE PLATES

When blocks are used as shown in Fig. 10.5, some accurate reference plane is required. Such a flat surface is known as a *surface plate*, and must be made with an accuracy comparable to that of the blocks themselves. In years past, carefully aged cast-iron plates with adequate ribbing on the reverse side were used. Such plates were prepared in sets of three, carefully ground and lapped together. When combinations of two are successively worked together, the three surfaces gradually approach the only possible surface common to all three—the true flat.

Machine-lapped and polished granite surface plates have largely replaced the hand-produced cast-iron type. Granite has several advantages. First, it is probably more nearly free from built-in residual stresses than any other material because it has had the advantage of a long period of time for relaxing. Hence, there is less tendency for it to warp when the plates are prepared. Secondly, should a tool or work piece be accidentally dropped on its surface, residual stresses are not induced, as they are in metals, causing warpage; the granite simply powders somewhat at the point of impact. Thirdly, granite does not corrode.

Optical flats, the uses of which are discussed later in Article 10.12, are very similar to surface plates except they are usually made of fused quartz and are generally much smaller in size.

10.6 TEMPERATURE PROBLEMS

Temperature differences or changes are major problems in accurate dimensional gaging. The coefficient of expansion for gage-block steels is about 6.4 microinches/inch·°F. Hence even a shift of one degree in temperature would cause a greater dimensional change in a one-inch gage block than the tolerance allowed for the A and AA gages. The standard gaging temperature has been established as 68°F (20°C).

Several solutions to the temperature problem are possible. First, the most obvious solution would be to use air-conditioned gaging rooms, with temperature maintained at 68°F. This is usually done when the volume of work warrants it. This is not the complete solution, however, for mere handling of the blocks causes thermal changes requiring up to 20 min to correct [2]. For this reason, use of insulating gloves and tweezers is recommended. In addition, care must be exercised to minimize radiated heat from light bulbs, etc. [3].

A constant-temperature bath of kerosene or some other noncorrosive liquid may be used to bring the blocks and work to the same temperature. They may be removed from the bath for comparison, or in extreme cases, measurement may be made with the items submerged.

On the other hand, if temperature control is not feasible, corrections may be used, based on existing conditions. A moment's thought will indicate that *if the gage blocks and work piece are of like materials, there will be no temperature error as long as the two parts are at the same temperature.* In by far the greatest number of applications, steel parts are gaged with steel gage blocks, and although there will probably be a slight difference in coefficients of expansion, and the gage and parts may be at slightly different temperatures, appreciable compensation exists and the problem is not always as great as suggested in the preceding several paragraphs.

Temperature corrections may be made by use of the following relations:

$$L = L_b[1 - (\Delta\alpha)(\Delta T)],$$
$$\Delta\alpha = (\alpha_p - \alpha_b), \tag{10.1}$$
$$\Delta T = (T_r - 68),$$

where

L = true length of dimension being gaged at 68°F, in.,

L_b = nominal length of gage blocks determined by summation of dimensions stamped thereon, in.,

α_p = coefficient of expansion of dimension being gaged, in./in·°F,

α_b = coefficient of expansion of gage-block material in./in·°F,

T_r = room temperature, °F (this assumes that both the part being gaged and the gage blocks are at room temperature).

In using the above relations, it is very necessary that proper signs be applied to $\Delta\alpha$ and ΔT.

Example 1. Let

$$L_b = 4.000 \text{ in.,}$$
$$\alpha_p = 7.0 \ \mu\text{in./in.}^\circ\text{F,}$$
$$\alpha_b = 6.4 \ \mu\text{in./in.}^\circ\text{F,}$$
$$T_r = 75^\circ\text{F.}$$

Substituting, we have

$$L = 4[1 - (0.6 \times 10^{-6})(7)]$$
$$= 3.999983 \text{ in.}$$

Example 2. Let

$$L_b = 9.7153 \text{ in.}$$
$$\alpha_p = 5.9 \ \mu\text{in./in.}^\circ\text{F,}$$
$$\alpha_b = 6.4 \ \mu\text{in./in.}^\circ\text{F,}$$
$$T_r = 88^\circ\text{F.}$$

Substituting, we have

$$L = 9.7153[1 - (-0.5 \times 10^{-6})(20)]$$
$$= 9.715397 \text{ in.}$$

10.7 USE OF GAGE BLOCKS WITH SPECIAL ACCESSORIES

Gage blocks are sometimes used with special accessories, including clamping devices for holding the blocks [2]. When so used, height gages, snap gages, dividers, pin gages, and the like, may be assembled, employing the basic gage blocks for establishing the essential dimensions. Use of devices of this type eliminates the necessity for transferring the dimension from the gage-block stack to the measuring device.

10.8 USE OF COMPARATORS

One of the primary applications of gage blocks is that of calibrating a device called a *comparator*. As its name indicates, a comparator is used to compare known and unknown dimensions. One form of mechanical comparator is shown in Fig. 10.6(a) and (b).

As an example of a comparator's use, suppose that the diameter of a plug gage is required. The nominal dimension may first be determined by use of an ordinary micrometer. Gage blocks would be stacked to the indicated rough dimension and placed on the comparator anvil, and the indicator would be adjusted on its support post until a zero reading is obtained. The gage blocks would then be removed and the part to be measured sub-

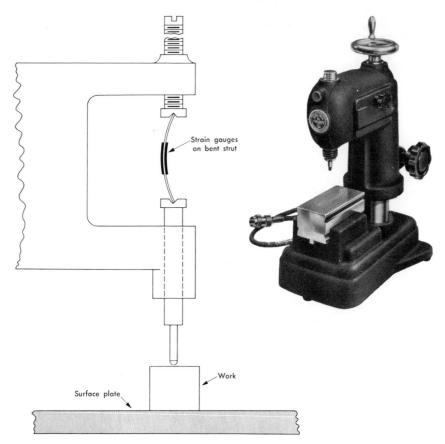

Fig. 10.6 (a) Brown and Sharpe strain gage comparator (readout meter not shown). (Courtesy Brown and Sharpe Mfg. Co., Providence, Rhode Island.) (b) Schematic of bent-column comparator sensing element.

stituted. A change in the indicator reading would show the difference between the unknown dimension and the height of the stack of gage blocks.

a) Types of comparators. Various types of comparators are available commercially, depending on the type of displacement-sensing element that is employed. A representative listing is as follows:

   I. Mechanical, employing reed-type amplifying linkage

 II. Electrical

     A. Using strain-gage elements
     B. Using variable-inductance elements
     C. Using variable capacitance

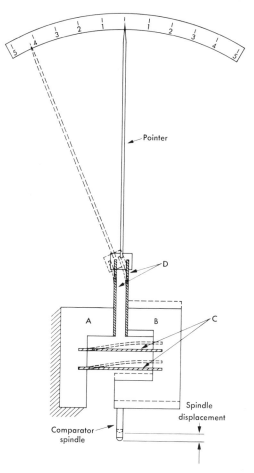

Fig. 10.7 Schematic of amplifying mechanism used in a mechanical comparator.

III. Pneumatic, employing variable orifices and differential pressures

These are discussed briefly in the following several paragraphs.

b) Mechanical comparators. Figure 10.7 illustrates the basic principles involved in amplifying linkages employed in most strictly mechanical comparators. Link *B* is constrained by thin metal flexure strips, or *reeds, C,* to move vertically relative to *A*, which is attached to, or is a part of, the housing. Because of their orientation relative to the motion, reeds *D* provide a large angular movement of the pointer. The scale may be calibrated (by means of gage blocks) to indicate any deviation from an initial setting. Comparators using this type of linkage have sensitivities in the order of 10 μin./scale division.

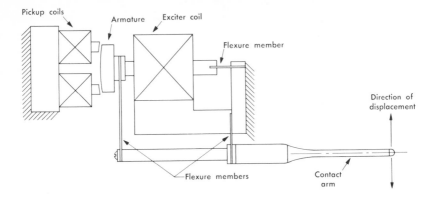

Fig. 10.8 Comparator sensing element using inductive principles. (Courtesy Cleveland Instrument Company, Cleveland, Ohio.)

c) Electrical comparators. Most of the displacement-sensitive electrical elements have been used for comparator transducers, with the exception of those that are only dynamically sensitive, such as the piezoelectric types. These include the resistance strain gage, inductive devices such as the linear variable-differential transformer, capacitance types, etc. Since it would be impractical to discuss all the many variations, two have been chosen to represent the area.

Strain-gage comparators. Figure 10.6a shows a comparator which uses strain gages attached to an elastic member as the displacement-sensitive transducer. Both mechanical and electrical amplification are employed through use of a bent column, as illustrated schematically in Fig. 10.6b. Sensitivities to 10 μin./division (0.0001 in. full scale) are possible with this instrument.

Inductive comparators. Figure 10.8 illustrates the sensing arrangement of a comparator employing an inductive transducer. Three coils are employed, an exciting coil and two pickup coils. The exciter coil and armature, energized from an a-c source, induce voltages in the pickup coils, which are connected in phase opposition. When the armature position, controlled by the contact finger, is in the electrical center, the output from the exciter coils is zero, and the instrument is in null balance.

When the instrument is used, the head is set to null with the contact finger bearing on a stack of gage blocks assembled to the desired dimension. The blocks are removed, and when work pieces are inserted, variation from the proper dimension is indicated on the calibrated indicator dial.

d) Pneumatic comparators. Pneumatic gaging is based on a double-orifice arrangement such as is illustrated in Fig. 10.9. Intermediate pressure, P_i, is dependent on the source pressure, P_s, and the pressure drops across orifices

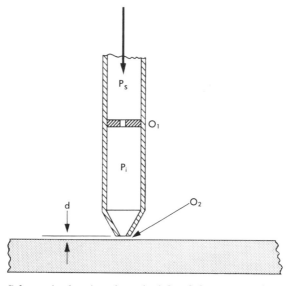

Fig. 10.9 Schematic showing the principle of the pneumatic comparator.

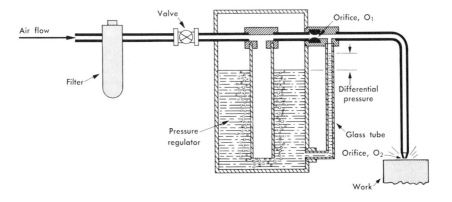

Fig. 10.10 Circuit diagram for a pneumatic comparator.

O_1 and O_2. The effective size of orifice O_2 may be varied by a change in distance, d. As d is changed, pressure P_i will change, and this change can be used as a measure of dimension, d. Figure 10.10 shows schematically how this arrangement may be used as a comparator.

Using a double orifice as shown in Fig. 10.11, Graneek [4] experimentally obtained the characteristic curve shown, for the following fixed values: $P_s = 15$ psia and the diameter of orifice $O_1 = 0.033$ in. It was found that

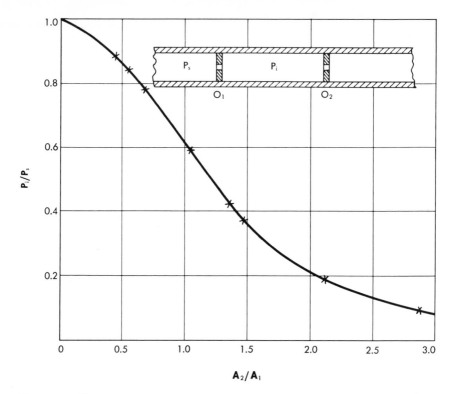

Fig. 10.11 Characteristics of the double orifice. (Courtesy *Engineering*, London.)

the curve agreed quite closely with the relation

$$\left(\frac{A_2}{A_1}\right)^2 = \frac{P_s}{P_i} - \frac{P_i}{P_s}, \qquad (10.2)$$

in which A_1 and A_2 are the areas of orifices O_1 and O_2, respectively. For
values of P_i/P_s between 0.4 and 0.9, the relation corresponds very nearly
to the straight line

$$\frac{P_i}{P_s} = 1.10 - 0.50\frac{A_2}{A_1}.$$

This indicates that for fixed values of P_s and A_1, P_i may be expected to
vary linearly with the value of A_2. In addition, the device may be made
quite sensitive, capable of measuring displacements of a few microinches.

10.9 OPTICAL METHODS

Optical methods applied to linear measurement may be divided into two
general areas as follows: (1) very accurate measurement of small dimension
or displacement by methods employing light-wave interference or through

the use of image magnification, and (2) measurement of large dimensions by use of alignment telescopes with accessories and by use of projection systems. Although such a classification can only be quite general, a dimension of about three feet or one meter may be suggested as a rough dividing line, with the realization that there will be overlapping in many cases.

10.10 MONOCHROMATIC LIGHT

The method of interferometry, employed for accurate measurement of small linear dimension, is described in subsequent articles. A required tool is a source of monochromatic (one color or wavelength) light.

There are various sources of monochromatic light. Optical filters may be employed singly or in combination to isolate narrow bands of approximately single-wavelength light. Or a prism may be employed to "break down" white light into its components and, in conjunction with a slit, to isolate a desired wavelength; however, both these methods are quite inefficient. Most practical sources rely on the electrical excitation of atoms of certain elements which radiate light at descrete wavelengths.

Possible sources include: mercury, mercury 198, cadmium, krypton, krypton 86, thallium, sodium, helium, and neon [16]. Means are provided for vaporizing the element, if not already gaseous, and to produce a visible light through the application of electrical potential, often at a high voltage and/or frequency. As has been seen, this method is employed with krypton 198 for producing the official length standard (Article 4.3).

Fig. 10.12 A commercially available monochromatic light source for applications to interferometry. (Courtesy The Van Keuren Company, Watertown, Mass.)

The common industrial standard is the helium lamp. This is obtained by means not unlike those used in the familiar neon sign. A tube is charged with helium and connected to a high-voltage source, which causes it to glow. The resulting light has a narrow range of wavelengths and is intense enough for practical use. The wavelength of this source is 23.2 μin. Figure 10.12 shows a commercially available helium light source. Table 10.2 gives the approximate wavelengths for the various primary colors, while wavelengths from several sources are given in Table 10.3.

TABLE 10.2

APPROXIMATE WAVELENGTHS
OF LIGHT OF THE VARIOUS
PRIMARY COLORS

Color	Range of wavelengths, microinches
Violet	15.7 to 16.7
Blue	16.7 to 19.3
Green	19.3 to 22.6
Yellow	22.6 to 23.6
Orange	23.6 to 25.4
Red	25.4 to 27.5

TABLE 10.3

WAVELENGTHS FROM SPECIFIC SOURCES

Source	Wavelengths, microinches	Fringe interval, microinches/fringe
Mercury isotope, Hg^{198}	21.5	10.75
Helium	23.2	11.6
Sodium	23.56	11.78
Krypton 86	23.85	11.92
Cadmium red	25.38	12.69

10.11 OPTICAL FLATS

An important accessory in the application of monochromatic light to measurement problems is the optical flat, Fig. 10.13, made from a transparent material such as strain-free glass. Probably the most common material for this purpose, however, is fused quartz. As the name indicates, one surface is

Fig. 10.13 A set of optical flats. (Courtesy The Van Keuren Company, Watertown, Mass.)

lapped and polished to a close flatness tolerance. Either square or circular flats are available in sizes ranging from 1 to 16 inches (or larger on special order) across a side or diameter. Thicknesses range from $\frac{1}{4}$ inch for the small flats to $2\frac{3}{4}$ in. for the large flats. Flats are manufactured to tolerances as follows: commercial, 8 μin.; working, 4 μin.; master, 2 μin.; reference, 1 μin.

10.12 APPLICATIONS OF MONOCHROMATIC LIGHT AND OPTICAL FLATS

An optical flat and a monochromatic light source may be employed to compare gage-block dimensions with unknown dimensions, i.e., they may be used together as a form of dimensional comparator. They may also be used to measure variation from flatness or to determine the contour of an almost flat surface. For these applications the principles involved are as follows.

When light waves are applied to measurement problems, principles of interferometry are used. In general, light waves from a single source may be caused to add or subtract, increasing or decreasing the light intensity, depending on the phase relation. The arrangement shown in Fig. 10.14 making use of an optical flat and a reflective surface, illustrates the basic principles.

Two requirements must be met: (1) *an air gap (a wedge) of varying thickness must exist between the two surfaces*, and (2) *the work surface must be reflective.*

As shown in Fig. 10.14, the light is reflected from both the working face of the flat and the work surface of the part being inspected. At the particular points where multiples of half-wavelengths occur, dark interference bands or

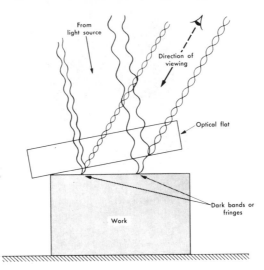

Fig. 10.14 Sketch illustrating the basic light-interference principle.

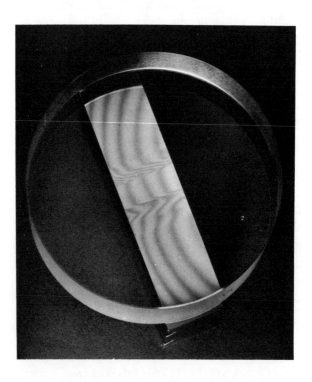

Fig. 10.15 Photograph showing a badly worn comparator anvil. (Courtesy The Van Keuren Company, Watertown, Mass.)

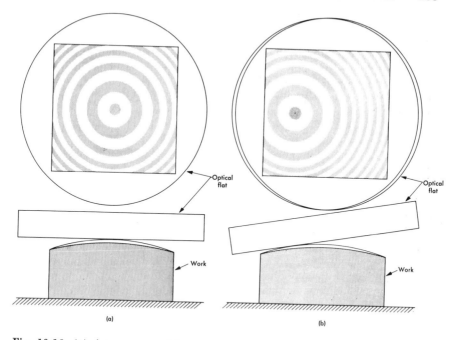

Fig. 10.16 (a) Appearance of interference fringes which occur when an optical flat is placed on a convex spherical surface. (b) Shift of fringe pattern caused by rocking an optical flat.

fringes will be observed. Figure 10.15 shows a typical pattern. It is seen that a fringe represents a locus of separation between work and flat of a definite integral number of half wavelengths of the light used. *Adjacent fringes may be interpreted, therefore, as representing contours of elevation differing by one-half wavelength.* We shall call this distance the *fringe interval*.

Point, or line, of contact. Another item of information that is generally desirable is the point of contact between the flat and the work surface. There will always be one, or possibly more, primary support points for the flat. Sometimes, instead of making contact at a point, the flat will make contact with the work along a line. This can be determined by gently rocking the flat on the surface. Actually, of course, this motion must be very slight in order to maintain an observable pattern, and only a very light pressure need be applied. The general rule is that as the flat is rocked, the point of contact is determined as that spot or line in the pattern which does not shift. Actually in some situations the point will move slightly, but in general it does not stray far from its starting point.

Figure 10.16(a) illustrates the pattern that would be observed if the work surface were spherically convex. The flat rests on the central high spot, and a fringe pattern of concentric circles results. It will be remembered that each adjacent fringe represents a change in elevation, or *fringe interval*, of one-half wavelength of the light used. For a helium source, this value would be 11.6 μin. See Table 10.3.

If one edge of the optical flat is pressed gently, it will rock on the high spot and the fringe pattern will shift into a form like that shown in Fig. 10.16(b). The high spot, indicated by the center of the concentric circles, will shift slightly. However, it remains as the primary center of the pattern. In addition, it should be noted *that many of the outer fringes run out at the edge of the work.*

On the other hand, suppose that the surface were spherically concave and that the flat rested on a line extending all the way around the edge, or at least on several points around the edge. If the edge of the optical flat is gently pressed down, it will be observed that although the central point may shift slightly, the edge points (or line) remain stationary and do not move as pressure is varied. From this simple test it can only be concluded that the surface is concave. Having determined the points of contact, we may now map the general contour.

We may put this on a more formal basis as follows. Let us define the term *fringe order* as the number we would assign a given set of fringes if we counted them in sequence, starting with a contact point as zero order. We may write the relation

$$\Delta d = (\text{fringe interval}) \times N$$

$$= \frac{\lambda}{2} N, \tag{10.3}$$

in which

Δd = difference in elevation between contact and the point in question, microinches (the contact point will always be the highest point, hence this difference in elevation will always be negative, or away from the face of the optical flat),

N = fringe order at the point in question,

λ = wavelength of the light source used, μin.

10.13 USE OF OPTICAL FLATS AND MONOCHROMATIC LIGHT FOR DIMENSIONAL COMPARISON

Suppose that the diameter of a plug gage (Article 10.2) is to be determined to an accuracy of a few microinches. Gage blocks and some form of comparator could be used. Or in place of the comparator, an optical flat and a monochromatic light source would serve. If the latter method were used, the

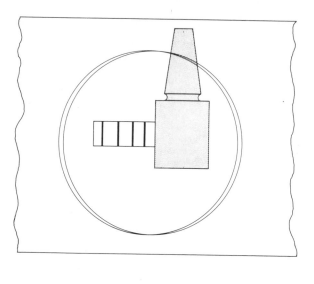

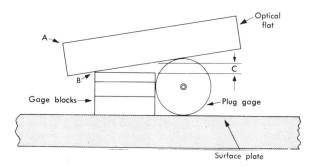

Fig. 10.17 Arrangement for measuring with gage blocks, optical flat, and monochromatic light source.

procedure would be as follows:

1. The gage diameter would first be approximately determined by use of a micrometer. An estimate of the ten-thousandths place would not be inappropriate.

2. Gage blocks would then be stacked to the "miked" dimensions. At this point it would be a good idea to check the micrometer against the blocks. If a discrepancy of half a thousandth or more is found, the gage-block stack should be corrected to the new value, accounting for the discrepancy.

3. The stack of blocks, along with the plug gage, would then be arranged as shown in Fig. 10.17, with an optical flat resting on top of the combination.

4. If the dimensions of the stack and of the plug are nearly the same, an observable fringe pattern will appear over the surface of the gage block.

Sometimes considerable patience is required to obtain a good pattern, and experience is certainly helpful. Probably the most important single factor in obtaining good results is cleanliness, because a small bit of lint or a foreign particle will be very large in comparison with the sensitivity of the inter-ference fringes. Ideally, a pattern such as is shown in the figure should finally be obtained.

5. Although Fig. 10.17 indicates that the plug dimension is greater than the stack dimension, this fact would have to be proved by determining the point of contact between flat and gage block. If the flat were gently pressed downward at point A, the fringes would either crowd more closely together (if the plug dimension is the larger) or spread apart (if the stack dimension is the larger). A moment's thought will confirm this if it is remembered that the difference in elevation between adjacent fringes is a fixed quantity. If the situation illustrated exists, then the fringes will crowd more closely together, and contact at B between block and flat will be indicated.

6. Now that the point of contact between flat and gage blocks has been determined, the difference in elevation between flat and blocks, C, may be determined. This is done by counting the number of fringes, say to the nearest $\frac{1}{5}$ fringe. This number, multiplied by the half-wavelength of the light used, gives the height of the flat above the block at C.

7. Simple application of similar triangles can now be used to determine the difference between the height of the stack of gage blocks and the plug-gage diameter.

10.14 THE INTERFEROMETER

Measurement of length to, say, one-half microinch is not commonly required in mechanical development work, but by this time the reader is undoubtedly aware that some accurate means must be available to establish the absolute length of gage blocks. In other words, how are gage blocks calibrated? We have already discussed methods whereby they may be compared with other gage blocks, but somehow a comparison must be made with a more basic standard. While various procedures may be used, some means for comparing the unknown length with a number of known wavelengths of light is invariably used.

Equipment used for this purpose is usually referred to as an interfer-ometer [3]. Its operation will be described by reference to a commercially available model. Figure 10.18 shows the external appearance of a Hilger interferometer, and Fig. 10.19 is a diagram illustrating its essential con-struction.

Let us say first that we would never use equipment of this type to determine nominal dimensions. Other methods would be much faster and more practical. Undoubtedly we will already have nominally measured the

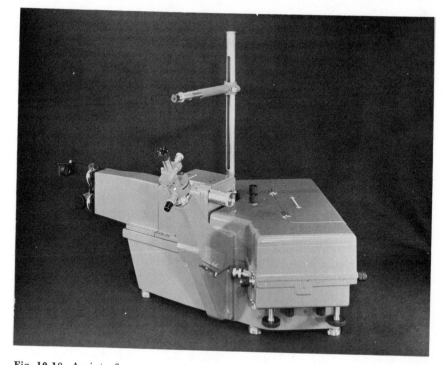

Fig. 10.18 An interferometer used to measure the length of a gage block in terms of wavelengths of light. (Courtesy Hilger and Watts, Ltd., London.)

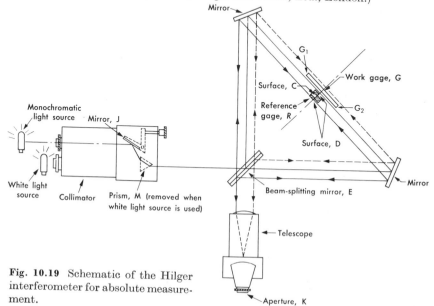

Fig. 10.19 Schematic of the Hilger interferometer for absolute measurement.

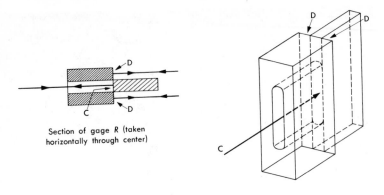

Fig. 10.20   Zero-thickness reference block.

gage in question. We are probably satisfied with our accuracy to at least
the nearest ten thousandth of an inch. The interferometer would only be
brought into use to reduce the error tolerance.

A necessary accessory to the instrument to be described is a special
gage block shown in Fig. 10.20. Although permanently assembled, it actually
consists of two blocks wrung together. The larger of the two blocks is made
with a central opening, as shown, through which one face of the smaller
block may be viewed. When assembled, a single plane, common to both
blocks, may be viewed from opposite directions, as indicated. This gage
will be referred to as the *reference* gage, and the two coplanar areas will be
designated by C and D, as shown in the figure.

A basic part of the interferometer is the beam-splitting mirror, E in
Fig. 10.19. This is a half-silvered mirror which approximately divides the
beam into equal parts of reflected and transmitted light. The beams are
then directed, in opposition, over the triangular course as indicated, and their
combination is viewed through aperture K. Reference gage R and work
gage G are positioned in the light path as shown.

Two light sources are provided, a white light source and a standard
source such as a krypton lamp, with means for introducing either without
disturbing adjustment of the other. White light is desirable because its zero
fringe results in complete absence of light (called the *achromatic fringe*) and
is quite prominent, while adjacent fringes are the colored spectral sequences,
becoming less and less well defined as the fringe order increases. This makes
positive identification of the zero fringe an easy matter.

In addition, provision is made so that different monochromatic components from the krypton source may be selected by means of mirror J,
rotation of which makes Prism M a selective filter.

In operation, the position of gage R is adjusted so that the distances
traveled by the two split light beams from E to R and back through E to K

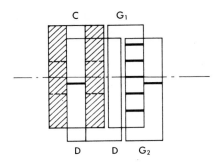

Fig. 10.21 View through eyepiece of Hilger interferometer (modified for purposes of explanation).

are equal. This is accomplished by using white light and positioning R such that the interfering beams resulting from reflection in each direction from R's common plane indicate zero fringe order as determined by the achromatic fringe. In a similar manner, gage G is adjusted so that faces G_1 and G_2 are equidistant from mirror E.

In addition to the interaction between light beams reflected from the two faces of G, on the one hand, or from the common plane of R on the other, there will also be interference between the reflected beams from R and G. Provisions are available to horizontally adjust the relative position of the images received at K from each of the two directions. By this means the images may be overlapped, providing a field of view at the eyepiece in the form shown in Fig. 10.21. (The two images are also shown with relative vertical displacement. This is done in order to clarify the diagram for the discussion. The images are normally in alignment vertically.)

Now that the faces G_1 and G_2 have been located equidistant from E and from K, and the common plane CD of R is centered in the light path, it is necessary to determine the distance from CD to one of the faces of the work gage, say to G_1. If this can be done in terms of light wavelengths, then the overall length of the work gage will be twice this value. This half-length is determined by interpretation of the fringes resulting from interference between beams reflected from G_1 and D.

By successively using different wavelengths from the krypton source, as selected by prism M, we may make measurements between portions of the central fringe on faces C and D and the fringes on the overlapping faces G_1 and D. The relative displacement is determined for each wavelength used, the number depending on the length to be measured and the accuracy required. Four or five are usually sufficient.

The method of deducing the length makes use of the fact that if a length is measured by a number of different scales, the units of which differ in length, the distance included before a set of all scale markings coincides will increase with the number of scales used. By a suitable selection of scales (wavelengths) and of the number of them used, the distance included before coincidence occurs can be increased to cover almost any range.

As pointed out previously, we begin by having a reasonably good idea of our unknown gage length. We have made measurements with more conventional equipment, and our only requirement is to determine where, within our initial error tolerance, the true dimension lies. This *initial* knowledge of approximate length, coupled with the information on *relative* fringe spacing for several different wavelengths, can be used to determine just where in the error range our gage dimension actually lies. Special tables or slide rules supplied with the instrument are used to make the final computation easy to perform. (See Problem 10.17.)

As greater dimensions are checked by this method, the fringes become less well defined; a length of about eight inches may be considered maximum without resorting to special procedures. Cooling the krypton lamp with liquid air increases definition, permitting larger dimensions to be measured; however, step measurement may also be used. In the latter case, a gage about five inches in length may be measured by the method outlined, after which it is substituted for the reference gage, and the procedure is repeated using this secondary reference. This method permits extending the procedure to a length of a meter or more.

Application of a laser light source considerably increases the upper limit of accurate dimensional measurement. The high-intensity coherent laser source has made feasible long-path interferometry to distances as great as 200 m [15]. (See also, Article 10.17.)

10.15 MEASURING MICROSCOPES

Figure 10.22 shows a section through a general-purpose low-power microscope. Basically the instrument consists of an objective cell containing the objective lens, an ocular cell containing the eye and field lenses, and a reticle mounting arrangement, all assembled in optical and body tubes. The ocular cell is adjustable in the optical tube, thereby allowing the eyepiece to be focused sharply on the reticle. The complete optical tube is adjustable in the body tube by means of a rack and pinion, for focusing the microscope on the work.

Aside from the necessary optical excellence required in the lens system, the heart of the measuring microscope lies in the reticle arrangement. The reticle itself may involve almost any type of plane outline, including scales, grids, and lines. Figure 10.23 illustrates several common forms. In use, the images of the reticle and the work are superimposed, making direct comparison possible. If a scale such as Fig. 10.23(a) is used and if the relation

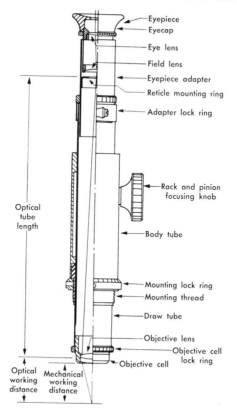

Fig. 10.22 Section through a simple low-power microscope. (Courtesy The Gaertner Scientific Corp., Chicago, Illinois.)

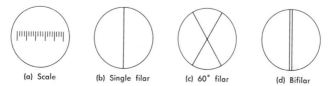

(a) Scale   (b) Single filar   (c) 60° filar   (d) Bifilar

Fig. 10.23 Examples of measuring-microscope reticles. (Courtesy The Gaertner Scientific Corp., Chicago, Illinois.)

between scale and work is known, the dimension may be determined by direct comparison.

Microscopes used for mechanical measurement are of relatively low power, usually less than 100× and often about 40×. They may be classified as follows: (1) fixed-scale, (2) filar, (3) traveling, (4) traveling-stage, and (5) draw-tube. The first two, fixed-scale and filar, are intended for measurement of relatively small dimensional magnitudes, from 0.050 to 0.200 in. in most cases.

Fig. 10.24 Filar measuring microscope. (Courtesy The Gaertner Scientific Corp., Chicago, Illinois.)

a) Fixed-scale microscopes. The fixed-scale measuring microscope uses reticles of the type shown in Fig. 10.23(a). After proper focusing has been accomplished, the scale is simply compared with the work dimension, and the number of scale units is thereby determined. The scale units, of course, must be translated into full-scale dimensions, i.e., the instrument must be calibrated. This is accomplished by focusing on a calibration scale. These are usually made of glass with etched scales. Typical calibration scales are: 100 divisions with each division 0.1 mm long, and 100 divisions with each division 0.004 in. long. Comparison of the calibration and reticle scales provides a positive calibration. Some microscopes are precalibrated and expected to maintain their calibration indefinitely. However, if the objective is changed or tampered with, recalibration will be necessary.

b) Filar microscopes. Filar microscopes make use of moving reticles. Actually in most cases a single or double hairline is moved by a fine-pitch screw thread, with the micrometer drum normally divided into 100 parts for subdividing the turns, as shown in Fig. 10.24. A total range of about 0.25 in. is common. The kind using the double hairline, referred to as a

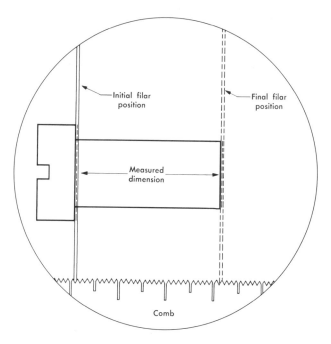

Fig. 10.25 View through eyepiece of a filar microscope, showing reference comb for indicating turns of the micrometer drum. (Courtesy The Gaertner Scientific Corp., Chicago, Illinois.)

bifilar type, is more common. In use, the double hairline is aligned with one extreme of the dimension, then moved to the other extreme, with the movement indicated by the micrometer drum. It is felt that the bifilar type is more easily used than the single-hairline type. A comparison of the views obtained by both the filar and the bifilar microscopes is shown in Fig. 10.23.

One of the problems in using a filar measuring microscope is to keep track of the number of turns of the micrometer screw. Two methods for accomplishing this are used. In the first case, a simple counter is attached to the microscope barrel, thus providing direct indication of the number of turns. The more common method is to use a built-in notched bar or *comb* in the field of view. This is shown in Fig. 10.25. Each notch on the comb corresponds to one complete turn of the micrometer wheel. Further minor and major divisions corresponding to 5 and 10 turns of the wheel are also provided. When the comb is used, a mental scale is applied. As an example, referring to Fig. 10.25, assume that for the initial position the micrometer wheel reads 85. The user might mentally designate the major divisions, reading to the right, as 0, 1000, 2000, 3000, etc. The initial reading would therefore be 085. The hairlines are then moved to the other extreme of the dimension being measured. Suppose now that the micrometer scale reads 27. Reference to the mental scale applied to the comb supplies the hundreds and thousands places, and the reading should therefore be 3127. The dimension then is 3127 less 85, or 3042 micrometer divisions. From previous calibration, each micrometer division has been determined to be, say, 0.000032 in. Therefore the actual dimension is 3042 × 0.000032, or 0.09734 in.

An example of a special application of the filar microscope is described in Article 16.10.

c) *Traveling and traveling-stage microscopes.* A traveling microscope is moved relative to the work by means of a fine-pitch lead screw, and the movement is measured in a manner similar to that used for the ordinary micrometer. In this case the microscope is employed merely to provide a magnified index. The traveling-stage type is similar, except that the work is moved relative to the microscope. In both cases the microscope simply serves as an index, and the micrometer arrangement is the measuring means. About 4 or 5 inches is the usual limit of movement, with a least count of 0.0001 in.

An instrument called a *tool-maker's* microscope is an elaborate version of the moving-stage microscope. Special illuminators are used, along with a protractor-type eyepiece.

d) *The draw-tube microscope.* This type uses a scale on the side of the optical tube to give a measure of the focusing position. The microscope is used to determine displacements in a direction along the optical axis. For example, the height of a step could be measured. The instrument would be focused on the first level, or elevation, and a reading made; then it would be moved to the second elevation and a second reading made. The difference in readings would be the height of the step.

Vernier scales are normally used, along with microscopes having very shallow depth of focus. A typical range of measurement is $1\frac{1}{2}$ in., with a least count of 0.005 in.

Focusing. Proper focusing of the measuring microscope is essential. First, the eyepiece is carefully adjusted on the reticle without regard to the work image. This is accomplished by sliding the ocular relative to the optical tube, up or down, until maximum sharpness is achieved. Next, the complete optical system is adjusted by means of the rack and pinion until the work is in sharp focus.

Positive check on proper focus may be had by checking for parallax. *When the eye is moved slightly from side to side, the relative positions of the reticle and work images should remain unchanged.* If the reticle image appears to move with respect to the work when this check is made, the focusing has not been properly done.

10.16 OPTICAL TOOLING

Precision measurement and alignment of parts of relatively large dimension became particularly important during World War II. Much of the development in this area came about by problems presented by the aircraft industry. Aircraft frame subcontractors were confronted with problems attending the assembly of large components such as wings, fuselage, engine mounts, tail surfaces, etc., each manufactured by different companies [7]. Missile assembly presents similar problems [8], as does the assembly of large machine tools, turbogenerators, and the like. Tolerances of a few thousandths of an inch in a number of feet are often specified (for example, 0.001 inch in 10 ft) [7]. In order to measure dimensions of this magnitude to the indicated accuracy, special gaging methods are required, resulting for the most part in the development and use of optical methods.

Equipment required. Special equipment available for optical tooling includes the following: (1) alignment telescopes, (2) collimators, (3) autocollimators, and (4) accessories. A simple alignment telescope is basically very similar to the familiar surveyor's transit; in fact, surveyors' transits are often employed. Basically, the instrument consists of a medium- to high-power telescope with a cross-hair reticle (other special reticles may be used) at the focal point of the eyepiece. The telescopes may be used in the same manner as the surveyor's transit for establishing datum lines and levels. They are particularly useful, however, when used in conjunction with a *collimator*.

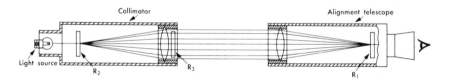

Fig. 10.26 Sketch showing arrangement of alignment telescope and collimator.

A collimator is simply a source of a bundle of *parallel* light rays. Essential parts of the device are a lens tube, a light source, and a lens system for projecting the bundle of rays. Also included are reticles whose images are projected by the collimator. Figure 10.26 presents a schematic showing the relative positions of collimator and telescope. Reticle R_2 is at the focal point of the collimator lens system, while reticle R_3 is in the collimated light beam. One important feature of the setup is that when reticle R_2 is in place, the observed image at the telescope is a function of *angular alignment only*, independent of lateral or transverse positioning. Another important feature is that when reticle R_3 is observed, its image is dependent only on *lateral position* and is independent of angular alignment. It is therefore possible

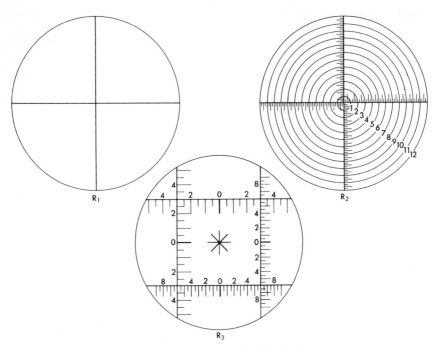

Fig. 10.27 Examples of reticle designs.

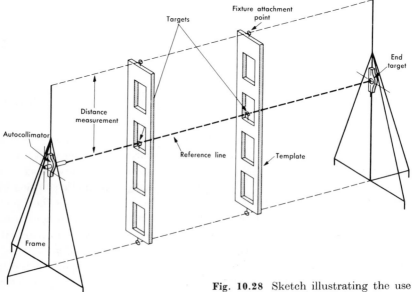

Fig. 10.28 Sketch illustrating the use of the autocollimator.

to first establish correct angular relation between the collimator and the telescope, and then to determine the magnitude of any lateral misalignment. Magnitudes are independent of distance of separation and are read from scales inscribed on the reticles (Fig. 10.27). Through use of optical means of this type, widely separated reference points may be established whose relative locations are accurately known. Such points may then be used for establishing shorter local dimensions by means of the more conventional methods.

Another form of instrument is the *autocollimator* (see Fig. 11.4). Basically, this device is a combined telescope and collimator which (a) projects a bundle of parallel light rays, and (b) employs the same lens system for viewing a reflected image. An important accessory is some form of mirror which is used as a target for reflecting the light beam. Figure 10.28 shows schematically how it may be used. The autocollimator-target combination

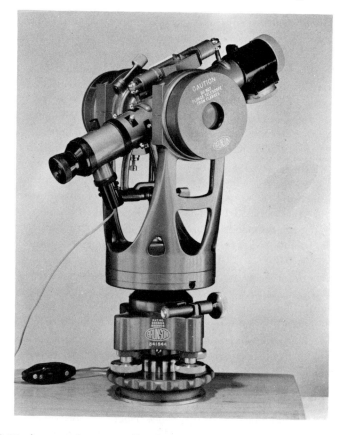

Fig. 10.29 A general-purpose "optical transit square."   (Courtesy Brunson Instrument Company, Kansas City, Missouri.)

provides an optical reference line to which important dimensions may be referred. Intermediate targets may be set up to provide reference points from which dimensional measurements may be made. Instruments of this type also provide for making direct horizontal and vertical measurements of about $\pm\frac{1}{8}$ in. with an accuracy of 0.001 in.

Figure 10.29 shows a general-purpose combined alignment telescope, collimator, and autocollimator featuring an open cross tube through the axis of the instrument, which may be used for establishing an orthogonal reference line.

Many accessories are available for the instruments discussed above. These include special mounts, reticles, optical squares, tooling bars and carriages, targets, etc.

10.17 LONG-PATH INTERFEROMETRY [17]

An important advance in optical tooling is represented by the laser interferometer. In its basic aspects it operates on the same principle as the interferometer employing ordinary or simple monochromatic light. Its important feature, however, lies in the use of a gas laser as the light source. This extends the useful range to distances as great as 200 in. as compared to roughly 10 in. for the common methods.

In the strictest sense, no light source has a single frequency or wavelength. For example, the Hg 198 source has a wavelength spread of approximately 0.005 Å (19.685 pico-in.) about a center wavelength of 5461 Å (μin.). Because of this lack of preciseness, fringes become more poorly defined as the fringe order increases, with reference to the achromatic. For this reason, in practical applications the usefulness of the common metrological light sources is limited to somewhere between 10 in. and one meter; the actual upper limit depends on methods employed to minimize the spread. For example, the use of krypton sources may be extended by application of cryogenic temperatures.

Use of the laser as a source provides light that is not only of a very precise wavelength, but also of a coherent nature (as opposed to randomly vibrating light). This makes possible the extension of practical interferometry into the optical tooling range of dimensions.

Figure 10.30(a) illustrates a simplified arrangement of a laser interferometer. Light from the source is divided by the beam splitter A, with one component reflected to the reference mirror B and the remainder to the moveable target C. Returning beams are recombined at A, where either reinforcement or cancelation takes place, depending on the phase relationship between the combining beams. Interference fringes are formed as previously described (Article 10.12). As the target mirror C is moved from one station to another, the distance in fringes is sensed with the output displayed in any

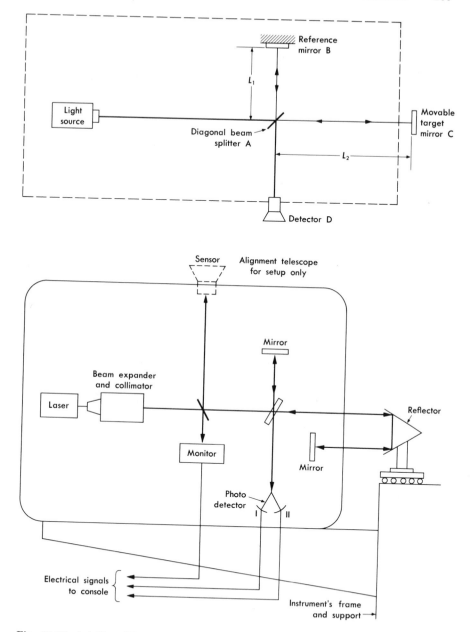

Fig. 10.30 (a) Simplified arrangement of the laser interferometer. (Courtesy: Airborne Instruments Laboratory, Division of Cutler-Hammer, Inc., Deer Park, N.Y.) (b) Schematic of a laser interferometer instrumented to provide direct readout of dimension. (Courtesy: Airborne Instrument Laboratory, Division of Cutler-Hammer, Inc., Deer Park, N.Y.)

convenient units of linear dimension. Through the use of prisms, dimensions along any orthogonal axes may be measured without disturbing the sensor position.

Figure 10.30(b) illustrates a more sophisticated arrangement. A photocell is employed as a detector, and its output is connected to appropriate analysis circuitry for determining sign (increasing and decreasing) and count.

10.18 SURFACE ROUGHNESS

Surface finish may be measured by many different methods, using several different units of measurement. Following is a list of some of the basic methods that have been used or suggested [9, 10, 11].

1. *Visual comparison* with a *standard* surface. This method is based on appearance, which involves more than the surface roughness.

2. *Tracer method*, which employs a stylus that is dragged across the surface. This is the most common method for obtaining quantitative results.

3. *Plastic-replica method*, wherein a soft, transparent, plastic film is pressed onto the surface, then stripped off. Light is then passed through the replica and measured. Refraction caused by the roughened surface reduces the transparency, and the intensity of transmitted light is used as the measure.

4. *Reflection of light* from the surface measured by a photocell.

5. *Magnified inspection*, using a binocular microscope or an electron microscope.

6. *Adsorption* of gas or liquid, wherein the magnitude of adsorption is used as the surface-roughness criterion. Radioactive materials have been used for providing a method of quantitative measurement.

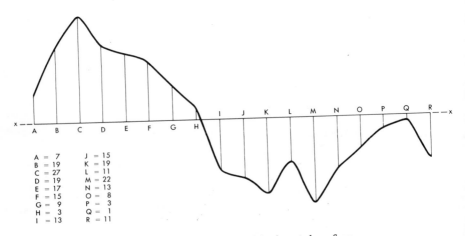

A = 7	J = 15	
B = 19	K = 19	
C = 27	L = 11	
D = 19	M = 22	
E = 17	N = 13	
F = 15	O = 8	
G = 9	P = 3	
H = 3	Q = 1	
I = 13	R = 11	

Fig. 10.31 Contour of a finished metal surface.

7. *Parallel-plane clearance.* Leakage of low-viscosity liquid or gas between the subject surface and a reference flat is used as the measure of roughness.

8. *Electron diffraction* has been proposed, but there are major drawbacks to its use.

9. *Electrolytic method.* The method assumes that the electrical capacitance is a function of the actual surface area, the rough surface providing a greater capacitance than a smooth surface.

10.19 UNITS OF ROUGHNESS MEASUREMENT

Suppose Fig. 10.31 represents a sample contour of a machined surface. The values listed thereon may be thought of as actual deviations of the surface

TABLE 10.4

CALCULATION OF MEAN ABSOLUTE HEIGHT AND
ROOT-MEAN-SQUARE AVERAGE

Position	Absolute elevation from x–x, microinches	Square of elevation
A	7	49
B	19	361
C	27	729
D	19	361
E	17	289
F	15	225
G	9	81
H	3	9
I	13	169
J	15	225
K	19	361
L	11	121
M	22	484
N	13	169
O	8	64
P	3	9
Q	1	1
R	11	121
	Total = 232	Total = 3828

Average absolute height $= 232/18 = 12.89$

Root-mean-square average $= \sqrt{3828/18} = 14.58$

from the reference plane x–x, which is located such that the sectional areas above and below the line are equal. These values are also listed in Table 10.4 as absolute values, and their average is calculated to be 12.89 μin. In addition, the root-mean-square (rms) average is calculated as 14.58 μin. It is also seen that the peak-to-peak height is 27 + 22, or 49 μin. All the values—the peak-to-peak height, the arithmetical average deviation, or the root-mean-square average—may be used as measures of roughness.

The American Standards Association specifies the arithmetical average deviation, defined by the equation

$$Y = 1/l \int_0^l |y|\, dx \tag{10.4}$$

as the standard unit for surface roughness [18]. In this equation,

Y = arithmetical average deviation,

y = ordinate of the curve profile from the center line,

l = length over which the average is taken.

The term "center line," as used in defining the distance "y," corresponds to "x–x" in Fig. 10.31. It is seen therefore that the value 12.89 μin. in our example is a practical evaluation of Eq. (10.4).

Many roughness measuring systems employing the tracer method yield the rms average. On a given surface, this value will be approximately 11% greater than the arithmetical average deviation [18]. The difference between the two values, however, is less than the normal variations from one piece to another and is commonly ignored.

An idea of the relative values for practical surface finishes may be had from Table 10.5.

TABLE 10.5

RELATIVE VALUES OF SURFACE FINISH

Common name for finish	rms roughness, microinches	Average peak-to-peak height, microinches	Usual tolerance specified for finished part, inches
Mirror	4	15	0.0002
Polished	8	28	0.0005
Ground	16	56	0.001
Smooth	32	118	0.002
Fine	63	220	0.003
semifine	125	455	0.004
Medium	250	875	0.007
Semirough	500	1750	0.013
Rough	1000	3500	0.025

Although the tracer method for measuring surface roughness is used more than any of the others, it does present several important problems. First, in order for the scriber to follow the contour of the surface, it should have as sharp a point as possible [12, 13]. Some form of conical point with a spherical end is most common. A point radius of 0.0005 in. is often used. Therefore the stylus will not always follow the true contour. If the surface irregularities are primarily what might be referred to as *wavy* or *rolling* hills and valleys of appreciable vertical radius, the stylus may indeed follow the actual contour. If the surface is rugged, on the other hand, the stylus will not extend fully into the valleys.

Second, the stylus will probably actually round off the peaks as it is dragged over the surface. This problem increases as the radius of the tip is decreased. It would seem, therefore, that there are two conflicting requirements with regard to stylus-tip radius. The marring of the surface will in part be a function of the material constants, which of course should have no bearing on the measure of surface roughness.

A third problem lies in the fact that the stylus can never inspect more than a very small percentage of the overall surface.

In spite of these problems, the tracer method is undoubtedly the most commonly used. Various forms of secondary transducers have been employed, including piezoelectric elements (Article 6.17), variable inductance (Article 6.13), variable reluctance (Article 6.15), and electronic transducers (Article 6.19). In each case the stylus motion is transferred to the transducer element, which converts it to an analogous electrical signal. Figure 10.32 illustrates typical construction. In this example the stylus motion is transmitted to the arm of an electronic transducer.

Readout devices may be any of the ordinary voltage indicators or recorders, such as the simple meter, an oscillograph, or CRO. The basic

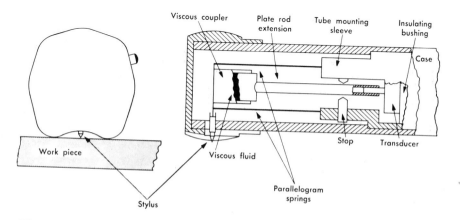

Fig. 10.32 Brush surface-analyzer pickup (shown in section). (Courtesy Brush Instruments, Division of Clevite Corp., Cleveland, Ohio.)

indicated value depends on the particular combination of transducing element and indicator employed. Piezoelectric variable-inductance and electronic transducers are basically displacement-sensitive, while the variable-reluctance type provides an output proportional to velocity. Outputs from the latter, however, may be integrated (Article 7.25b), thereby providing displacement information.

An ordinary VTVM provides a convenient and inexpensive indicator. Outputs from both the piezoelectric and electronic pickups are of sufficient level to provide adequate drive without additional amplification. In addition, if the meter has reasonable frequency-response characteristics, the speed with which the displacement-type pickups are moved over the work need not be carefully controlled.

Of course, the meter, because of its inherent characteristics, indicates rms amplitudes. When some form of oscillograph or oscilloscope is used, peak-to-peak values may be indicated or also recorded. Because of the lower cost, increased portability, and simple operating technique, combinations of displacement-type transducers (the piezoelectric or electronic) and meter indicators are the most popular.

10.20 DISPLACEMENT TRANSDUCERS

In concluding this chapter on measurement of displacement, it may be appropriate to review very briefly the basic transducer elements that are displacement-sensitive. Detailed discussions of these devices were included in Chapter 6.

The following elements are basically displacement-sensitive:

1. Resistance potentiometers (Article 6.9),
2. Differential transformers (Article 6.14),
3. Variable-inductance transducers (Article 6.13),
4. Capacitance transducers (Article 6.16),
5. Piezoelectric transducers (Article 6.17),
6. Electronic transducers (Article 6.19),
7. Ionization transducers (Article 6.20).

Usually variable-inductance, capacitance, piezoelectric, and electronic transducers are suitable only for small displacements (a few microinches to perhaps $\frac{1}{4}$ in.). The differential transformer and ionization transducers may be used over intermediate ranges, say a few microinches to several inches. While resistance potentiometers are not as sensitive to small displacements as most of the others, there is practically no limit on the maximum displacement for which they may be used [14]. With the exception of the piezoelectric type, all may be used for both static and dynamic displacements.

SUGGESTED READINGS

ASME Performance Test Code, PTC 19.14-1958, *Linear Measurements.*

Busch, T., *Fundamentals of Dimensional Metrology.*   Albany, N.Y. Delmar Publishers, Inc., 1964.

Busch, T., Laboratory Experiments in *Fundamentals of Dimensional Metrology.* Albany, N.Y. Delmar Publishers, Inc., 1965.

Greve, J. W., and F. W. Wilson (Eds), *Handbook of Industrial Metrology.* Englewood Cliffs, N.J.: Prentice-Hall, Inc., 1967.

National Bureau of Standards: *Dimensional Metrology,* Subject Classified with Abstracts., U.S. Dept. of Commerce Misc. Publication 265, 1966.

National Bureau of Standards, *Precision Measurement and Calibration.* Handbook 77, U.S. Dept. of Commerce Vol. III, *Optics, Metrology and Radiation,* 1961.

PROBLEMS

10.1 A manufacturer of dimensional measuring apparatus advertises, "Every-thing made . . . must be measured." How many exceptions to this statement can you list?

10.2 Lack of space prevents discussion of a number of the devices listed in Table 10.1. Obtain a catalog of machinist's tools and check on the form and use of any with which you are not familiar.

10.3 Obtain an ordinary "ten-thousandths" micrometer. Select various objects having dimensions that fall within the measuring range of the micrometer. Have various persons measure the selected dimensions to the nearest 0.0001 in. How consistent are the readings?

10.4 Gage blocks and a sensitive comparator are used to accurately measure a linear dimension, which is indicated as *1.718565 in.* However, the measured part is of aluminum ($\alpha_a = 11.6$ μ-in./in. °F), the blocks are of steel ($\alpha_s = 6.4$ μ-in./in. °F), and the measurement is made at 90°F. What is the standardized dimension of the part (at 68°F)?

10.5 Applying elementary theory from the mechanics of materials, analyze the action of the mechanical comparator mechanism shown in Figs. 10.7 and 10.33. Assume that the spindle displacement Δy is quite small compared to the overall dimensions of the mechanism.

10.6 Show, by drawing a sketch, the setup for measuring the diameter of a 2-in. length of nominal O.D. 0.040 in. hypodermic tubing. Use gage blocks and a monochromatic light source of wavelength equal to 23.2 μ in. Given that the actual diameter is 0.0407 in., determine the number of interference fringes counted for your setup, and indicate the actual gage blocks of wearing surface $\frac{1}{2} \times 2$ in. that you will use.

10.7 Figure 10.34 illustrates an arrangement of gage blocks, optical flat, and surface plate for measuring the dimension of a hex bar across flats. If the true dimensions are shown, what number of interference fringes would be seen if a helium light source were used?

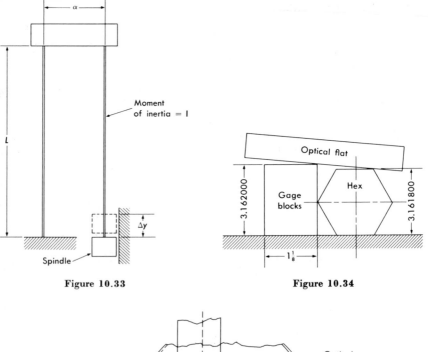

Figure 10.33

Figure 10.34

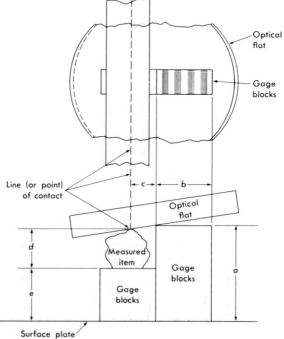

Figure 10.35

VARIOUS COMBINATIONS OF VALUES FOR THE PARAMETERS DEFINED IN FIG. 10–35 ARE GIVEN IN THE TABLE BELOW. DETERMINE THE MISSING NUMBERS.

Problem Number	Geometry of measured item and material	a	b	c	d	e	Is $a > (d + e)$?	Light source	Fringes	Temp., °F
				Inches						
10.8	Rectangular: Carbon steel	1.7839	0.875	1.250	?	0.0	No	Helium	$6\frac{1}{2}$	68
10.9	Cylindrical: Carbon steel	1.7619	1.250	d/2	?	1.4387	Yes	Helium	8	68
10.10	Cylindrical: Aluminum	2.7814	1.000	d/2	2.7805	0	...	Krypton 86	?	68
10.11	Cylindrical: Aluminum	3.7892	1.250	d/2	?	0	No	Hg 198	$15\frac{1}{2}$	90
10.12	Spherical: Brass	1.3470	0.875	d/2	?	1.2750	Yes	Cad. Red	12	75
10.13	Rectangular: Stainless steel	6.3400	1.125	1.500	6.3400*	0	?	Helium	?	98
10.14	Rectangular: Carbon steel	1.7839	0.875	1.250	?	0	Yes	Helium	?	98
10.15	Cylindrical: Carbon steel	1.7619	1.250	0.400	?	1.4387	Yes	Helium	8	68

* at 68 °F

Note: Use the following values for thermal coefficients of linear expansion, α, μ in./in.–°F; gage block steel, 6. 4; carbon steel, 6. 7; stainless steel, 9. 4; aluminum, 13. 0; brass, 10.0.

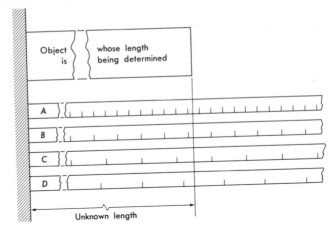

Figure 10.36

10.16 Investigate the theory, construction, operation, and historical background of the Michelson interferometer. Physics texts, general encyclopedias, reference sources on optical theory, etc., should be consulted. Write a concise report covering your findings.

10.17 This exercise is presented to assist in understanding the use of an interferometer and light waves of various wavelengths to determine the absolute length of a gage block. The situation described below is analogous to the method discussed in Article 10.14.

a) Assume that a linear distance is to be measured whose beginning, or zero point, is in general unavailable to the metrologist, but whose nominal length is known in very approximate terms, to be somewhere between 150 and 200 linear units.

b) Assume that four different tapes A, B, C, and D have been placed beside the unknown length for the purpose of comparison, and that the metrologist knows that the "zero" of each tape coincides with the beginning point of the unknown length, even though that point is unavailable for inspection.

c) Assume that each tape is divided into increments by uniformly spaced markings, but that there are no identifying numerals of any kind on the tapes. Assume further that the increments on each of the tapes, A, B, C, and D are known to be 2, 4, 6, and 7 units in length, respectively.

d) If Fig. 10.36 illustrates the only readout available to the metrologist, what must he conclude is the magnitude of the unknown length, to the nearest half-unit?

CHAPTER 11

STRAIN MEASUREMENT

11.1 INTRODUCTION

Strain gages are used for either of two purposes: (1) to determine the state of strain existing at a point on a loaded member for the purpose of stress analysis, or (2) to act as a strain-sensitive transducer element calibrated in terms of quantities such as force, pressure, displacement, acceleration, or the like, for the purpose of measuring the magnitude of the input quantity.

For the first application, accurate and reliable relations between the gage output and the actual strain input must be predictable, whereas in the second cast strain measurement *per se* is incidental, and the device is calibrated directly in terms of the primary input quantity.

Strain is a function of displacement or deformation. It is a percentage displacement rather than total displacement. It is the change in a dimension brought about by load application, divided by the initial dimension, or, as shown in Fig. 11.1,

$$\epsilon_a = \frac{L_2 - L_1}{L_1}, \tag{11.1}$$

in which

ϵ_a = axial strain,

L_1 = linear dimension or gage length, in.,

L_2 = final strained linear dimension, in.

Correctly, the term *unit strain* should be used for the quantity above, and is usually intended when the word *strain* is used alone. Throughout the following discussion, when the word *strain* is used, the quantity defined by Eq. (11.1) will be meant. If the net change in a dimension is required, the term *total strain* will be used.

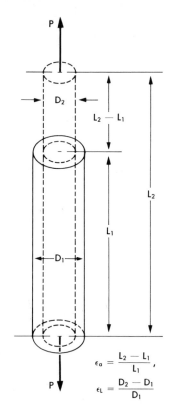

$$\epsilon_a = \frac{L_2 - L_1}{L_1},$$

$$\epsilon_L = \frac{D_2 - D_1}{D_1}$$

Fig. 11.1 Defining relations for axial and lateral strain.

Because the quantity *strain*, as applied to most engineering materials, is a very small number, it is commonly multiplied by one million; the resulting quantity is termed *microstrain*.*

The stress-strain relation for a uniaxial condition, such as exists in a simple tension-test specimen or at the outer fiber of a beam, is expressed by

$$E = \frac{\sigma_a}{\epsilon_a}, \tag{11.2}$$

where

E = Young's modulus,

σ_a = uniaxial stress,

ϵ_a = strain in direction of stress, in./in.

---

* Considerable use of the term *microstrain* will be made throughout this chapter and elsewhere in the book. For convenience, the abbreviation μs will often be employed.

This relation is linear, i.e., E is a constant for most materials as long as the stress is kept below the elastic limit.

When a member is subject to simple uniaxial stress in the elastic range (Fig. 11.1), lateral strain results in accordance with the following relation:

$$\nu = -\frac{\epsilon_L}{\epsilon_a}, \tag{11.2a}$$

where

$$\nu = \text{Poisson's ratio,}$$

$$\epsilon_L = \text{lateral strain.}$$

Equation (11.2a) is the defining equation for Poisson's ratio, but is valid only for the single-directional stress condition. When a *general stress situation* exists, with normal (tensile or compressive) stresses occurring in more than one direction, other stress-strain relations must be written. Such a condition as this occurs in the shell of a simple pressure vessel. These relations are discussed in Article 11.18b and also in Appendix A.

11.2 STRAIN GRADIENTS

A strain is not always uniform over the gage length for which it is measured. The rate of change, called the *strain gradient*, may be defined as

$$\beta = \frac{d\epsilon}{dL}. \tag{11.2b}$$

If the strain gradient is constant, the average unit strain corresponding to that given in Eq. (11.1) will be equal to the strain existing at the geo-

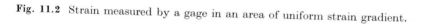

Fig. 11.2 Strain measured by a gage in an area of uniform strain gradient.

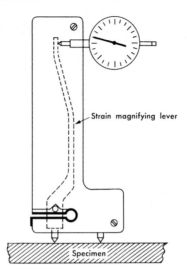

Fig. 11.3 Berry-type strain gage.

metric center of the gage length. See Fig. 11.2. This is the condition existing on the outer fiber of most beams subject to concentrated loading. Therefore the actual gage length used in a situation of this sort is not theoretically important.

On the other hand, if the gradient is not constant, Eq. (11.1) yields the average strain over the gage length and for high gradients may provide misleading data [56]. In such cases the *average* strain approaches the *point* strain as the gage length is reduced, and the shortest practical gage length should be used.

11.3 MECHANICAL AND MECHANICAL-OPTICAL GAGES

Until about 1930, all strain gages were essentially mechanical or optical in nature, and the term *extensometer* [49] was quite generally applied. Lever systems were employed which magnified the displacements measured over gage lengths as great as 10 or more inches. The strain was then determined by dividing the measured displacement by the gage length. Several of these gages, such as that shown schematically in Fig. 11.3, incorporated dial gages for part of the magnification.

Optical strain gages often use some form of light lever. One of the more successful is the Tuckerman strain gage [1, 2] shown in Fig. 11.4(a) and (b). This instrument consists of two parts, an autocollimator and an extensometer. An autocollimator incorporates both the source of a bundle of parallel rays of light and an optical system with reticle for determining the deflection

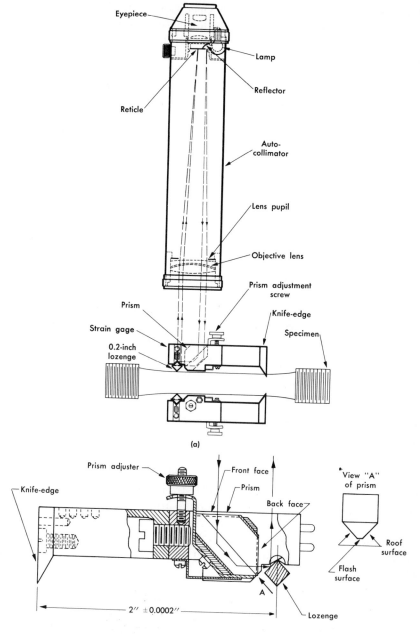

Fig. 11.4 (a) Section through the Tuckerman optical strain gage, consisting of an autocollimator and extensometer. (b) Section through the Tuckerman extensometer. (Both figures courtesy of American Instrument Company, Inc., Silver Spring, Maryland.)

of a reflected ray (see Article 10.16). Operation of the Tuckerman gage is based on the rotation of a mirror, which is actually one face of a tungsten carbide rocker, referred to by the manufacturer* as the *lozenge*. As the distance between the fixed and rocking knife-edges changes, the resulting rotation of the lozenge deflects the light beam (which has its origin in the autocollimator) back to the measuring reticle viewed through the autocollimator eyepiece. The relative movement is calibrated as strain. An important feature of this sort of instrument is that there need not be a fixed relationship between the positions of the autocollimator and the extensometer. In fact, readings may be made while the autocollimator is held by hand.

Interferometry (Articles 10.10 and 10.17) has also been employed for strain measurement [3]. However, the extreme sensitivity of the methods make the technique difficult to use for ordinary engineering applications. A simple arrangement was used by Vose [4] for measuring the lateral strains in a photoelastic model. Fringe order was determined through use of a low-power telescope.

Gages of the types described above are satisfactory for static strain measurement, but suffer from the obstacles inherent in all mechanical systems (Articles 7.2 through 7.8) if used for dynamic measurement. Poor dynamic response and magnifications limited by friction and inertia restrict their use to measurements of static strains in structures and for physical testing with slow-speed testing machines. In addition, relatively large gage lengths are required, up to one inch or even greater. This reduces the gage's usefulness where large strain gradients are encountered.

11.4 THE PHOTOELASTIC STRAIN GAGE†

A unique, self-contained strain gage employing photoelastic principles,‡ is illustrated in Fig. 11.5. The gage is a small, integrated reflection-type polariscope combined with a plate of photoelastically sensitive material [45, 46]. During manufacture residual strains are frozen into the photoelastic plate, producing visible interference fringes. When external strain is applied, the strain component along the gage axis causes a proportional shift of the fringes, in one direction for positive strain and in the other direction for negative strain. The strain magnitude may then be calculated from the relation

$$\epsilon = SF \times \Delta d, \tag{11.3}$$

---

* American Instrument Company, Silver Spring, Maryland.
† Commercially available from Baldwin-Lima-Hamilton Corporation, Waltham, Massachusetts, and the Budd Company, Phoenixville, Pennsylvania.
‡ For detailed discussion of photoelastic methods, see the following works [40, 41, 42, 43, 44].

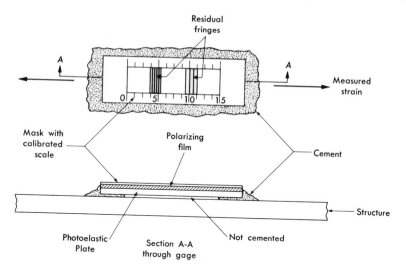

Fig. 11.5 Uniaxial or rectangular photoelastic strain gage.

where we find that

ϵ = strain, μs,

SF = strain factor (provided by the manufacturer)

Δd = displacement of the fringe in terms of calibrated scale, divisions.

Although primarily intended for static strain measurement, the use of synchronized stroboscopic lighting or high-speed motion pictures permits measurement of dynamic strains.

The important advantage of the photoelastic strain gage lies in the simplicity of its application. No auxiliary equipment or special training is required. The disadvantages include its appreciable gage length (about $\frac{3}{4}$ inch), limited resolution (in the order of 50 μs), limited operating temperature (120°F maximum), requirement for direct reading, necessity of a near-flat mounting surface, and a relatively limited range (3500–7000 μs, depending on the type of adhesive employed).

A second type of biaxial photoelastic strain gage employing the same principles of operation, but of circular geometry, is primarily intended for the determination of principal stress directions. When employed in a uniform strain field, a symmetrical fringe pattern is produced with axes corresponding to the principal stress directions. This provides valuable information for further experimental work. As will be seen later, knowing the directions of the principal axes permits the use of simpler bonded-type gages and considerably easier data reduction (see Article 11.18b).

Still another type of photoelastic gage should be noted at this point. Its construction and application result in a *stress* readout, rather than a strain readout; hence it is termed a photoelastic *stress* gage. See Reference [47] for further details.

11.5 ELECTRICAL STRAIN GAGES

As we have already seen, electrical methods yield advantages for dynamic measurements that no other systems provide (Article 6.7). This observation also holds for strain measurement; hence most of the discussion that follows in this chapter refers to strain measurement by electrical methods.

Electrical strain gages have been devised which employ simple resistive, piezoresistive, capacitive, inductive, piezoelectric, and photoelectric principles. The resistive types are by far the most popular and will be discussed in considerable detail in later sections of this chapter. They have advantages, primarily of size and mass, over the other types of electrical gages. On the other hand, strain-sensitive gaging elements used in calibrated devices for measuring other mechanical quantities are often of the inductive type, while the capacitive kind is used more for special-purpose applications. Inductive [5, 6] and capacitive [7, 8] gages are usually more rugged than resistive ones and better able to maintain calibration over a long period of time.

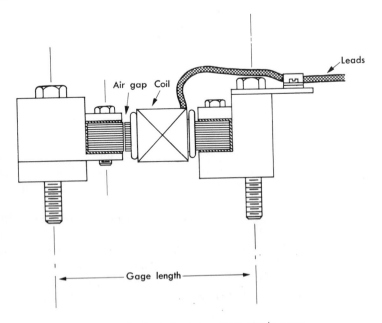

Fig. 11.6 An inductance-type strain gage.

Figure 11.6 shows an inductive strain gage for general-purpose application. Gages such as this are used primarily for permanent installations (they are bolted in place), rather than for general experimental development work. Torque meters often make use of strain gages in one form or another. Inductive meters have also been used [9], and Fig. 6.17 illustrates the principle employed by a capacitive type. Basically both are forms of strain gages.

Of somewhat limited application are barium-titanate piezoelectric elements used for strain measurement [10, 11]. Wafers about 0.01 in. thick, with suitable electrodes, are cemented directly to the test item. The voltage output that results when the gage is deformed along with the test item is used as a measure of the strain. These gages are limited to dynamic inputs and are rather difficult to calibrate. Also, the gages are equally sensitive to strain in both lateral directions. Their primary advantages result from their high output and good dynamic response.

11.6 THE ELECTRICAL RESISTANCE STRAIN GAGE

In 1856 Lord Kelvin described to the Royal Philosophical Society the results of experiments in which he demonstrated that the resistances of copper wire and iron wire change when subject to strain. He made use of a Wheatstone bridge with a galvanometer as indicator [12]. Although the principle had been proved, no practical use was made of the discovery for many years.

Probably the first wire resistance strain gage was that made by Carlson in 1931 [13]. It was of the unbonded type. A force of one pound was required to deflect the gage 0.0005 inch; however, the gage was sensitive to 10 pounds per square inch of stress in concrete. What was probably the first bonded gage was used by Bloach [14]. This consisted of a carbon film applied to the surface of the member in which the strain was to be measured.

In 1938 Edward Simmons made use of a bonded wire gage in a study of stress-strain relations under tension impact [15]. His application consisted of 14 feet of No. 40 constantan wire cemented to the four faces of a steel bar. Glyptal was used as a binder, and the wire was protected by tape. The basic idea (U.S. Patent No. 2,292,549) was to make the cement bond stronger than the wire. At about the same time, Ruge of M.I.T. conceived the idea of making a preassembly of the gage by mounting the wire between thin pieces of paper. Simmons and Ruge were later jointly credited with inventing the gage, and the "SR" in the SR-4 trademark* honors the co-inventors. The basic gage is constructed according to the general arrangement shown in Fig. 11.7.

---

* Baldwin-Lima-Hamilton Corporation, Waltham, Massachusetts.

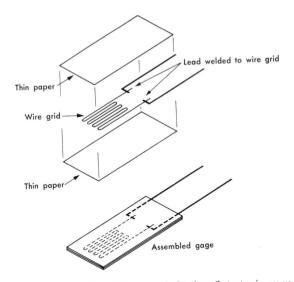

Fig. 11.7 Construction of the bonded wire, flat strain gage.

The unbonded resistance-type gage. The *unbonded* resistance element is often used as a secondary electromechanical element in such transducers as pressure sensors, accelerometers, etc. Usually four separate filaments are connected electrically in the form of a Wheatstone bridge and arranged mechanically so that filaments in adjacent bridge arms are subject to strains of opposite sign. Figures 13.12 and 16.9 illustrate applications of this type. The construction of the unbonded gage must provide a built-in prestrain in the grids greater than the maximum compressive strain to be sensed.

The foil strain gage (Fig. 11.8). During the 1950's considerable attention was given to the *foil-type* strain gage. Much of the credit for development and acceptance of this type of gage belongs to William T. Bean of Detroit [48]. Currently (1969), the foil gage has largely displaced the wire gage. The common form consists of a thin foil grid element on some type of support. Thin paper is sometimes used for this purpose, but an epoxy film is most common. Epoxy filled with fiber glass is employed for higher temperatures (see Article 11.20e), and a removable temporary support is sometimes provided for still higher temperatures. The gages are manufactured by printed circuit techniques. Hence, the configurations are not limited to the constant cross section of ordinary wire. This permits the production of many complicated configurations which are otherwise difficult or impossible to manufacture.

The Semiconductor gage. During the period 1954 to 1957, important advances in strain-gage technology were made through studies of the piezoresistive properties of silicon and germanium [50, 51]. This work resulted in the

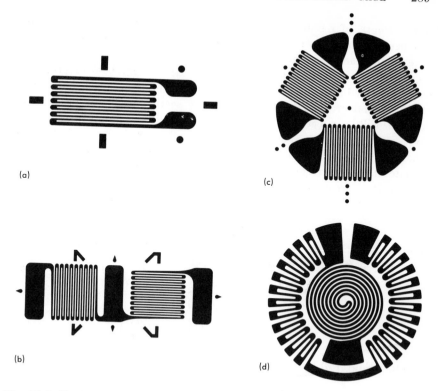

(a)

(b)

(c)

(d)

Fig. 11.8 Typical foil-type strain gages illustrating the following types: (a) single element, (b) two-element rosette, (c) three-element rosette, (d) one example of many different special-purpose gages. The one shown here is for use on pressurized diaphragms. (Courtesy the Budd Co., Instrument Division, Phoenixville, Pa.)

practical semiconductor strain gage.* The primary feature of the S-C gage lies in its extreme sensitivity. It is approximately 65 to 130 times more sensitive to strain than is the metallic gage. This gain in sensitivity, however, is not without certain disadvantages which will be discussed in later articles (see Article 11.14).

11.7 THE METALLIC RESISTANCE STRAIN GAGE

The theory of operation of the metallic resistance gage is relatively simple. When a length of wire (or foil) is mechanically stretched, a *longer* length of *smaller* conductor results and the electrical resistance is normally increased.

---

* Throughout the remainder of this book the word "semiconductor" or the abbreviation "S-C," will be employed to designate the germanium- or silicon-type piezoresistive gage. The common wire or foil gage will be referred to as the "metallic" or simply the "resistive" gage.

If the length of resistance element is intimately attached to a strained member in such a way that it will also be strained, then the measured change in resistance can be calibrated in terms of the strain.

A general relation between the electrical and mechanical properties may be derived as follows: Assume an initial conductor length L, having a cross-sectional area CD^2. (In general the section need not be circular; hence, D will be a sectional dimension and C will be a proportionality constant. If the section is square, $C = 1$; if it is circular, $C = \pi/4$, etc.) If the conductor is strained axially in tension thereby causing an increase in length, the lateral dimension should reduce as a function of Poisson's ratio.

We will start with the relation (Eq. 6.2)

$$R = \frac{\rho L}{A} = \frac{\rho L}{CD^2}. \tag{11.4}$$

If we assume that the conductor is strained, each of the quantities in Eq. (11.4) may change. Differentiating Eq. (11.4) gives

$$dR = \frac{CD^2(Ld\rho + \rho dL) - 2C\rho\, DL\, dD}{(CD^2)^2}$$

$$= \frac{1}{CD^2}\left((Ld\rho + \rho dL) - 2\rho L\frac{dD}{D}\right). \tag{11.5}$$

Dividing Eq. (11.5) by Eq. (11.4) yields

$$\frac{dR}{R} = \frac{dL}{L} - 2\frac{dD}{D} + \frac{d\rho}{\rho}, \tag{11.6}$$

which may be written

$$\frac{dR/R}{dL/L} = 1 - 2\frac{dD/D}{dL/L} + \frac{d\rho/\rho}{dL/L}. \tag{11.6a}$$

Now

$$\frac{dL}{L} = \epsilon_a = \text{axial strain,}$$

$$\frac{dD}{D} = \epsilon_L = \text{lateral strain,}$$

and

$$\nu = \text{Poisson's ratio} = -\frac{dD/D}{dL/L}.$$

Making this substitution gives us the basic relation for what is known as the *gage factor*, for which we shall use the symbol F.

$$F = \frac{dR/R}{dL/L} = \frac{dR/R}{\epsilon_a} = 1 + 2\nu + \frac{d\rho/\rho}{dL/L}. \tag{11.7}$$

This is basic for the resistance strain gage.

TABLE 11.1
REPRESENTATIVE PROPERTIES OF VARIOUS GRID MATERIALS

Grid Material	Composition	Approx. Gage Factor, F	Resistivity		Approximate Temperature Coefficient of Resistance ppm/°F	Maximum Operating Temp., °F (approx)
			Approximate Microhm-cm	Ohms per mil-foot		
Nichrome V*	80% Ni; 20% Cr	2.0	108	650	220	2000
Constantan*; Copel*; Advance*	45% Ni; 55% Cu	2.0	49	290	6	900
Isoelastic*	36% Ni; 8% Cr; 0.5% Mo; Fe remainder	3.5	112	680	260	...
Karma*	74% Ni; 20% Cr 3% Al; 3% Fe	2.4	130	800	10	1500
Manganin*	4% Ni; 12% Mn; 84% Cu	0.47	48	260	6	...
Platinum-Iridium	95% Pt; 5% Ir	5.1	24	137	700	2000
Monel*	67% Ni; 33% Cu	1.9	42	240	1100	...
Nickle		−12†	7.8	45	3300	...
Platinum		4.8	10	60	1670	...

* Trade names
† Varies widely with cold work

Considerable work must still be done before the relation between actual gage performance and Eq. (11.7) is adequately explained [16, 17]. One of the major problems is to accurately determine the true physical properties of the filament materials in the sizes used in strain gages. One-mil wire (0.001 inch) is about the largest wire used for commercial strain-gage construction. Most foil elements are considerably thinner. Just what are the values of Poisson's ratio and the resistivity for filaments of these dimensions?

Assuming for the moment that resistivity should remain constant with strain, then according to Eq. (11.7) the gage factor should be a function of Poisson's ratio alone, and in the elastic range should not vary much from $1 + (2)(0.3) = 1.6$. Table 11.1 lists typical values for various materials. Obviously, more than Poisson's ratio must be involved, and if resistivity is the only other variable, apparently its effect is not consistent for all materials. Notice the value of the gage factor for nickel. The negative value indicates that the stretched wire with increased length and decreased diameter (assuming elastic conditions) actually exhibits a reduced resistance.

In spite of our incomplete knowledge of the physical mechanism involved, the factor for metallic gages is essentially a *constant* in the usual range of required strains, and its value, determined experimentally, is reasonably consistent for a given material.

By rewriting Eq. (11.7) and replacing the differential by an incremental resistance change, the following equation is obtained:

$$\epsilon = \frac{1}{F} \frac{\Delta R}{R}. \tag{11.8}$$

In practical application, value of F and R are supplied by the gage manufacturer, and the user determines ΔR corresponding to the input situation he is measuring. This is the fundamental procedure for using resistance strain gages.

11.8 SELECTION AND INSTALLATION FACTORS FOR BONDED METALLIC STRAIN GAGES

Five important factors influence metallic-gage characteristics and application:

1. Grid material and construction

2. Backing material

3. Bonding material and method

4. Gage protection

5. Gage configuration.

About the only opportunity the average strain gage user has to control gage design is through selection of such gages as are made available to him by commercial sources. The variations are many, however, differing most significantly in types of configurations, but also in grid and backing materials and lead arrangements. In the following paragraphs, we shall discuss some of these factors.

a) *The grid.* Selection of grid materials is based on a compromise of the following desirable factors:

1. High gage factor, F.
2. High resistivity, ρ.
3. Low temperature sensitivity.
4. High electrical stability.
5. High yield point.
6. High endurance limit.
7. Good workability.
8. Good solderability or weldability.
9. Low hysteresis.
10. Low thermal emf when joined with other materials.
11. Good corrosion resistance.

Temperature sensitivity is one of the most worrisome factors in the use of resistance strain gages. In the majority of applications, compensation is provided in the electrical circuitry; however, this does not always eliminate the problem. Two factors are involved: (1) the differential expansion existing between the grid support and the grid proper, resulting in a strain which the gage is unable to distinguish from load strain, and (2) the change in resistivity ρ, with temperature change   The importance of these factors is made apparent by inspection of columns 7 and 8 of Table 11.1, which list the strain increments that may be caused by temperature change in the uncompensated gage. This is commonly termed *apparent strain*.

Thermal emf superimposed on gage output obviously must be avoided if d-c circuitry is employed. For a-c circuitry this factor would be of little importance. Corrosion at a junction between grid and lead could conceivably result in a miniature rectifier, which would be more serious in an a-c than in a d-c circuit.

Another factor of relatively recent discovery is the influence of strong magnetic fields on gage performance. In particular, grid materials of high nickel content are susceptible to the effects of magnetostriction and magneto-resistivity [18, 19, 20].

Table 11.1 lists several possible grid materials and some of the properties influencing their use for strain gages. Commercial gages are usually of constantan or isoelastic. The former provides a relatively low temperature coefficient along with reasonable gage factors. Isoelastic gages are some

40 times more sensitive to temperature than are constantan gages. However, they have appreciably higher output, along with generally good characteristics otherwise. They are therefore made available primarily for dynamic application where the short time of strain variation minimizes the temperature problem.

The gage factor listed for nickel is of particular interest, not only because of its relatively high value, but also because of its negative sign. It should be noted, however, that the value of F for nickel varies over a relatively wide range, depending on how it is processed. Cold working has a rather marked effect on the strain- and temperature-related characteristics of nickel and its alloys, and this is taken advantage of to produce special temperature self-compensating gages (see Article 11.12).

b) Backing materials. The strain-gage grid is normally supported on some form of *backing* material. This not only provides the necessary electrical insulation between grid and tested material, but also provides a convenient carriage for handling the unmounted gage. Certain types of gages intended for high-temperature applications employ a *temporary* backing which is removed when the grid is mounted. In this case, at the time of installation the grid is embedded in a special ceramic material which provides the necessary electrical insulation and high-temperature adhesion (see Article 11.20e).

Desirable characteristics for backing materials include:

1. Minimum thickness consistent with other factors.

2. High mechanical strength.

3. High dielectric strength.

4. Minimum temperature restrictions.

5. Good adherence to cements used.

6. Nonhygroscopic.

Common backing materials include thin paper, phenolic-impregnated paper, epoxy-type plastic films and epoxy-impregnated fiber glass. Most foil gages intended for a moderate range of temperatures (-100 to $200°F$) employ an epoxy film backing. Table 11.2 lists commonly recommended temperature ranges.

c) Bonding materials and methods. Strain gages normally are attached to the test item by some form of cement or adhesive. A multitude of bonding materials are available, which require various detailed techniques for their use. The wide range makes it necessary to give these bonding materials only a very general coverage here. The user is referred to the suppliers for more detailed instructions. We can, however, consider general types and ranges of application.

TABLE 11.2

GENERAL RECOMMENDATIONS FOR STRAIN GAGE BACKING
MATERIALS AND ADHESIVES

Grid and Backing materials	Recommended adhesive	Permissible Temperature Range, °F.
Wire or foil on paper	Nitrocellulose (e.g. Duco)	−300 to 180
Foil on Epoxy	Cyanoacrylate (e.g. Eastman 910)	−100 to 200
Wire or foil on impregnated paper	Phenolic (e.g. Bakelite)	−400 to 350
Foil on phenol impregnated fiber glass	Phenolic	−400 to 400
Strippable foil or wire	Ceramic	−400 to 750 (to 1800 for short-time dynamic tests)
Free filament wire	Ceramic	−400 to 1200 (to 2000 for short-time dynamic tests)

Strain-gage adhesives normally fall into one of the following categories: cellulose, phenolic, epoxy, or ceramic. Table 11.2 summarizes common applications.

In general terms, the following are desirable characteristics of strain-gage adhesives:

1. High mechanical strength.

2. High creep resistance.

3. High dielectric strength.

4. Minimum temperature restrictions.

5. Good adherence.

6. Minimum moisture attraction.

7. Ease of application.

8. The capacity to set up fast.

The mechanical strength of both the backing and the cement is important, since they must transmit the force required to strain the grid. The dielectric strength is also important because this along with moisture absorption determines the electrical leakage resistance to the ground. The recommended minimum resistance between grid and test structure is 50 megohms, with higher values desired. For high stability, resistances to ground should be 1000 megohms or more [21].

Moisture attraction by cement and backing material must be minimized. As moisture is absorbed, the cement swells, causing straining of the grid and hence large zero shifts. This factor is most important when installations prohibit rechecks of gage zeros. According to Boiten [22] any cement providing good adherence to metal must *inherently* be of a type which will attract moisture. If this is accepted, it would indicate that all bonded strain-gage installations should be moisture-proofed. This is indeed desirable, except for short tests or tests permitting careful zero checking.

No particular difficulty should be experienced in mounting strain gages if the manufacturer's recommended techniques are carefully followed. However, we may make one observation which is universally applicable. *Cleanliness* is an absolute requirement if consistently satisfactory results are to be expected. The mounting area must be cleaned of all corrosion, paint, etc., and bare base material must be exposed. All traces of greasy film must be removed. Several of the gage suppliers offer kits of cleaning materials along with instructions for their use. These are very satisfactory.

Not only must the test surface be perfectly clean, but the gage must also be free from any trace of oily film. Many of the modern gages are supplied with a temporary, strippable backing which is removed immediately as the gage is mounted, thus helping to ensure a clean gage surface.

Nitrocellulose cement, used with paper-backed gages, requires 6 to 8 hours to set up, under moderate clamping pressure. An infrared heat lamp may also be used to accelerate the process. Other cements, such as the phenolic type, require greater pressures (100–200 psi) and a cycle of elevated temperature curing (to as high as 350°F) for proper bond.

Most of the epoxy cements will set up at room temperatures; however accelerated bonding can be achieved through the application of heat (to a temperature as high as 250–350°). Ceramic cements require curing at temperatures at least as high as the maximum temperature of use.

For simple strain-gage use over a modest temperature range (−100–200°F) a most useful cement is the cyanoacrylate type (Eastman 910). Mounting methods are simple and convenient (requiring no temperature cure and a clamping pressure of approximately 15 psi for a minute or two). The installation is ready for use almost as quickly as leads can be attached.

d) Gage protection. Most gage installations are not complete until provision is made to protect the gage from ambient conditions. The latter may include mechanical abuse, moisture, oil, dust and dirt, and the like. Of particular importance is moisture-proofing.

Once again, gage suppliers provide recommended materials for this purpose, including petroleum waxes, silicone resins, epoxy preparations, and rubberized brushing compounds. A variety of materials is necessary because of the many types of protection required—from such things as hot oil, immersion in water, liquified gases, etc. An extreme requirement for gage protection is found in the case of gages mounted on the exterior of a ship or submarine hull for the purpose of sea trials [52, 53, 54]. Special methods of protection are employed including rubber boots vulcanized over the gages.

The common practice is to mount the gage, make electrical connections and then, as a final step, to apply the protective coating by brushing or spreading the liquid, paste, or gel preparation over the complete installation. If paper-base gages are employed, it is usually desirable to ensure a low initial moisture content by heating the gage (with a heat lamp or hand-held hair drier) immediately before applying the protective coating.

e) Gage configurations and sizes. Gages may have one or more elements (Fig. 11.8). When employed for stress analysis, the single-element gage is applied to the uniaxial stress condition; the two-element rosette is applied to the biaxial condition when either the principal axes or the axes of interest are known, and the three-element rosette is applied when a biaxial stress condition is completely unknown (see Article 11.18 for a more complete discussion).

Selection of a gage size must be a compromise based on consideration of several factors: axial and transverse strain gradients, and power dissipation. It is desirable to minimize a gage dimension in the direction of greatest strain gradient, either axial or transverse. High strain gradients occur not only near points of stress concentration but also in cases of dynamic loading where propagation of stress waves is being studied. On the other hand, it is sometimes desirable to "average out" the local stress variations that may be caused by nonhomogeneity of material or local, uncontrollable dimensional deviations. An example of the latter would be handling defects or instances of out-of roundness* in a thin-walled tube which introduce local bending strains. In this case a long gage length may be desirable.

Of course resistance strain gages are passive transducers requiring external power. The energizing current causes ohmic heating which must be dissipated, largely to the structure upon which the gage is mounted. Foil gages bonded to metal can safely handle about 5 watts per square inch. An approximate maximum current for wire gages is 35 ma. Because sensitivity

---

* In one instance measurable, but not obvious, out-of-roundness of a pressure vessel with a 30-inch diameter and $\frac{1}{4}$-inch wall caused a strain difference of some 35% at points 90° apart around a free meridian.

TABLE 11.3

TYPICAL SINGLE ELEMENT STRAIN GAGES*

Item	Nominal resistance, ohms	Nominal gage factor	Length of grid, inches	Width of grid, inches	Grid material and form	Backing	Remarks
1.	120	2	$\frac{13}{16}$	$\frac{9}{64}$	Constantan wire	Paper	Original commercial gage
2.	300	2	6	$\frac{1}{32}$	Constantan wire	Paper	Extra long gage
3.	60	1.7	$\frac{1}{16}$	$\frac{1}{16}$	Constantan wire	Paper	Short gage length
4.	500	3.5	1	$\frac{3}{16}$	Isoelastic wire	Paper	
5.	240	2	$\frac{1}{4}$	$\frac{3}{32}$	Constantan wire	Phenolic	
6.	120	2.1	1	0.31	Constantan foil	Epoxy	
7.	120	2.0	0.015	0.020	Constantan foil	Epoxy	Very short gage length
8.	750	2.1	6	0.74	Constantan foil	Epoxy	Long gage length, high resis.
9.	350	2.1	0.12	0.21	Constantan foil	Phenolic	
10.	350	2.2	0.5	0.5	Karma foil	Fiber glass epoxy	
11.	120	2.1	0.50	0.29	Nichrome V foil	Strippable	For elevated temp. use.
12.	120	2.2	$\frac{1}{2}$	$\frac{1}{16}$	Nichrome V foil	Weldable	For elevated temp. use.

* Although the gages listed above are not identified with specific manufacturers, the list does approximate the range of single-element, metallic types presently available. There is a multitude of variations, however, and new gages are being developed at a rapid rate, hence, manufacturers' lists must be consulted for specific applications.

is, among other things, a function of energizing voltage (hence, power for a given resistance), it is obvious that the larger gage is inherently capable of the greater sensitivity. It may be said, therefore, that it is advantageous to use the largest gage compatible with the strain conditions being measured. An additional dividend gained by using the larger gage is considerably greater ease of installation.

Table 11.3 lists a small but representative sampling of available gages.

11.9 CIRCUITRY FOR THE METALLIC STRAIN GAGE

When the sensitivity of a metallic resistance gage is considered, its versatility and reliability are truly amazing. The basic relation as expressed by Eq. (11.8) is

$$\epsilon = \frac{1}{F} \frac{\Delta R_g}{R_g}. \tag{11.9}$$

Typical gage constants are

$$F = 2.0, \qquad R_g = 120 \text{ ohms.}$$

Strains of one μs (0.000001 in./in.) are detectable with commercial equipment; hence, the corresponding resistance change that must be measured in the gage will be

$$\Delta R_g = F R_g \epsilon = (2)(120)(0.000001) = 0.00024 \text{ ohm.}$$

This amounts to a resistance change of 0.0002%. Obviously, to measure changes as small as this, instrumentation more sensitive than the ordinary ohmmeter will be required.

Two circuit arrangements are used for this purpose: the simple voltage-dividing potentiometer or ballast circuit (Article 7.15), and the Wheatstone bridge (Article 7.18). Some form of bridge arrangement is by far the more generally useful.

11.10 THE STRAIN-GAGE BALLAST CIRCUIT

Figure 11.10 illustrates a simple strain-gage ballast arrangement. Using Eq. (7.33) and substituting R_g for kR_t, we may write

$$e_o = e_i \frac{R_g}{R_b + R_g}$$

and

$$de_o = \frac{e_i R_b \, dR_g}{(R_b + R_g)^2} = \frac{e_i R_b R_g}{(R_b + R_g)^2} \frac{dR_g}{R_g}.$$

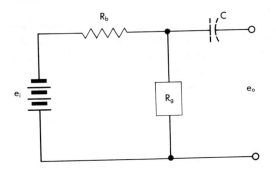

Fig. 11.9   Ballast circuit for use with strain gages.

From Eq. (11.9),

$$de_o = \frac{e_i R_b R_g}{(R_b + R_g)^2} F\epsilon,$$ (11.10)

where

e_i = exciting voltage,

e_o = voltage output,

R_b = ballast resistance, ohms,

R_g = strain-gage resistance, ohms,

F = gage factor,

ϵ = strain, in./in.

Some of the limitations inherent in this circuit may be demonstrated by the following example. Let

$$R_b = R_g = 120 \text{ ohms.}$$

A resistance of 120 ohms is common in a strain gage, and it will be recalled that in Article 7.15 we showed that equal ballast and transducer resistances provide maximum sensitivity. Also, let

$$e_i = 8 \text{ volts.}$$

Let

$$F = 2.0$$

which is a common value. Then

$$e_o = 8\left(\frac{120}{120 + 120}\right) = 4 \text{ volts,}$$

$$de_o = \frac{8 \times 120 \times 120 \times 2 \times \epsilon}{(120 + 120)^2} = 4\epsilon.$$

If our indicator is to provide an indication for strain of, say, one micro-strain, it must sense a 4-microvolt variation in 4 V, or 0.00010%. This severe requirement practically eliminates the ballast circuit for *static* strain work. We may use it, however, in certain cases for dynamic strain measurement when any static strain component may be ignored. If a capacitor is inserted in an output lead, the d-c exciting voltage is blocked and only the variable component is allowed to pass (Fig. 11.9). Temperature compensation is not provided; however, when only transient strains are of interest, this is usually of no importance.

11.11　THE STRAIN-GAGE BRIDGE CIRCUIT

A resistance-bridge arrangement is particularly convenient for use with strain gages because it may be easily adjusted to a null for zero strain, and it provides means for effectively reducing or eliminating the temperature effects previously discussed (Article 11.8a). Figure 11.10 shows a minimum bridge arrangement, where arm 1 consists of the strain-sensitive gage mounted on the test item. Arm 2 is formed by a similar gage mounted on a piece of unstrained material as nearly like the test material as possible and placed near the test location so that the temperature will be the same. Arms 3 and 4 may simply be fixed resistors selected for good stability, plus portions of slide-wire resistance, D, required for balancing the bridge.

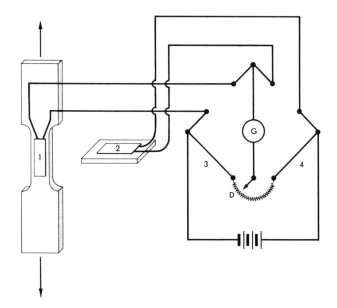

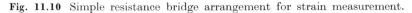

Fig. 11.10 Simple resistance bridge arrangement for strain measurement.

If we assume a voltage-sensitive deflection bridge with all initial resistances nominally equal, using Eq. (7.44) we have

$$\frac{\Delta e_o}{e_i} = \frac{\Delta R_1/R}{4 + 2(\Delta R_1/R)} .$$

In addition,

$$\epsilon = \frac{1}{F} \frac{\Delta R_1}{R} \quad \text{or} \quad \Delta R = FR\epsilon.$$

Then

$$\Delta e_o = \frac{e_i F\epsilon}{4 + 2F\epsilon} .$$

For $e_i = 8$ volts and $F = 2$,

$$\Delta e_o = \frac{8 \times 2 \times \epsilon}{4 + (2)(2)\epsilon} .$$

If we neglect the second term in the denominator, which is normally negligible, then

$$\Delta e_o = e_o = 4\epsilon \text{ volts}$$

or for $\epsilon = 1$ microstrain, $e_o = 4$ microvolts.

We see that under similar conditions the output increment for the bridge and ballast arrangements is the same. The tremendous advantage that the bridge possesses, however, is that the incremental output is not superimposed on a large fixed-voltage component. Another important advantage, which is discussed in Article 11.12, is that temperature compensation is easily attained through the use of a bridge circuit incorporating a "dummy" or compensating gage.

a) Bridges with two and four arms sensitive to strain. In many cases bridge configuration permits the use of more than one arm for measurement. This is particularly true where a known relation exists between two strains, notably the case of bending. For a beam section symmetrical about the neutral axis, it is known that the tensile and compressive strains are equal except for sign. In this case, both gages 1 and 2 may be used for strain measurement. This is done by mounting gage 1 on the tensile side of the beam and mounting gage 2 on the compressive side, as shown in Fig. 11.11. The resistance changes in the gages will then be alike but of opposite sign, and a double bridge output will be realized. Bridge *unbalance* may be indicated by the following relation:

$$\frac{R_1 + \Delta R_1}{R_2 - \Delta R_2} \neq \frac{R_3}{R_4} .$$

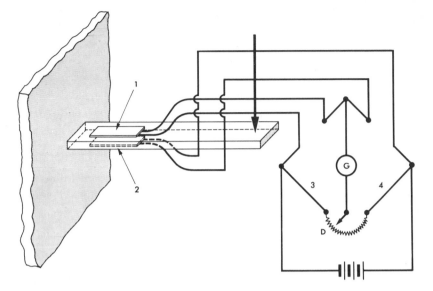

Fig. 11.11 Bridge arrangement with two gages sensitive to strain.

Here ΔR_1 and ΔR_2 represent the resistance changes due to strain, one caused by tensile strain and the other by compressive strain.

This may be carried further, and all four arms of the bridge made strain-sensitive, thereby *quadrupling* the output that would be obtained if only a single gage were used. In this case gages 1 and 4 would be mounted to record like strain (say tension) and 2 and 3 to record the opposite type. Bridge unbalance would then be represented by the following equation:

$$\frac{R_1 + \Delta R_1}{R_2 - \Delta R_2} \neq \frac{R_3 - \Delta R_3}{R_4 + \Delta R_4}.$$

Bridge circuits of these kinds may be used either as null-balance bridges or as deflection bridges (Article 7.18). In the former the slide-wire movement becomes the indicated measure of strain. This is most useful for strain-indicating devices used for static measurement. Most dynamic strain-measuring systems, however, use a voltage- or current-sensitive deflection bridge. After initial balance is accomplished, the output, amplified as necessary, is used to deflect an indicator, such as a cathode-ray oscilloscope beam, or a recorder of the stylus or light-beam types.

b) The bridge constant. At this point we introduce the term *bridge constant*, which we shall define by the following equation:

$$k = A/B, \tag{11.11}$$

where

k = bridge constant,

A = actual bridge output,

B = output from the bridge if only the single gage,
       sensing maximum strain, were effective.

In the example illustrated in Fig. 11.11, the bridge constant would be 2. This is true because the bridge provides an output double that which would be had if only gage 1 were strain-sensitive. If all four gages were used, quadrupling the output, the bridge constant would be 4. In certain other cases (Article 11.19), gages may be mounted sensitive to lateral strains which are functions of Poisson's ratio. In such cases bridge constants of 1.3 and 2.6 (for Poisson's ratio = 0.3) are common.

11.12 TEMPERATURE COMPENSATION

As already intimated, resistive-type strain gages are normally quite sensitive to temperature. Both the differential expansion between grid and the tested material and the temperature coefficient of the resistivity of the grid material contribute to the problem. It has been shown (Table 11.1) that the temperature effect is large enough to require careful consideration. Temperature effects may be handled by (1) cancellation or compensation, or (2) evaluation as a part of the data-reduction problem.

The former method is commonly applied to both the semiconductor and the metallic types of gages while the latter is primarily applied to the S-C types (Article 11.14).

Compensation may be provided (a) through the use of adjacent-arm balancing or compensating gage or gages, (b) by means of self-compensation, and (c) through the use of special external controlling circuitry.

a) *The adjacent-arm compensating gage.* Consider bridge configurations such as those shown in Figs. 11.10 and 11.11. Initial electrical balance is had when

$$R_1/R_2 = R_3/R_4.$$

If the gages in arms 1 and 2 are *alike* and *mounted on similar materials*, and if both gages experience the same resistance shift, ΔR_t, caused by temperature change, then

$$\frac{R_1 + \Delta R_t}{R_2 + \Delta R_t} = \frac{R_3}{R_4}.$$

It is seen that the bridge remains in balance and the output unaffected by the change in temperature. When the compensating gage is employed merely to complete the bridge and to balance out the temperature component, it is often referred to as the "dummy" gage.

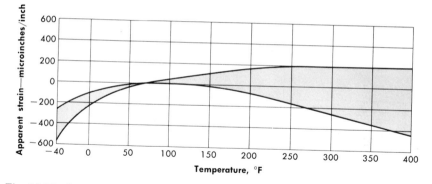

Fig. 11.12 Approximate range of apparent strain vs. temperature for a typical "selected melt" gage mounted on the appropriate material (e.g., steel at 6 ppm/deg. F.)

b) Self-temperature compensation. In certain cases it may be difficult or impossible to obtain temperature compensation by means of an adjacent-arm compensating or dummy gage. For example, temperature gradients in the test part may be sufficiently great to make it impossible to hold any two gages at similar temperatures. Or, in certain instances, it may be desirable to use the ballast rather than the bridge circuit, thereby eliminating the possibility of adjacent-arm compensation. Situations of this sort make *self-compensation* highly desirable.

The two general types of self-compensated gages available are the *selected-melt* gage and the *dual-element* gage. The former is based on the discovery that through proper manipulation of alloy and processing, particularly through cold-working, some control over the temperature sensitivity of the grid material may be exercised. Through this approach grid materials (both wire and foil) may be prepared which show very low apparent strain versus temperature change over certain temperature ranges when the gage is mounted on a particular test material. Figure 11.12 shows typical characteristics of selected-melt gages compensated for use with a material having a coefficient of expansion of 6 ppm/deg. F, which corresponds to the coefficient of expansion of most carbon steels. In this case, practical compensation is accomplished over a temperature range of approximately 50–250°F. Other gages may be compensated for different thermal expansions and temperature ranges. These curves give some idea of the degree of control that may be had through manipulation of the grid material.

The second approach to self-compensation makes use of two wire elements connected in series in one gage assembly. The two elements have different temperature characteristics and are selected so that the net temperature-induced strain is minimized when the gage is mounted on the specified

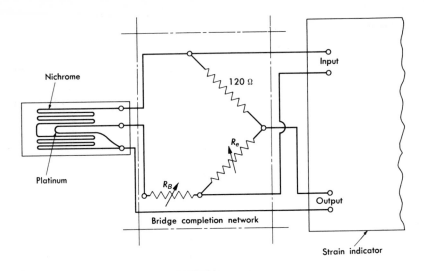

Fig. 11.13 "Universally" temperature-compensated strain gage. (U.S. Patent 2,672,048) (Courtesy: BLH Electronics, Inc., a subsidiary of Baldwin-Lima-Hamilton Corp., Waltham, Mass.)

test material. In general, the performance of this type of gage is similar to that of the selected-melt gage depicted in Figure 11.12.

Neither the selected-melt nor the dual-element gage has a distinctive outward appearance. One company employs color-coded backings to assist in identifying gages of different specifications.

c) *The universally temperature-compensated strain gage.* Figure 11.13 shows a special two-element gage, which when coupled with an appropriate bridge-completion network provides widely adjustable temperature-response characteristics. The gage consists of a Nichrome foil element and a platinum wire element arranged as shown. By varying the parameters of the external network, compensation for different test materials and temperature ranges may be had. Specific instructions for determining network components are supplied by the manufacturer.*

This discussion of temperature self-compensating methods should not be permitted to overshadow the adjacent-arm or dummy-gage method. The latter technique is so effective and simple that it should be employed whenever possible. It is, of course, quite permissible to combine self-compensating gages and the adjacent-arm or dummy methods in a single test setup, thereby ensuring even greater cancellation of temperature effects.

---

* Baldwin-Lima-Hamilton Corporation, Waltham, Mass.

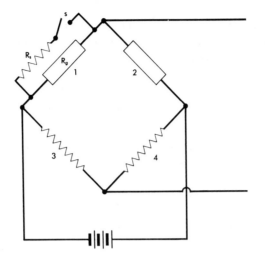

Fig. 11.14 Bridge employing a shunt resistance for calibration.

11.13 CALIBRATION

Ideally, calibration of any measuring system consists of introducing an accurately known sample of the variable that is to be measured and then observing the system's response. This ideal cannot often be realized in bonded resistance strain-gage work because of the nature of the transducer. Normally, the gage is bonded to a test item for the simple reason that the strains (or stresses) are unknown. Once bonded, the gage can hardly be transferred to a *known* strain situation for calibration. Of course, this is not necessarily the case if the gage or gages are employed as secondary transducers applied to an appropriate elastic member for the purpose of measuring force, pressure, torque, etc. In cases of this sort, it may be perfectly feasible to introduce known inputs and carry out satisfactory calibrations. When the gage is used for the purpose of experimentally determining strains, however, some other approach to the calibration problem is required.

Resistance strain gages are manufactured under carefully controlled conditions, and the gage factor for each lot of gages is provided by the manufacturer within an indicated tolerance of about $\pm 0.2\%$. Knowing the gage factor and gage resistance makes possible a simple method for calibrating any resistance strain-gage system. The method consists of determining the system's response to the introduction of a known small resistance change at the gage and of calculating an equivalent strain therefrom. The resistance change is introduced by shunting a relatively high-value precision resistance across the gage as shown in Fig. 11.14. When switch S is closed, the resistance of bridge arm 1 is changed by a small amount, as determined by the following calculations.

Let

$$R_g = \text{gage resistance, ohms,}$$

$$R_s = \text{shunt resistance, ohms.}$$

Then the resistance of arm 1 before the switch is closed $= R_g$, and the resistance of arm 1 after the switch is closed $= (R_g R_s)/(R_g + R_s)$, as determined for parallel resistances. Therefore the change in resistance is:

$$\Delta R = R_g - \frac{R_g R_s}{(R_g + R_s)} = \frac{R_g^2}{(R_g + R_s)}.$$

Now, to determine the equivalent strain, we may use the relation given by Eq. (11.9),

$$\epsilon = \frac{1}{F} \frac{\Delta R_g}{R_g}.$$

By substituting ΔR for ΔR_g, the equivalent strain is found to be

$$\epsilon_e = \frac{1}{F} \left(\frac{R_g}{R_g + R_s} \right). \tag{11.12}$$

Example 1. Suppose that

$$R_g = 120 \text{ ohms,}$$

$$F = 2.1,$$

$$R_s = 100 \text{ K (i.e., 100,000 ohms).}$$

What equivalent strain will be indicated when the shunt resistance is connected across the gage?

Solution. From Eq. (11.12),

$$\epsilon_e = \frac{1}{2.1} \left(\frac{120}{100,000 + 120} \right) = 0.00057 = 570 \, \mu s.$$

Example 2. What shunt resistor should be used to provide an equivalent strain of 100 μs, using the same gage as was used in Example 1?

Solution. Solving Eq. (11.12) for R_s,

$$R_s = \frac{R_g}{F \epsilon_e} - R_g.$$

Substituting,

$$R_s = \frac{120}{2.1 \times 0.00010} - 120 = 572,000 - 120 = 571,880 \text{ ohms.}$$

Dynamic calibration is sometimes provided by replacing the manual calibration switch with an electrically driven switch, often referred to as a *chopper*, which makes and breaks the contact 60 or 100 times per second. When displayed on a CRO screen or recorded, the trace obtained is found to be a square wave. The *step* in the trace represents the equivalent strain calculated from Eq. (11.12).

There are other methods of electrical calibration. One system replaces the strain-gage bridge with a substitute load, initially adjusted to equal the bridge load [23]. A series resistance is then employed for calibration. Another method injects an accurately known voltage into the bridge network.

11.14 THE SEMICONDUCTOR, OR PIEZORESISTIVE-TYPE, STRAIN GAGE

A thin sliver of silicon, typically 0.005 by 0.0005 inches in cross section, is the common transducer element of the semiconductor, or S-C, strain gage. Effective lengths may range from roughly 0.050 upward to $\frac{1}{2}$ inch and the sensitive elements may be either backed or unbacked. Ultrasonic techniques are employed to cut the S-C gage elements from single crystals of heavily doped silicon. Figure 11.15 illustrates a typical, commercially available S-C gage. Essentially the same types of backing, bonding materials, and mounting techniques as used for metallic gages are suitable for S-C gages.

Fig. 11.15 An example of a semi-conductor strain gage. (Courtesy: BLH Electronics, Inc., a subsidiary of Baldwin-Lima-Hamilton Corp., Waltham, Mass.)

The major advantage of the silicon S-C gage is its high gage factor, currently approximating 130, but which may be increased through further development. This represents a very decided improvement in sensitivity compared to the $2-3\frac{1}{2}$ exhibited by the ordinary metallic element.

The increased sensitivity, however, is also accompanied by comparative disadvantages, namely:

a) the output of the S-C element is inherently nonlinear with strain;

b) strain sensitivity is markedly temperature dependent;

c) It is somewhat more fragile than the corresponding wire or foil element, though it can be bent to a radius as small as $\frac{1}{8}$ inch;

d) the strain range of the S-C gage is roughly limited to 3000 to 10,000 μs (dependent upon the specific gage type), as compared to an upper limit of 100,000 μs for some metallic resistance gages;

e) the semiconductor gage is considerably more expensive than the ordinary metallic gage;

f) because of the high sensitivity of the S-C element, the nonlinearity of the simple Wheatstone bridge (See Figs. 7.15 and 7.16) cannot always be ignored, as is normally done when conventional metallic-element gages are used. This may or may not necessitate special instrumentation (see Article 11.14b).

11.14a THEORETICAL RELATIONSHIPS

Semiconductor-gage behavior may be described by the following relation [48]:

$$\frac{\Delta R}{R_0} = \left(\frac{T_0}{T}\right)(F_0)\epsilon + \left(\frac{T_0}{T}\right)^2 (C_0)\epsilon^2, \qquad (11.13)$$

where

ΔR = change in gage resistance, ohms,

R_0 = resistance of the unstrained gage element at temperature T_0,

T_0 = reference temperature, degrees Kelvin commonly taken as 298° (or 25°C or 77°F),

T = temperature, deg. K,

F_0 = gage factor = $(\Delta R/R_0)/\epsilon$ = the slope of the $\Delta R/R_0$ versus ϵ curve at *zero strain* (see Fig. 11.17),

C_0 = a constant for a particular gage,

= a multiplying factor which, along with the temperature, describes the nonlinearity of the gage.

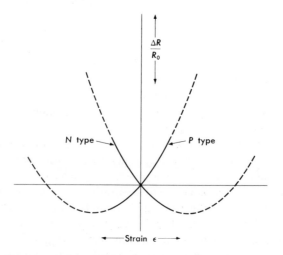

Fig. 11.16 Characteristic strain-resistance sensitivity of silicon semiconductor strain gage material (solid portion of curves indicate practical ranges).

Both F_0 and C_0 are supplied by the manufacturer, but note that F_0 is the value for the original unbacked gage element and that the value may be altered both by the manufacturing process of mounting the element on its backing and by the application process of bonding the gage.

It is seen that, in terms of strain, the variation of gage sensitivity is parabolic and that increased temperature T reduces both sensitivity and nonlinearity. By control of the impurities (doping) it is possible to obtain gages with a wide variation of characteristics. Figure 11.16 illustrates two basic types, the P or *positive* type and the N or *negative* type. The term "positive" or "negative" refers to the general slope of the curves over the practical strain ranges, i.e., in the vicinity of zero strain.

11.14b PRACTICAL ASPECTS

It is obvious that the greatest advantage of the semiconductor gage is that its sensitivity is greater than that of the ordinary metallic gage. This factor becomes particularly useful (a) when the range of strains is small enough to make the metallic gage of limited usefulness, and (b) when the required gage output is greater than usual.

Small strain range may correspond to small load increments, to "stiff" transducer elements, or to both. For example, a cell employed for measuring rolling mill loads requires a "stiff" design; otherwise the dimensional tolerance of the mill product becomes difficult to maintain. For this reason the load cell would be designed for minimum deflection of the elastic element and a high output secondary transducer would be desirable.

Greater than normal outputs may be used to advantage in certain control or telemetry applications, or merely for the purpose of adapting simplified or low-sensitivity readout devices.

The two cases (a) and (b) above place considerably different requirements on the circuitry and data-reduction methods that may be employed. For example, the high sensitivity of the S-C gage generally limits direct use of the ordinary strain-measuring indicators and amplifiers intended for metallic gages. There are two reasons for this: (1) such instruments do not have the necessary measurement span required by the S-C gage, and (2) their design is usually based on the simple Wheatstone bridge, whose nonlinearity becomes excessive when used with the S-C gage. Both of these factors, however, are no more critical for the S-C than for the metallic application, provided the ranges of actual resistance changes for the two are comparable. We see, then, that in general the common metallic-strain gage instrumentation may be used with S-C gages over the range of resistance change for which the instruments are designed. For gages of the same nominal resistance the instrument's range of strain measurement is reduced by the factor $F_{\text{S-C}}/F$ metallic.

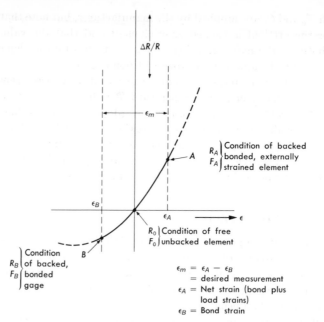

Fig. 11.17 Diagram illustrating the resistance-strain relationships between a free, a backed, and a backed and bonded grid of the positive S-C type.

When large outputs are desired (perhaps to take advantage of less expensive, less sensitive readout instrumentation) then the inherent nonlinearity of the bridge, as commonly employed, may be a nuisance. One solution is to replace the constant voltage exciting source with a constant current source. The advantages of doing this are discussed in Article 7.18c Also see Chapter 7, Ref. [25].

One manufacturer* refers basic gage constants to the characteristics of the *free, unbacked* silicon element. Both gage resistance R_0 and gage factor F_0 (slope of the $\Delta R/R_0$ versus ϵ curve at zero strain), are determined for the *free* element.† If the free element is then mounted on a backing, both the element resistance and its gage factor are altered. The resulting resistance is termed the *packaged* resistance R_p and may be provided by the manufacturer along with the other data. Further changes occur when the gage is bonded to the test structure, yielding a *bonded resistance* R_B and a *bonded gage factor* F_B. In general, shrinkage of the bonding agents places the

---

* Baldwin-Lima-Hamilton Corporation, Waltham, Massachusetts.
† Each free element is mechanically strained and gage factors determined experimentally.

element in a compressed state of strain. It is seen then that initial test zeros may be sufficiently different from the basic reference zeros supplied by the manufacturer as to make corrections necessary. This, of course, is a penalty that must be paid for the increased sensitivity obtained. Similar conditions apply to metallic-gage applications but low sensitivities make the discrepancies insignificant. Realization of S-C gage potentials, however, requires that this discrepancy be taken into consideration.

Figure 11.17 illustrates the resistance-strain relationships of the three conditions discussed above. Values for R_0, F_0 and C_0 are commonly provided by the manufacturer.

Although Eq. (11.13) expresses the relationship between strain, temperature, and resistance change for S-C type gages, direct application of the equation for the purpose of data reduction is difficult. Practical strain evaluation is most easily accomplished through the use of precalculated tables relating resistance, temperature, and strain. Gage characteristics vary sufficiently, however, as to rule out "universal tables" such as the familiar thermocouple tables. It is necessary for the manufacturer to supply the required tabulations for each lot of gages.

Semiconductor strain gages are still in a state of active development, and industry-wide methods of strain evaluation are not as yet solidified. For these reasons, the reader is referred to instruction and data sheets of the specific manufacturer for further details, especially for information on data reduction.

11.15 COMMERCIAL STRAIN-MEASURING SYSTEMS

Four types of commercial strain-gage instrumentation will be discussed in this article. They are representative of four functionally different systems, as follows:

1. A basic strain indicator useful for single-channel, static readings.

2. A single-channel system for dynamic measurement, using a cathode-ray oscilloscope for output indication.

3. A stylus-and-paper oscillographic-type recording system.

4. A light-beam and photographic-paper recording oscillographic system.

Each of the systems selected for discussion is typical of many commercially available systems in that category, and all of them are intended for use with metallic gages. Commercial systems specifically designed for semiconductor gages, however, are not generally available at this time.

a) A basic indicator for static strains. Figure 11.18 illustrates a commercial indicator primarily intended for static strain measurements. Basically the instrument is an oscillator-excited resistance bridge provided with coarse

Fig. 11.18 Baldwin Type-N strain-gage indicator.   (Courtesy Baldwin-Lima-Hamilton Corp., Waltham, Mass.)

Fig. 11.19 Ellis Model BAM-1 strain gage bridge and transistorized amplifier-meter. (Courtesy: Ellis Associates, Pelham, N.Y.)

and fine adjustments for balancing. After amplification, the bridge output is fed to a phase-sensitive demodulating circuit which supplies the balance meter with a d-c voltage proportional to the bridge unbalance. The meter indicates a positive or negative unbalance, depending on the sign of the unbalancing strain. The system is used as a simple null-balance resistance bridge, with the balancing adjustments calibrated in terms of strain in microstrains.

b) Strain-gage bridge and amplifier for use with cathode-ray oscilloscope. Figure 11.19 shows a versatile single-channel strain-gage bridge and amplifier suitable for static and dynamic measurements. Basically the instrument consists of a dc-energized resistance bridge connected to a simple transistorized amplifier. A simplified schematic diagram of the circuit is shown in Fig. 11.20. Shunting resistors are provided for calibration (Article 11.13), and either meter or CRO readout may be employed, depending primarily on whether the input is static or dynamic.

c) Multichannel stylus-and-paper strain-recording systems. Figure 8.15 illustrates a typical oscillographic strain-recording system. The equipment shown is an eight-channel model; however, assemblies of this kind are available with as few as one channel. These systems are quite versatile: various preamplifiers may be had, depending on the form of input signal. For strain-gage work, a carrier amplifier (Article 7.23d) is recommended. Figure 11.21 shows the system arrangement in block form. Frequency

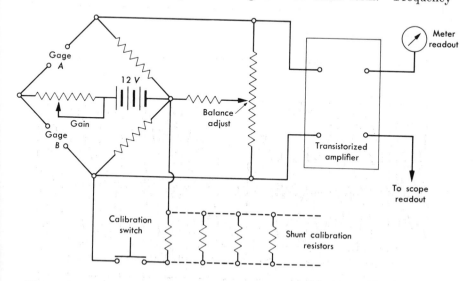

Fig. 11.20 Schematic circuit diagram (simplified) of Ellis Bridge amplifier and meter (BAM-1) for use with resistance-type strain gages.

response of the system shown is flat from d-c to about 70 Hz and down about 20% at 100 Hz. With proper damping (supplied by the driving amplifier), square-wave overshoot is about 4%, and 90% of final amplitude deflection is reached in 0.01 sec.

d) *Multichannel light-beam oscillographic system.* Another commercially available strain-measuring system employs an oscillographic recorder. Mirror-type galvanometers (Article 8.10) are used in conjunction with moving photographic paper or film for recording. Strain-gage bridge configurations are commonly used, powered either from external d-c or a-c sources or by an

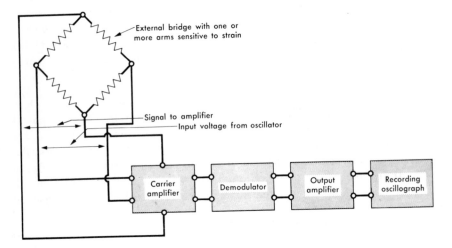

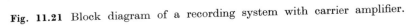

Fig. 11.21 Block diagram of a recording system with carrier amplifier.

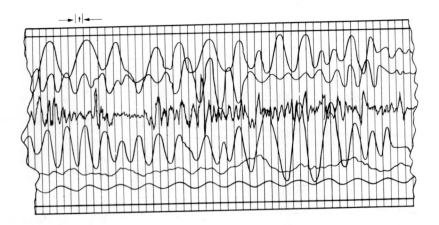

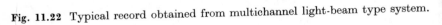

Fig. 11.22 Typical record obtained from multichannel light-beam type system.

a-c carrier arrangement. Figure 8.17 shows an example of this type of system, and Fig. 11.22 shows a typical record. These systems are available with as many as 50 separate channels.

Galvanometers used in these systems may be had with frequency response flat within $\pm 5\%$ to 4800 Hz (Table 8.1). Their use extends the frequency range of measurable strains far above that of the stylus recorder. Many frequencies from mechanical sources fall within the useful range of the light-beam oscillographic recording systems. It will be recalled, however, that wide frequency range can only be obtained at the expense of sensitivity (see Article 8.11).

11.16 STRAIN-GAGE SWITCHING

Mechanical development problems often require the use of many gages mounted throughout the test item, and simultaneous or nearly simultaneous readings are often necessary. Of course, if the data must be recorded at precisely the same instant, it will be necessary to provide separate channels for each gage involved. However, many times steady-state conditions may be maintained or the test cycle repeated, and readings may be made in succession until all the data have been recorded. In other cases, the budget may prohibit duplication of the required instrumentation for simultaneous multiple readings or recordings, and it becomes desirable to switch from gage to gage, taking data in sequence.

Two basic switching arrangements are possible when resistance bridges are used. They are *intrabridge* switching and *interbridge* switching. Figure 11.23 illustrates the first arrangement, in which various gages are switched

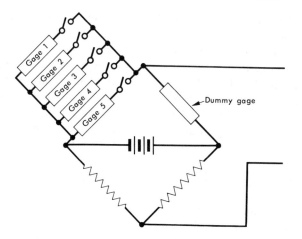

Fig. 11.23 Intrabridge switching, where gages are switched into and out of a bridge.

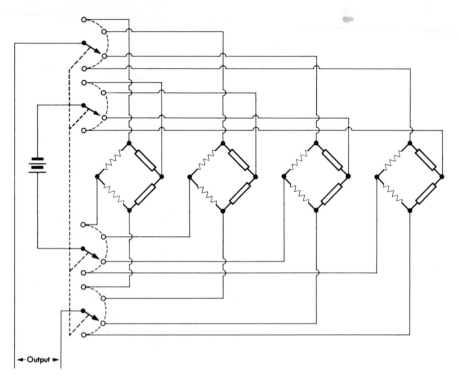

Fig. 11.24 Interbridge switching, where complete bridges are switched into or out of the circuit.

into and out of a single bridge circuit. Figure 11.24 illustrates the second, in which switch connections are entirely outside the bridges.

When metallic gages and intrabridge switching are simultaneously employed, variations in switch resistance can be a very annoying problem. As was shown in Article 11.11, the resistance change due to strain of the metallic gage is very small. It is quite possible that the change in switch resistance from one "switching" to the next may be of the same order of magnitude as the quantity of interest [24]. Unless extreme care is used and the highest quality of switch components employed, this method of switching may prove entirely impractical. It is readily apparent that this problem is reduced in inverse proportion to the ratio of gage factors when the semi-conductor gage is used.

For really trouble-free strain-gage switching, the method illustrated in Fig. 11.24 is recommended. Here all switch contacts are placed completely outside the bridge ring, and complete bridges are switched into and out of the measuring system. Of course, the same variations in switch resistance occur as do in intrabridge switching, but in this case they do not alter bridge balance. Their effect is to alter bridge sensitivity slightly, although this is not normally measurable. The disadvantage in the latter method lies in the more complicated installation. This disadvantage must be weighed against the possibly questionable data yielded by the simpler setup.

11.17 USE OF STRAIN GAGES ON ROTATING SHAFTS

Strain-gage information may be conducted from rotating shafts in at least three different ways: (1) by direct connection, (2) by telemetering, and (3) by use of slip rings. When a shaft rotates slowly enough and when only a sampling of data is required, direct connections may be made between the gages and the remainder of the measuring system. Sufficient lead length is provided, and the cable is permitted to wrap itself onto the shaft. In fact, the available time may be doubled with a given length of cable if it is first wrapped on the shaft so that the shaft rotation causes it to unwrap and then to wrap up again in the opposite direction. If the machine cannot be stopped quickly enough as the end of the cable is approached, a fast or automatic disconnecting arrangement may be provided. This actually need be no more than soldered connections that can be quickly peeled off. Shielded cable should be used to minimize reactive effects resulting from the coil of cable on the shaft. This technique is somewhat limited, of course, but should not be overlooked, because it is quite workable at slow speeds and avoids many of the problems inherent in the other methods.

A second method is that of actually transmitting the strain-gage information through the use of a radio-frequency transmitter mounted on the shaft and picking up the signal by means of a receiver placed nearby. This method has been used successfully [25], and the procedure and equipment will undoubtedly be perfected for more general usage. There is commercially available transistorized FM equipment of relatively small size in which the frequency change of the transmitter RF is a function of the strain. Such a system is quite practical when the added cost can be justified.

Undoubtedly the most common method for obtaining strain-gage information from rotating shafts is through the use of slip rings. Slip-ring problems are similar to switching problems, as discussed in the previous article, except that additional variables make the problem more difficult. Such factors as ring and brush wear and changing contact temperatures make it imperative that the full bridge be used at the test point and that the slip rings be introduced external to the bridge as shown in Fig. 11.25.

Commercial slip-ring assemblies are available whose performances are quite satisfactory. Their use, however, presents a problem that is often difficult to solve. The assembly is normally self-contained, consisting of brush supports and a shaft with rings mounted between two bearings. The construction requires that the rings be used at a free end of a shaft,

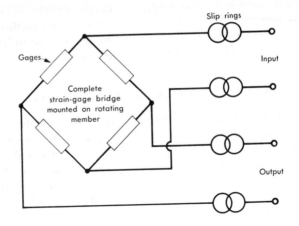

Fig. 11.25 Slip rings used external to the bridge.

which more often than not is separated from the test point by some form of bearing. This presents the problem of getting the leads from the gage located on one side of a bearing to the slip rings located on the opposite side. It is necessary to feed the leads through the shaft in some manner, which is not always convenient. Where this presents no particular problem, the commercially available slip-ring assemblies are practical and also probably the most inexpensive solution to the problem.

For certain applications it may be necessary to construct special-purpose rings for the problem at hand. In this case the experiences of others may be helpful [26, 27, 28, 29, 30, 31, 55].

11.18 STRESS-STRAIN RELATIONSHIPS

As previously stated, strain gages are usually used for one of two reasons: to determine stress conditions through strain measurements, or to act as secondary transducers calibrated in terms of such quantities as force, pressure, displacement, and the like. In either case, intelligent use of strain gages demands a good concept of stress-strain relationships. Knowledge of the *plane*, rather than of the general three-dimensional case, is usually sufficient for strain-gage work because it is only in the very unusual situation that a strain gage is mounted anywhere except on the *unloaded surface* of a stressed member. For a review of the plane stress problem, the reader is directed to Appendix A.

a) The simple uniaxial stress situation. In bending, or in a tension or compression member, the unloaded outer fiber is subject to a uniaxial stress situation. However, this condition results in a triaxial strain condition, because we know that there will be lateral strain in addition to the strain in the direction of stress. Because of the simplicity of the ordinary tensile (or compressive) situation and its prevalence (see Fig. 11.1), the *fundamental stress-strain relation* is based on it. Young's modulus is *defined* by the relation expressed by Eq. (11.2), and Poisson's ratio is *defined* by Eq. (11.2a). It is important to realize that both these definitions *are made on the basis of the simple, one-direction stress system.*

For situations of this sort, calculation of stress from strain measurements is quite simple. The stress is determined merely by multiplying the strain, measured in the axial direction in microstrains, by the modulus of elasticity for the test material in millions of psi.

Example 1. Suppose the tensile member in Fig. 11.10 is of aluminum having a modulus of elasticity equal to 10 million psi and the strain measured by the gage is 325 μs. What axial stress exists at the gage?

Solution

$$\sigma_a = E\epsilon_a = (10) \times (325) = 3250 \text{ psi.}$$

Example 2. A strain gage is mounted on a beam as shown in Fig. 11.11. The beam is of steel having an estimated modulus of elasticity of 29.5×10^6 psi. If the strain measured by the gage is 195 μs, what stress exists at the gage centerline?

Solution

$$\sigma_b = E\epsilon = (29.5) \times (10)^6 \times (195) \times (10)^{-6} = 5752 \text{ psi.}$$

b) The biaxial stress situation. Often gages are used at locations subject to stresses in more than one direction. If the test point is on a free surface, as is usually the case, the condition is termed *biaxial*. A good example of this condition exists on the outer surface, or shell, of a cylindrical pressure vessel. In this case, we know there are *hoop* stresses, acting circumferentially, tending to open up a longitudinal seam. There are also longitudinal stresses tending to blow the heads off. The situation may be represented as shown in Fig. 11.26.

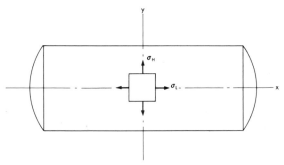

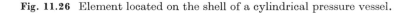

Fig. 11.26 Element located on the shell of a cylindrical pressure vessel.

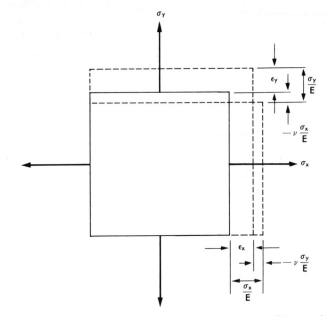

Fig. 11.27 An element taken from a biaxially stressed condition, with normal stresses shown.

Let us consider a general element subject to stresses which we shall call σ_x and σ_y, as shown in Fig. 11.27. Suppose that the stresses σ_x and σ_y are applied one at a time. If σ_x is applied first, there will be a strain in the x-direction equal to σ_x/E. At the same time, because of the Poisson's ratio effect, there will be a strain in the y-direction equal to $-\nu\sigma_x/E$.

Now suppose that the stress in the y-direction, σ_y, is applied. This stress will result in a y-strain of σ_y/E and an x-strain equal to $-\nu\sigma_y/E$. The net strains, *which are the strains that a strain gage would sense*, are expressed by the relations*

$$\epsilon_x = \frac{1}{E}(\sigma_x - \nu\sigma_y) \quad \text{and} \quad \epsilon_y = \frac{1}{E}(\sigma_y - \nu\sigma_x). \quad (11.14)$$

---

* When a stress σ_z exists, acting in the third orthogonal direction, it is seen that the more general three-dimensional relations are:

$$\epsilon_x = \frac{1}{E}\left[\sigma_x - \nu(\sigma_y + \sigma_z)\right],$$

$$\epsilon_y = \frac{1}{E}\left[\sigma_y - \nu(\sigma_z + \sigma_x)\right],$$

$$\epsilon_z = \frac{1}{E}\left[\sigma_z - \nu(\sigma_x + \sigma_y)\right]. \quad (11.16)$$

If these relations are solved simultaneously for σ_x and σ_y, we obtain the equations

$$\sigma_x = \frac{E(\epsilon_x + \nu\epsilon_y)}{1 - \nu^2} \quad \text{and} \quad \sigma_y = \frac{E(\epsilon_y + \nu\epsilon_x)}{1 - \nu^2}. \tag{11.15}$$

Assume now that we have a general, unknown, plane-stress problem for which we wish to determine the stresses in given orthogonal (right-angled) directions. Equation (11.15) shows us that in addition to knowing the modulus of elasticity and Poisson's ratio for the material, we must measure two strains in the x- and y-directions. Even though only a single stress is desired, it will be necessary to measure both orthogonal strains.

Example 3. Suppose we wish to determine, by strain measurement, the stresses on the outer surface of a cylindrical pressure-vessel in the circumferential or hoop direction. The modulus of elasticity of the material is 15×10^6 psi, and Poisson's ratio is 0.28. By strain-gage measurement, the hoop and longitudinal strains are determined to be

$$\epsilon_h = 425 \ \mu s \quad \text{and} \quad \epsilon_L = 115 \ \mu s.$$

What hoop stress corresponds to these values?

Solution

$$\sigma_h = \frac{E(\epsilon_h + \nu\epsilon_L)}{1 - \nu^2}$$

$$= \frac{15(425 + 0.28 \times 115)}{1 - (0.28)^2} = 7440 \text{ psi}.$$

Although we may not be directly interested, we have the necessary information to determine the longitudinal stress also, as follows:

$$\sigma_L = \frac{E(\epsilon_L + \nu\epsilon_h)}{1 - \nu^2}$$

$$= \frac{15(115 + 0.28 \times 425)}{1 - (0.28)^2} = 3810 \text{ psi}.$$

It may be noted that the 2-to-1 stress ratio traditionally expected for the thin-walled pressure vessel does not yield a like ratio of strains. The strain ratio is more nearly 4 to 1.

Use of Eq. (11.15) permits us to determine the stresses in two orthogonal directions. However, this information gives the *complete* stress-strain picture only when the two right-angled directions coincide with the *principal directions* (see Appendix A). If we do not know the principal directions, it would only be by chance that our gages would yield maximum stress. In

TABLE 11.4
STRESS-STRAIN RELATIONS FOR ROSETTE GAGES*†

Type of rosette	Rectangular	Equiangular (delta)	T-delta
Principal strains, ϵ_p, ϵ_q	$\dfrac{1}{2}[\epsilon_a + \epsilon_c]$ $\pm \sqrt{2(\epsilon_a - \epsilon_b)^2 + 2(\epsilon_b - \epsilon_c)^2}$	$\dfrac{1}{3}[\epsilon_a + \epsilon_b + \epsilon_c]$ $\pm \sqrt{2(\epsilon_a - \epsilon_b)^2 + 2(\epsilon_b - \epsilon_c)^2 + 2(\epsilon_c - \epsilon_a)^2}$	$\dfrac{1}{2}[\epsilon_a + \epsilon_d]$ $\pm \sqrt{(\epsilon_a - \epsilon_d)^2 + \frac{4}{3}(\epsilon_b - \epsilon_c)^2}$
Principal stresses, σ_1, σ_2	$\dfrac{E}{2}\left[\dfrac{\epsilon_a + \epsilon_c}{1-\nu} \pm \dfrac{1}{1+\nu}\right.$ $\left.\sqrt{2(\epsilon_a - \epsilon_b)^2 + 2(\epsilon_b - \epsilon_c)^2}\right]$	$\dfrac{E}{3}\left[\dfrac{\epsilon_a + \epsilon_b + \epsilon_c}{1-\nu} \pm \dfrac{1}{1+\nu}\right.$ $\left.\sqrt{2(\epsilon_a - \epsilon_b)^2 + 2(\epsilon_b - \epsilon_c)^2 + 2(\epsilon_c - \epsilon_a)^2}\right]$	$\dfrac{E}{2}\left[\dfrac{\epsilon_a + \epsilon_d}{1-\nu} \pm \dfrac{1}{1+\nu}\right.$ $\left.\sqrt{(\epsilon_a - \epsilon_d)^2 + \frac{4}{3}(\epsilon_b - \epsilon_c)^2}\right]$
Maximum shear, τ_{max}	$\dfrac{E}{2(1+\nu)}$ $\sqrt{2(\epsilon_a - \epsilon_b)^2 + 2(\epsilon_b - \epsilon_c)^2}$	$\dfrac{E}{3(1+\nu)}$ $\sqrt{2(\epsilon_a - \epsilon_b)^2 + 2(\epsilon_b - \epsilon_c)^2 + 2(\epsilon_c - \epsilon_a)^2}$	$\dfrac{E}{2(1+\nu)}$ $\sqrt{(\epsilon_a - \epsilon_d)^2 + \frac{4}{3}(\epsilon_b - \epsilon_c)^2}$
$\tan 2\theta$	$\dfrac{2\epsilon_b - \epsilon_a - \epsilon_c}{\epsilon_a - \epsilon_c}$	$\dfrac{\sqrt{3}(\epsilon_c - \epsilon_b)}{(2\epsilon_a - \epsilon_b - \epsilon_c)}$	$\dfrac{2}{\sqrt{3}}\dfrac{(\epsilon_c - \epsilon_b)}{(\epsilon_a - \epsilon_d)}$
$0 < \theta < +90°$	$\epsilon_b > \dfrac{\epsilon_a + \epsilon_c}{2}$	$\epsilon_c > \epsilon_b$	$\epsilon_c > \epsilon_b$

* References: [32, 33, 34, 35, 36, 37].   † *Note:* θ = angle of reference, measured positive in the counterclockwise direction from the a-axis of the rosette to the axis of the algebraically larger stress.

general, if a plane stress condition is completely unknown, at least three strain measurements must be made, and it is necessary to use some form of three-element rosette (see Article 11.8e, also Fig. 11.8). From the strain data secured in the three directions, the complete stress-strain picture is obtained.

Stress-strain relations for rosette gages are given in Table 11.4.

While only three strain measurements are necessary to completely define a stress situation, the T-delta rosette, which includes a fourth gage element, is sometimes used to advantage for the following reasons:

1. The fourth gage may be used as a check on the results obtained from the other three elements.

2. If the principal directions are approximately known, gage d may be aligned with the estimated direction. Then, if the readings from gages b and c are of about the same magnitude, it is known that the estimate is reasonably correct, and the principal stresses may be calculated directly from Eq. (11.15), greatly simplifying the arithmetic. If the estimate of direction turns out to be incorrect, complete data is still available for use in the equations from Table 11.4.

3. If the four readings are used in the T-delta equations in Table 11.4. an averaging effect results in better accuracy than if only three readings are used.

In spite of the advantages of the T-delta rosette, the rectangular one is probably the most popular, with the equiangular (delta) kind receiving second greatest use.

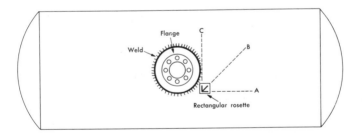

Fig. 11.28 Rosette installation near a pressure-vessel nozzle.

Example 4. Figure 11.28 illustrates a rectangular rosette used to determine the stress situation near a pressure-vessel nozzle. For thin-walled vessels, the assumption that principal directions correspond to the hoop and longitudinal directions is valid for the shell areas removed from discontinuities. Near an opening, however, the stress condition is completely unknown, and a rosette with at least three elements must be used.

Let us assume that the rosette provides the following data:

$$\epsilon_a = 72 \ \mu s, \qquad \epsilon_b = 120 \ \mu s, \qquad \epsilon_c = 248 \ \mu s.$$

In addition, we shall say that

$$\nu = 0.3, \qquad E = 30 \times 10^6 \ \text{psi}.$$

Note: A study of the equation forms in Table 11.4 shows that for each case, the principal strain, the principal stress, and maximum shear relations involve similar radical terms. Therefore, in evaluating rosette data, it is convenient to calculate the value of the radical as the first step. It will also be noted that the second term in the principal stress relations is equal to the shear stress; thus arithmetical manipulations may be kept to a minimum if the shear stress is calculated before the principal stresses are determined. Hence,

$$\sqrt{2(\epsilon_a - \epsilon_b)^2 + 2(\epsilon_b - \epsilon_c)^2}$$
$$= \sqrt{2(72 - 120)^2 + 2(120 - 248)^2}$$
$$= 193 \ \mu s$$

and

$$\epsilon_1 = \tfrac{1}{2}[72 + 248 + 193] = 256\tfrac{1}{2} \ \mu s$$

$$\epsilon_2 = \tfrac{1}{2}[72 + 248 - 193] = 63\tfrac{1}{2} \ \mu s$$

$$\tau_{\max} = \left(\frac{30 \times 10^6}{2(1 + 0.3)}\right)(193) = 2230 \ \text{psi},$$

$$\sigma_1 = \frac{30}{2}\left(\frac{72 + 248}{0.7}\right) + 2230$$

$$= 6860 + 2230 = 9090 \ \text{psi},$$

$$\sigma_2 = 6860 - 2230 = 4630 \ \text{psi}.$$

To determine the principal planes,

$$\tan 2\theta = \frac{(2\epsilon_b - \epsilon_a - \epsilon_c)}{(\epsilon_a - \epsilon_c)}$$

$$= \frac{(2 \times 120) - 72 - 250}{(72 - 150)} = 0.46,$$

$$2\theta = 24.7° \qquad \text{or} \qquad 204.7°,$$

or

$$\theta = 12.3° \qquad \text{or} \qquad 102.3°,$$

measured counterclockwise from the axis of element *A*. We must test for the

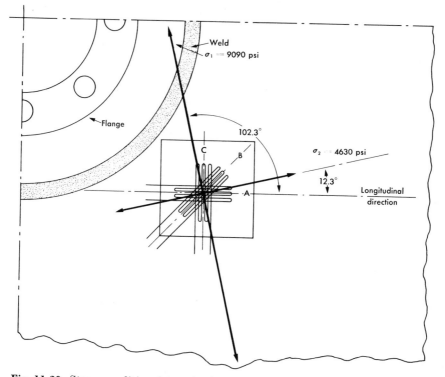

Fig. 11.29 Stress condition determined from data obtained by the rosette shown in Fig. 11.28.

proper quadrant as follows (see the last line in Table 11.4):

$$\frac{\epsilon_a + \epsilon_c}{2} = \frac{72 + 248}{2} = 160,$$

which is greater than ϵ_b. Therefore the axis of maximum principal stress does *not* fall between 0° and 90°. Hence, $\theta = 102.3°$. Figure 11.29 illustrates this condition.

c) Computational aids. When a test installation makes use of many rosette gages and tests are conducted for many different load conditions, the mass of data so quickly accumulated produces a computation problem whose magnitude is often staggering. For this reason, much effort has been devoted to the development of fast data-reduction methods and special computers. One approach has been to develop aids in the form of nomographic charts and various graphical constructions. In other cases, mechanical or electrical computing means have been developed. For detailed information in this area, the reader is referred to the *Proceedings of the Society for Experimental Stress Analysis* and to References 33, 34, 35, 36, 37 and 57 in the bibliography.

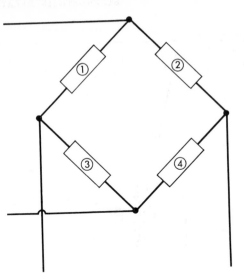

Fig. 11.30 Symbol standards for examples of Article 11.19.

11.19 PROPER ORIENTATION OF GAGES* AND INTERPRETATION OF RESULTS

In a given situation it is often possible to place gages in several different arrangements to obtain the desired data. Often there is a "best" way, however, and in certain cases unwanted components may be canceled by proper orientation. For example, it is often desirable to eliminate unintentional bending when direct axial loading is of primary interest. Or, perhaps only the bending component in a shaft is desired, and torsional-stress pickup must be eliminated.

The following discussion should be helpful in determining the proper orientation of gages and interpretation of results in a given problem. We will assume a *standard* bridge arrangement, as shown in Fig. 11.30, and gages will be numbered in the examples in accordance with this standard. When fewer than four measuring gages are indicated, it will be assumed that the remaining arms consist of fixed resistances, either as a part of the internal circuitry of the measuring system or provided externally in the bridge arrangement.

a) For axial loading

Example A. The simplest application uses a single measuring gage with an external compensating gage, as shown in Fig. 11.10. This arrangement measures strain at the test point, which may include a bending component due to eccentric loading. Temperature compensation is provided by the dummy gage in arm 2 of the bridge. Bridge constant = 1 (see Article 11.11b).

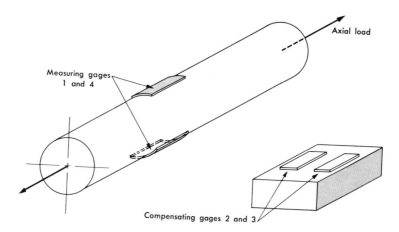

Fig. 11.31 Two gages arranged to measure axial load but to cancel bending.

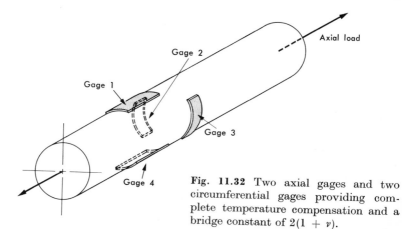

Fig. 11.32 Two axial gages and two circumferential gages providing complete temperature compensation and a bridge constant of $2(1 + \nu)$.

Example B. Figure 11.31 shows two measuring gages used in opposite arms of the bridge. If bending is present, its effect on gages 1 and 4 will be equal but of opposite sign. The bending component will therefore be canceled. Compensating gages 2 and 3 will eliminate temperature effects provided they are mounted on a material identical to that of the loaded member and the two members are kept at the same temperature. Bridge constant = 2.

Example C. In this bridge (Fig. 11.32) gages 1 and 4 are mounted in the axial direction and gages 2 and 3, often referred to as "Poisson's ratio gages," are mounted laterally. For a tensile load, as shown, gages 1 and 4 will react to the tensile strain, while gages 2 and 3 will sense a compressive strain resulting from the lateral contraction of the member. If Poisson's

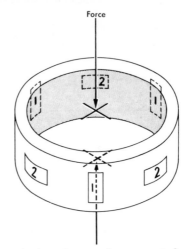

Fig. 11.33 Load cell employing three series-connected axial gages and three series-connected Poisson-ratio gages.

ratio is 0.3, the output of this arrangement will be 2.6 times the output of that for Example A and 1.3 times as great as that for Example B. Temperature compensation will be obtained and bending components eliminated. Bridge constant $= 2(1 + \nu)$, where $\nu =$ Poisson's ratio.

Example D. Figure 11.33 illustrates a load cell arrangement wherein gages are used in series. The three gages numbered 1 are connected in series and placed in the number 1 arm of the bridge, while the three gages numbered 2 are also connected in series and are placed in the number 2 arm of the bridge. This arrangement provides temperature compensation and eliminates any effect of eccentric loading.

At first glance it might be thought that the three gages in series would provide an output three times as great as that from a single gage under like conditions. Such is not the case, for it will be recalled that it is the percentage change in resistance, or dR/R, that counts, not dR alone. It is true that the resistance change for one arm, in this case, is three times what it would be for a single gage, but so also is the total resistance three times as great. Therefore the only advantage gained is that of *averaging* to eliminate incorrect readings resulting from eccentric loading. The remaining two arms (not shown in the figure) may be made up of either inactive strain gages or fixed resistors. Bridge constant $= (1 + \nu)$.

b) Bending

Example E. Figure 11.34 shows the most simple arrangement for measuring a bending stress. A single gage in bridge arm 1 is aligned on the beam as shown and a dummy gage is used for temperature compensation. In this

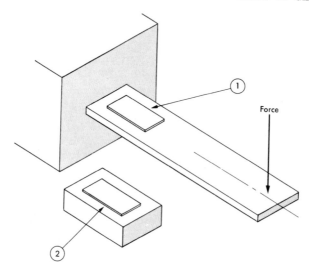

Fig. 11.34 Simple arrangement for measuring bending strain. Gage 2 provides temperature compensation.

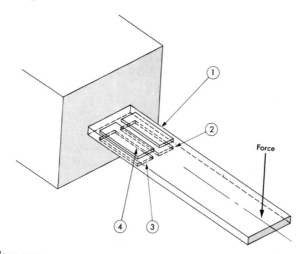

Fig. 11.35 Four-gage arrangement for measuring bending strain. Temperature compensation is provided, and strains from axial loads are eliminated.

case, the measuring gage is also sensitive to any axial load component. Bridge constant = 1.

Example F. Two- or four-arm bridges may be used for measuring bending strains, as shown in Figs. 11.11 and 11.35. In both instances temperature compensation is provided, and strain components caused by axial loads are canceled. (a) Bridge constant = 2. (b) Bridge constant = 4.

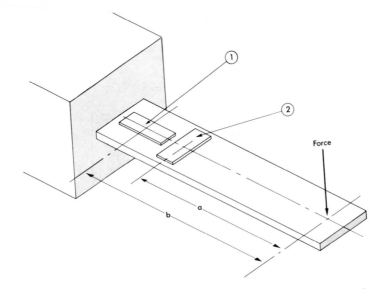

Fig. 11.36 Strain-gage configuration for measuring bending strain from one side only.

Example G. Figure 11.36 shows an arrangement that could be used if only one side of the bending member were available for gaging. Gage 1 would be mounted axially and gage 2 transversely. If the moments at the gage lines of 1 and 2 were the same, then the bridge output would be increased about 30% (Poisson's ratio) more than that in Example E above. In general, however, gage 2 would be mounted at a point of bending moment different from that of gage 1, and this fact would have to be accounted for. Temperature compensation would be provided; however, the effect of any axial load would not be canceled. Bridge constant $= [1 + (a/b)\nu]$.

c) Torsion

Example H. Figure 11.37 illustrates gage mounting that is sensitive to torque and bending but insensitive to axial loads. Bridge constant $= 2$.

Example I. By placing two gages on the same side of the shaft oriented as shown in Fig. 11.38, an arrangement is obtained that is sensitive to torque, but not to axial loads, and relatively insensitive to bending. If the bending moment at the gage lines of both gages were the same, the bending strains would be canceled. Otherwise only the difference in the moments at the two points would be indicated. Bridge constant $= 2$.

Example J. Figure 11.39 shows gage orientation that provides maximum sensitivity to torsion, eliminates strains caused by axial loading, and

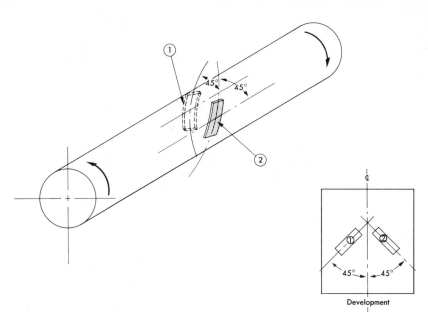

Fig. 11.37 Two-gage configuration for measuring torsion and providing temperature compensation.

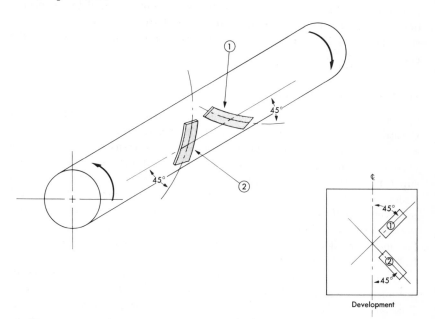

Fig. 11.38 Two-gage arrangement sensitive to torque and relatively insensitive to bending.

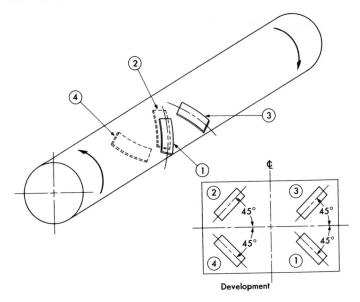

Fig. 11.39 Configuration sensitive to torque and insensitive to temperature, axial load, or bending.

minimizes bending strains. Bending effects may not be completely canceled, depending on the moment gradient along the shaft. It is therefore desirable that the gages be placed as closely together axially as possible. Bridge constant = 4.

11.20 SPECIAL PROBLEMS

a) Cross sensitivity. Wire strain-gage grids are arranged with the majority of the strain-sensitive filament aligned with the sensitive axis of the gage. However, unavoidably, a part of the grid wire is aligned transversely. The transverse portion of the grid senses the strain in that direction, and this effect is superimposed upon the longitudinal output. This is known as *cross sensitivity.* Unless it is corrected, it introduces an error in the output. The error is relatively small, seldom exceeding 2 or 3% in the worst cases, and the overall accuracy of many applications does not warrant accounting for it. For more detailed consideration of the problem, the reader is referred to References 32, 58, and 59.

Special gage designs minimize the problem by making the transverse portions of the grid much heavier in section than the longitudinal portions. This is accomplished very nicely in the case of "printed circuit" foil-type gages because of the unrestricted variation in cross section that is possible. A European design, designated as the *G-H strain gage* [37], minimizes cross sensitivity by fabricating the grid of two wire sizes. The transverse elements are of a relatively large-diameter wire, while the longitudinal strain-sensitive elements are of small section. In both examples, the heavier transverse elements are not appreciably strained, simply because the cement is incapable of supplying the required force.

b) Plastic strains and the post-yield gage. The average commercial strain gage will behave elastically to strain magnitudes as high as 2 to 3%. This represents a surprising performance when it is realized that the corresponding uniaxial elastic stress in steel would be almost 1,000,000 psi (if elastic conditions in the steel were maintained). It is not very great, however, when viewed by the engineer seeking strain information beyond the yield point. When mild steel is the strained material, strains as great as 10 to 15% may occur immediately following attainment of the elastic limit, before the stress again begins to climb above the yield stress. Hence, the usable strain range of the common resistance gage is quickly exceeded.

Gages known as *post-yield* gages have been developed, extending the usable range to approximately 10%. Grid material in very ductile condition is used, which is literally caused to flow with the strain in the test material. The primary problem, of course, in developing an "elastic-plastic" grid is to obtain a gage factor that is the same under both conditions. Data reduction presents special problems and for coverage of this aspect, the reader is referred to References 60 and 61.

c) Fatigue applications of resistance strain gages. Strain gages are subject to fatigue failure in the same manner as are other engineering structures. The same factors are involved in determining their fatigue endurance. In general, the vulnerable point is the discontinuity formed at the juncture of the grid proper and the lead wire to which the user makes connection. Of course, as with any fatigue problem, strain level is the most important factor in determining life.

An improvement in the endurance of wire gages has been obtained by the use of what are known as *dual-lead* gages. An intermediate-size wire formed into a loop is interposed between the grid and the lead wire. This reduces the magnitude of the discontinuity at the joint by providing what may be thought of as a stepped fillet. The construction markedly increases the endurance limit of the gage.

It has also been shown that isoelastic grid material performs better under fatigue conditions than does constantan and also that the carrier material is an important factor.

Figure 11.40 illustrates the effects of most of the factors discussed above.

d) Cryogenic temperature applications. Extreme cryogenic temperatures often cause relatively unpredictable performances of resistance strain gages. Adhesives and backings become glass-hard and quite brittle. While the

11.20

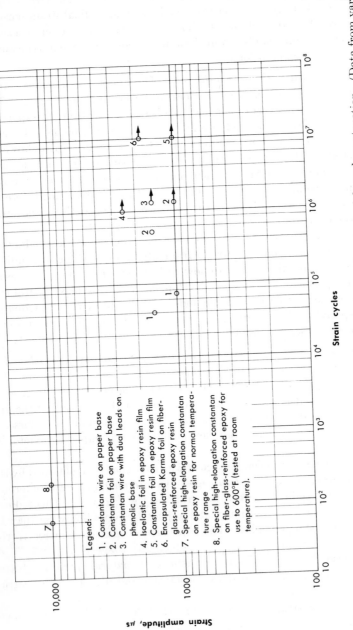

Fig. 11.40 Relationship of endurance limit to strain level for gages of various materials and construction. (Data from various sources including manufacturers' literature.) ◯→ indicates tests ended before failure.

mechanical properties of certain grid materials are drastically curtailed, those of others remain only slightly affected. Large changes in resistivities may be encountered with the effective values dependent to a great degree on trace elements and previous mechanical working of the materials. Much work is being conducted in this area and the state-of-the-art is rapidly changing. Even if all the temperature-related properties were known, however, there would still remain the difficult problem of either controlling the temperature or measuring it.

Telinde reports on a comprehensive evaluation of strain-gage use at temperatures as low as −452°F [62]. His work favors Karma as a grid material, supported on fiber-glass reinforced epoxy.

e) High-temperature applications. Maximum temperatures for short-period use of paper, epoxy, and glass-filled phenolic-base gages with appropriate cements are about 180°, 250° and 600°F, respectively. Primary limiting factors are decomposition of cement and carrier materials. At these temperatures grid materials present no particular problems. For applications at higher temperatures (to 1800°F) some form of ceramic-base insulation must be employed. The grid may be either of the strippable support, free-element type with the bonding as described below, or the gage may be of the "weldable" type.

Use of the free-element-type gage involves "constructing" the gage on the spot. Either brushable or flame-sprayed ceramic bonding materials are employed. Application of the former consists of laying down an insulating

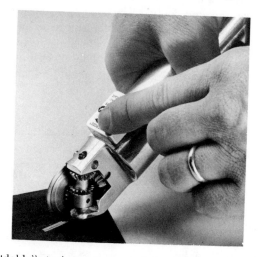

Fig. 11.41 "Weldable" strain gages consist of a ceramic insulated strain element encapsulated within a metal casing. The illustration shows a gage being attached to a flat surface through use of a "Rollectrode" welding apparatus. (Courtesy: Microdot Inc., South Pasadena, Calif.)

coating upon which the free-element grid is secured with more cement. The process demands considerable skill and carefully controlled baking or curing-temperature cycling.

Flame spraying involves the use of a plasma-type, oxyacetylene gun [63, 64]. Molten particles of ceramic are propelled onto the test surface and employed as both the cementing and insulating material for bonding the grid element to the test item. In both cases, leads must be attached by welding to provide the necessary high-temperature properties to the connections. Lead-wire temperature-resistance variations may also present problems.

It is obvious that considerable technique must be developed to satisfactorily use either of these types.

The "weldable" gage consists of a resistance strain element surrounded by a ceramic-type insulation and encapsulated within a metal casing (see Fig. 11.41). The gage is applied by spot-welding the edges to the test member [65].

A novel laser-based extensometer usable at 3500°F or higher and having an overall accuracy of ± 0.0002 in. over a 0.3 gage length is described in Reference [66].

f) Creep. Creep in the bond between gage and test surface is a factor sometimes ignored in strain-gage work. This problem is approximately diametrically opposite to the fatigue problem in that it is of importance only in static strain testing, primarily of the long-duration variety. For example, residual stresses are sometimes determined by measuring the dimensional relaxation as stressed material is removed. In this case, the strain is applied to the gage once and once only. The loading cycle cannot be repeated. Under these circumstances, gage creep will result in direct errors equal to the magnitude of the creep. If the load can be slowly cycled, the creep will appear as a hysterysis loop in the results. This effect is a function of several things, but is primarily determined by the strain level and the cement used for bonding.

TABLE 11.5

Cement	Creep, μs		
	After 10 hr	After 20 hr	After 30 hr
Bakelite	5	7	8
Duco	12	18	21
Armstrong A-1 epon with activator A	8	9	10

The following work will indicate the order of magnitudes that are encountered. C. H. Betts [39] of the Canadian Bureau of Mines has observed creep in the order of 35 μs at a strain level of 2400 μs over a period of 700 hours. Approximately 75% of this value took place during the first hour. R. F. Jordan [39a] of the General Electric Company observed the creep values shown in Table 11.5 corresponding to an initial strain level of 1500 μs.

SUGGESTED READINGS

Dally, J. W., and W. F. Riley, *Fundamental Stress Analysis*, New York. McGraw-Hill Book Co., 1965.

Dean, M. III, and R. D. Douglas, *Semiconductor and Conventional Strain Gages*. New York: Academic Press, 1962.

Dove, R. C., and P. H. Adams, *Experimental Stress Analysis and Motion Measurement*. Columbus, Ohio: Charles E. Merrill Books, Inc., 1964.

Durrelli, A. J., *Applied Stress Analysis*, Englewood Cliffs, N.J.: Prentice-Hall, Inc., 1967.

Hendry, A. W., *Elements of Experimental Stress Analysis*, New York: Pergamon Press, 1964.

Hetenyi, M., *Handbook of Experimental Stress Analysis*, New York: John Wiley & Sons, Inc., 1950.

Holister, G. S., *Experimental Stress Analysis*, Cambridge, England: Cambridge University Press, 1967.

Perry, C. C., and H. R. Lissner, *The Strain Gage Primer*, 2nd Ed. New York: McGraw-Hill Book Co., 1962.

PROBLEMS

11.1 A single strain gage is mounted to measure the axial strain in a simple tensile member. If the recorded strain is 380 μs, what is the axial stress
   a) if the member is of steel?
   b) if the member is of aluminum?

11.2 Prove that the readout from a bonded strain gage, subjected to a linear-strain gradient, corresponds to the strain at the center of the gage length (see Fig. 11.2).

11.3 A strain gage is centered along the length of a simply supported beam carrying a centrally positioned concentrated load. The beam is three gage lengths long. The strain in the beam will vary over the gage length, and the gage will indicate a mean strain. What correction factor should be applied to the gage reading to give maximum strain?

11.4 A foil strain gage has a gage factor of 1.90 and a resistance of 120.5 ohms. If a calibration resistor of 390,000 ohms is shunted across the gage, what equivalent strain should be indicated?

11.5 Two gages are mounted on a simple tensile specimen of steel. One is mounted axially and the other transversely. The two gages are connected in adjacent arms of a bridge. If the total bridge readout (based on single gage calibration) is 716 μs, what is the axial stress?

11.6 A simple tension member (minimum sectional area = 0.200 in.^{2}) is subjected to an axial load of 5100 lbs. Strains of 1620 and -455 microstrain are measured in the axial and transverse directions, respectively. Assuming that elastic conditions prevail, determine the values of Young's modulus and Poisson's ratio.

11.7 Four strain gages are mounted on a simple flat tensile specimen arranged for complete temperature compensation and maximum strain sensitivity when connected in the four arms of a bridge circuit. An 800,000-ohm calibration resistor is shunted across one of the gages. Gage resistances are each 118 ohms and the gage factors are 1.22.

   If the strain indicator readout is 150 divisions when the calibration switch is thrown and 240 divisions when the load is applied, what is the axial strain in the specimen?

11.8 What calibration resistor should be used in a strain-gage system to provide an equivalent strain of 150 μs? Gage resistance = 312 ohms and the gage factor is 3.50.

11.9 A two-element rosette is mounted on the shell of a cylindrical pressure vessel at a point well removed from any stress concentration or discontinuity. The gages are aligned with the circumferential and longitudinal directions. If the vessel is of steel and the circumferential and longitudinal strains are measured as 625 and 132 μs, respectively, what are the corresponding stresses? Sketch Mohr's circle for stresses. What is the ratio of circumferential to longitudinal stress? What is the ratio of strains?

11.10 A four-arm strain-gage bridge is mounted on a steel shaft (assume $E = 30 \times 10^6$ psi and $\nu = 0.3$), so that two gages sense maximum tensile strain and the other two sense maximum compressive strain. The gages are arranged to yield maximum strain output due to torque.
   a) Show how the gages should be mounted on the shaft and connected in a bridge circuit.
   b) If the reading obtained by means of a "direct reading" strain indicator (such as the BLH Type N, Fig. 11.18) is 1500 μs, what torsional stress will the shaft be subject to?

11.11 (a) through (j). Employing the strain data given in Table 11.6, calculate the principal strains, principal stresses, maximum shear stress, and principal stress directions, and sketch Mohr's circle of stresses. Also sketch an element similar to that shown in Fig. 11.29. Assume the stressed material is steel ($E = 30 \times 10^6$ psi and $\nu = 0.29$), and that a rectangular rosette is employed.

TABLE 11.6

Problem	ϵ_a	ϵ_b	ϵ_c
(a)	620	−200	410
(b)	1300	516	0
(c)	860	−1240	980
(d)	210	470	770
(e)	−816	0	0
(f)	2080	−1710	−865
(g)	−1100	−1440	−835
(h)	ϵ	ϵ	ϵ
(i)	−ϵ	ϵ	ϵ
(j)	−ϵ	−ϵ	ϵ

11.12 (a) through (j). Repeat Problem 11.11, substituting aluminum ($E = 10 \times 10^6$ psi and $\nu = 0.28$) for steel.

11.13 (a) through (j). Repeat Problem 11.11, assuming an equiangular rosette.

11.14 (a) through (j). Repeat Problem 11.12, assuming an equiangular rosette.

11.15 Strain readouts from a rectangular rosette are

$$\epsilon_a = 620, \qquad \epsilon_b = -200, \qquad \epsilon_c = 410.$$

Assume that under the above conditions, a delta rosette is mounted and that its a element is aligned with the direction of the a element of the above rectangular rosette. What readouts should be expected from the delta gage? (*Hint;* See Appendix A, for Mohr's circle for strain.)

11.16 Repeat Problem 11.15, using the data selected from Table 11.6.

11.17 In order to determine the power transmitted by a steel shaft with a 3-in. diameter, four strain gages are mounted as shown in Fig. 11.39. They are connected as a four-arm bridge, and the output is fed to a recording oscillograph. The gage resistance is 121.5 ohms, and G.F. is 1.95. The calibration resistor, R_s, has a value of 250,000 ohms. Figures 11.42 (a) and (b) show the calibration and strain records, respectively. Chart speed was 100 mm/sec.

Determine the extreme and mean values of the transmitted horsepower.

11.18 Four strain gages are mounted on a beam to provide maximum sensitivity to bending. All gages have a gage factor of 1.98 and a resistance of 121

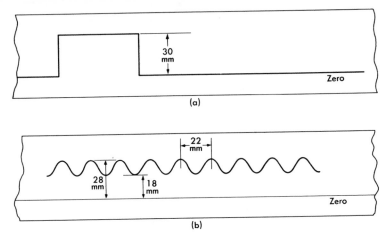

Figure 11.42

ohms. A 750,000 precision resistor is shunted across one of the gages for calibration.

If the strain-indicating meter reads 180 divisions when the calibration switch is closed and 310 divisions when a load is applied to the beam, what bending strain is measured?

11.19 Four strain gages, arranged for maximum sensitivity, are mounted on a shaft in torsion. The gages have a resistance of 121 ohms and a gage factor of 2.04. When a 750,000-ohm resistor is shunted across one gage for calibration, an oscillograph displacement of 2.2 cm is recorded. If straining the shaft causes a deflection of 3.2 cm, what maximum strain will be indicated? What will be the stress if the shaft is of steel?

11.20 A pressure transducer consists of a straight piece of thin walltube with strain gages mounted on the outer surface of the tube. The tube is of steel for which $E = 30,000,000$ psi and Poisson's ratio $= 0.3$. Two gages are mounted on the tube to sense maximum strain and two gages are mounted as compensating gages on an unstressed piece of steel of the same composition as the tube.

a) Show how the gages should be mounted on the tube and how the four gages should be connected in the bridge for maximum output.

b) If the reading obtained by a strain indicator is 1200 μs, and assuming the usual 2/1 stress ratio exists, determine the maximum stress on the outer surface of the tube. (It should be noted that the strain ratio will *not* be 2/1, but that the ratio of strains can be calculated.)

CHAPTER 12

MEASUREMENT OF FORCE AND TORQUE

12.1 MASS, WEIGHT, AND FORCE

Along with displacement and time, *force* and *mass* are physical quantities fundamental to engineering. Basically, mass may be defined as a measure of quantity of matter, and force may be defined as a measure of the gravitational attraction between masses. Weight is defined as the gravitational attraction between the mass of a given item and a specific second mass, the earth. Gravitational attraction between two bodies varies directly as the product of the magnitudes of the two masses and inversely as the square of the relative displacement of their centers of gravity. Hence weight will vary from point to point over the earth's surface. Mass, or quantity of matter, on the other hand, is invariable.

Mass and weight are related through Newton's second law of motion, which may be stated as follows: *The acceleration of a particle is directly proportional to and in the same direction as the resultant applied force.* This may be expressed as

$$\frac{F_1}{a_1} = \frac{F_2}{a_2}, \tag{12.1}$$

when F_1 and F_2 are two different forces applied to a given particle of mass, and a_1 and a_2 are the resulting accelerations.

A most convenient force to apply is the earth's gravitational attraction for the body or particle, which is the weight, W. Then the resulting acceleration is that of the falling body *in vacuo* at the particular location. Both the weight and the gravitational acceleration will vary from location to location. Their ratio, however, remains constant and proportional to the mass, M, of the particle. Hence we may write

$$\frac{F_1}{a_1} = \frac{F_2}{a_2} = \frac{W}{g} = CM, \tag{12.1a}$$

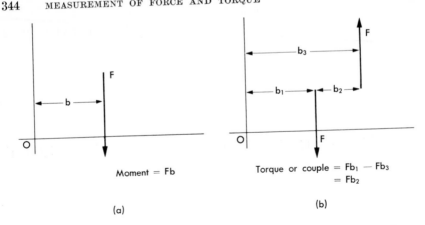

Moment = Fb

(a)

Torque or couple = Fb₁ — Fb₃
= Fb₂

(b)

Fig. 12.1   (a) Definition of moment.   (b) Definition of torque or couple.

where C is a proportionality factor whose value depends upon the units used. For the usual engineering applications, C is equal to unity.

As may be seen from Eq. (12.1a), the ratio W/g is *not* indispensable to the application of Newton's second law. Any ratio F/a, where F is an applied force and a is the resulting acceleration, is just as valid as W/g. The ratio W/g is particularly convenient, however, because g is relatively constant over the surface of the earth and need not be measured for each and every application. The required proportionality may be established simply by determining gravitational force (weighing) alone, rather than by measuring both a force and resulting acceleration. For most engineering applications, the standard value of g is 32.1739 ft/sec², or 386.087 in./sec².

Force, in addition to its effect along its line of action, may exert a turning effect relative to any axis other than those intersecting the line of action. Such a turning effect is variously termed *torque*, *moment*, or *couple*, depending upon the manner in which it is produced. The term *moment* is applied to conditions such as those illustrated in Fig. 12.1(a), while the terms *torque* and *couple* are applied to conditions involving counterbalancing forces, such as shown in Fig. 12.1(b).

Mass standards. As stated previously (Article 4.4), the fundamental unit of mass is the kilogram, equal to the mass of the International Prototype Kilogram located at Sèvres, France. A gram is defined as a mass equal to one-thousandth of the mass of the International Prototype Kilogram. The commonly used avoirdupois* pound is 0.453,592,37 kilograms, as agreed to in 1959 (Article 4.4). Various classifications and tolerances for laboratory standards are recommended by the National Bureau of Standards [1] [21].

---

* From the French, meaning "goods of weight."

12.2 MEASURING METHODS

As in other areas of measurement, there are two basic approaches to the problem of force and weight measurement [2]: (a) direct comparison, and (b) indirect comparison through use of calibrated transducers. Directly comparative methods employ some form of beam balance using a null-balance technique. If the beam neither amplifies nor attenuates, the comparison is *direct*. The simple analytical balance is of this type. Often, however, as in the case of a platform scale, the force is attenuated through a system of levers so that a smaller weight may be used to *balance* the unknown, with the variable in this case being the magnitude of attenuation. This method requires calibration of the system.

It should be pointed out that, strictly speaking, the beam balancing methods *compare masses*, whereas the calibrated transducers sense gravitational attraction, or weight.

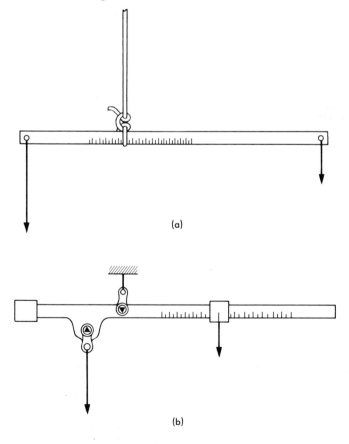

(a)

(b)

Fig. 12.2 (a) Danish steelyard. (b) Roman steelyard.

12.3 MECHANICAL WEIGHING SYSTEMS

Mechanical weighing systems originated in Egypt, and were probably used as early as 5000 B.C. [3]. The earliest devices were of the cord and *equal-arm* type, traditionally employed to symbolize justice. *Unequal-arm* balances were apparently first used in the form shown in Fig. 12.2(a). This device, called a Danish steelyard, was described by Aristotle (384–322 B.C.) in his *Mechanics*. Balance is accomplished by moving the beam through the loop of cord, which acts as the fulcrum point, until balance is obtained. A later unequal-arm balance, the Roman steelyard, which employed fixed pivot points and movable balance weights, is still in use today [Fig. 12.2(b)].

a) The analytical balance. Probably the simplest weight or force measuring system is the ordinary equal-arm beam balance (Fig. 12.3). Basically this is a device which operates on the principle of *moment comparison*. The moment produced by the unknown weight, mass, or force is compared with that produced by a known mass. When null balance is obtained, the two weights are equal, provided the two arm lengths are identical. A check on arm equivalence may easily be made by simply interchanging the two weights. If balance was initially achieved and if it is maintained after exchanging the weights, it can only be concluded that the weights are equal, as also are the

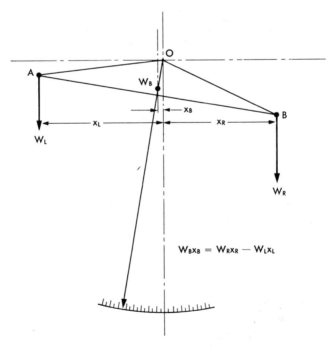

Fig. 12.3 Requirement for equilibrium for an analytical balance.

arm lengths. This method for checking the true null of a system is known as the method of *symmetry*.

A common example of the equal-arm balance is the analytical scale used principally in chemistry and physics. Devices of this type have been constructed with capacities as high as 400 lb, having sensitivities of 0.0002 lb [4]. In smaller sizes the analytical balance may be constructed to have sensitivities of 0.001 mg. Some of the factors governing operation of this type of balance were discussed in Article 2.5.

b) Multiple-lever systems. When large weights are to be measured, neither the equal-arm nor the simple unequal-arm balance is adequate. In such cases, multiple-lever systems, shown schematically in Fig. 12.4, are often

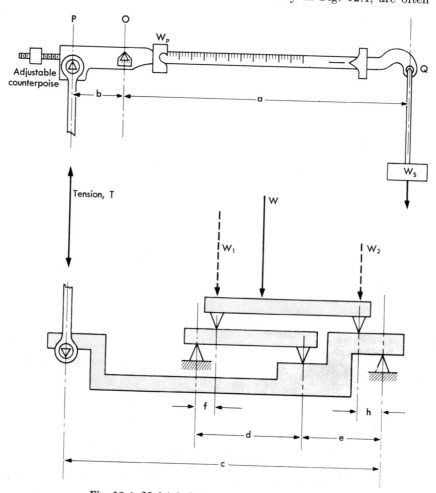

Fig. 12.4 Multiple-lever system for weighing.

employed. With such systems, large weights W may be measured in terms of much smaller weights W_p and W_s. Weight W_p is termed the *poise weight* and W_s is the *pan weight*. An adjustable counterpoise is used to obtain an initial zero balance.

We will assume for the moment that W_p is at the zero beam graduation, that the counterpoise is adjusted for initial balance, and that W_1 and W_2 may be substituted for W. With W on the scale platform and balanced by a pan weight W_s, we may write the relations

$$T \times b = W_s \times a \tag{12.2}$$

and

$$T \times c = W_1 \frac{f}{d} e + W_2 h. \tag{12.3}$$

Now if we proportion the linkage such that

$$\frac{h}{e} = \frac{f}{d},$$

then

$$T \times c = h(W_1 + W_2) = hW. \tag{12.3a}$$

From this we see that W may be placed anywhere on the platform and that its position relative to the platform knife-edges is immaterial.

Solving for T in Eqs. (12.2) and (12.3a) and equating, yields

$$\frac{W_s a}{b} = \frac{W h}{c}$$

or

$$W = \frac{a}{b} \frac{c}{h} W_s = R W_s. \tag{12.4}$$

The constant

$$R = \frac{a}{b} \frac{c}{h}$$

is the scale *multiplication ratio*.

Now if the beam is divided with a scale of u lb/in., then a poise movement of v inches should produce the same result as a weight W_p placed on the pan at the end of the beam. Hence,

$$W_p v = u v a \qquad \text{or} \qquad u = \frac{a}{W_p}.$$

This relation determines the required scale divisions on the beam for any poise weight W_p.

Dynamic response of a scale of this sort is a function of the natural frequency and damping. The natural frequency will be a function of the

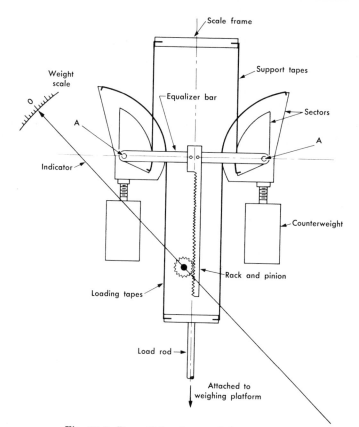

Fig. 12.5 Essentials of a pendulum scale.

moving masses, multiplication ratio, and the restoring forces. The latter are determined by the relative vertical placement of the pivot points, primarily those of the balance beam, O, P, and Q. If O is below a line drawn from P to Q, then the beam will be unstable, and balance will be unattainable. Pivot O is normally above line PQ, and as the distance above the line is increased, the natural frequency and sensitivity are both reduced.

c) The pendulum force-measuring mechanism. Another type of moment-comparison device used for measurement of force and weight is shown in Fig. 12.5. This is often referred to as a *pendulum scale*. Basically, the pendulum mechanism is a force-measuring device of the multiple-lever type, with the fixed-length levers replaced by ribbon- or tape-connected sectors. The input, either a direct force or a force proportional to weight and transmitted from a suitable platform, is applied to the load rod. As the load is applied, the sectors rotate about points A, as shown, moving the counter-

TABLE 12.1

Elastic element	Deflection equation	Deflection constant K_1, lb/in.
A F = Load, lb L = Length, in. A = Cross-sectional area, in.^{2} y = Deflection at load, in. E = Young's modulus, psi	$y = \dfrac{FL}{AE}$	$K = \dfrac{AE}{L}$
B F = Load, lb L = Length, in. E = Young's modulus, psi I = Moment of inertia, in.^{4}	$y = \dfrac{1}{48}\dfrac{FL^3}{EI}$	$K = \dfrac{48EI}{L^3}$
C F = Load, lb L = Length, in. E = Young's modulus, psi I = Moment of inertia, in.^{4}	$y = \dfrac{1}{3}\dfrac{FL^3}{EI}$	$K = \dfrac{3EI}{L^3}$

	Diagram	Symbols	Deflection	Stiffness
D		F = Load, lb D_m = Mean coil dia., in. N = Number of coils E_s = Shear modulus, psi D_w = Wire dia., in.	$y = \dfrac{8FD_m^3 N}{E_s D_w^4}$	$K = \dfrac{E_s D_w^5}{8 D_m^3 N}$
E		F = Load, lb D = Dia. of ring, in. E = Young's modulus, psi I = Moment of inertia of section about centroidal axis of bending section, in.[4]	$y = \dfrac{1}{16}\left(\dfrac{\pi}{2}-\dfrac{4}{\pi}\right)\dfrac{FD^3}{EI}$	$K = \dfrac{16}{(\pi/2)-(4/\pi)}\left(\dfrac{EI}{D^3}\right)$
F		K_1 = Defl. const. of member 1, lb/in. K_2 = Defl. const. of member 2, lb/in. F = Load, lb	$y = \dfrac{F}{K_1 + K_2}$	$K = K_1 + K_2$
G		K_1 = Defl. const. of member 1, lb/in. K_2 = Defl. const. of member 2, lb/in. F = Load, lb	$y = F\left(\dfrac{1}{K_1}+\dfrac{1}{K_2}\right)$	$K = \dfrac{1}{(1/K_1)+(1/K_2)}$

weights outward. This movement increases the counterweight effective
moment until the load and balance moments are equalized. Motion of the
equalizer bar is converted to indicator movement by a rack and pinion, the
sector outlines being proportioned to provide a linear dial scale. This device
may be applied to many different force-measuring systems, including dynamometers (Article 12.7).

12.4 ELASTIC TRANSDUCERS

Many force-transducing systems make use of some mechanical elastic member
or combination of members. Application of load to the member results in an
analogous deflection, usually linear. The deflection is then observed directly
and used as a measure of force or load, or a secondary transducer is used to
convert the displacement into another form of output, often electrical.

Most force-resisting elastic members adhere to the relation

$$K = \frac{F}{y}, \tag{12.5}$$

in which

$$F = \text{applied load, lb,}$$

$$y = \text{resulting deflection, in.,}$$

$$K = \text{deflection constant, lb/in.}$$

To determine the value of the deflection constant of an element, it is only
necessary to write the deflection equation, and if the deflection is a linear
function of the load, K may be found. Table 12.1 lists representative relations
indicating the general form.

Design detail of the detector-transducer element is largely a function of
capacity, required sensitivity, the nature of any secondary transducer, and
depends on whether the input is static or dynamic. Although it would be
impossible to discuss all situations, there are several general factors we may
consider.

It is normally desirable that the detector-transducer be as sensitive as
possible; i.e., maximum output per unit input should be obtained. This
would require an elastic member that deflects considerably under load,
indicating as low a value of K as possible. There are usually conflicting
factors, however, with the final design being a compromise. For example,
if we were to measure rolling-mill loads by placing cells between the screw-
down and bearing blocks, our application could scarcely tolerate a *springy*
load cell, i.e., one that deflected considerably under load. It would be necessary to construct a stiff cell at the expense of elastic sensitivity and then
attempt to make up for the loss by using as sensitive a secondary transducer
as possible.

Another factor involving sensitivity is response time, or time required to come to equilibrium. This is a function of both damping and natural frequency (see Article 2.6). Fast response corresponds to high natural frequency, requiring a stiff elastic member.

Stress, also, may be a limiting factor in any loaded member. It is especially important that the stresses remain below the elastic limit, not only in gross section, but also at every isolated point. In this respect residual stresses are often of significance. While load stresses may be well below the elastic limit for the material, it is possible that when they are added to *locked-in* stresses, the total may be too great. Even though such a situation occurs only at a single isolated point, hysteresis and nonlinearity will result.

Accuracy and repeatability are absolute requirements, and the factors discussed in Chapter 2 apply. Manufacturing tolerances are yet another factor of importance in design and application of elastic load elements. These are discussed in some detail in Article 7.8.

a) Calibration adjustment. Various calibration adjustments may be made to account for variation in characteristics of elastic load members. Sometimes a simple check at the time of assembly and the selection of one of several standard scale graduations may suffice. In the example of a weighing spring design in Article 7.8, errors of almost $\pm 12\%$ were found to be possible from quite reasonable dimensional tolerances. Optimum maximum deflection was 4 in. By providing three different face plates, A, B, and C, for the *nominal*, the *strong*, and the *weak* ranges, the error could be reduced to a maximum of about $\pm 4\%$. This scheme is often used, not only for load-measuring devices, but for all varieties of inexpensive instruments employing a scale. It does not provide for calibration adjustment in use, however.

When coil springs are used, means are sometimes provided to adjust the number of effective coils through use of an end connection that may be screwed into or out of the spring, thereby changing the number of *active* coils and hence the stiffness of the spring. In other cases, the springs may be purposely overdesigned with regard to stress, and the number of coils specified so that in no case may the tolerances add up to give a spring that is too flexible. Then at the time of assembly the springs are buffed on a wheel to obtain the required deflection constant.

If a secondary transducer is used, we may be able to provide for calibration by making adjustment in its characteristics. As an example, we could use a voltage-dividing potentiometer to sense the load deflection of the spring just discussed. We might do this to provide remote indication or recording. A circuit arrangement could be used in which an adjustable series resistor would be employed to provide calibration for the complete system.

Figure 12.6 illustrates a precision force indicator in which the spring system is held to an overall tolerance of $\pm 0.2\%$ although the tolerances

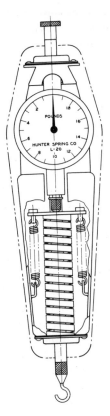

Fig. 12.6 A precision force indicator. (Courtesy Hunter Spring Company, Lansdale, Pa.)

on the individual items are no closer than $\pm 2\%$. This is accomplished by use of two small *compensator* springs selected at the time of assembly. Although it adds no more than 10% to the total spring stiffness, its selection can be made to correct for quite small overall variations from nominal.

While the above discussion is primarily in terms of load-sensing elements of the coil-spring type, the principles involved are applicable to most elastic force-sensing elements of other designs. Proving rings and load cell systems to be discussed in subsequent articles are all susceptible to mechanical and electrical adjustment of the kinds discussed above.

b) The proving ring. This device has long been the *standard* for calibrating tensile-testing machines and is, in general, the means whereby accurate measurement of large static loads may be obtained. Figure 12.7 shows the construction of a compression-type ring. Capacities usually fall in the range of from 300 to 300,000 lb [5].

Here, again, deflection is used as the measure of applied load, with the deflection measured by means of a precision micrometer. Repeatable micrometer settings are obtained with the aid of a vibrating reed. In use, the reed A is plucked (electrically driven reeds are also available), and the microm-

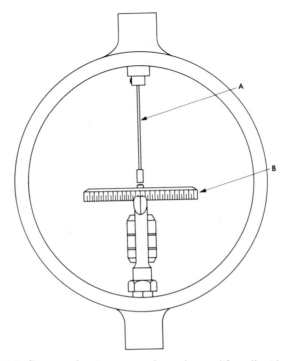

Fig. 12.7 Compression-type proving ring with vibrating reed.

eter spindle B is advanced until contact is indicated by the marked damping of the vibration. Although different operators may obtain somewhat different individual readings, consistent differences in readings still will be obtained provided both zero and loaded readings are made by the same person. With 40 to 64 micrometer threads per inch, readings may be made to one- or two-hundred thousandths of an inch [5].

The equation given in Table 12.1 for circular rings is derived with the assumption that the radial thickness of the ring is small compared with the radius. Most proving rings are made of section with appreciable radial thickness. However, Timoshenko [6] shows that use of the thin-ring rather than the thick-ring relations introduces errors of only about 4% for a ratio of section thickness to radius of $\frac{1}{2}$. Increased stiffness in the order of 25% is introduced by the effects of integral bosses [5]. It is, therefore, apparent that use of the simpler thin-ring equation is normally justified.

Stresses may be calculated from the bending moments M determined by the relation [6]

$$M = \frac{PR}{2}\left(\cos \phi - \frac{2}{\pi}\right). \tag{12.6}$$

Symbols correspond to those shown in Fig. 12.8.

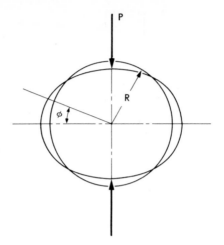

Fig. 12.8 Ring loaded diametrically in compression.

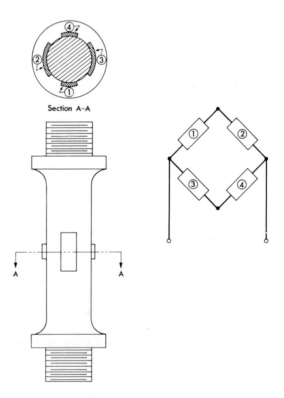

Fig. 12.9 Tension-compression resistance strain-gage load cell.

c) Strain-gage load cells. Instead of using total deflection as a measure of load, the strain-gage load cell measures load in terms of *unit* strain. Resistance gages are very suitable for this purpose (see Chapter 11). One of the many possible forms of elastic member is selected, and the gages are mounted to provide maximum output. If the loads to be measured are large, the direct tensile-compressive member may be used. If the loads are small, strain amplification provided by bending may be employed to advantage.

Figure 12.9 illustrates the arrangement for a tensile-compressive cell using all four gages sensitive to strain and providing temperature compensation for the gages. The bridge constant (Article 11.11b) in this case will be $2(1 + \nu)$ where ν is Poisson's ratio for the material. Compression cells of this sort have been used with a capacity of 3 million pounds [8]. Simple beam arrangements may also be used, as illustrated in Figs. 11.11 and 11.35.

Figures 12.10(a) and (b) illustrate proving-ring strain-gage load cells. In Fig. 12.10(a) the bridge output is a function of the bending strains only, the axial components being canceled in the bridge arrangement. By mounting the gages as shown in Fig. 12.10(b), somewhat greater sensitivity may be obtained because the output includes both the bending and axial components sensed by gages 1 and 4.

d) Temperature sensitivity. The sensitivity of elastic load-cell elements is affected by temperature variation. This change is caused by two factors: variation in Young's modulus and altered dimensions. Variation in Young's modulus is the more important of the two effects, amounting to roughly

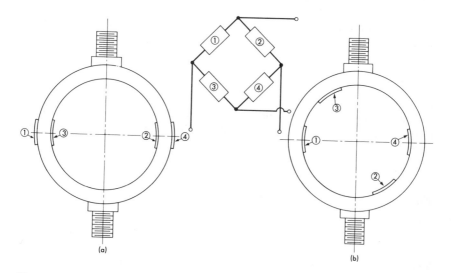

(a)   (b)

Fig. 12.10 Two arrangements of circular-shaped load cells employing resistance strain gages as secondary transducers.

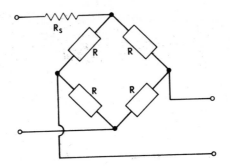

Fig. 12.11 Schematic diagram of a strain-gage bridge with a compensation resistor.

$2\frac{1}{2}\%$ per 100°F. On the other hand, the increase in cross-sectional area of a tension member of steel will amount to only about 0.15% per 100°F change.

Obviously, when accuracies of $\pm\frac{1}{2}\%$ are desired, as provided by certain commercial cells, a means of compensation, particularly for variation in Young's modulus, must be supplied. When resistance-strain gages are used as secondary transducers, this is accomplished electrically by causing the bridge's electrical sensitivity to change in the opposite direction to the modulus effect [9]. As temperature increases, the deflection constant for the elastic element decreases; it becomes more *springy*, and deflects a greater amount for a given load. This increased sensitivity is offset by reducing the sensitivity of the strain-gage bridge through use of a thermally sensitive compensating resistance element, R_s, as shown in Fig. 12.11.

As discussed in Article 7.18e, the introduction of a resistance in an input-lead reduces the electrical sensitivity of an equal-arm bridge by the factor expressed as follows:

$$n = \frac{1}{1 + (R_s/R)}.$$

Requirements for compensation may be analyzed through use of the relation for the initially balanced equal-arm bridge, Eq. (7.44). If we assume

$$2\frac{\Delta R}{R} \ll 4,$$

Eq. (7.44) may be modified to read

$$\frac{\Delta e_o}{e_i} = \frac{k}{4}\frac{\Delta R}{R}.$$

This is true, particularly for a *strain-gage bridge* for which $\Delta R/R$ is always small. A bridge constant, k, is included to account for use of more than one active gage. If all four gages are equally active, $k = 4$. For the arrangement shown in Fig. 12.9, $k = 2(1 + v)$, where v is Poisson's ratio. If we account for the compensating resistor, the equation will then read

$$\frac{\Delta e_o}{e_i} = \frac{k}{4} \frac{\Delta R}{R} \left[\frac{1}{1 + (R_s/R)} \right]. \qquad (12.7)$$

Rewriting Eq. (11.8),

$$\epsilon = \left(\frac{1}{F} \right) \left(\frac{\Delta R}{R} \right),$$

and from the definition of Young's modulus, E, Eq. (11.2),

$$P = EA\epsilon,$$

we may solve for sensitivity,

$$\frac{\Delta e_o}{P} = \left(\frac{e_i}{4} \right) \left(\frac{FRk}{A} \right) \left[\frac{1}{E(R + R_s)} \right]. \qquad (12.8)$$

If it is assumed that the gages are arranged for compensation of resistance variation with temperature and that the gage factors F remain unchanged with temperature, and, further, that any change in the cross-sectional area of the elastic member may be neglected, then complete compensation will be accomplished if the quantity $E(R + R_s)$ remains constant with temperature.

Using Eqs. (7.20) and (7.28), we may write

$$E(R + R_s) = E(1 + c \, \Delta T)[R + R_s(1 + b \, \Delta T)], \qquad (12.9)$$

from which we find

$$\frac{R_s}{R} = - \frac{c}{b + c} . \qquad (12.10)$$

This indicates that temperature compensation may possibly be accomplished through proper balancing of the temperature coefficients of Young's modulus, c, and electrical resistivity, b. Because c is usually negative (see Table 7.1) and because the resistances cannot be negative, it follows that

$$b > -c.$$

In addition, we may write [see Eq. (6.2)]

$$R_s = \rho \frac{L}{A} = -R \left(\frac{c}{b + c} \right), \qquad (12.11)$$

from which

$$L = - \frac{RA}{\rho} \left(\frac{c}{b + c} \right). \qquad (12.11a)$$

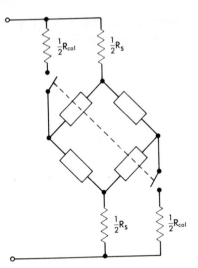

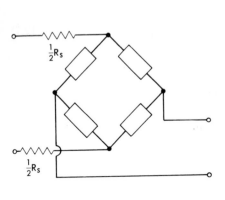

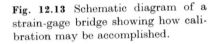

Fig. 12.12 Strain-gage bridge with two compensation resistors.

Fig. 12.13 Schematic diagram of a strain-gage bridge showing how calibration may be accomplished.

From these relations, specific requirements for compensation may be derived. After a resistance material, usually in the form of wire, is selected, the required length may be determined through use of Eq. (12.11a).

Although a single resistor would serve, commercial cells normally use two modulus resistors, as shown in Fig. 12.12. This assures proper connections regardless of instrumentation and also permits electrical calibration of the gages by shunt resistances as described in Article 11.13. It is necessary, however, to use two calibration resistors as shown in Fig. 12.13. If each resistor is considered as one-half the total calibration resistance, then the relation given, Eq. (11.12), will remain legitimate.

12.5 BALLISTIC WEIGHING

Theoretically, if a mass is suddenly applied to a resisting member having a linear load-deflection characteristic, the dynamic deflection will be exactly twice the final static deflection. This is true so long as damping is absent. This fact may be used as the basis for a weighing system [2]. The basic equation for a system of this type is

$$\frac{W}{g}\frac{d^2y}{dt^2} + Ky = W, \qquad (12.12)$$

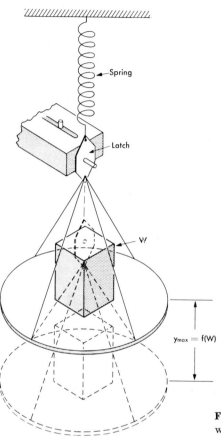

Fig. 12.14 A ballistic weighing system.

in which W = weight, lb,

K = deflection constant, lb/in.,

g = gravity acceleration = 386 in./sec²,

y = deflection, in.,

t = time, sec.

A solution of this equation is

$$y = \frac{W}{K}(1 - \cos \omega t), \qquad (12.12a)$$

for which the maximum value is

$$y_0 = \frac{2W}{K} = 2y_{\text{static}} \qquad (12.13)$$

when

$$t = \frac{\pi}{\omega}.$$ (12.13a)

The period of oscillation will be

$$T = 2\pi\sqrt{W/Kg}.$$ (12.14)

In operation, the platform is locked (Fig. 12.14), then the weight to be measured is put in place, the system is unlocked, and the maximum excursion is measured. If damping is minimized, the maximum displacement will be linearly proportional to the weight and can be used to measure the weight. Of course, the system is useful only for weight measurement and cannot be used to measure force.

12.6 HYDRAULIC AND PNEUMATIC SYSTEMS

If a force is applied to one side of a piston or diaphragm, and a pressure, either hydraulic or pneumatic, is applied to the other side, some particular value of pressure will be necessary to exactly balance the force. This is the principle upon which hydraulic and pneumatic load cells are based.

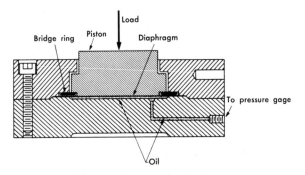

Fig. 12.15 Section through a hydraulic load cell.

For hydraulic systems, conventional piston and cylinder arrangements may be employed. However, the friction between piston and cylinder wall and required packings and seals is unpredictable, making good accuracy difficult to obtain. Use of a *floating* piston with a diaphragm-type seal practically eliminates this variable.

Figure 12.15 shows a hydraulic cell in section. This is similar to the type employed in some materials-testing machines. The piston does not actually contact a cylinder wall in the usual sense, but a thin elastic diaphragm, or bridge ring, of steel is used as the positive seal, which allows small piston movement. Mechanical stops prevent the seal from being overstrained.

When force acts on the piston, the resulting oil pressure is transmitted to some form of pressure-sensing system such as the simple Bourdon gage. If the system is completely filled with fluid, very small transfer or flow will be required. Piston movement may be less than 0.002 in. at full capacity. In this respect, at least, the system will have good dynamic response; however, overall response will be determined very largely by the response of the pressure-sensing element.

Very high capacities and accuracies are possible with cells of this type. Capacities to 5,000,000 lb and accuracies in the order to $\pm\frac{1}{2}\%$ of reading or $\pm\frac{1}{10}\%$ of capacity, whichever is greater, have been attained [10]. Since hydraulic cells are somewhat sensitive to temperature change, provision should be made for adjusting the zero setting. Temperature changes during the measuring process cause errors of about $\frac{1}{4}\%$ per 10°F change [10].

Pneumatic load cells [11, 12, 13] are quite similar to hydraulic cells in that the applied load is balanced by a pressure acting over a resisting area, with the pressure becoming a measure of the applied load. However, in addition to using air rather than liquid as the pressurized medium, these cells differ from the hydraulic ones in several other important respects.

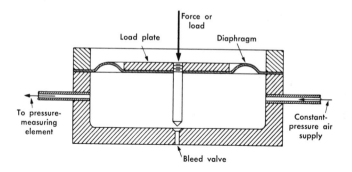

Fig. 12.16 Section through a pneumatic load cell.

Pneumatic load cells commonly employ diaphragms of a flexible material rather than pistons, and they are designed to automatically regulate the balancing pressure. A typical arrangement is shown in Fig. 12.16. Air pressure is supplied to one side of the diaphragm and allowed to escape through a position-controlling *bleed* valve. The pressure under the diaphragm, therefore, is controlled both by source pressure and bleed-valve position. The diaphragm seeks the position that will result in just the proper air pressure to support the load. This, of course, assumes that the supply pressure is great enough so that its value multiplied by the effective area will at least support the load.

It is seen that as the load changes magnitude, the measuring diaphragm must change its position slightly. Unless care is used in the design, a non-linearity may result, the cause of which may be made clear by reference to Fig. 12.17(a). As the diaphragm moves, the portion between the load plate and the fixed housing will alter position as shown. If it is assumed that the diaphragm is of a perfectly flexible material, incapable of transmitting any but tensile forces, then the division of vertical load components transferred to housing and load plate will occur at points A or A', depending on dia-phragm position. It is seen then that the effective area will change, depending on the geometry of this portion of the diaphragm. If a complete semicircular roll is provided, as shown in Fig. 12.17(b), this effect will be minimized.

Since simple pneumatic cells may tend to be dynamically unstable, most commercial types provide some form of viscous damper to minimize this tendency. Also, additional chambers and diaphragms may be added to provide for *tare* adjustment.

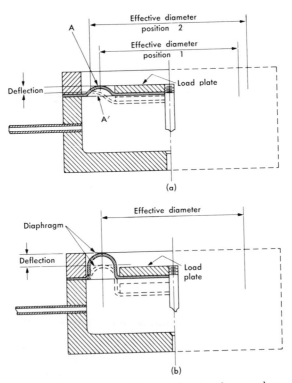

Fig. 12.17 (a) Section through a diaphragm showing how a change in effective area may take place. (b) When sufficient "roll" is provided, the effective area remains constant.

Single-unit capacities to 80,000 lb may be had, and by use of parallel units practically any total load or force may be measured. Errors as small as 0.1% of full scale may be expected.

12.7 TORQUE MEASUREMENT

Torque measurement is often associated with determination of mechanical power, either power required to operate a machine or power developed by the machine. In this connection, torque-measuring devices are commonly referred to as *dynamometers*. When so applied, both torque and angular speed must be determined. Another important reason for measuring torque is to obtain load information necessary for stress or deflection analysis.

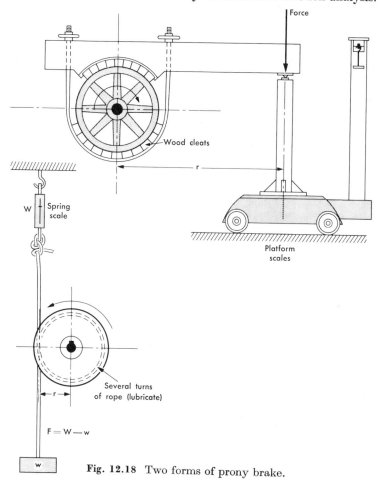

Fig. 12.18 Two forms of prony brake.

There are three basic types of torque-measuring apparatus, namely, absorption, driving, and transmission dynamometers. *Absorption dynamometers* dissipate mechanical energy as torque is measured; hence they are particularly useful for measuring power or torque developed by power sources such as engines or electric motors. *Driving dynamometers*, as their name indicates, both measure torque or power and also supply energy to operate the tested devices. They are, therefore, useful in determining performance characteristics of such things as pumps and compressors. *Transmission dynamometers* may be thought of as passive devices placed at an appropriate location within a machine or between machines, simply for the purpose of sensing the torque at that location. They neither add to nor subtract from the transmitted energy or power, and are sometimes referred to as *torque meters*.

a) Mechanical and hydraulic dynamometers. Probably the simplest type of absorption dynamometer is the familiar *prony brake*, which is strictly a mechanical device depending on dry friction for converting the mechanical energy into heat. There are many different forms, two of which are shown in Fig. 12.18.

Another form of dynamometer operating on similar principles is the *water brake*, which uses fluid friction rather than dry friction for dissipating the input energy. Figure 12.19 shows this type of dynamometer in its simplest

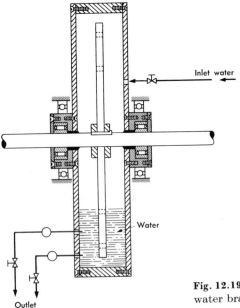

Inlet water

Water

Outlet

Fig. 12.19 Section through a typical water brake.

form. Capacity is a function of two factors, speed and water level. Power absorption is approximately a function of the *cube* of the speed, and the absorption at a given speed may be controlled by adjustment of the water level in the housing. This type of dynamometer may be made in considerably larger capacities than the simple prony brake because the heat generated may be easily removed by circulating the water into and out of the casing. Trunnion bearings support the dynamometer housing, allowing it freedom to rotate except for restraint imposed by a reaction arm.

In each of the above devices the power-absorbing element tends to rotate with the input shaft of the driving machine. In the case of the prony brake, the absorbing element is the complete brake assembly, while for the water brake it is the housing. In each case such rotation is constrained by a force-measuring device, such as some form of scales or load cell, placed at the end of a reaction arm of radius r. By measuring the force at the known radius, the torque T may be computed by the simple relation

$$T = Fr. \tag{12.15}$$

If the angular speed of the driver is known, power may be determined from the relation

$$P = 2\pi(T) \text{ (rpm)} \tag{12.16}$$

or, in terms of horsepower,

$$\text{bhp} = \frac{2\pi Fr \text{ (rpm)}}{33,000}, \tag{12.16a}$$

where

$T =$ torque, ft·lb,

$F =$ force measured at radius r, lb,

$r =$ length of reaction arm, ft,

$P =$ power, ft·lb/min,

bph $=$ brake horsepower,

rpm $=$ revolutions per minute.

b) Electric dynamometers. Almost any form of rotating electric machine can be used as a driving dynamometer, or as an absorption dynamometer, or as both. Of course, those designed especially for the purpose are most convenient to use. Four possibilities may be listed as follows: (1) eddy-current dynamometers, (2) d-c dynamometers or generators, (3) d-c motors and generators, (4) a-c motors and generators.

Eddy-current dynamometers are strictly of the absorption type. They are incapable of driving a test machine such as a pump or compressor; hence they are only useful for measuring the power from a source such as an internal combustion engine or electric motor.

The eddy-current dynamometer is based on the following principles. When a conducting material moves through a magnetic flux field, voltage is generated, which causes current to flow. If the conductor is a wire forming a part of a complete circuit, current will be caused to flow through that circuit, and with some form of commutating device a form of a-c or d-c generator may be the result. If the conductor is simply an isolated piece of material, such as a short bar of metal, and not a part of a complete circuit as generally recognized, voltages will still be induced. However, only local currents may flow in practically short-circuit paths within the bar itself. These currents, called eddy currents, become dissipated in the form of heat.

An eddy-current dynamometer consists of a metal disk or wheel which is rotated in the flux of a magnetic field. The field is produced by field elements or coils excited by an external source and attached to the dynamometer housing, which is mounted in trunnion bearings. As the disk turns, eddy currents are generated, and the reaction with the magnetic field tends to rotate the complete housing in the trunnion bearings. Torque is measured in the same manner as for the water brake, and Eqs. (12.15), (12.16), and (12.16a) are applicable. Load is controlled by adjusting the field current. As with the water brake, the mechanical energy is converted to heat energy,

Fig. 12.20 An example of a general-purpose electric dynamometer. (Courtesy General Electric Company, Schenectady, N.Y.)

presenting the problem of satisfactory dissipation. Most eddy-current dynamometers must employ water cooling. Particular advantages of this type are the comparatively *small size* for a given capacity and characteristics permitting *good control at low rotating speeds.*

Undoubtedly the most versatile of all types is the *cradled d-c dyna-mometer,* shown in Fig. 12.20. This type of machine is usable both as an absorption and as a driving dynamometer in capacities to 5000 hp. Basically the device is a d-c motor generator with suitable controls to permit operation in either mode. When used as an absorption dynamometer, it performs as a d-c generator and the input mechanical energy is converted to electrical energy, which is dissipated in resistance racks. This latter feature is important, for unlike the eddy-current dynamometer, the heat is dissipated external to the machine. Cradling in trunnion bearings permits the determination of reaction torque and the direct application of Eqs. (12.15), (12.16), and (12.16a). Provision is made for measuring torque in either direction, depending on the direction of rotation and mode of operation. As a driving dynamometer, the device is used as a d-c motor, which presents a problem in certain instances of obtaining an adequate source of d-c power for this purpose. Use of either an a-c motor-driven d-c generator set or a rectified source is required. *Ease of control* and *good performance at low speeds* are features of this type of machine.

Ordinary *electric motors* or *generators* may be adapted for use in dynamometry. This is more feasible when d-c rather than a-c machinery is employed. Cradling the motor or generator may be used for either driving or absorbing applications, respectively. By measuring torque reaction and speed, power may be computed. This, of course, requires special effort in designing and fabricating a minimum-friction arrangement. Adjustment of driving speed or absorption load could be provided through control of field current. Load-cell mounting may be used.

Knowledge of motor or generator characteristics versus speed presents another approach. If a d-c generator is used as an *absorption dynamometer,* then

$$\text{hp absorbed} = \frac{1.341 \times \text{kilowatts}}{\text{efficiency}} = \frac{1.341 \times (e)(i)}{1000 \,(\text{efficiency})} . \qquad (12.17)$$

Likewise, if a d-c motor is used as a *driving dynamometer,*

$$\text{hp input} = 1.341 \times \text{kilowatts} \times \text{efficiency} = \frac{1.341 \times (e)(i) \,(\text{efficiency})}{1000} ,$$

$$(12.17a)$$

where e = volts and i = amperes. In both cases,

$$\text{Torque} = \frac{33,000 \text{ hp}}{2\pi \,(\text{rpm})} . \qquad (12.18)$$

Both e and i may be measured separately, or a wattmeter may be employed and the electrical power measured directly.

In many applications, only approximate results may be required, in which case *typical* motor or generator efficiencies supplied by the manufacturer should suffice. For more accurate results, some form of dynamometer would be required to determine the efficiencies for the particular machine to be used. The use of a-c motors or generators, while feasible, is considerably more difficult and will not be discussed here. For methods of application, the reader is referred to the references for this chapter, particularly Ref. 14. In any case, application of *general-purpose* electrical rotating machinery to dynamometry must be considered special and will not yield as satisfactory results as equipment particularly designed for the purpose.

12.8 TRANSMISSION DYNAMOMETERS

As mentioned earlier, transmission dynamometers may be thought of as passive devices neither appreciably adding to nor subtracting from the energy involved in the test system. Various devices have been used for this purpose, including gear train arrangements and belt or chain devices.

Any gear box producing a speed change is subjected to a reaction torque equal to the difference between the input and output torques. When the reaction torque of a cradled gear box is measured, a function of either input or output torque may be obtained.

Belt or chain arrangements, in which reaction is a function of the difference between the tight and loose tensions, may also be used. Torque at either main pulley is also a function of the difference between the tight and loose tensions; hence the measured reaction may be calibrated in terms of torque, from which, with speed information, power may be determined. Mechanical losses introduced by arrangements of these types, combined with general awkwardness and cost, make them rather unsatisfactory except for an occasional special application.

More common forms of transmission dynamometers are based on calibrated measurement of unit or total strains in elastic load-carrying members. A popular dynamometer of the elastic type employs bonded strain gages applied to a section of torque-transmitting shaft [15, 16], as shown in Fig. (11.39). Such a dynamometer, often referred to as a *torque meter*, is used as a coupling between driving and driven machines, or between any two portions of a machine. Gages are applied as discussed in Article 11.19c (see also Example 3 in Article A.6 of Appendix A). A complete four-arm bridge is employed, incorporating modulus gages to minimize temperature sensitivity (Article 12.4d). Electrical connections are made through slip rings, with means provided to lift the brushes when they are not in use, thereby minimiz-

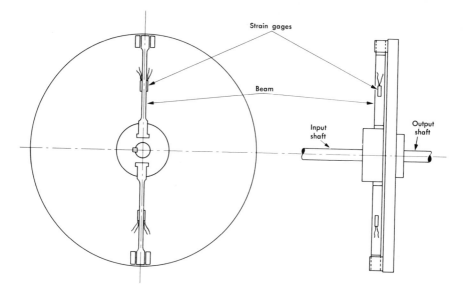

Fig. 12.21 Transmission dynamometer which employs beams and strain gages for sensing torque.

ing wear. Any of the common strain-gage indicators or recorders are usable to interpret the output. Dynamometers of this type are commercially available in capacities of 100 to 30,000 in.·lb. Accuracies to $\frac{1}{4}\%$ are claimed.

In most cases resistance strain-gage transducers are most sensitive when bending strains can be employed. Figure 12.21 suggests methods whereby torsion may be converted to bending for measurement.

Slip rings are subject to wear and may present annoying maintenance problems when permanent installations are required. For this reason many attempts have been made to devise electrical torque meters which do not require direct electrical connection to the moving shaft. Inductive [17, 18] and capacitive [19] transducers (see Fig. 6.17) have been used to accomplish this.

In addition to temperature sensitivity resulting from variation in elastic constants, further variation may be caused in the inductive type by change in magnetic constants with temperature. This may be compensated by resistors in a manner similar to that used for strain-gage load cells (Article 12.4d).

These types are relatively expensive, and cannot be considered general-purpose instruments. However, in permanent installations they provide the advantage of long service without maintenance problems.

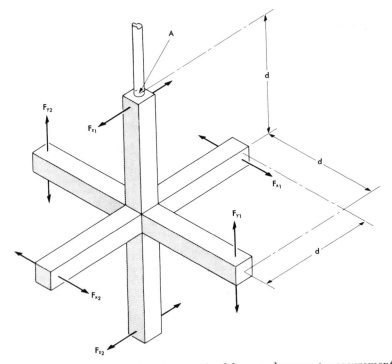

Fig. 12.22 Cross-type resolver for combined force and moment measurement.

12.9 COMBINED FORCE AND MOMENT MEASUREMENT

Certain special situations require combined force and moment measurement. One example is the measurement of forces and moments on a wind-tunnel model. Another is related to the determination of forces and moments caused by pressure and thermal expansions in a power-piping system [20]. In situations such as these, determination of three force and three moment components will completely define the condition. Usually the force and moment components are referred to a set of orthogonal axes, x, y, and z.

One of many possible systems will be described briefly. It is applicable to determination of forces and moments in models of power-pipe systems. The method employs a cross-shaped resolver arranged as shown in Fig. 12.22. A model piping-system terminal point is attached to the resolver at point A. Forces required to maintain equilibrium are then applied at the cross extremities. These forces may be designated, F_{x_1}, F_{x_2}, F_{y_1}, F_{y_2}, F_{z_1}, and F_{z_2}. Inspection shows that

$$R_x = F_{x_1} + F_{x_2},$$
$$R_y = F_{y_1} + F_{y_2},$$
$$R_z = F_{z_1} + F_{z_2},$$

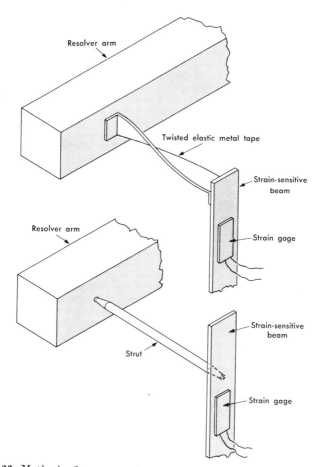

Fig. 12.23 Methods for connecting resolvers to force-sensing elements.

and also that

$$M_x = d(F_{z_1} - F_{z_2}),$$
$$M_y = d(F_{x_1} - F_{x_2}),$$
$$M_z = d(F_{y_1} - F_{y_2}).$$

Actual measurement of the forces may be made by use of any of the various force transducers that have been discussed. In practice, small cantilever beams with attached resistance-type strain gages have been used. Forces between the resolver and the beams may be transmitted by short compression struts, or twisted elastic tapes (Fig. 12.23), thereby permitting freedom of movement in all but the load-measuring direction.

If the relative stiffness of the transducer to the test subject is great enough, the small movement resulting from deflection of the transducer under load may be ignored. Otherwise means must be incorporated to adjust for transducer deflection in order to return point A to its initial unloaded position.

SUGGESTED READINGS

ASME Performance Test Code, PTC 19.7-1961 *Measurement of Shaft Horsepower.*

ASME Performance Test Code, PTC 19.5; 101964 *Weighing Scales.*

National Bureau of Standards: *Precision Measurement and Calibration*, Handbook 77, U.S. Dept. of Commerce, 1961; Vol. III, *Optics, Metrology and Radiation.*

PROBLEMS

12.1 A simple circular-sectioned column with a diameter of $1\frac{1}{2}$ in. of phosphor bronze is used as the basis for a compression-type load-measuring device. Two differential transformers are employed as secondary transducers, one on each side of the column, arranged to average out the effects of eccentric loading. The effective length of the column is 2 in. Calculate the error caused by dimensional and modulus changes that are introduced by a 50°F change in temperature. Consider the elastic member only, disregarding any component of error which may be due to the secondary transducers.

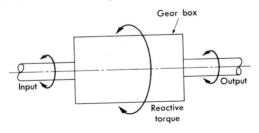

Figure 12.24

12.2 Assume a simple gear box with the input shaft speed R times the speed of the colinear output shaft (see Fig. 12.24).

a) If frictional losses are neglected, show that the reactive torque required to hold the box may be found from the equation

$$\text{Reactive torque} = 33{,}000\ (\text{hp})\ (R \pm 1)/(2\pi)\ (\text{Input rpm}).$$

b) Derive a comparable relationship including an efficiency factor.

c) It is seen that by application of proper instrumentation, a gearbox of this type may be employed as a dynamometer. Suggest one method for doing this.

12.3 Small toys are available, powered by d-c motors deriving their energy from ordinary flashlight batteries. The motors measure approximately $1\frac{1}{4}$ in. diameter and are about $1\frac{1}{2}$ in. long. Output shaft diameter is $\frac{1}{8}$ in.

a) From the above description of size, propose procedures for determining: (1) horsepower versus speed from "no-load" to "stalling load" conditions, and (2) efficiencies for the above speed range.

b) Discuss points which you feel might cause problems in the method you are proposing.

12.4 Obtain a motor such as the one described in Problem 12.3 above and carry through with your proposed tests. Estimate the accuracies of your tests.

12.5 Apply Problems 12.3 and 12.4 above to a model airplane engine.

CHAPTER 13

MEASUREMENT OF PRESSURE

13.1 INTRODUCTION

Pressure is the force exerted by a medium, usually a fluid, on a unit area. It differs from normal stress only in the mode of application. In engineering it is most commonly expressed in terms of pounds per square inch (psi). Measuring devices usually register it as a differential pressure, i.e., the difference between two pressures, with atmospheric pressure as the common reference. This is termed *gage pressure*, psig. Figure 13.1 illustrates the relationships.

Pressure is often equated to the unit of force exerted by a column of fluid, such as mercury or water. For example, the atmospheric *standard pressure* is specified as 14.696 psi. This is identical to the unit force that a column of mercury 760 mm (29.921 in.) in height exerts at its base. Therefore, it is common to refer to standard atmospheric pressure as 760 mm Hg. It is obvious that fundamentally the unit of pressure is neither millimeters nor inches and that these units have meaning only when employed in the proper context.

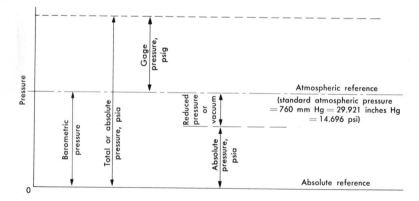

Fig. 13.1 Relations between absolute, gage, and barometric pressures.

A negative gage pressure is often referred to as a *vacuum*. In addition to the units mentioned above, the following are commonly used for evaluating low pressure:

$$1 \text{ millibar} = 1000 \text{ dyne/cm}^2 = 14.5 \times 10^{-3} \text{ psi,}$$

$$1 \text{ micron} = 10^{-6} \text{ m Hg} = 19.34 \times 10^{-6} \text{ psi,}$$

$$1 \text{ torr} = 1 \text{ mm Hg} = 1000 \text{ microns} = 19.34 \times 10^{-3} \text{ psi.}$$

Extremely high pressure is often designated in terms of atmospheres (atm):

$$1 \text{ atm} = 14.696 \text{ psi.}$$

When the fluid is in equilibrium, the pressures at a point are identical in all directions and independent of orientation. This is referred to as a *static pressure*. When pressure gradients occur within a continuous body of fluid, the attempt to restore equilibrium results in fluid flow from regions of higher pressure to regions of lower pressure. In this case, total pressures are no longer independent of direction.

Velocity and impact pressures. Various pressure components exist in a flowing fluid. If we attempt to use a small tube or probe for sampling the pressure in an air duct, we find that the results depend on how the tube is oriented. If the tube or probe is aligned so that the flow impacts against the tube opening as shown at A in Fig. 13.2, we obtain one result; if it is positioned as shown at B, we obtain another result.

Probe A senses a *total or stagnation pressure*, while tap B senses only the *static* component of pressure. Static pressure may be thought of as the pressure one would sense if moving along with the stream, while total pressure may be defined as the pressure that would be obtained if the

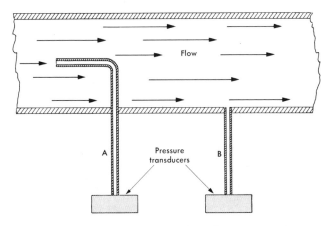

Fig. 13.2 Impact-pressure and static-pressure tubes.

stream were brought to rest isentropically. If we take the difference between the two pressures, we obtain the pressure due to the fluid motion, referred to as the *velocity pressure*, or

velocity pressure = total pressure − static pressure.

We see, therefore, that to properly obtain and interpret pressure information it is necessary to account for flow conditions. Conversely, to properly interpret flow measurements, consideration must be given to the pressure situation. With the above factors in mind, we shall proceed to consider some of the methods used to measure pressure.

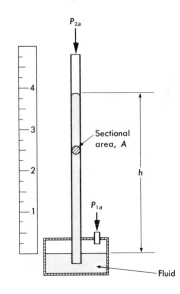

Fig. 13.3 Well-type manometer.

13.2 PRESSURE-MEASURING SYSTEMS

Pressure-measuring systems probably vary over a greater range of complexity than any other type of measuring system. On the one hand, the ordinary manometer (Fig. 13.3) is one of the most elementary measuring devices imaginable. It is simple, inexpensive, and relatively free from error, yet it may be arranged to almost any degree of sensitivity. Its major disadvantages are in the areas of certain pressure ranges and its poor dynamic response. It is not very practical for measuring pressures greater than, say, 100 psi, and it is incapable of following any but slowly changing pressures. Another familiar pressure-measuring device, the common Bourdon-tube gage, is quite useful over a wide pressure range, but only for static or slowly changing pressures.

In general, it can be said that when the pressure is *dynamic*, some form of pressure-measuring system utilizing electromechanical transducer methods is required. A major portion of this chapter will be devoted to discussing applications of devices of this kind.

In accounting for the dynamic response of a pressure-measuring system, the instrumentation and the application must be considered as a whole. The response is not determined by the isolated physical properties of the instrument components alone, but must include the mass-elastic-damping effects of the pressurized media and conducting passageways.

As an example, a diaphragm-type pickup may be employed for measuring the pressure at a specific point on an aircraft skin. In such an application, it is often undesirable to place the diaphragm flush with the aircraft surface. Possibly the size of the diaphragm is too great in comparison with the pressure gradients existing; or perhaps flush mounting would disturb the surface to too great a degree; or it may be necessary to mount the pickup internally to protect it from large temperature variations. In such cases, the pressure would be conducted to the sensing element of the pickup through a passageway, and a small space or cavity would exist over the diaphragm. The passageway and cavity become, in essence, an integral part of the transducer, and the mass-elastic-damping properties contribute to the determination of the overall response of the system. It is obvious that it would be insufficient to know only the transducer characteristics.

Ideally, a pressure pickup should be insensitive to temperature change and acceleration; friction should be minimized, and any that is unavoidable should be predictable. Damping should remain constant for all operating conditions. These items will be discussed in more detail later in the chapter.

13.3 PRESSURE-MEASURING TRANSDUCERS

Often pressure is measured by transducing its effect to a deflection through use of a pressurized area and either a gravitational or elastic restraining element. A comprehensive classification of basic pressure-measuring methods is difficult to make. However, the following should suffice for our purposes.

I. Gravitational types
    A. Liquid columns
    B. Pistons or loose diaphragms, and weights

II. Direct-acting elastic types
    A. Unsymmetrically loaded tubes
    B. Symmetrically loaded tubes
    C. Elastic diaphragms
    D. Bellows
    E. Bulk compression

III. Indirect-acting elastic type
    Piston with elastic restraining member

13.4 GRAVITATIONAL TRANSDUCERS

The simple well-type manometer (Fig. 13.3) is one of the most elementary forms of pressure-measuring devices. A force-equilibrium expression for the net liquid column is

$$P_{1a}A - P_{2a}A = Ah\delta \tag{13.1}$$

or

$$P_{1a} - P_{2a} = P_{1d} = h\delta, \tag{13.1a}$$

where

$$P_{1a}, P_{2a} = \text{applied absolute pressures,}$$

$$P_{1d} = \text{differential pressure or difference}$$
$$\text{in pressure between points 1 and 2,}$$

$$\delta = \text{unit density of the fluid.}$$

In practice, pressure P_{2a} is commonly atmospheric and,

$$P_{1a} - P_{\text{atm}} = P_{1g} = h\delta, \tag{13.2}$$

where

$$P_{1g} = \text{gage pressure at point 1.}$$

It is seen that *because the fluid density is involved, accurate work will require consideration of temperature variation;* the manometer will possess a certain amount of temperature sensitivity.

When the applied absolute pressure P_{2a} is made to be zero, and P_{1a} is atmospheric, the ordinary barometer is obtained. In this case, the fluid is usually mercury.

Figure 13.4 illustrates the functioning of the simple U-tube manometer. Pressures are applied to both legs of the U, and the manometer fluid is displaced until force equilibrium is attained. Pressures P_{1a} and P_{2a} are transmitted to the manometer legs through some fluid of density δ_t, while the manometer fluid has some greater density, δ_m. In general it is seen that

$$P_{1a} - P_{2a} = h(\delta_m - \delta_t). \tag{13.3}$$

As an example, suppose the manometer fluids are water and mercury, respectively. This situation might occur when a manometer is employed to measure the differential pressure across a venturi meter through which water is flowing. Then

$$P_{1a} - P_{2a} = h(13.6 - 1)(0.0361),$$

where the pressures are in pounds per square inch, 13.6 and 1 are the specific gravities of mercury and water, respectively, and 0.0361 lb/in.[3] is the density of water under standard conditions.

When the relative density of the transmitting fluid is small enough to be ignored, then Eq. (13.3) reverts to Eq. (13.1a).

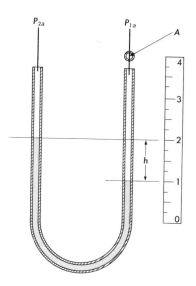

Fig. 13.4 Simple U-tube manometer.

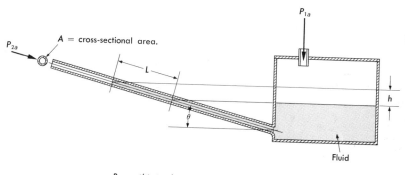

$$P_{1a} = \delta(L \sin \theta) + P_{2a} \quad \text{(Eq. 13.4)}$$

Fig. 13.5 Inclined-type manometer.

To obtain displacement amplification, various schemes may be applied to manometers, such as inclining one or both legs (Fig. 13.5), or employing combinations of liquids (Fig. 13.6).

Figure 13.7 illustrates the familiar dead-weight tester commonly used as a source of static pressure for calibration purposes but which is basically a pressure-producing and pressure-measuring device. When the applied weights and piston area are known, the resulting pressure may be readily calculated.

Figure 13.8 illustrates the principle of operation of the inverted bell pressure-measuring system. In this case, the force exerted by the pressure

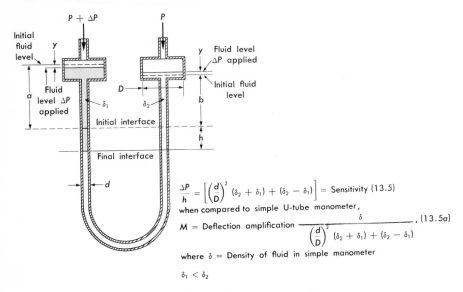

$$\frac{\Delta P}{h} = \left[\left(\frac{d}{D}\right)^2 (\delta_2 + \delta_1) + (\delta_2 - \delta_1)\right] = \text{Sensitivity (13.5)}$$

when compared to simple U-tube manometer,

$$M = \text{Deflection amplification} \quad \frac{\delta}{\left(\dfrac{d}{D}\right)^2 (\delta_2 + \delta_1) + (\delta_2 - \delta_1)}, \ (13.5a)$$

where δ = Density of fluid in simple manometer

$$\delta_1 < \delta_2$$

Fig. 13.6 Two-fluid type manometer.

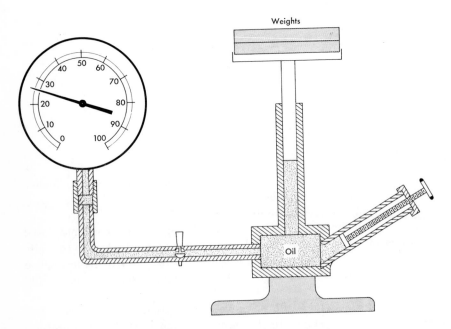

Fig. 13.7 A dead-weight tester.

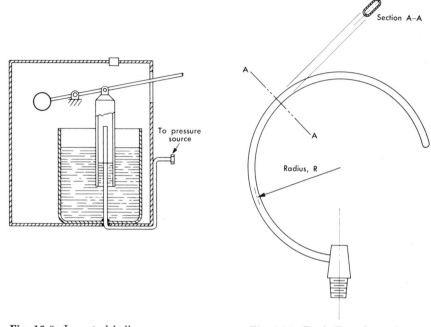

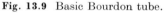

Fig. 13.8 Inverted bell pressure-measuring device.

Fig. 13.9 Basic Bourdon tube.

against the inner top of the bell is balanced against the net weight of the bell. The net weight depends on the depth of immersion, and as the pressure varies, the bell rises or falls according to pressure magnitude. The primary application of this device is for actuating industrial pressure recorders and controllers. Of course, all gravitational-type pressure transducers are sensitive to the local value of gravity acceleration.

13.5 ELASTIC TRANSDUCERS

Elastic elements operate on the principle that the deflection or deformation accompanying a balance of pressure and elastic forces may be used as a measure of pressure. A familiar example is the ordinary Bourdon tube. A tube, normally of oval section, is initially coiled into a circular arc of radius R, as shown in Fig. 13.9. The included angle of the arc is usually less than $360°$; however, in some cases, when increased sensitivity is desired, the tube may be formed into a helix of several turns.

    As a pressure is applied to the tube, the oval section tends to round out, becoming more circular in section. The inner and outer arc lengths will remain approximately equal to their original lengths, and hence the only recourse for the tube is to uncoil. In the simple pressure gage, the movement of the end of the tube is communicated through linkage and gearing to a pointer whose movement over a scale becomes a measure of pressure. Rigorous treatment of the mechanics of Bourdon-tube action is complex, and only approximate analyses have been made [4, 5].

13.6 ELASTIC DIAPHRAGMS

Many dynamic pressure-measuring apparatus employ an elastic diaphragm as the primary pressure transducer. Such diaphragms may be either flat or corrugated; the flat type [Fig. 13.10(a)] is often used in conjunction with electrical secondary transducers whose sensitivity permits quite small diaphragm deflections, while the corrugated type [Fig. 13.10(b)] is particularly useful when larger deflections are required.

Diaphragm displacement may be transmitted by mechanical means to some form of indicator, perhaps a pointer and scale as is used in the familiar aneroid barometer. For engineering measurements, particularly when dynamic results are required, diaphragm motion is more often sensed by some form of electrical secondary transducer, whose principle of operation may be resistive (Fig. 13.12), capacitive (Fig. 6.18), or inductive (Fig. 13.13). The output from the secondary transducer is then processed by appropriate intermediate devices and fed to an indicator, recorder, or controller.

Diaphragm design for pressure transducers generally involves all the following requirements to some degree:

1. Dimensions and total load must be compatible with physical properties of the material used.

2. Flexibility must be such as to provide the sensitivity required by the secondary transducer.

3. Volume of displacement should be minimized to provide reasonable dynamic response.

4. Natural frequency of the diaphragm should be sufficiently high to provide satisfactory frequency response.

5. Output should be linear.

a) *Flat metal diaphragms.* Deflection of flat metal diaphragms is limited either by stress requirements or by deviation from linearity. It has been found that as a general rule the maximum deflection that can be tolerated maintaining a linear pressure-displacement relation, is about 30% of the diaphragm thickness [6, 7].

In certain cases secondary transducers require physical connection with the diaphragm at its center. This is usually true when mechanical linkages are employed and is also necessary for certain types of electrical secondary transducers. In addition, auxiliary spring force is sometimes

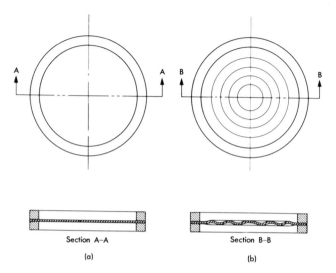

Section A–A   Section B–B

(a)   (b)

Fig. 13.10 (a) Flat diaphragm. (b) Corrugated diaphragm.

introduced to increase the diaphragm deflection constant. These requirements make necessary some form of boss or reinforcement at the center of
the diaphragm face, which reduces diaphragm flexibility and complicates
theoretical design analysis.

When a central connection is made, a concentrated force F will normally
be applied. In general, therefore, the diaphragm may be simultaneously
subjected to two deflection forces, the distributed pressure load and a
central concentrated force. Design relationships for the fixed-edge, pressurized diaphragm may be found in Reference [7]; for diaphragms with central
bosses, in Reference [7a].

Calculations for diaphragm dimensions should not be relied upon as
representing more than a rough guide for design purposes. There are several
factors that cannot be accurately predicted. Among these are: (1) the rigidity
of the outer supporting ring and inner boss, which is never as complete as
assumed, and (2) the material physical properties, which are seldom accurately known. In addition, an undesirable characteristic of simple flat diaphragms that is often encountered is a nonlinearity referred to as "oil
canning." The term is derived from the action of the bottom of a simple oil
can when it is pressed. A slight unintentional dimpling in the assembly of a
flat-diaphragm pressure pickup is difficult to eliminate unless special precautions are taken. In addition, oil canning may be aggravated by differential
expansions due to changing ambient conditions. It is desirable, therefore,
to construct a pressure cell from materials having the same coefficient of
expansion. Even this, however, may not always solve the problem because

temperature gradients within the instrument itself may result in a different expansion. One solution to this problem is obtained by using a stretched or *radially* preloaded diaphragm [8]. Theoretical solutions for the radially preloaded diaphragm are considerably more involved than those for the simple flat type. An approach to the problem is presented in the cited reference [8]. Another solution to the oil-canning problem is to employ a small external spring load to *bias* the diaphragm. This, of course, adds mass and thereby sacrifices dynamic response. In all cases, care must be exercised to minimize undesirable temperature effects.

b) Corrugated diaphragms. Corrugated diaphragms are normally used in larger diameters than the flat types. Corrugations permit increased linear deflections and reduced stresses. Since the larger size and deflection reduce the dynamic response of the corrugated diaphragms as compared with the flat type, they are more commonly used in static applications.

Adding convolutions to a diaphragm increases the complexity of the theoretical design approach. Grover and Bell [9] have used brittle coatings as means for evaluating approximate theoretical solutions for stresses.

Two corrugated diaphragms are often joined at their edges to provide what is referred to as a *pressure capsule*. This is the type commonly used in aneroid barometers.

Metal bellows are sometimes used as pressure-sensing elements. Bellows are generally useful for pressure ranges from about $\frac{1}{2}$ psi to 150 psi full scale. Hysteresis and zero shift are somewhat greater problems with this type of element than with most of the others.

13.7 SECONDARY TRANSDUCERS USED WITH DIAPHRAGMS

Most electromechanical transducer principles have been applied to diaphragm pressure pickups. The following examples are only representative of many possible variations.

a) *Use of resistance strain gages with flat diaphragms.* An obvious approach is to simply apply strain gages directly to a diaphragm surface and calibrate the measured strain in terms of pressure. One drawback of this method that is often encountered is the small physical area available for mounting the gages; for this reason, gages with short gage lengths must be employed.

Special spiral grids have been used [6, 10], but they have exhibited less reliability than ordinary gages. When used, grids are mounted in the central area of the diaphragm, with the elements in tension (see Fig. 11.8d).

Wenk [6] has found that a satisfactory method for mounting strain gages is the one illustrated in Fig. 13.11. When pressure is applied to the side opposite the gages, the central gage is subject to tension while the outer gage

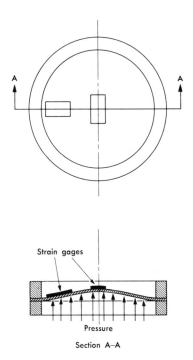

Fig. 13.11 Location of strain gages on flat diaphragm.

senses compression. The two gages may be used in adjacent bridge arms, thereby adding their individual outputs and simultaneously providing temperature compensation.

A special-purpose bonded strain gage is used in the Norwood* pressure transducer. In this case a diaphragm is used to load a *strain tube* as a compression or column member. The load causes the tube to both shorten and increase in diameter. Two strain windings, one with the effective grid placed axially and the other wound circumferentially, form the two active arms in a resistance bridge. Change in resistance caused by pressure change is calibrated in terms of pressure. A feature of the gage is the preformed diaphragm having catenary section between supports. This shape provides several distinct advantages over a flat diaphragm: only tensile strains are produced in the diaphragm, permitting thicknesses as small as one-eighth that of a flat diaphragm for the same purpose; because of the thinner section, essentially all the load restraint is provided by the sensing tube, and changes in temperature have negligible effect on its flexibility. Gages of this type are made in sizes having ranges from a lower range of −15 to 400 psi,

---

* Control Engineering Corporation, Norwood, Conn.

Fig. 13.12 Pressure-gage assembly which employs unbonded resistance strain elements as the secondary transducer. (Courtesy Consolidated Electrodynamics Corporation, Pasadena, California.)

to an upper range of −15 to 15,000 psi. Forced air or water-cooling permit their use to 5000°F in certain applications. Low mass and high stiffness provide resonant frequencies in the order of 45,000 Hz. This permits pressure measurements within ±1% at operating frequencies as high as 20,000 Hz.

Unbonded resistance gages are also used. Statham* gages employ sensing elements having the same general design as their strain gages and accelerometers (Fig. 16.9). Flat diaphragms are used to convert the applied pressure to displacement. Gages of this type are available covering pressure ranges from 0–0.05 psig to 0–10,000 psig.

Another method for straining unbonded resistance elements is shown in Fig. 13.12. This device employs a four-legged flexure linkage referred to as a "star spring" [11]. Pressure applied to a diaphragm (not shown) displaces the central socket (extreme top in the photograph), thereby stretching two resistance elements of 0.005 in. in diameter wound around the spring assembly. These two active elements, along with two inactive elements wound around fixed posts, are connected to form a conventional resistance bridge. Gages employing this principle are available† in absolute, differential, or gage pressure ranges from 0–1.0 to 0–5000 psi for temperatures ranging from −65 to 600°F.

b) Inductive types. Variable inductance has also been successfully employed as a form of secondary transducer used with a diaphragm [8]. Figure 13.13 illustrates one arrangement of this sort. Flexing of the diaphragm due to

---

* Statham Laboratories, Beverly Hills, Calif.

† Consolidated Electrodynamics Corporation, Pasadena, Calif.

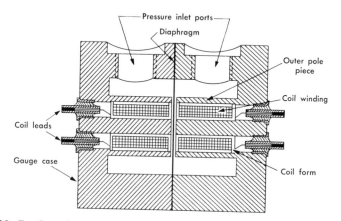

Fig. 13.13 Section through a differential pressure cell which uses an inductive secondary transducer. (Courtesy Ultradyne Inc., Albuquerque, New Mexico.)

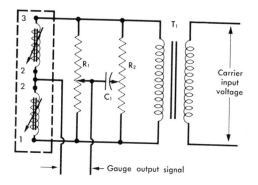

Fig. 13.14 Circuit diagram for the pressure cell shown in Fig. 13.13.

applied pressure causes it to move toward one pole piece and away from the other, thereby altering the relative inductances. An inductive bridge circuit may be employed, as shown in Fig. 13.14. Standard laboratory equipment, such as an oscilloscope or VTVM, as well as recorders, may be used to display the gage output. Available ranges are from 0–1.0 to 0–100 psi.*

c) Other types of secondary transducers. Flexing diaphragms have been used to alter capacitance as a means of producing an electrical output (see Fig. 6.18). This method is not so common as those previously discussed, primarily because of low sensitivity and the problems accompanying the requirement for relatively high carrier frequencies.

---

* Ultradyne Engineering Laboratories, Inc., Albuquerque, New Mexico.

Another method that has been employed with some success* uses an electromechanical resonant system consisting of a fine wire under tensile load vibrating at its natural frequency. One end of the wire is connected to the center of a pressure-sensing diaphragm, which varies the wire tension, depending on the applied pressure. Small permanent magnets provide a magnetic field in which the wire vibrates, causing an a-c potential to be developed in the wire. After amplification, a portion of this voltage is fed back to energize driving coils that maintain the vibration. Output *frequency* is the measure of pressure.

Still another system employs the electrokinetic potential (see Article 6.21) developed by a fluid flowing through a porous disk as a measure of pressure. The device consists of a capsule formed by two flexible diaphragms, with the electrokinetic disk dividing the chamber into two parts. Input pressure is applied to one diaphragm only. The resulting displacement forces fluid through the disk, thereby generating an electrical potential that may be used as a measure of pressure. Inherently, the pickup is suitable for dynamic pressures only. A commercially available model† is usable over a pressure range of 10^{-4} to 100 psig at frequencies from 4 Hz to 15,000 Hz (down 3 db at each end).

13.8 STRAIN-GAGE PRESSURE CELLS

Any form of container will be strained when pressurized. Sensing the resulting strain with an appropriate secondary transducer, such as a bonded-wire strain gage, will provide a measure of the applied pressure. The term *pressure cell* has gradually become applied to this type of pressure-sensing device, and various forms of elastic *containers* or cells have been devised.

For low pressures, a pinched tube may be used (Fig. 13.15). This supplies a bending action as the tube tends to round out. Gages may be placed diametrically opposite on the flattened faces, as shown, with two unstressed temperature-compensating gages mounted elsewhere. This arrangement completes the electrical bridge. Cells of this general design are commercially available.‡

Probably the simplest form of strain-gage pressure transducer is a cylindrical tube such as shown in Fig. 13.16. In this application two active gages mounted in the hoop direction may be used for pressure sensing, along with two temperature-compensating gages mounted in an unstrained location. Temperature-compensating gages are shown mounted on a separate

---

* Byron Jackson Co., Los Angeles, Calif.
† Consolidated Electrodynamics Corporation, Pasadena, Calif.
‡ Baldwin-Lima-Hamilton Corp.

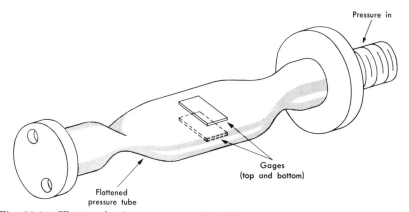

Fig. 13.15 Flattened-tube pressure cell which employs resistance strain gages as secondary transducers.

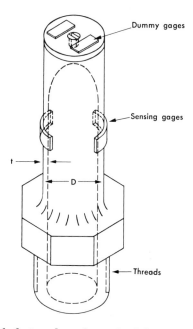

Fig. 13.16 Cylindrical type of pressure cell.

disk fastened to the end of the cell. Design relationships may be found in most mechanical design texts [12].

The sensitivity of a pair of circumferentially mounted strain gages (Fig. 13.16) with gage factor F is expressed by the relationship [32]:

$$\frac{\Delta R}{P_i} = \frac{2FR\,d^2}{E}\left[\frac{2-\nu}{D^2-d^2}\right] \tag{13.6}$$

where

$$\Delta R = \text{strain-gage resistance change, ohms,}$$
$$R = \text{nominal gage resistance, ohms,}$$
$$P_i = \text{internal pressure, psi,}$$
$$d = \text{inside diameter of cylinder, inches,}$$
$$D = \text{outside diameter of cylinder, inches,}$$
$$E = \text{Young's modulus, psi,}$$
$$\nu = \text{Poisson's ratio.}$$

The bridge constant, 2, appears because two circumferential gages are assumed. If a single strain-sensitive gage is to be employed, the sensitivity will be one-half that given by Eq. (13.6). Of course, these relations are true only if elastic conditions are maintained and if the gages are located so as to be unaffected by end restraints.

Improved frequency response may be obtained for a cell of this type by minimizing the internal volume. This may be accomplished by use of a solid "filler" such as a plug, which will reduce the flow into and out of the cell with pressure variation.

By employing a double-walled tube with gages mounted externally on the outer shell and internally on the inner shell, full advantage of the four arms of a resistance bridge may be obtained. In this case the outer gages are subject to tensile strains while the inner gages are subject to compression.

Figure 13.17 shows the electrical circuitry used for a transducer of this type. Gage M is a modulus gage, discussed in Article 12.4d, used to compen-

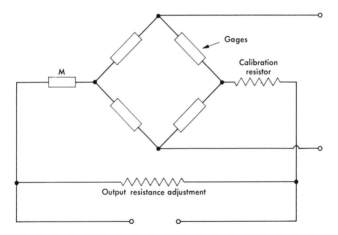

Fig. 13.17 Strain-gage circuitry for pressure cells employing a modulus gage.

sate for variation in Young's modulus with temperature. The calibration and output resistors are adjusted to provide predetermined bridge resistance and calibration.

13.9 MEASUREMENT OF HIGH PRESSURES

The high-pressure range has been defined as beginning at 10,000 psi and extending upward to the limit of present techniques, which is in the order of a quarter of a million psi [13]. Conventional pressure-measuring devices, such as strain-gage pressure cells and Bourdon-tube gages, may be used at pressures as high as 50,000 to 100,000 psi. Bourdon tubes for such pressures are nearly round in section and have a high ratio of wall thickness to diameter. They are, therefore, quite stiff, and the deflection per turn is small. For this reason, high-pressure Bourdon tubes are often made of a number of turns.

Electrical resistance pressure gages. Very high pressures may be measured by electrical resistance gages, which make use of the resistance change brought about by direct application of pressure to the electrical conductor itself. The sensing element consists of a loosely wound coil of relatively fine wire. When pressure is applied, the bulk-compression effect results in an electrical resistance change that may be calibrated in terms of the applied pressure.

Figure 13.18 shows a bulk modulus gage in section. The sensing element does not actually contact the process medium, but is separated therefrom by a kerosene-filled bellows. One end of the sensing coil is connected to a central terminal, as shown, while the other end is grounded, thereby completing the necessary electrical circuit.

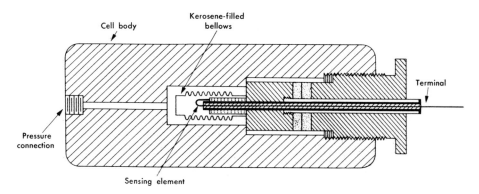

Fig. 13.18 Section through a bulk-modulus pressure gage.

Although Eq. (11.6) was written with a somewhat different application in mind, it also applies to the situation being discussed. Rewrite this relation,

$$\frac{dR}{R} = \frac{dL}{L} - 2\frac{dD}{D} + \frac{d\rho}{\rho}, \tag{13.7}$$

in which

R = electrical resistance,

L = length of conductor,

D = a sectional dimension,

ρ = resistivity.

The wire will be subject to a biaxial stress condition because the ends, in providing electrical continuity, will generally not be subject to pressure. Using relations of the form expressed by Eqs. (11.14), and assuming that $\sigma_x = \sigma_y = -P$ and $\sigma_z = 0$, we may write

$$\epsilon_x = \epsilon_y = \frac{dD}{D} = -\frac{P}{E}(1 - \nu) \tag{13.8}$$

and

$$\epsilon_z = \frac{dL}{L} = \frac{2\nu P}{E}. \tag{13.9}$$

Combining the above relations,

$$\frac{dR}{R} = \frac{2P}{E} + \frac{d\rho}{\rho} \tag{13.10}$$

or

$$\frac{dR/R}{P} = \frac{2}{E} + \frac{d\rho/\rho}{P}. \tag{13.11}$$

Two metals are commonly used for resistance gages, manganin and an alloy of gold and 2.1% chromium. Both metals provide linear outputs with the following sensitivities: 1.692×10^{-7} and 0.673×10^{-7} ohm/ohm·psi for manganin and the gold alloy, respectively. Although the former possesses the greater pressure sensitivity, final selection must also be based on temperature sensitivity. Whereas manganin exhibits a resistance change of about 0.2% for the temperature range of 70–180°F, the corresponding change for the gold alloy is in the order of 0.01% [14]. Because of the difference, the gold alloy is generally preferred. The lower output is compensated for by greater electrical amplification.

13.10 MEASUREMENT OF LOW PRESSURES

Pressures may or may not be referred to the atmospheric datum as depicted in Fig. 13.1. We know, of course, that a *positive* magnitude of absolute pressure exists at all times. It is impossible to reach the absolute zero value.

Atmospheric pressure, however, serves as a reference, and in general, pressures below atmospheric may be termed low pressures or vacuums.

A common unit of low pressure is the *micron*, which is one millionth of a meter (0.001 mm) of mercury column. *Very low* pressure may be defined as any below 1 mm of mercury, and an *ultralow* pressure as any less than a millimicron (10^{-3} micron). The torr is also employed (Article 13.1).

There are two basic methods for measuring low pressure: (1) *direct* measurement resulting in a displacement caused by the action of force, and (2) *indirect* or *inferential* methods wherein pressure is determined through the measurement of certain other pressure-controlled properties, such as volume, thermal conductivity, etc. Devices included in the first category would be spiral Bourdon tubes, flat and corrugated diaphragms, capsules, and various forms of manometers. Since these have been discussed in the preceding pages, they need not be discussed further here except to say that their use is generally limited to a lowest pressure value of about 10 mm of mercury. For measure-

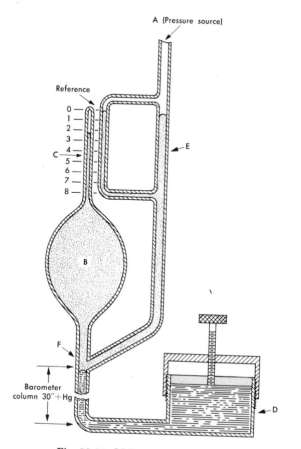

Fig. 13.19 McLeod vacuum gage.

ment of pressures below this value, one of the inferential methods is normally dictated.

a) The McLeod gage. Operation of the McLeod gage is based on Boyle's fundamental relation,

$$P_1 = \frac{P_2 V_2}{V_1}, \qquad (13.12)$$

where P_1 and P_2 are pressures at initial and final conditions, respectively, and V_1 and V_2 are volumes at corresponding conditions. By compressing a known volume of the low pressure gas to a higher pressure and measuring the resulting volume and pressure, one can calculate the initial pressure.

Figure 13.19 illustrates the basic construction and operation of the McLeod gage. Measurement is made as follows. The unknown pressure source is connected at point A, and the mercury level is adjusted to fill the volume represented by the darker shading. Under these conditions the unknown pressure fills the bulb B and capillary C. Mercury is then forced out of the reservoir D, up into the bulb and reference column E. When the mercury level reaches the cutoff point F, a *known* volume of gas is trapped in the bulb and capillary. The mercury level is then further raised until it reaches a zero reference point in E. Under these conditions the volume remaining in the capillary is read directly from the scale, and the difference in heights of the two columns is the measure of the trapped pressure. The initial pressure may then be calculated by use of Boyle's law.

Pressure of gases containing vapors cannot normally be measured with a McLeod gage, for the reason that the compression will cause condensation. By use of instruments of different ranges, a total pressure range of from about 0.01 micron to 50 mm of mercury may be measured with this type of gage.

b) Thermal conductivity gages. The temperature of a given wire through which an electric current is flowing will depend on three factors, the magnitude of the current, the resistivity, and the rate at which the heat is dissipated. The latter will be largely dependent on the conductivity of the surrounding media. As the density of a given media is reduced, its conductivity will also reduce and the wire will become hotter for a given current flow.

This is the basis for two different forms of gages for measurement of low pressures. Both employ a heated filament, but differ in the means for measuring the temperature of the wire. A single platinum filament enclosed in a chamber is employed by the *Pirani gage.* As the surrounding pressure changes, the filament temperature, and hence its resistance, also changes. The resistance change is measured by use of a resistance bridge which is calibrated in terms of pressure, as shown in Fig. 13.20(a). A compensating cell is employed to minimize variations caused by ambient temperature changes.

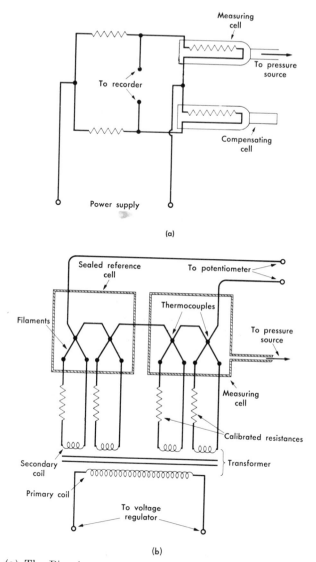

(a)

(b)

Fig. 13.20 (a) The Pirani-type thermal conductivity gage. (b) Thermocouple-type conductivity gage.

A second gage also depending on thermal conductivity is of the thermo-couple type. In this case the filament temperatures are measured directly by means of thermocouples welded directly to them. Filaments and thermo-couples are arranged in two chambers, as shown schematically in Fig. 13.20(b). When conditions in both the measuring and reference chambers are the same, no thermocouple current will flow. When the pressure in the measuring chamber is altered, changed conductivity will cause a change temperature, which will then be indicated by a thermocouple current.

In both cases the gages must be calibrated for a definite pressurized media, for the conductivity is also dependent upon this factor. Gages of these types are useful in the range of 1 to 1000 microns.

c) *Ionization gages.* For measurement of extremely low pressures, an ionization gage, which is usable to pressures down to 0.000001 micron (one billionth of a millimeter of mercury), is employed. The maximum pressure for which an ionization gage may be used is about 1 micron. An ionization cell for pressure measurement is very similar to the ordinary triode electronic tube (Fig. 7.21). It possesses a heated filament, a positively biased grid and a negatively biased plate, in an envelope evacuated by the pressure to be measured. The grid draws electrons from the heated filament, and collision between them and gas molecules causes ionization of the molecules. The positively charged molecules are then attracted to the plate of the tube, causing a current flow in the external circuit, which is a function of the gas pressure.

Disadvantages of the heated-filament ionization gage are: (1) excessive pressure (above 1 or 2 microns) will cause rapid deterioration of the filament and a short life, and (2) the electron bombardment is a function of filament temperature, therefore requiring careful control of filament current. Another form of ionization gage minimizes these disadvantages by substituting a radioactive source of alpha particles for the heated filament.

13.11 DYNAMIC CHARACTERISTICS OF PRESSURE-MEASURING SYSTEMS

Basic pressure-measuring transducers are driven, damped, spring-mass systems whose isolated dynamic characteristics are theoretically similar to the generalized systems discussed in Chapter 2. In application, however, the actual dynamic characteristics of the complete pressure-measuring system are usually controlled more by factors extraneous to the basic pickup than by the pickup characteristics alone. In other words, overall dynamic performance is determined less by the transducer than by the manner in which it is inserted into the complete system.

When the pickup is used to measure a dynamic air or gas pressure, system damping will be determined to a considerable extent by factors external to the pickup. The extraneous pneumatic circuitry will have frequency characteristics of its own, affecting system response. When liquid pressures are measured, the effective sprung mass of the system will necessarily include some portion of the liquid mass. In addition, the elasticity of any conducting tubing will act to change the overall spring constant. Connecting tubing and unavoidable cavities in the pneumatic or hydraulic

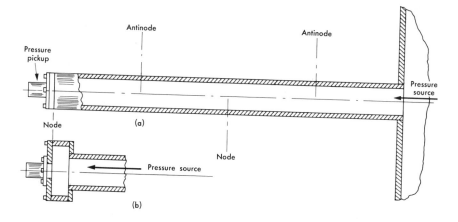

Fig. 13.21 (a) Gas-filled pressure-measuring system. (b) Gas-filled pressure-measuring system with cavity.

circuitry introduce losses and phase lags, causing differences between measured and applied pressures. Much theoretical work has been done in an attempt to evaluate these effects [15, 16, 17, 18, 19]. Each application, however, must be weighed on its own individual merits; for this reason only a general summary of some of the factors involved is practical in this discussion.

a) Gas-filled systems. As outlined above, the response of a pressure-measuring system involves more than the pickup characteristics alone. The complete system, including the method of conducting the pressure variation to the pickup, must be accounted for. In many applications it is necessary to transmit the pressure through some form of passageway or connecting tube. Figure 13.21 illustrates typical cases.

If the pressurized media is a gas, such as air, acoustical resonances may occur in the same manner in which the air in an organ pipe resonates. If sympathetic driving frequencies are present, nodes and antinodes will occur, as shown in the figure. A node, characterized by a point of zero air motion, will occur at the blocked end. (This assumes that the displacement of the pressure-sensing element, such as a diaphragm, is negligible.) Maximum pressure variation takes place at this point. Maximum oscillatory motion will occur at the antinodes, and the distance between adjacent nodes and antinodes equals one-fourth the wavelength of the resonating frequency. Theoretical resonant frequencies may be determined from the relation

$$f = \frac{C}{4L}(2n - 1), \tag{13.13}$$

in which

f = resonant frequencies (including both fundamental and harmonics), Hz,

C = velocity of sound in pressurized media, ft/sec,

L = length of the connecting tube, ft,

n = any positive integer. (It will be noted from the equation that only odd harmonics occur.)

In many cases a cavity is required at the pickup end to adapt the instrument to the tubing, as shown in Fig. 13.21(b). If we assume that the medium is a gas, and that the elasticity of the containing system, including the pickup device, is relatively stiff compared with that of the gas, we have what is known as a Helmholtz resonator. The column of gas with its weight and elasticity form a spring-mass system having an acoustical resonance whose fundamental frequency may be expressed by the relation [20]

$$f = \frac{C}{2\pi}\sqrt{\frac{a}{V(L + \frac{1}{2}\sqrt{\pi a})}}, \tag{13.14}$$

where

a = cross-sectional area of the connecting tube, ft^2,

V = net internal volume of the cavity, excluding volume of tube, ft^3.

By proper configuration and proportioning, connecting systems of this type may be used for acoustical filtering. In certain applications, quite sensitive or fragile sensing devices are required to measure small differential pressures. At the same time, high-energy pressure cycles may be present at frequencies above the range of interest and of sufficient intensity to cause pickup failure. This situation is often present in aircraft testing.

Figure 13.22 shows a connecting arrangement designed as a low-pass filter, flat to around 40 Hz [21]. Typical frequency-response curves for this design at various pressure amplitudes are shown in Fig. 13.22(b).

b) Liquid-filled systems. When a pressure-measuring system is filled with liquid rather than a gas, a considerably different situation is presented. The liquid becomes a major part of the total sprung mass, thereby becoming a significant factor in determining the natural frequency of the system. If a single degree of freedom is assumed,

$$f_n = \frac{1}{2\pi}\sqrt{\frac{kg}{W}}, \tag{13.15}$$

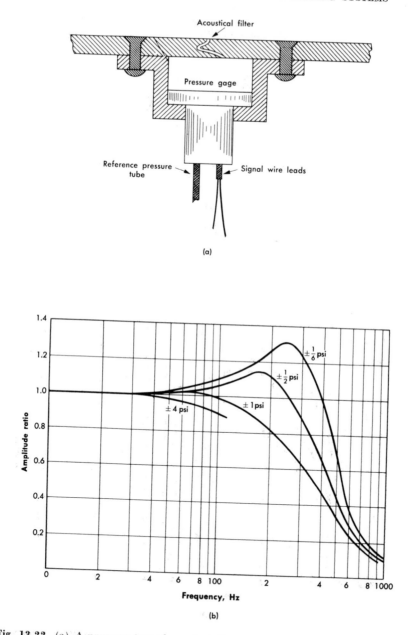

(a)

(b)

Fig. 13.22 (a) A pressure-transducer installation employing an acoustical filter. (b) Typical frequency response relations for a system such as shown in (a). (Both reprinted from January, 1957, issue of *Control Engineering*, by permission of McGraw-Hill Publishing Company, Inc.)

in which

f_n = natural frequency, Hz,

W = equivalent weight of moving mass, lb

  = $W_1 + W_2$,

W_1 = weight of moving transducer elements, lb,

W_2 = equivalent moving weight of liquid column, lb.

By simplified analysis, White has determined the following approximate relation for the effective weight of the liquid column [22]:

$$W_2 = \tfrac{4}{3} \delta a L \left(\frac{A}{a}\right)^2,$$ (13.16)

in which

δ = fluid density, lb/in³,

a = area of tube, in²,

L = length of tube, in.

A = effective area of transducer sensing element, in².

It will be noted that A is the *effective* area, which is not necessarily equal to the actual diaphragm or bellows area, but may be defined by the following relation:

$$A = \frac{\Delta v}{\Delta y},$$ (13.17)

where

Δv = volume change accompanying sensing-element deflection, in³,

Δy = significant displacement of sensing element, in.

By substitution,

$$f_n = \frac{1}{2\pi}\sqrt{\frac{kg}{W_1 + (4/3)\,\delta a L (A/a)^2}}.$$ (13.18)

In many cases the equivalent weight of the liquid, W_2, is of considerably greater magnitude than W_1, and the latter may be ignored without introducing appreciable discrepancy. By so doing and substituting:

$$a = \frac{\pi D^2}{4}, \qquad D = \text{tubing I.D.},$$

$$f_n = \frac{D}{8A}\sqrt{\frac{3kg}{\pi \delta L}}.$$ (13.19)

As mentioned before, pressure pickups involve spring-restrained masses in the same manner as do galvanometers and seismic-type accelerometers, and therefore good frequency response is obtainable only in a frequency range well below the natural frequency of the measuring system itself.* For this reason it is desirable that the pressure-measuring system have as high a natural frequency as is consistent with required sensitivity and installation requirements. Inspection of Eq. (13.19) indicates that the diameter of the connecting tube should be as large as practical and that its length should be minimized.

In addition, it has been shown that optimum performance for systems of this general type requires damping in rather definite amounts. White [22] gives the following relation for the damping ratio ζ of a system of the sort being discussed:

$$\zeta = \frac{4\pi L\nu(A/a)^2}{\sqrt{kW/g}} \tag{13.20}$$

$$= \frac{4\pi L\nu(A/a)^2}{\sqrt{k/g[W_1 + (4/3)\,\delta a L(A/a)^2]}}, \tag{13.21}$$

where ν = viscosity of the fluid. If we ignore W_1 and insert $a = (\pi D^2)/4$, we may write the equation as

$$\zeta = \frac{16\nu A}{D^3}\sqrt{\frac{3Lg}{\pi k\delta}}. \tag{13.22}$$

13.12 CALIBRATION METHODS

Static calibration of pressure gages presents no particular problems unless the upper pressure limits are unusually high. The familiar dead-weight tester, Fig. 13.7, may be used to accurately supply reference pressures with which transducer outputs may be compared. Testers of this type are useful to pressures as high as 10,000 psi, and by use of special designs, this limit may be extended to 100,000 psi.

While static calibration is desirable, pickups used for dynamic measurement should also receive some form of dynamic calibration. Dynamic calibration problems consist of two: (1) obtaining a satisfactory source of pressure, either periodic or pulsed, and (2) reliably determining the true pressure-time relation produced by such a source. These two problems will be discussed in the next few pages.

a) *Pressure sources for dynamic calibration.* Methods for obtaining dynamic calibration pressures are classified in the following tabulation. It must be noted, however, that simply providing a dynamic pressure falls short of

---

* The reasons are discussed in Articles 2.9, 8.11, 16.6, and 16.7 and will not be repeated here.

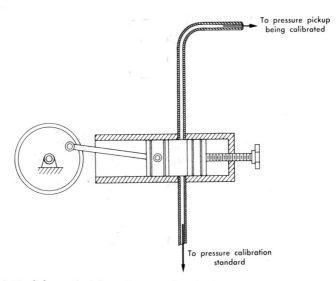

Fig. 13.23 Schematic of a piston and cylinder steady-state pressure source.

providing a dynamic standard. To be a standard, the precise pressure-time relation must be known. Sources of dynamic pressure are as follows:

I. Steady-state periodic sources
   A. Piston and chamber
   B. Cam-controlled jet
   C. Acoustic resonator
   D. Siren disk

II. Transient sources
   A. Quick-release valve
   B. Burst diaphragm
   C. Closed bomb
   D. Shock tube

b) Steady-state methods. One source of steady-state periodic calibration pressure is simply an ordinary piston and cylinder arrangement, shown schematically in Fig. 13.23 [23]. If the piston stroke is fixed, pressure amplitude may be varied by adjusting the cylinder volume. Although such a system is normally special-purpose, existing equipment, such as a CFR* engine, may be adapted for this use. Amplitude and frequency ranges will depend on the mechanical design; however, peak pressures of 1000 psi and frequencies to 100 Hz may be obtained.

    A method very similar to this is employed for the purpose of microphone calibration. In this case required pressure amplitudes are quite small, and instead of the piston's being driven with a mechanical linkage, an electromagnetic system is used [24]. Piston excursion may be determined by the technique described in Article 16.11. This suggests the possibility of using

---

* Cooperative Fuels Research.

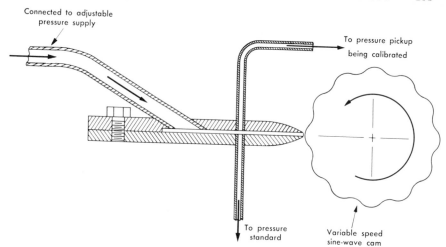

Connected to adjustable
pressure supply

To pressure pickup
being calibrated

To pressure
standard

Variable speed
sine-wave cam

Fig. 13.24 Jet and cam steady-state pressure source.

vibration test shakers (Article 16.17) as a source of piston motion for this method of calibration [25].

A variation of the piston source is to drive a diaphragm, bellows, or Bourdon tube. The latter has been successfully used by flexing it by means of an eccentric attached to the end of the tube through a link or connecting rod [26].

Figure 13.24 illustrates another method for obtaining a steady-state periodic pressure. A source of this type has been used to 3000 Hz with amplitudes to 1 psi [8]. A variation of this method employs a motor-driven siren-type disk having a series of holes drilled in it so as to alternately vent a pressure source to atmosphere and then shut it off [27].

Another system that has been used is to employ a variable-speed motor to drive a pressure transmitter by means of a circular cam [28]. The transmitter is essentially an adjustable servo valve that controls the output from a constant-pressure source.

Steady-state sinusoidal pressure generators consisting of an acoustically driven resonant system (see Article 13.11a) have been used. Hylkema and Bowersox [18] obtained pressure fluctuations in the order of 0.5 psi rms, using a 40-inch long pipe energized with a 35-watt loudspeaker type of driving unit. Usable frequencies to about 2000 Hz in integral multiples of the fundamental were obtained.

All the methods suggested above simply supply sources of pressure variation, but in themselves do not provide means for determining magnitudes or time characteristics. They are particularly useful, however, for comparing pickups having unknown characteristics with those of proven performance.

c) Transient methods. Steady-state periodic sources used to determine dynamic characteristics of pressure transducers are limited by amplitude and frequency that can be produced. High amplitudes and steady-state frequencies are difficult to obtain simultaneously. For this reason it is necessary to resort to some form of step function in order to determine high-frequency response of pressure transducers in the higher amplitude ranges.

Various methods are employed to produce the necessary pulse. One of the simplest is to use a fast-acting valve between a source of hydraulic pressure and the pickup. Rise times, from 0 to 90% of full pressure, of 10 milliseconds are reported [29].

Pressure steps may also be obtained through use of bursting diaphragms. Two chambers are separated by a thin plastic diaphragm or plate whose failure is mechanically induced by a plunger or knife. It has been found that a pressure drop, rather than a rise, produces a more nearly ideal step function. Drop time in the order of $\frac{1}{4}$ millisecond has been obtained [18].

Still another source of stepped-pressure function is the closed bomb, in which a pressure generator such as a dynamite cap is exploded. Peak pressure is controlled by net internal volume, and pressure steps as high as 700 psi in 0.3 milliseconds have been obtained [18].

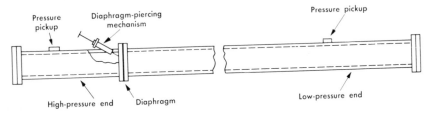

Fig. 13.25 Basic shock tube.

Undoubtedly the so-called *shock tube* provides the nearest thing to a transient pressure "standard." Construction of a shock tube is quite simple consisting of a long tube, closed at both ends, separated into two chambers by a diaphragm, as shown in Fig. 13.25. A pressure differential is built up across the diaphragm, and the diaphragm is burst, either directly by the pressure differential or initiated by means of an externally controlled probe, or *dagger*. Rupturing of the diaphragm causes a pressure discontinuity, or *shock wave*, to travel into the region of the lower pressure and a rarefaction wave to travel through the chamber of initially higher pressure. The reduced pressure wave is reflected from the end of the chamber and follows the stepped

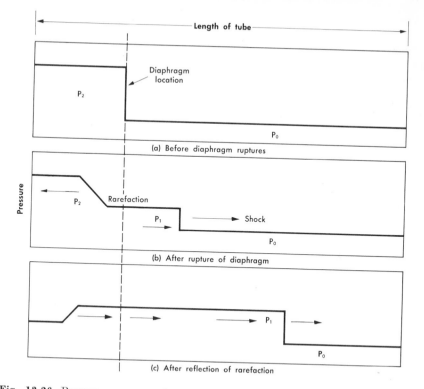

Fig. 13.26 Pressure sequence in a shock tube before and immediately after diaphragm is ruptured. Abscissa represents longitudinal axis of shock tube.

pressure down the tube at a velocity which is higher because it is added to the velocity already possessed by the gas particles from the pressure step. Figure 13.26 illustrates the sequence of events immediately following the bursting of the diaphragm.

A relationship between pressures and shock-wave velocity may be expressed as follows [30]:

$$P_1/P_0 = 1 + [2k/(k + 1)][M_0^2 - 1], \qquad (13.23)$$

in which

$P_1 =$ intermediate transient pressure,

$P_0 =$ lower initial pressure,

$k =$ ratio of specific heats,

$M_0 =$ Mach number corresponding to the lower initial conditions.

It is seen, then, that if the gas properties are known, measurement of the propagation velocity will be sufficient to determine the magnitude of the pressure pulse. Propagation velocity may be determined from information supplied by accurately positioned pressure pickups in the wall of the tube. By this means, a known transient pressure pulse may be applied to a pressure transducer or to a complete pressure-measuring system simply by mounting the pickup in the wall of the shock tube. The response characteristics, as determined in this manner, may then be used to calculate the general response of the device or system over a spectrum of frequencies [31]. The methods for doing this, however, are beyond the scope of this book, and the large amount of computation required is particularly adaptable to the modern digital computer.

Concluding remarks. An attempt has been made in the foregoing pages to introduce the reader to some of the problems attending accurate experimental determination of pressure. It is realized that many approaches to the problem have been omitted and that in certain respects the coverage has been brief and somewhat superficial. Such is a penalty that must be paid in assembling a book of this nature. For more detailed discussions, the reader is referred to the references for this chapter.

SUGGESTED READINGS

ASME Performance Test Code, PTC 19.2-1964 *Pressure Measurement.*

National Bureau of Standards Circular 558, *Bibliography and Index on Dynamic Pressure Measurement*, Washington: U.S. Department of Commerce, 1955.

National Bureau of Standards Monograph 67, *Methods for the Dynamic Calibration of Pressure Transducers*, Washington: U.S. Department of Commerce, 1963.

PROBLEMS

13.1 Determine the factors for converting pressure in psi, to "head" in (a) feet of water, (b) inches of mercury.

13.2 Standard atmospheric pressure is 14.696 psia. What are the equivalents in (a) tons per square foot, (b) feet of water, (c) millibars, (d) microns, and (e) torr?

13.3 A television tube has a face area that is approximately $18 \times 22\frac{1}{2}$ inches in size. If the internal pressure is 10^{-3} microns, what net force is exerted on the face by standard atmospheric pressure?

13.4 Investigate the practical pressure-measuring ranges of the manometer, the bourdon-type gage, and the strain-gage-type pressure cell.

13.5 A simple U-tube manometer (total length of water column $= 15$ in.), was used across an orifice to measure the output air flow from a large single-acting reciprocating compressor. As the machine was being brought up to speed, the meter behaved quite normally until suddenly the water in the manometer began to oscillate violently and was drawn into the air stream. Can you estimate the rpm of the compressor when this happened? What remedy do you suggest for the metering difficulty?

13.6 Express the ratio of sensitivities of an inclined manometer to that of a simple manometer in terms of angle θ (see Fig. 13.5). In order to make an inclined manometer six times as sensitive as a simple manometer, what should be the angle of incline?

13.7 Note that the equation in Fig. 13.5 is based on a moving datum, namely, the liquid level in the reservoir. Derive an equation for the differential pressure based on the movement of the liquid in the inclined column only.

13.8 Confirm Eq. (13.4), Fig. 13.5.

13.9 Confirm Eqs. (13.5) and (13.5a), Fig. 13.6.

13.10 A two-fluid manometer as shown in Fig. 13.6 employs a combination of kerosene (sp. gr. $= 0.80$) and alcohol-diluted water (sp. gr. $= 0.83$). $d = \frac{1}{4}$ in. and $D = 2$ in. What amplification ratio is obtained with this arrangement as compared to a simple water manometer? What error would be introduced if the ratio of diameters were ignored?

13.11 Equation (13.5) may be approximated by $\Delta P/h = \delta_2 - \delta_1$ if the term $(d/D)(\delta_2 + \delta_1)$ is ignored. Assuming that the percent error introduced by omitting this term is no greater than ϵ and that $r = \delta_1/\delta_2$, show that the necessary relationship is

$$\frac{d}{D} \leq \sqrt{\frac{(1-r)\epsilon}{(100 - \epsilon)(r + 1)}}.$$

13.12 A suction tube is permanently installed in a large water tank, as shown in Fig. 13.27. Assuming that the only access to the tank is via the tube, devise a simple arrangement employing a water manometer, whereby the manometer column directly represents the head h in the tank. You may

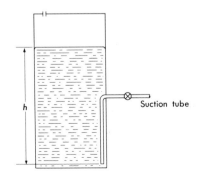

Suction tube

Figure 13.27

assume that the readings will only be taken intermittently, during any pumping standby periods.

13.13 Assume that cans of beer (or Pepsi, Coke, hair spray, or shaving cream) are under pressure. Outline a test procedure whereby the pressure-versus-temperature relationship existing in a randomly selected can of beer (or appropriate substitute) might be experimentally determined. Assume that the temperature range of interest corresponds to the extremes which might be experienced under normal conditions (say, from the back of a beer truck on a hot summer's day to the inside of a refrigerator).

After establishing what appears to be a satisfactory methodology, procede with a test [33].

CHAPTER 14

MEASUREMENT OF FLUID FLOW

14.1 INTRODUCTION

Accurate measurement of flow presents many and varied problems. The flowing medium may be liquid, gaseous, a granular solid, or any combination of these. The flow may be laminar or turbulent, steady-state or transient. In addition, there are several very different *basic* approaches to the problem of flow measurement. This section, therefore, will only present an outline of some of the more important aspects of the general topic.

Flow measurement methods may be categorized according to device or method as follows:

I. Primary or quantity methods

    A. Weight or volume tanks, burettes, etc.

    B. Positive-displacement meters

II. Secondary or rate devices

    A. Obstruction meters

        1. The venturi

        2. Flow nozzles

        3. Orifices

        4. Variable-area meters

    B. Velocity probes

        1. Total-pressure probes

        2. Static-pressure probes

        3. Direction-sensing probes

    C. Special methods

        1. Turbine-type meters

        2. Thermal or hot-wire meters

        3. Magnetic flowmeters

        4. Sonic flowmeters

        5. Mass flowmeters

The above outline does not exhaust the list of flow-measuring systems, but does attempt to include those of primary interest to the mechanical engineer. Application of some of the methods listed is so obvious that only passing note will be made of them. This is particularly true of *quantity* methods. Weight tanks are especially useful for steady-state calibration of liquid flowmeters, and no particular problems are connected with their use.

There are many forms or variations of displacement meters. Common examples are the water and gas meters used by suppliers to establish charges for services. Basically, displacement meters are hydraulic or pneumatic motors whose cycles of motion are recorded by some form of counter. Only such energy from the stream is absorbed as is necessary to overcome the friction in the device, and this is manifested by a pressure drop between inlet and outlet. Most of the configurations used for motors have been applied to metering. These include reciprocating and oscillating pistons, vane arrangements, including the nutating (or nodding) disk, helical screw devices, etc.

14.2 FLOW CHARACTERISTICS

When fluids move through uniform conduits at very low velocities, the motions of individual particles are generally along lines paralleling the conduit walls. Actual particle velocity is greatest at the center and theoretically zero at the wall, with the velocity distribution as shown in Fig. 14.1(a). Plots of individual loci are called *streamlines*, and the flow is termed *laminar* or *viscous*.

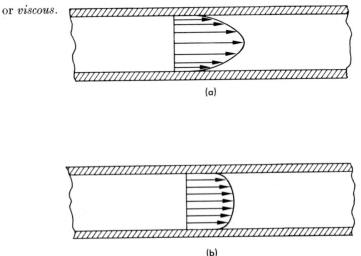

(a)

(b)

Fig. 14.1 Velocity distribution for (a) laminar flow in a pipe or tube, (b) turbulent flow in a pipe or tube.

As the flow rate is increased, a point is reached when the particle motion becomes more random and complex. Although this change in the nature of flow may appear to occur at a very definite velocity, careful observation will show that the change is somewhat gradual over a relatively narrow range of velocities. The *approximate* velocity at which the change occurs is called the *critical velocity*, and the flow at higher rates is referred to as *turbulent*. The corresponding velocity distribution across a circular tube is shown in Fig. 14.1(b).

It has been found that the critical velocity is a function of several factors that may be put in a dimensionless form called the Reynolds number, R_D,* as follows:

$$R_D = D\rho V/\mu, \tag{14.1}$$

where

$D =$ a sectional dimension of the fluid stream (normally the diameter if the conduit is a pipe of circular section), ft,

$\rho =$ mass density of the fluid, slugs/ft³,

$V =$ fluid velocity, ft/sec,

$\mu =$ absolute viscosity of the fluid, lb-sec/ft².

Of course, consistent dimensional units must be used, usually those that are indicated, although hours are sometimes used instead of seconds.

It has been shown by many investigators [1] that below the critical-velocity range, friction loss in pipes is a function of R_D only, while for turbulent flow, the Reynolds number combined with surface roughness determines the losses. The critical-velocity range for pipes usually falls between 2000 and 2300.

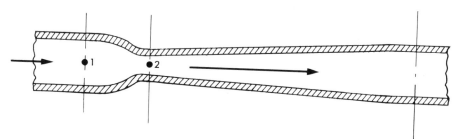

Fig. 14.2 Section through a restriction in a pipe or tube.

---

* The subscript D is employed to indicate nominal pipe diameter. When Reynold's number is based, for example, on the throat diameter of a venturi or a nozzle, the lower-case d is commonly employed.

Bernoulli's equation for the flow of incompressible fluids between points 1 and 2 in Fig. 14.2 may be written

$$P_1/\gamma_1 - P_2/\gamma_2 = (V_2^2 - V_1^2)/2g + Z_2 - Z_1, \qquad (14.2)$$

in which

P = pressure, lb/ft^2,

γ = specific weight, lb/ft^3,

V = linear velocity, ft/sec,

Z = elevation, ft,

G = acceleration due to gravity, 32.2 ft/sec^2.

As written above, the relationship assumes that there is no mechanical work done on or by the fluid and that there is no heat transferred to or from the fluid as it passes between points 1 and 2. This equation provides the basis for evaluating the operation of flow-measuring devices generally classified as *obstruction meters*.

14.3 OBSTRUCTION METERS

Figure 14.3 shows three common forms of obstruction meters, the venturi, the flow nozzle, and the orifice. In each case the basic meter acts as an obstacle placed in the path of the flowing fluid, causing localized changes in velocity. Concurrently with velocity change, there will be pressure change, as illustrated in the figure. At points of maximum restriction, hence maximum velocity, minimum pressures are found. A certain portion of this pressure drop becomes irrecoverable; therefore, the output pressure will always be less than the input pressure. This is indicated in the figure, which shows the venturi, with its guided re-expansion, to be the most efficient. Losses in the order of 30–40% of the differential pressure occur through the orifice meter.

a) Obstruction meters for incompressible fluids. For incompressible fluids, $\gamma_1 = \gamma_2 = \gamma$ and $Q = A_1V_1 = A_2V_2$, where Q represents the quantity of flowing fluid per unit time in cubic feet per second. Substituting $V_1 = (A_2/A_1)V_2$ in Eq. (14.2), we obtain

$$P_1 - P_2 = \frac{V_2^2 \gamma}{2g}\left[1 - \left(\frac{A_2}{A_1}\right)^2\right] \qquad (14.3)$$

and

$$Q_{\text{ideal}} = A_2V_2 = \frac{A_2}{\sqrt{1 - (A_2/A_1)^2}}\sqrt{2g\left(\frac{P_1 - P_2}{\gamma}\right)}. \qquad (14.4)$$

For a given meter, A_1 and A_2 have definite values, and it is often convenient

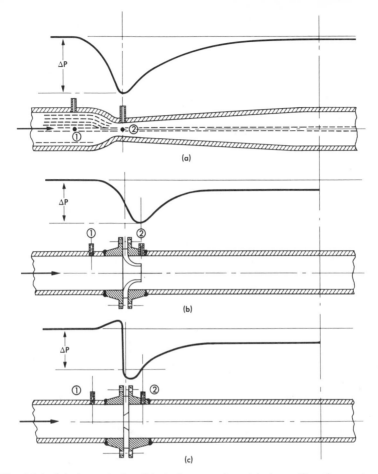

Fig. 14.3 (a) A venturi. (b) A flow-nozzle. (c) An orifice flowmeter.

to calculate

$$M = \frac{1}{\sqrt{1 - (A_2/A_1)^2}} .$$ (14.5)

This quantity is termed the *velocity-of-approach* factor.

Two other factors used with obstruction meters are the *discharge coefficient*, C, and the *flow coefficient*, K. These may be defined as follows:

$$C = \frac{Q_{\text{actual}}}{Q_{\text{ideal}}}$$

and

$$K = CM.$$

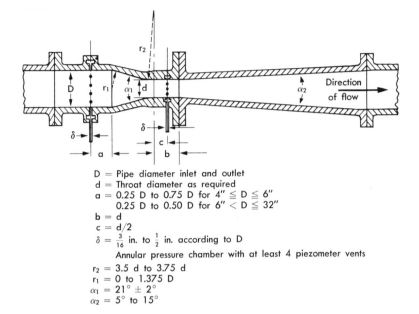

D = Pipe diameter inlet and outlet
d = Throat diameter as required
a = 0.25 D to 0.75 D for 4″ ≤ D ≤ 6″
    0.25 D to 0.50 D for 6″ < D ≤ 32″
b = d
c = d/2
$\delta = \frac{3}{16}$ in. to $\frac{1}{2}$ in. according to D
    Annular pressure chamber with at least 4 piezometer vents
r_2 = 3.5 d to 3.75 d
r_1 = 0 to 1.375 D
α_1 = 21° ± 2°
α_2 = 5° to 15°

Fig. 14.4 Recommended proportions of Herschel-type venturi tubes. (*ASME Power Test Codes*, "Instruments and Apparatus," Part 5, Chapter 4, 1959.)

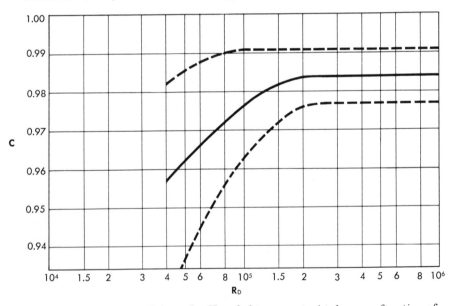

Fig. 14.5 Discharge coefficients for Herschel-type venturi tubes as a function of the Reynolds number. Applicable to values of diameter ratio from 0.25 to 0.75 in pipes two inches and larger. The tolerance limits are shown by dashed lines. Velocity of approach factor not included. (*ASME Power Test Codes*, "Instruments and Apparatus," Part 5, Chapter 4, 1959.)

The discharge coefficient C is the factor which accounts for losses through the meter, while the flow coefficient, K, is used as a matter of convenience, combining the loss factor with the meter constants. The quantities C and M are usually applied to venturis, while the combined term, K, is used with nozzles and orifices. Therefore, we may write for venturi meters,

$$Q_{\text{actual}} = CMA_2\sqrt{2g/\gamma}\,\sqrt{P_1 - P_2} \tag{14.6a}$$

and for nozzles and orifices,

$$Q_{\text{actual}} = KA_2\sqrt{2g/\gamma}\,\sqrt{P_1 - P_2}. \tag{14.6b}$$

b) Venturi characteristics. Venturi proportions are not standardized. However, the dimensional ranges shown in Fig. (14.4) include most cases. Typical relations between discharge coefficients and Reynolds numbers is shown in Fig. 14.5.

c) Flow-nozzle characteristics. Figure 14.6 illustrates examples of several "standard" types of flow nozzles. The approach curve must be proportioned to prevent separation between the flow and the wall, and the parallel section is used to ensure that the flow fills the throat. Typical flow characteristics are shown in Fig. 14.7 ($\beta = d/D$).

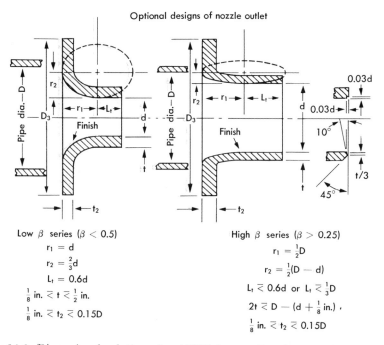

Fig. 14.6 Dimensional relations for ASME long-radius flow nozzles. (*ASME Power Test Codes*, "Instruments and Apparatus," Part 5, Chapter 4, 1959.)

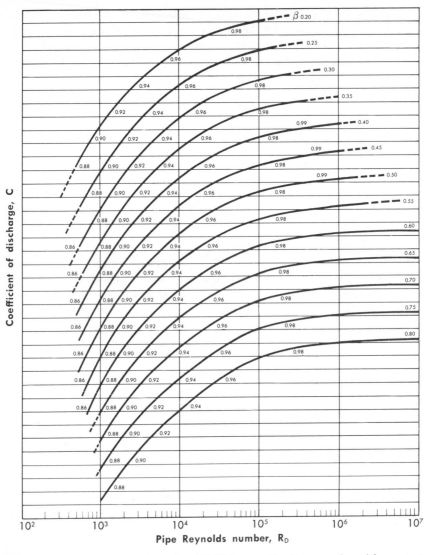

Fig. 14.7 Discharge coefficients for ASME long-radius flow nozzles with pressure taps located one pipe diameter preceding and one-half pipe diameter following the inlet face. Velocity of approach factor not included. (*ASME Power Test Codes*, "Instruments and Apparatus," Part 5, Chapter 4, 1959.)

d) Orifice characteristics. The primary variables in the use of flat-plate orifices are ratio of orifice to pipe diameter, tap locations, and characteristics of orifice sections. Various configurations of bevels and rounded edges are sometimes employed in seeking particular performance characteristics, especially constant coefficients at low Reynolds numbers. Figure 14.8(a) illustrates a typical orifice installation and recommended range of tap

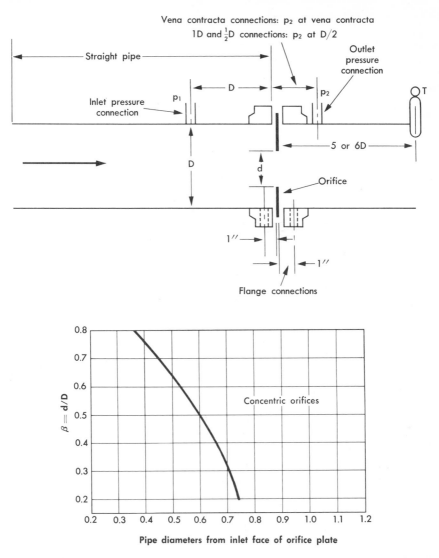

Fig. 14.8 (a) Location of pressure taps used with concentric, thin-plate, square-edge orifices. (b) Location of outlet pressure connections for vena contracta taps. (*ASME Power Test Codes*, "Instruments and Apparatus," Part 5, Chapter 4, 1959.)

locations. Figure 14.8(b) may be used for determining the approximate location of the *vena contracta* (smallest stream diameter). Figure 14.9 shows typical orifice flow coefficients ($\beta = d/D$).

e) Relative merits of the venturi, flow nozzle, and orifice. High accuracy, good pressure recovery, and resistance to abrasion are the primary advan-

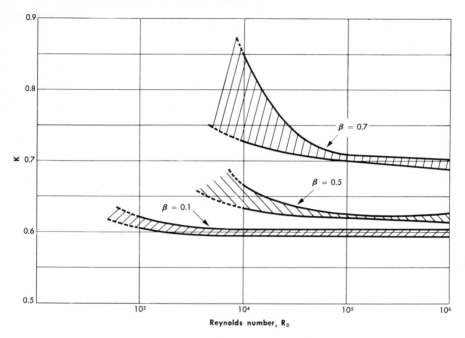

Fig. 14.9 Ranges of flow coefficients for flat-plate orifices. (From data in *ASME Power Test Codes*, "Instruments and Apparatus," Part 5, Chapter 4, 1959.)

tages of the venturi. These are offset, however, by considerably greater cost and space requirements as compared with the orifice and nozzle. The orifice is inexpensive, and may often be installed between existing pipe flanges. However, its pressure recovery is poor, and it is especially susceptible to inaccuracies resulting from wear and abrasion. It may also be damaged by pressure transients because of its lower physical strength. The flow nozzle possesses the advantages of the venturi, except that it has lower pressure recovery, plus the added advantage of shorter physical length. It is expensive compared with the orifice and is relatively difficult to install properly.

Example 1. A venturi designed in accordance with the specifications of Fig. 14.4 is placed in an 8-inch-diameter line passing 500 gallons of water per minute. If the throat diameter is 4 in. what differential pressure may be expected across the pressure taps? Assume a water temperature of 70°F.

Solution:

$$Q = 500 \text{ gpm} = 66.8 \text{ ft}^3/\text{min} = 1.11 \text{ ft}^3/\text{sec},$$
$$A_1 = 0.348 \text{ ft}^2, \qquad A_2 = 0.087 \text{ ft}^2,$$
$$V_1 = Q/A_1 = 3.2 \text{ ft/sec}, \qquad V_2 = Q/A_2 = 12.8 \text{ ft/sec}.$$

For water at 70°F (see Table D.3, Appendix D),

$$\rho = 62.3/32.2 = 1.93 \text{ slug/ft}^3,$$

$$\mu = 2.02 \times 10^{-5} \text{ slug/ft-sec},$$

$$R_D = D\rho V/\mu = (8/12)(1.93)(3.2)/(2.02 \times 10^{-5}) = 2.10 \times 10^5.$$

From Fig. 14.5, $C = 0.984$:

$$M = 1/\sqrt{1 - (A_2/A_1)^2} = 1.033.$$

Substituting in Eq. (14.6a), we obtain

$$1.11 = (0.984)(1.033)(0.087)\sqrt{(2 \times 32.2/62.3)} \sqrt{P_1 - P_2},$$

$$P_1 - P_2 = 152.5 \text{ lb/ft}^2 = 1.272 \text{ psi}.$$

14.4 OBSTRUCTION METERS FOR COMPRESSIBLE FLUIDS

When compressible fluids flow through obstruction meters of the types discussed in the previous article, the specific weight does not remain constant during the process; that is, $\gamma_1 \neq \gamma_2$. The usual practice is to base the energy relation, Eq. (14.2), on the specific weight at condition 1 (Fig. 14.2) and to introduce an *expansion factor*, Y, as follows:

$$W = KA_2\gamma_1 Y \sqrt{2g\frac{P_1 - P_2}{\gamma_1}}, \tag{14.7}$$

where W = flow rate, lb/sec.

The expansion factor, Y, may be determined theoretically for nozzles and venturis and experimentally for orifice meters. Theoretical values may be calculated from the following relations [2]

$$Y = \left\{ \left(\frac{P_2}{P_1}\right)^{2/k} \left(\frac{k}{k-1}\right) \left[\frac{1 - (P_2/P_1)^{(k-1)/k}}{1 - (P_2/P_1)}\right]\right.$$
$$\left. \times \left[\frac{1 - (A_2/A_1)^2}{1 - (A_2/A_1)^2(P_2/P_1)^{2/k}}\right]\right\}^{1/2}, \tag{14.8}$$

in which

$$k = \frac{\text{specific heat at constant pressure}}{\text{specific heat a constant volume}}.$$

For square-edged orifices, an empirical relation has been developed [3], which is expressed as follows:

$$Y = 1 - \left[0.41 + 0.35\left(\frac{A_2}{A_1}\right)^2\right]\frac{P_1 - P_2}{kP_1}. \tag{14.8a}$$

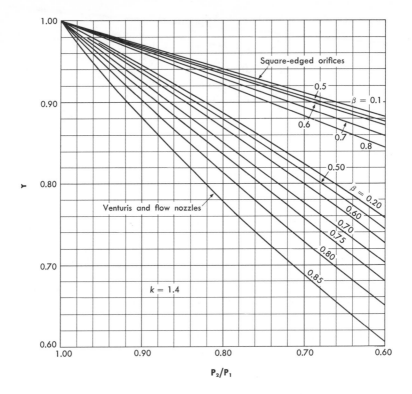

Fig. 14.10 Expansion factors Y versus pressure ratio. Upper curves are for square-edged concentric orifices, and lower are for venturis and flow nozzles.

Expansion factors, Y, for venturis and orifices, for $k = 1.4$, are shown plotted against pressure ratio in Fig. 14.10 ($\beta = d/D$).

Example 2. A sharp-edged concentric orifice is used to measure the flow of 90°F air through a 10-in. diameter circular-sectioned duct. Estimate the flow rate if the differential pressure between *vena contracta* taps is 8.5 in. of 0.92 specific gravity oil, the upstream pressure P_1 is 60 pisa, and $\beta = 0.6$.

$$W = KA_2 \gamma_1 Y \sqrt{2g(P_1 - P_2)/\gamma_1},$$
$$A_1 = 0.546 \text{ ft}^2, \qquad A_2 = 0.196 \text{ ft}^2, \qquad A_2/A_1 = 0.36,$$
$$K = CM.$$

From Fig. 14.9, we estimate that $C = 0.62$; then

$$M = 1/\sqrt{1 - (A_2/A_1)^2} = 1.07,$$
$$K = 0.62 \times 1.07 = 0.664.$$

From Table D.4, Appendix D,

$$\gamma_1 = 0.0722(60/14.7) = 0.295 \text{ lb/ft}^3,$$
$$P_1 - P_2 = (8.5/12)(62.4)(0.92) = 40.7 \text{ lb/ft}^2 \qquad \text{or} \qquad 0.282 \text{ psi},$$
$$(P_1 - P_2)/P_1 = 0.282/60 = 0.0047.$$

From Eq. (14.8a),

$$Y = 1 - [0.41 + (0.35)(0.36)^2](0.0047/1.4) = 0.998,$$
$$W = (0.664)(28.27/144)(0.295)(0.998)\sqrt{(64.4)(40.7)/(0.295)}$$
$$= 3.61 \text{ lb/sec}.$$

To check our estimate of C,

$$R_D = D\rho V/\mu,$$
$$D = 0.833 \text{ ft}, \qquad V = W/A\gamma = 3.61/0.546 \times 0.295 = 22.4 \text{ ft/sec},$$
$$\rho = 0.295/32.2 = 0.00915 \text{ slug/ft}^3,$$
$$\mu = 0.390 \times 10^{-6} \text{ (at 14.7 psia) (use } 0.392 \times 10^{-6} \text{ as per Note 1 Table D.4)},$$
$$R_D = (0.833)(0.00915)(22.4)/(0.392)(10^{-6})$$
$$= 0.436 \times 10^6.$$

Inspection of Fig. 14.9 shows that the original estimate of C is sufficiently good to make a refinement in the calculations unnecessary.

14.5 THE VARIABLE-AREA METER

A major disadvantage of the common forms of obstruction meters (the venturi, the orifice, and the nozzle) is that the pressure drop varies as the square of the flow rate [Eq. (14.6)]. This means that if these meters are to be used over a wide range of flow rates, pressure-measuring equipment of very wide range will be required. In general, if the range is accommodated, accuracy at low flow rates will be poor. One solution would be to use two (or more) pressure-measuring systems: one for low flow rates and another for high rates.

A device whose indication is essentially linear with flow rate is shown in section in Fig. 14.11. This instrument is a variable-area meter, commonly called a *rotameter*. Two parts are essential, the float and the tapered tube in which the float is free to move. The term "float is somewhat a misnomer, because it must be heavier than the liquid it displaces. As flow takes place upward through the tube, four forces act on the float: a downward gravity force, an upward buoyant force, pressure, and viscous drag forces.

For a given rate of flow, the float assumes a position in the tube where the forces acting on it are in equilibrium. Through careful design, the effects of changing viscosity or density may be minimized, leaving only the pressure force as a variable. The latter is dependent on flow rate and the annular area between it and the tube. Hence its position will be determined by the flow rate alone. A basic equation for the rotameter has been developed [4] in the following form:

$$Q = A_w C \left[\frac{2gv_f(\rho_f - \rho_w)}{A_f \rho_w} \right]^{1/2}, \quad (14.9)$$

where

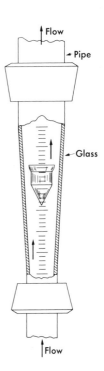

Q = volumetric rate of flow,

v_f = volume of float,

g = acceleration due to gravity,

ρ_f = float density,

ρ_w = liquid density,

A_f = area of float,

C = discharge coefficient,

A_w = area of annular orifice
    = $\pi/4[(D + by)^2 - d^2]$,

D = effective diameter of tube depending on position of float,

b = change in tube diameter per unit change in height,

d = maximum diameter of float,

y = height of float above zero position.

Fig. 14.11 A rotameter.

Certain disadvantages of the rotameter as compared with the other forms of obstruction meter are: the meter must be installed in a vertical position; the float may not be visible when opaque fluids are used; they cannot be used with liquids carrying large percentages of solids in suspension; for high pressures or temperatures, they are expensive. Advantages include: a uniform flow scale over the range of the instrument, with the pressure loss fixed at all flow rates; the capacity may be changed with relative ease by changing float and/or tube; many corrosive fluids may be handled without complication; the condition of flow is readily visible.

14.6   MEASUREMENTS OF FLUID VELOCITIES

Flow is generally proportional to some flow velocity; hence, by measuring the velocity, a measure of flow is obtained. Often velocity *per se* is desired, particularly velocity *relative* to a fluid. An example of the latter is an aircraft moving through the air. The following articles will deal with measurement of absolute and relative velocities of fluids.

14.7 PRESSURE PROBES

Point measurement of pressure is accomplished by use of tubes joining the location in question with some form of pressure transducer. A *probe* or sampling device is intended insofar as possible to obtain a reliable and interpretable indication of the pressure at the signal source. Therein lies a difficulty, however, for the mere presence of the probe will alter, to some extent, the quantity being measured.

A common reason for desiring point-pressure information is to determine flow conditions. Flowing media may be gaseous or liquid in a symmetrical conduit or pipe or in a more complex situation such as in a jet engine or compressor.

There are many different types of pressure probes, with the selection depending on the information required, space available, pressure gradients, and constancy of flow magnitude and direction. Basically, pressure probes measure either of two different pressures or some combination thereof. In Article. 13.1 we briefly discussed *static* and *total* pressures and indicated that the difference is a result of the flow velocity. This may be expressed as

$$P_t = P_s + P_v.$$ (14.10)

Referring to Eq. (14.2), for incompressible fluids we may write

$$P_t = P_s + \frac{\gamma V^2}{2g},$$ (14.11)

in which

P_t = total pressure,

P_s = static pressure,

P_v = velocity pressure,

γ = specific weight of the fluid,

V = velocity,

g = acceleration of gravity.

Solving for velocity, we obtain

$$V = \sqrt{\frac{2g(P_t - P_s)}{\gamma}}.$$ (14.11a)

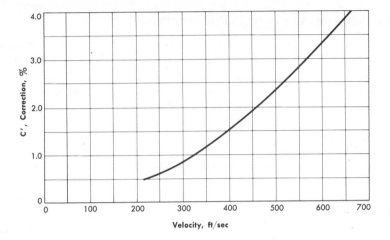

Fig. 14.12 Approximate velocity corrections for fluid compressibility based on air at atmospheric pressure.

From this we see that velocity may be determined simply by measuring the difference between the total and static pressures.

Actually, Eqs. (14.11) and (14.11a) should only be applied to incompressible fluids. However, the effect of compressibility is usually handled through use of a correction factor rather than more correct but unwieldy equations. Figure 14.12 provides correction factors applied as follows [3].

$$V = (1 - C')\sqrt{\frac{2g(P_t - P_s)}{\gamma}}. \tag{14.11b}$$

Example 3. A pitot-static tube is employed to determine the velocity of air at the center of a pipe. Static pressure is 18 psia, the air temperature is 80°F, and a differential pressure of 3.8 in. of water is measured. What is the air velocity in ft/sec? Assume that $C' = 0.02$.

Solution. From Table D.4, at 14.7 psia,

$$\gamma_{80} = 0.0735 \text{ lb/ft}^3.$$

At 18 psia,

$$\gamma_{80} = 0.0735(18/14.7) = 0.090 \text{ lb/ft}^3.$$
$$\Delta P = P_t - P_s = (3.8 \times 144)/(12 \times 2.31)$$
$$= 19.75 \text{ lb/ft}^2.$$

Using Eq. (14.11b), we obtain

$$V = 0.98\sqrt{(64.4 \times 19.75)/0.090}$$
$$= 116.5 \text{ ft/sec.}$$

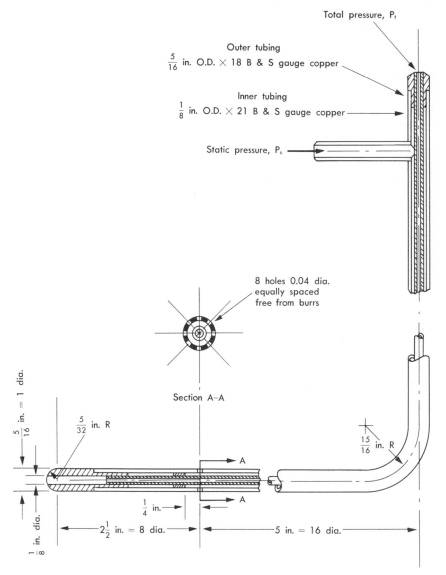

Fig. 14.13 A pitot-static tube.

When velocity is used to measure flow rate, consideration must be given to the velocity distribution across the channel or conduit. A mean may be found by traversing the area to determine the velocity profile, from which the average may be calculated, or a multiplication constant may be determined by calibration (see Problem 14.15).

a) Total-pressure probes. Obtaining a measure of total or impact pressure is usually somewhat easier than getting a good measure of static pressure. This is true except for something such as an open jet, when a barometer reading may be used for the static component. The simple pitot tube (named for Henri Pitot) shown at *A* in Fig. 13.2 is usually adequate for determining impact pressure. More often, however, the pitot tube is combined with static openings, constructed as shown in Fig. 14.13. This is known as a pitot-static tube, or sometimes as a Prandtl-Pitot tube. For steady-flow conditions, a simple differential manometer, often of the inclined type, suffices for pressure measurement, and $P_t - P_s$ is determined directly. When variable conditions exist, some form of pressure transducer, such as one of the diaphragm types, may be used. Of course, care must be exercised in providing adequate response, particularly in the connecting tubing (see Article 13.11).

A major problem in the use of an ordinary pitot-static tube is to obtain proper alignment of the tube with flow direction. The angle formed between the probe axis and the flow streamline at the pressure opening is termed the

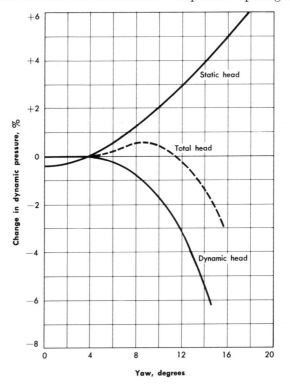

Fig. 14.14 Yaw sensitivity of a standard pitot-static tube.   (Courtesy The Airflo Instrument Company, Glastonbury, Connecticut.)

yaw angle. This should be zero, but in many situations the yaw angle may not be constant: the flow may be neither fixed in magnitude nor in direction. In such cases, yaw sensitivity is very important. The pitot-static tube is particularly sensitive to yaw, as shown in Fig. 14.14. Although sensitivity is influenced by orientation of both impact and static openings, the latter probably has the greater effect.

The Kiel tube, designed to measure total or impact pressure only (there are no static openings), is shown in Fig. 14.15. It consists of an impact tube surrounded by what is essentially a venturi. The curve demonstrates the striking insensitivity of this type to variations in yaw. Modifications of the Kiel tube employ a cylindrical duct, beveled at each end, rather than the streamlined venturi. This appears to have little effect on the performance and makes the construction much less expensive.

b) Static-pressure probes. Static-pressure probes have been used in many different forms [5]. Ideally, the simple opening with axis normal to flow direction should be satisfactory. However, slight burrs or yaw introduce

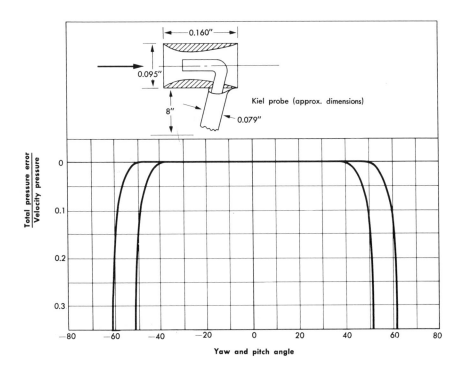

Fig. 14.15 Kiel-type total-pressure tube and plot of yaw sensitivity. (Courtesy The Airflo Instrument Company, Glastonbury, Connecticut.)

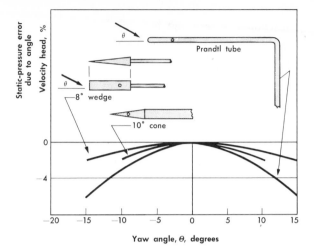

Fig. 14.16 Angle characteristics of certain static-pressure-sensing elements. (Courtesy Instrument Society of America.)

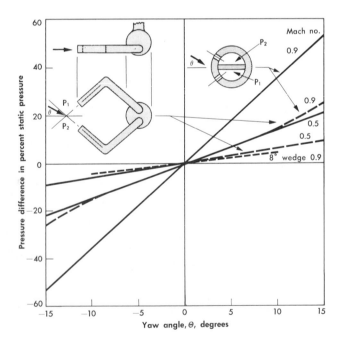

Fig. 14.17 Special direction-sensing elements and their yaw characteristics. (Courtesy Instrument Society of America.)

appreciable errors. As mentioned previously, in many situations yaw angle may be continually changing. For these reasons, special static-pressure probes may be used. Figure 14.16 shows several probes of this type and corresponding yaw sensitivities.

As mentioned earlier, the mere presence of the probe in a pressure-flow situation alters the parameters to be measured. Probes interact with other probes, with their own supports, and with duct or conduit walls. Such interaction is primarily a function of geometry and relative dimensional proportions; it is also a function of Mach number. Much work has been conducted in this area, and the interested reader is referred to References [6] and [7] for a review of some of these efforts.

c) Direction-sensing probes. Figure 14.17 illustrates two forms of direction-sensing or yaw-angle probes. Each of these probes employ two impact tubes. In each case the probe is placed transverse to flow and is rotatable around its axis. The angular position of the probe is then adjusted until the pressures sensed by the openings are equal. When this is the case, the flow direction will correspond to the bisector of the angle between the openings. Probes are also available with a third opening midway between the other two. The additional hole, when properly aligned, senses maximum impact pressure.

14.8 SPECIAL FLOW-MEASURING METHODS AND DEVICES

While the foregoing discussion covers the common methods of flow measurement, there are many additional methods of relatively specialized nature. In this article we shall consider some of the more important types.

a) Turbine-type meters. The familiar anemometer employed by weather stations to measure wind velocity is a simple form of free-stream turbine meter. Somewhat similar rotating-wheel flowmeters are used by civil engineers to measure water flow in rivers and streams [8]. Both the cup-type rotors and the propeller types are used for this purpose. In each case the number of turns of the wheel per unit time is counted and used as a measure of the flow rate.

Figure 14.18 illustrates a modern adaptation of these methods to the measure of flow in tubes and pipes. Wheel motion, proportional to flow rate, is sensed by a reluctance-type pickup coil. A permanent magnet is encased in the rotor body, and each time a wheel blade passes the pole of the coil, change in permeability of the magnetic circuit produces a voltage pulse at the output terminal. The pulse rate may then be indicated by a frequency meter, displayed on a CRO screen, or counted by some form of EPUT meter. Frequency converters are also available which convert flowmeter pulses to a proportional d-c output, permitting use of simple

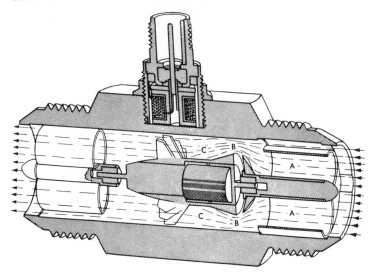

Fig. 14.18 Potter Aeronautical Company turbine-type flowmeter. (Courtesy Potter Aeronautical Company, Newark, N.J.)

meters for indication. Accuracies within $\pm\frac{1}{2}\%$ are claimed for these devices within specific flow range. Available sizes cover the range from $\frac{1}{8}$ inch to 8 inches. Transient response is good; time constants for stepped pressure pulses are in the order of 2 to 12 milliseconds [9].

In addition to bearing maintenance, a major problem inherent in this type of meter is reduced accuracy at low flow rates. Maximum to minimum capacities vary from about 8 to 1 for small meters to about 40 to 1 for the large sizes.

b) Thermal methods. When an electrically heated wire is placed in a flowing stream (assumed gaseous), heat will be transferred between the two, depending on a number of factors, including the flow rate. This type of flow-sensing element is called the *hot-wire anemometer*. The element consists of a short length of fine wire stretched between two supports, such as shown in Fig. 14.19. Two methods are employed to measure flow. The first technique employs a constant current passing through the sensing wire. Variation in flow results in changed wire temperature, hence changed resistance, which thereby becomes a measure of flow. The second technique employs a servo system to maintain wire resistance, hence wire temperature. In this case, a change in flow results in a corresponding change in electrical current. The latter is then interpreted as a flow analog. The two methods are termed *constant-current* and *constant-temperature*, respectively.

When the hot wire is placed in a flowing stream, heat will be transferred from the wire, primarily by convection. Radiation and conduction are

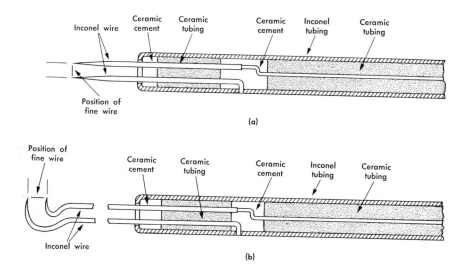

Fig. 14.19 Two forms of hot-wire, high-temperature anemometer probes. (a) Wire mounted normal to probe axis. (b) Wire mounted parallel to probe axis.

normally negligible. The rate of heat transfer will depend upon a number of factors included in the three dimensionless quantities, the Nusselt, Reynolds, and Prandtl numbers. Governing relations have been written as follows [10, 11, 12]:

$$\frac{\text{Power/(unit length)}}{\text{Temperature difference}} = \frac{i^2 R}{T_w - T_a} = A + B\sqrt{\rho V}, \qquad (14.12)$$

where A and B are constants and

$i =$ instantaneous current,

$R =$ resistance of wire per unit length,

$T_w =$ temperature of wire

$T_a =$ ambient temperature,

$\rho =$ density of gas,

$V =$ free-stream velocity.

When the flow past the wire is varying with time, the sensing-element response will lag behind the actual fluctuations because of the heat capacity of the wire. To a considerable extent, it is possible to compensate for lag of this type in the electrical circuitry. This is illustrated in Fig. 14.20.

When a constant-current system is employed, compensation is achieved by means of a passive network (Article 7.25c) of inductance and resistance, or of capacitance and resistance, or through use of a suitable transformer.

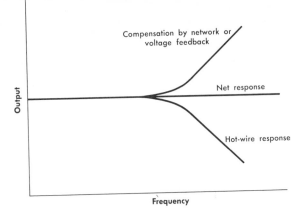

Fig. 14.20 Effect of compensation on hot-wire anemometer response characteristics.

However, compensation must be adjusted for the particular flow condition being measured and, in general, is usable only for fluctuations having magnitudes up to about 15% of the mean stream velocity.

In the constant-temperature anemometer, compensation forms an inherent part of the basic system. The sensing element is incorporated in an electrical bridge whose unbalance is employed as a measure of current required to maintain wire temperature. Fluctuations of the required current thereby become a measure of the flow variations. An advantage of this system is that a wide range of conditions does not affect the ability of the system to provide flat response. Another very practical advantage is that the system provides inherent protection against wire burn-out.

c) *Magnetic flowmeters.* Magnetic flowmeters are based on Faraday's law of induced voltage, expressed by the relation [13]

$$e = Blv \times 10^{-8}, \qquad (14.13)$$

where

$$e = \text{induced voltage, volts,}$$

$$B = \text{flux density, gauss,}$$

$$l = \text{length of conductor, cm,}$$

$$v = \text{velocity of conductor, cm/sec.}$$

Basic flowmeter arrangement is as shown in Fig. 14.21. The flowing medium is passed through a pipe, a short section of which is subjected to a transverse magnetic flux. Fluid motion relative to the field causes a voltage to be induced proportional to the fluid velocity. This emf is detected by

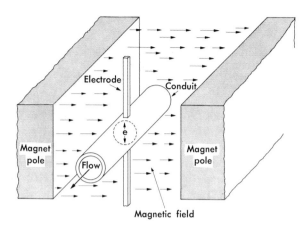

Fig. 14.21 Schematic showing operation of a magnetic flowmeter.

electrodes placed in the conduit walls. Either an alternating or direct magnetic flux may be employed. However, if amplification of the output is required, the advantage lies with the alternating field.

Two basic types have been developed. In the first, the fluid need be only slightly electrically conductive, and the conduit must be of glass or some similar nonconducting material. The electrodes are placed flush with the inner conduit surfaces making direct contact with the flowing fluid. Output voltage is quite low, and an alternating magnetic field is used for amplification and to eliminate polarization problems. Special circuitry is required to separate the no-flow output from the signal caused by flow [14].

A second form of magnetic flowmeter is primarily intended for use with highly conductive fluids such as liquid metals. This meter operates on the same basic principle but may employ electrically conducting materials for the conduit. Stainless steel is commonly used. A permanent magnet supplies the necessary flux, and the electrodes may be simply attached to diametrically opposite points on the *outside* of the pipe. This provides for easy installation at any time and at any point along the pipe. The output of this type is sufficient to drive ordinary commercial indicators or recorders, and zero output for nonflow conditions is an added advantage [15].

d) An ultrasonic flowmeter [16]. Another type of noncontacting rate meter for measuring flow in pipes employs ultrasonic waves. Similar piezoelectric or magnetostrictive transducers are placed externally on the conduit a few inches apart. One serves as a 100-kHz energy source and the other as the pickup. As the wave travels from source to pickup, its normal velocity in stationary fluid will be either increased or decreased by the fluid velocity, depending on the relative motions. In order to minimize errors, the functioning of the two transducers is reversed 10 times per second. In this manner, phase shifts due to both addition and subtraction of the velocities are employed. The relative phase shift is used as the measure of flow rate.

14.9 PREDICTABILITY OF FLOW-METER PERFORMANCE

The foregoing discussion of flow-measuring devices and procedures is based to a great extent upon practical information assembled and published by the American Society of Mechanical Engineers (see general references at the end of this chapter). Tables and charts of coefficients for various and precisely prescribed metering methods have evolved through the years from accumulated experiences of many different people and interested commercial and research organizations. This material, stemming from committee actions, probably forms as useful and reliable a guide to flow measurement as is available, and when diligently adhered to, yields very satisfactory and practical results. It is recognized, however, that there must be a "limit of accuracy" applied to the values of any generally expressed discharge or flow coefficients or expansion factors and that the charts can only be accurate within a specific tolerance range, and further, that such tolerances can only be approximately determined.

The tolerance ranges in Table 14.1 have been abstracted from ASME Performance Test Code, PTC 19.5, 4-1959, *Flow Measurement*, and apply to coefficient and factor data presented therein. From this tabulation it is seen

TABLE 14.1

TOLERANCE RANGES FOR THEORETICAL FLOW-MEASUREMENT
COEFFICIENTS AND FACTORS

Type of element	Tolerances (in percent of coefficient or factor)	
	For discharge or flow coefficients	For expansion factors
Venturi tubes, Herschel type	$\pm$ 0.75 to $\pm$ 2.25	0.0 to $\pm$ 0.25
Flow nozzles, long radius	$\pm$ 0.95 to $\pm$ 1.50	$\pm$ 0.25 to $\pm$ 1.0
Square-edge concentric orifices	$\pm$ 0.50 to $\pm$ 1.00	0.0 to $+$ 2.0

Adapted from ASME Performance Test Code, PTC 19.5, 4–1959, *Flow Measurement*. Reproduced by permission.

that the coefficients may be quite accurately predicted. Of course, it must be remembered that these figures do not include errors of observation, errors of subsequent signal conditioning apparatus, etc. For more accurate work, calibration of individual meters along with associated instrumentation is required.

14.10 CALIBRATION OF FLOW-MEASURING DEVICES

Facilities for producing standardized flows are required for flow-meter calibration. Fluid at known rates of flow must be passed through the meter and the rate compared with the meter readout. When the basic flow input is determined through measurement of time and either linear dimensions (volumetric flow) or weight (mass flow), the procedure may be termed *primary calibration*. After receiving a primary calibration, a meter may then be employed as a *secondary standard* for standardizing other meters through *comparative calibration*.

Primary calibration is usually carried out at a constant flow rate with the procedure consisting of an integration or summing of the total flow for a predetermined period of time. Volumetric displacement of a liquid may be measured in terms of the liquid level in a carefully measured tank or container. For a gas, at moderate rates, volume may be determined through use of an inverted bell-type *gasometer* or "meter prover." Primary calibration in terms of mass is commonly accomplished by means of the familiar *weigh tank* in which the liquid is collected and weighed. Although the latter method is normally employed only for liquids, with proper facilities it may also be used for calibration with gases [18].

Primary calibration is usually considered only at relatively low flow rates; however, on occasion, the cost of large-scale facilities can be justified by the importance of the application. An unusual system adapted to liquid hydrogen flow calibrations, at rates to 7000 gallons per minute, is described in Reference [19]. Two 50,000-gallon dewars and an interconnecting flow loop were employed and volume rates were determined by means of liquid-level gages. Calibration errors of no more than 4% were claimed for this unusual application.

Secondary calibration may be either *direct* or *indirect*. Direct secondary calibration is accomplished by simply placing a secondary standard in series with the meter to be calibrated and comparing their respective readouts over the desired range of flow rates. Turbine-type meters are particularly useful as secondary standards for "field" calibration of orifice or venturi meters [20]. It is clear that this procedure requires careful consideration of meter installations, minimizing interactions or other forms of disturbances such as might be caused by nearby line obstructions, e.g., elbows, tees, etc.

Indirect calibration is based on the equivalencies of two different meters. The requirement for similarity is had through maintenance of equal Reynolds numbers, or

$$D_1\rho_1 V_1/\mu_1 = D_2\rho_2 V_2/\mu_2,$$

where the subscripts 1 and 2 refer to the "standard" and the meter to be calibrated, respectively. The practical significance lies in the fact that, provided that similarity is maintained, discharge coefficients of the two meters will be directly comparable.

It is seen then that for geometrically similar meters it is possible to predict the performance of one meter on the basis of the experimental performance of another. Meters which are small in physical size may be employed to determine the discharge coefficient of large meters. Indeed, coefficients for one fluid may be determined through test runs with another fluid, provided that similarity is maintained through Reynold's numbers. However, when a liquid is employed to calibrate a meter intended for a gas, corrections for density and expansion must also be made.

SUGGESTED READINGS

ASME Performance Test Code, PTC 19.5; 2-1966, *Volumetric* (*Displacement*) *Meters*.

ASME Performance Test Code, PTC 19.5; 3-1965, *Fluid Velocity Measurement*.

ASME Performance Test Code, PTC 19.5; 4-1959, *Flow Measurement*.

ASME Performance Test Code, PTC 19.5; 5-1966, *Special Methods of Flow Measurement*.

ASME, *Fluid Meters, Their Theory and Application*, 1959.

Cusick, C. F., *Flow Meter Engineering Handbook*, 3rd Ed., Philadelphia: Minneapolis-Honeywell Regulator Co., Brown Instrument Div., 1961.

Spink, L. K., *Principles and Practice of Flow Meter Engineering*, 8th Ed., Foxboro, Mass.: The Foxboro Co., 1958.

Katys, G. P., *Continuous Measurement of Unsteady Flow*, New York: The Macmillan Co., 1964.

Linford, A., *Flow Measurement and Meters*, 2nd. Ed., London: E. & F. N. Spon Ltd., 1961.

PROBLEMS

14.1 Water at 95°F flows through a pipe with an inside diameter of 4 in. at an average velocity of 16 ft/sec. Calculate the value of R_D (see Appendix D for necessary data).

14.2 If Problem 14.1 had specified a 4-in. schedule 140 pipe, what would be the resulting R_D? (Check any general engineering handbook for the significance of pipe "schedule" numbers.)

14.3 Show that Eq. (14.6b) may be written as follows:

$$Q = KA_2 \sqrt{2gh},$$

where h = differential pressure across the meter measured in "head" of the flowing fluid.

14.4 A sharp-edged concentric orifice is employed to measure the flow of water in a line with a 4-in. diameter. If $\beta = 0.4$, what differential pressure may be expected if the flow rate is 12 ft³/min and the water temperature is 50°F?

14.5 Oil (sp. gr. at 70° = 0.93) is to be metered with a 2 × 1 venturi (2-in.-diameter pipe, 1-in.-diameter throat). Maximum flow rate will not exceed 30 gpm. Ambient temperature will range from 40 to 100°F. In specifying a manometer for measuring the differential pressure, what maximum magnitude of differential pressure must be accommodated?

14.6 A long-radius nozzle is to be selected for the measurement of water flow under the following conditions: Pipe size equals 7-in. IPS (ID = 7.023 in.), flow rate = 350 gpm, and average water temperature = 60°F. Available instrumentation makes it desirable that the nozzle provide a differential pressure head of about 48 in. of water under the above conditions. What approximate value of β should be specified? (Note: IPS refers to "iron pipe size.")

14.7 A given orifice employed for metering water is calibrated at 70°F over a relatively narrow range of flow rates. Given that $Q_T/Q_{70} = (1 + \xi)$ = ratio of the rate at temperature T to the rate at 70°, make a plot of ξ versus T over the range of 40 to 120°F. Assume that $R_D \approx 10^5$. Could the plot be used over a "large" range of flow rates? Explain.

14.8 Repeat Problem 14.7 with air as the flowing medium.

14.9 A venturi with a 15-in. throat is employed to meter 60°F air in a 24-in.-diameter duct. If the differenɟial pressure measured across *vena contracta* taps is 3.25 in. of water and the upstream pressure is 18 psia, what is the flow rate in lb/sec? In ft³/min?

14.10 Direct secondary calibration is employed to determine the flow coefficient for an orifice to meter the flow of nitrogen. For an orifice inlet pressure of 25 psia, the flow rate determined by the primary meter is 9.0 lb/min at 68°F. If the differential pressure across the orifice being calibrated is 3.1 in. of water, the conduit diameter is 4 in. and $\beta = 0.5$, what is the value of $(K \times Y)$? Use $\gamma = 0.0726$ lb/ft³ for nitrogen at 68°F and 14.7 psia.

14.11 What special consideration must be given to the readout from an obstruction-type meter installed in a run of pipe whose flow axis is inclined relative to the horizontal?

14.12 A pitot-static tube is employed to measure the velocity of 50°F water in an open conduit. If a differential pressure of 2 in. of water is measured, what is the corresponding flow velocity?

14.13 A pitot-static tube is employed to measure the velocity of an aircraft. If air temperature and pressure are 40°F and 13.2 psia, respectively, what is the aircraft velocity in miles-per-hour if the differential pressure is 17.5 in. of water?

14.14 The factor $(1 - C')$ is applied as a practical correction for adapting Eq. (14.11a) to compressible fluids. More accurately [21],

$$V^2 = 2g\left(\frac{k}{k-1}\right)\left(\frac{P_s}{\gamma}\right)\left[\left(\frac{P_t - P_s}{P_s}\right)^{(k-1)/k} - 1\right].$$

Employing the data of Problem 14.13 and the above relationship, determine the velocity of the aircraft. Use $k = 1.4$.

14.15 The velocity profile for turbulent flow in a smooth pipe is sometimes given as [17]:

$$V_r/V_{center} = [1 - (2r/D)]^{1/n},$$

where $D =$ pipe diameter,
             $r =$ radial coordinate from center of pipe,

and n ranges in value from about 6 to 10, depending on Reynold's number. For $n = 8$, determine the value of r at which a pitot tube should be placed to provide the velocity V_{ave}, such that $Q = AV_{ave}$.

14.16 A one-fifth size, geometrically similar model is made of the orifice described in Example 2. If water at 70°F is employed to experimentally determine the flow coefficient, to what velocity should the flow be adjusted for dynamic similarity?

CHAPTER 15

TEMPERATURE MEASUREMENTS

15.1 INTRODUCTION

Temperature is a manifestation of the molecular kinetic energy within a body and is readily perceived by the human nervous system. It is a fundamental property in much the same sense as mass, dimension, and time and hence is more difficult to define than are derived quantities. Various definitions have been proposed, including: "A condition of a body by virtue of which heat is transferred to or from other bodies," [1] and "A quantity whose difference is proportional to the work from a Carnot engine operating between a hot source and a cold receiver" [2]. The latter definition assumes some basis for the relation between *hot* and *cold*. For purposes of discussion in this chapter, the concept that temperature is an indication of intensity of molecular activity will suffice.

Temperature cannot be measured by use of basic standards for direct comparison. It can only be determined through use of some form of standardized calibrated device or system. Its measurement is not based on a tangible entity such as the International Meter or the International Kilogram, but rather, on ideas and instructions, as outlined in Article 4.6. When energy in the form of heat is introduced to or extracted from a body, altered molecular activity will be made apparent as a temperature change. Various primary effects may accompany the temperature change, any of which may be employed for the purpose of temperature measurement. These include (1) change in physical state, (2) change in chemical state, (3) altered physical dimensions, (4) change in electrical properties, (5) change in radiating ability.

The first two possibilities are seldom used for direct temperature measurement. As we have already seen, however, in Chapter 4, actual *establishment* of temperature standards is based on changes in physical state, i.e., freezing or melting and boiling or condensing. Ignition temperatures for combustible materials [3] determining the minimum temperature at which a change in chemical state may be sustained could, perhaps, be used as a limited method

TABLE 15.1

APPROXIMATE RANGE AND ACCURACY OF VARIOUS
TEMPERATURE-MEASURING ELEMENTS

Type	Range, °F	Accuracy, °F
Glass thermometers		
Mercury filled	−38 to 760	0.5 to 2
Mercury and nitrogen filled	−38 to 1000	0.5 to 10
Alcohol filled	−95 to 150	1 to 2
Pressure-gage thermometers		
Vapor-pressure type	20 to 400	2 to 10
Liquid or gas filled	−200 to 1000	2 to 10
Bimetallic thermometer	−100 to 1000	0.5 to 25
Thermocouples		
Base metal	−300 to 2000	0.5 to 20
Precious metal	−300 to 2800	0.5 to 20
Resistance thermometer	−400 to 1800	0.005 to 5
Thermistors	−150 to 500	*
Pyrometers		
Optical	1400 up	20 for blackbody conditions
Radiation	1000 up	20 to 30 for blackbody conditions
Fusion	1100 to 3600	As low as 20 or 30 under optimum conditions

* Depends on aging.

for determining temperature magnitude. It is mentioned here, however, merely as a possibility rather than as a practical method.

The use of change in dimensions that accompany a temperature change as a measuring method is well known and, of course, forms the basis of operation for the common liquid-in-glass, as well as gas and bimetal, thermometers. Electrical methods include means based on change in electrical conductivity and the well-known thermoelectric effects that produce an electromotive force (emf) at the junction of two dissimilar metals. Another temperature-measuring method, one that makes use of the energy radiated from a hot body, is the basis of operation of the optical and radiation pyrometers.

Table 15.1 outlines the range and accuracy of representative temperature-measuring methods. Table 15.1(a), p. 444, shows an expanded summary of the characteristics of temperature sensors, primarily those which yield an electrical output.

15.2 USE OF BIMATERIALS

a) Liquid-in-glass thermometers. The ordinary thermometer is an example of the liquid-in-glass type. Essential elements consist of a relatively large bulb at the lower end, a capillary tube with scale, and liquid filling both the bulb and a portion of the capillary. In addition, a smaller bulb is usually incorporated at the upper end to serve as a safety reservoir when the intended temperature range is exceeded. As the temperature is raised, the greater expansion of the liquid compared with that of the glass causes it to rise in the capillary or stem of the thermometer, and the height of rise is used as a measure of the temperature. The volume enclosed in the stem above the liquid may either contain a vacuum or be filled with air or other gas. For the higher temperature ranges, an inert gas at a carefully controlled initial pressure is introduced in this volume, thereby raising the boiling point of the liquid and increasing the total useful range. In addition, it is claimed that such pressure minimizes column separation.

There are several desirable properties for a liquid used in a glass thermometer, as follows:

1. The temperature-dimensional relation should be linear, permitting a linear instrument scale. An example of a poor liquid in this respect is water at temperatures in the region immediately above its freezing point, for it will be remembered that in this range water actually expands with drop in temperature.

2. The liquid should have as large a coefficient of expansion as possible. For this reason, alcohol is better than mercury. Its larger expansion makes possible larger capillary bores, and hence provides easier reading.

3. The liquid should accommodate a reasonably large temperature range without change of state. Mercury is limited at the low-temperature end by its freezing point ($-37.97°F$), and the spirits are limited at their high-temperature ends because of boiling.

4. The liquid should be clearly visible when drawn into a fine thread. Mercury is inherently good in this regard, while alcohol is usable only if dye is added.

5. Preferably, the liquid should not adhere to the capillary walls. When rapid temperature drops occur, any film remaining on the wall of the tube will cause a reading that is too low. In this respect, mercury is better than alcohol.

Within its capabilities, mercury is undoubtedly the best liquid for liquid-in-glass thermometers and is generally used in the higher-grade instruments. Alcohol is usually satisfactory. Other liquids are also used, primarily for the purpose of extending the useful ranges to lower temperatures.

b) Calibration and stem correction. High-grade liquid-in-glass thermometers are made with the scale etched directly upon the thermometer stem, thereby

TABLE 15.1(a)

CHARACTERISTICS OF TEMPERATURE SENSORS

Sensor type	Nominal* drift	Nominal minimum† recorder span	Repeatability	Instrument compensation required	Max. dist. between sensor and receiver
Thermal elements	Can be zero—usually < 1%/yr.	50°F	¼% span	Yes—Hg No—Others	200'
Nickel resistance thermometer	0.5°F/yr.	< 20°F	0.1°F		< 1000' without special compensation. Can go to 5000'; copper wire OK
Platinum resistance thermometer	< 0.1°F/yr.	< 5°F 1 MV/of output from bridge available	0.05°F	No	Can go to 5000'; copper wire OK
Thermistors	< 0.1°F/yr. aged. Up to 2°F/yr. unaged.	< 2°F	0.05°F–0.2°F		Can go to 5000'; copper wire OK
Quartz crystals	Very stable	Digital readout to 0.001°F			1000'
Germanium sensors		1.5–10°K	0.005–0.010°K		200–300'

Type	Drift		Accuracy	Compensation/extension	Lead wire
Copper-constantan (T) thermocouples	T/C Drift is widely subject to type atmos., thermal cycle, gage size, etc. It can be less than 1°F/year or more than 20°F	150°F	On the order of 0.2°F for all T/C's base and noble metal	Yes, also exten. lead wires from sensor to instrument	Generally < 5000' with potentiometer type receivers. Millivoltmeters limited by external ohms. Normally use special lead wire
Iron-constantan (J)' thermocouples		125°F			
Chromel alumel‡ (K) thermocouples		150°F			
Chromel†-constantan (E) thermocouples		100°F			
Noble metal (R&S) thermocouples		500°F			
Tungsten-rhenium thermocouples		500–1000°F	Very good		
Radiation pyrometer		Large because high temperature application		No, compensation part of sensor	Same as for T/C

This table is reproduced by permission of Honeywell, Inc., Industrial Division, Fort Washington, Pa.

* Drift figures are based on no mistreatment of sensor—overheating, wrong atmosphere, too much current, etc. Other extreme figures given do not necessarily apply simultaneously.

† Figure for thermocouples based on potentiometer type recorder or indicator. Resistance thermometers and transistors based upon Wheatstone Bridge type Recorders or Indicators.

‡ Trademark, Hoskins Manufacturing Co.

TABLE 15.1(a) (continued)

Sensor type	Nominal minimum size	Scale linearity	Atmosphere environment	Weak points	Strong points
Thermal elements	$\frac{3}{8}'' \times 2''$	Linear except vapor systems	Depends on bulb material	Integral bulb and instrument bulky	Good overall system economy
Nickel resistance thermometer	Commercially $\frac{3}{8}''$ dia.	Expanded at low temperature	Element should be protected in liquid and corrosive atmosphere	Modest high temp. limit. Must have protective cover to prevent mechanical damage	Good stability. Fairly narrow spans
Platinum resistance thermometer	$0.084''$ diameter	Excellent—nearly linear	Element should be protected in liquid and corrosive atmosphere	Slightly more expensive than T/C or thermistor. More subject to mechanical damage	Higher signal output than T/C. Best stability. Int'l temp. std. Between $-183°$C to $+630°$C. Used primarily for precision measurements. Fast response. Good sensitivity.
Thermistors	$0.014''$ diameter	Poor—expands at low end (negative temp. coefficient type)	Practically any	Very nonlinear. High drift possible if not aged correctly	High signal output than RTB or T/C. For Narrow spans, small size. Fast response. Excellent sensitivity.
Quartz crystals	$\frac{3}{8}'' \times 2''$	Excellent		Expensive—requires oscillatory circuitry	Extreme sensitivity $10^{-9}°$F claimed
Germanium sensors	$\frac{1}{8}'' \times \frac{3}{8}''$	Very nonlinear	Cryogenic atmospheres		Used in cryogenic temperature measurements—more practical than paramagnetic salts

Type	Measurement / Size	Linearity	Atmosphere	General	Disadvantages	Advantages
Copper-constantan (T) thermocouples		Good but crowds at low end	Oxidizing or reducing	For all T/C's relatively wide spans req'd; cold junction compensation req'd. Linearity generally less than RTB	Relatively low max. temp.	High resis. to corrosion from moisture. Good in med. low temp. range
Iron-constantan (J) thermocouples		Good—nearly linear in range 300–800°F	Reducing			Good in reducing atmos. Most economical T/C
Chromel alumel‡ (K) thermocouples	Nearly a point measurement. Industrial applications limited to about 30 ga. wire and $\frac{1}{8}''$ dia. protecting tube	Good—most linear of all T/C's	Oxidizing		More expensive than T or J	Good in oxidizing atmos.—most linear T/C
Chromel‡-constantan (E) thermocouples		Good	Oxidizing		Larger drift	Highest emf/degree — Small size, fast response
Noble metal (R&S) thermocouples		Good at high temp. Poor below 1000°F	Types R&S oxidizing—other types inert		More expensive than K	Protected types R&S T/C good in oxidizing atmospheres
Tungsten-rhenium thermocouples		Same as noble metal T/C's	Vacuum or clean inert gas		Cannot be used in oxidizing atmos. Brittle—hard to handle	Highest temperature T/C
Radiation pyrometer	$\frac{1}{8}''$ dia. target	Poor—varies as temp. to fourth power	Clear sighting path to target required		Poor linearity. More expensive than T/C	Requires no contact with material being measured

making it mechanically impossible to shift the scale relative to the stem. The care with which the scale is laid out depends upon the intended accuracy of the instrument (and to a large extent governs its cost). The process of establishing *bench marks* from which a scale is determined is known as "pointing" [4], and two or more *marks* or *points* are required. In spite of intentions, a particular thermometer will exhibit some degree of nonlinearity. This may be caused by nonlinear temperature-dimension characteristics of liquid or glass, or by the nonuniformity of the bore of the column. In the simplest case, two points may be established, such as the freezing and boiling points of water, and equal divisions used to interpolate (and extrapolate) the complete scale. For a more accurate scale, additional points would be employed, and sometimes as many as five are used. Calibration points for this purpose are obtained through use of fixed temperature points, as discussed in Article 15.11.

Greatest sensitivity to temperature is at the bulb, where the largest volume of liquid is contained; however, all portions of a glass thermometer are temperature-sensitive. With temperature variation, the stem and any upper bulb will also change dimensions, thereby altering the available liquid space and hence the thermometer reading. For this reason, if greatest accuracy is to be attained, it is necessary to prescribe how a glass thermometer is to be subjected to the temperature. Maximum control is obtained when the complete thermometer is entirely immersed in a uniform temperature medium. This is not often possible, especially when the medium is liquid. A common practice, therefore, is to calibrate the thermometer for a given partial immersion, with the proper depth of immersion indicated by a scribed line around the stem. Thermometer accuracy is then prescribed for this condition only. This, of course, does not insure absolute uniformity, because the upper portion of the stem is still subject to some variation in ambient conditions.

When the immersion employed is different from that used for calibration, an *estimate* of the correct reading may be obtained from the following relation (for mercury-in-glass thermometers only):

$$T = T_1 + 0.00009T'(T_1 - T_2), \tag{15.1}$$

where

$T = $ correct temperature, °F,

$T_1 = $ actual thermometer reading, °F,

$T_2 = $ ambient temperature surrounding emergent stem (this may be determined by attaching a second thermometer to the stem of the main thermometer), °F,

$T' = $ degrees of thread emergence to be corrected, °F.

T' is determined as follows: For a *total immersion thermometer*, T' should be

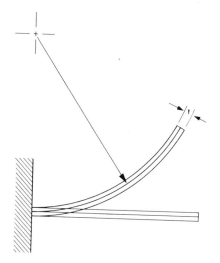

Fig. 15.1 Diagram illustrating deflection of a bimetal strip caused by temperature change.

the actual length of the thread of mercury which is emerging, measured in scale degrees. For the *partial immersion thermometer*, T' should be the number of scale degrees between scribed calibration immersion line and the actual point of emergence. When the thermometer is immersed to too great a depth, the value of T' will be *negative*.

The constant 0.00009 in Eq. (15.1) represents the approximate coefficient of differential expansion between mercury and glass.

Another factor influencing liquid-in-glass thermometer calibration is variation in the applied pressure, particularly pressure applied to the bulb. The resulting elastic deformation causes displacement of the column, hence incorrect reading. Normal variation in atmospheric pressure is not usually of importance, except for the most precise work. However, if the thermometer is subjected to system pressures of higher values, considerable error may be introduced.

c) Bimetal temperature-sensing elements. When two metal strips having different coefficients of expansion are brazed together, a change in temperature will cause a free deflection of the assembly, as indicated in Fig. 15.1. Such bimetal strips are the basis for many control devices and are also used to some extent for temperature measurement. The following relation [5] may be used to determine the radius of curvature of such a strip, which was initially flat at temperature T_0:

$$r = \frac{t\{3(1+m)^2 + (1+mn)[m^2 + (1/mn)]\}}{6(\alpha_2 - \alpha_1)(T - T_0)(1+m)^2}. \tag{15.2}$$

In this equation,

r = radius of curvature at temperature T, in.,

t = total thickness of strip, in.,

m = ratio of thicknesses of low-expansion to high-expansion components,

n = ratio of modulus of elasticity of low-expansion to high-expansion components,

α_1 = lower coefficient of expansion, in./in. °F,

α_2 = higher coefficient of expansion, in./in. °F,

T = temperature, °F,

T_0 = initial temperature, °F.

The above relation may be used to determine the deflection curve caused by temperature, upon which could be superimposed any deflection caused by concentrated or distributed load.

Thermometers with bimetallic temperature-sensitive elements are often used because of their ruggedness and also because their particular form is sometimes more convenient.

15.3 PRESSURE THERMOMETERS

Figure 15.2 illustrates the principle of operation of the *pressure thermometer*. Essential parts of the arrangement are: bulb A, tube B, pressure-sensing gage C, and some sort of filling medium. Pressure thermometers are termed *liquid-filled*, *gas-filled*, or *vapor-filled*, depending on whether the filling medium is completely liquid, completely gaseous, or a combination of liquid and its vapor. A primary advantage of these thermometers is that they provide sufficient force output to permit the direct driving of recording and controlling devices. In such applications the pressure type of temperature-sensing system is usually less costly than other systems. Tubes as long as 200 ft may be used successfully.

Expansion (or contraction) of bulb A and the contained fluid or gas, caused by temperature change, alters the volume and pressure in the system. In the case of the liquid-filled system, the sensing device C acts primarily as a differential volume indicator, with the volume increment serving as an analog of temperature. For the gas- or vapor-filled systems, the sensing device serves primarily as a pressure indicator, with the pressure providing the measure of temperature. In both cases, of course, both pressure and volume change.

Ideally the tube or capillary should serve simply as a connecting link between the bulb and the indicator. When liquid- or gas-filled systems

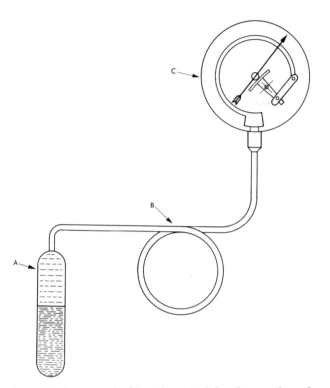

Fig. 15.2 Schematic diagram showing the principle of operation of a pressure thermometer.

are used, the tube and its filling are also temperature-sensitive, and any difference from calibration conditions along the tube introduces output error. This error is reduced by increasing the ratio of bulb volume to tube volume. Unfortunately, increasing bulb size reduces the time response of a system, which may introduce problems of another nature. On the other hand, reducing tube size, within reason, does not degrade response particularly because, in any case, flow rate is negligible. Another source of error which should not be overlooked is any pressure gradient resulting from difference in elevation of bulb and indicator not accounted for by calibration.

Temperature along the tube is not a factor for vapor-pressure systems, however, as long as a free liquid surface exists in the bulb. In this case, Dalton's law for vapors applies, which states that if both phases (liquid and vapor) are present, only one pressure is possible for a given temperature. This is an important advantage of the vapor-pressure system. In many cases, though, the tube in this type of system will be filled with liquid, and hence the system is susceptible to error caused by elevation difference.

Application limits are approximately as shown in the following tabulation:

Type of pressure thermometer	Approx. overall range of usefulness, °F	Maximum approx. range for a given instrument, °F
Liquid filled	−150 to 1000 (depending on liquid)	400 to 800 (depending on liquid)
Gas filled	−400 to 1000	800
Vapor filled	−300 to 700 (depending on vapor)	250

15.4 THERMOELECTRIC THERMOMETRY

Several temperature-sensitive electrical elements are available as measuring means. Of primary importance are thermal emf and both positive and negative variation in resistance with temperature. These are discussed in the following articles.

15.5 THERMORESISTIVE ELEMENTS

We have already seen that the electrical resistance of most materials varies with temperature. This was discussed in Article 7.9. In Articles 11.8a, and 11.12 we found this to supply a troublesome extraneous input in the application of strain gages. It can only follow that this relation, which proves so worrisome when unwanted, should be the basis for a good method of temperature measurement.

Traditionally, resistance elements sensitive to temperature are made of one of the metals generally considered a good conductor of electricity. Examples are nickel, copper, platinum, and silver. Temperature-measuring devices employing elements of this type are commonly referred to as *resistance thermometers*. Of more recent origin are elements made from semiconducting materials having large negative coefficients of resistance. Such materials are usually some combination of metallic oxide of cobalt, manganese, and nickel. These devices are called *thermistors*.

One important difference between these two kinds of material is that, whereas the resistance change in the thermometer element is small and positive (increasing temperature causes increased resistance), that of the thermistor is relatively large and negative. In addition, the thermometer type provides practically a linear temperature-resistance relation, while that of the thermistor is nonlinear. Still another important difference lies in the temperature ranges over which each may be used. The practical operating range for the thermistor lies between approximately −150 to 500°F. The range for the resistance thermometer is much greater, being from about −400 to 1800°F. Finally, the metal resistance elements are considered to be more time-stable than the oxides; hence they provide better reproducibility with lower hysteresis.

a) Resistance thermometers. Evidence of the importance and reliability of the resistance thermometer may be had by recalling that the International Temperature Scale of 1948 specifies use of this instrument for interpolation over a very important portion of the scale. A platinum resistance thermometer is employed to divide the interval between the oxygen point (−297.35 °F) and the antimony point (1166.9°F). (See Article 4.6.)

Certain properties are desirable in material used for resistance-thermometer elements. The material should have a resistivity permitting fabrication in convenient sizes without excessive bulk which would produce poor time response. In addition, its thermal coefficient of resistivity should be high and as constant as possible, thereby providing an approximately linear output of reasonable magnitude. The material should be corrosion-resistant and should not undergo phase changes in the temperature ranges of interest. Finally, it should be available in conditions providing reproducible and consistent results. In regard to this last requirement, it has been found that to produce precision resistance thermometers, great care must be exercised in minimizing residual strains, requiring careful heat treatment subsequent to forming.

As is usually the case in such matters, there is no universally acceptable material for resistance-thermometer elements. Several materials are commonly employed, the choice depending on the compromises that may be accepted. Although the actual resistance-temperature relation must be determined experimentally, for most metals the following empirical equation holds very closely:

$$R_t = R_0(1 + AT + BT^2),$$ (15.3)

where

R_t = resistance at temperature T,

R_0 = resistance at 0°F,

T = temperature, °F,

A and B = constants depending on material.

Undoubtedly platinum, nickel, and copper are the materials most commonly used, although others, such as tungsten, silver, and iron, have also been employed.

Figure 15.3 shows the general construction of two forms of resistance thermometer. In Fig. 15.3(a) the element consists of a number of turns of resistance wire wrapped around a solid-silver core. Heat is transmitted quickly from the end flange through the core to the winding. Figure 15.3(b)

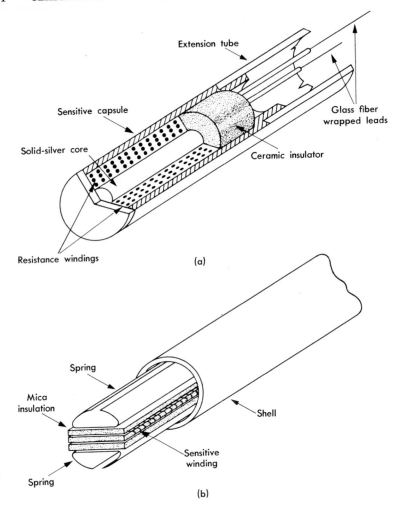

Fig. 15.3 Sections illustrating construction of two types of resistance thermometer elements. (a) Courtesy of The Foxboro Company, Foxboro, Mass. (b) Courtesy of Thomas A. Edison, Inc., West Orange, N.J.

illustrates another construction in which the wire is wrapped around a mica strip and sandwiched between two additional mica strips. Resistance thermometers of the sort illustrated may be used directly, enclosed only by the basic tube or sleeve. However, when permanent installations are made and when corrosion or mechanical protection is advisable, a *well* or *socket* may be employed. Such an arrangement is shown in section in Fig. 15.4. Table 15.2 describes characteristics of several typical, commercially available, resistance-thermometer elements.

TABLE 15.2

TYPICAL PROPERTIES OF RESISTANCE-THERMOMETER ELEMENTS

Type of element	Case material	Temperature range, °F	Resistance, ohms	Sensitivity, ohms/°F (approx.)	Limits of error, °F	Response,* sec
Platinum (laboratory)	Pyrex glass	−300 to 1000	25 at 32°F	0.05	±0.02	—
Platinum (industrial)	Stainless Steel	−325 to 250† 0 to 1000‡	25 at 32°F 25 at 32°F	—	±1.5 ±3	10 to 30 10 to 30
Copper	Brass	−100 to 250	10 at 77°F	0.2	±0.75	20 to 60
Nickel	Brass	32 to 250	100 at 68°F	—	±0.5	20 to 60

* Time required to detect 90% of any temperature change in water moving at 1 fps. The lower value is for the thermometer in a case only, while the higher value is for the thermometer in a protective well.

† Low range.

‡ High range.

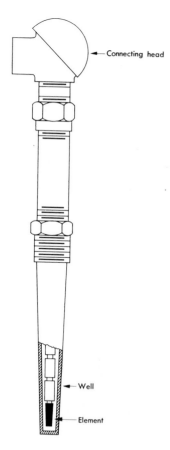

Temperature-sensitive resistance elements similar in construction to the resistance-type strain gages are available; both foil* and wire† grids are used. They differ from strain gages in that the elements are relatively sensitive to temperature and insensitive to strain. In addition, the circuitry employed is arranged to enhance the temperature sensitivity. When properly used, common strain gage instrumentation may be employed for readout. Continuous monitoring of temperatures as high as 500°F is possible.

Fig. 15.4 Installation assembly for an industrial-type resistance thermometer.

---

* William T. Bean, Detroit, Michigan.

† Baldwin-Lima-Hamilton Corporation, Waltham, Massachusetts.

b) Instrumentation for resistance thermometry. Some form of electrical bridge is normally used to measure the resistance change in these thermometers. However, particular attention must be given to the manner in which the thermometer is connected into the bridge. Leads of some length appropriate to the situation are normally required, and any resistance change therein due to any cause, including temperature, may be credited to the thermometer element. It is desirable, therefore, that the lead resistance be kept as low as possible relative to the element resistance. In addition, some modification may be employed, providing lead compensation.

Figure 15.5 illustrates three different arrangements used to minimize lead error. Inspection of the diagrams indicates that arms AD and DC each contain the same lead lengths. Therefore, if the leads have identical properties to begin with and are subject to like ambient conditions, the effects they introduce will cancel. In each case the battery and galvanometer may be interchanged without affecting balance. When the Siemen's arrangement is used, however, no current will be carried by the center lead at balance, as shown. This may be considered an advantage. The Callender arrangement is quite useful when thermometers are used in both arms AD and DC to provide an output proportional to temperature differential between the two thermometers. The four-lead arrangement is used in the same way as the one with three leads. Provision is made, however, for using any combination of three, thereby permitting checking for unequal lead

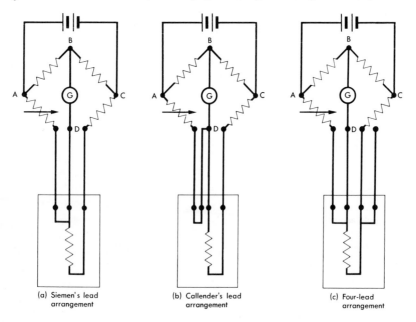

(a) Siemen's lead
arrangement

(b) Callender's lead
arrangement

(c) Four-lead
arrangement

Fig. 15.5 Three methods for compensating lead resistance.

resistance. By averaging readings, more accurate results are possible. Some form of this arrangement is used where highest accuracies are desired.

The usual practice is to use the bridge in the null-balance form, but the deflection bridge may also be used (see Article 7.18). In general, the null-balance arrangement is limited to measurement of static or slowly changing temperatures, while the deflection bridge is used for more rapidly changing inputs. Dynamic changes are most conveniently recorded rather than simply indicated, and for this purpose either the self-balancing or the deflection types may be used, depending upon time rate of temperature change.

When a resistance bridge is employed for measurement, current will necessarily flow through each bridge arm. An error may, therefore, be introduced, caused by I^2R heating. For resistance thermometers such an error will be of opposite sign to that caused by conduction and radiation from the element (Article 15.10a), and in general it will be small because the gross effects in individual arms will be largely balanced by similar effects in the other arms. An estimate of the overall error resulting from ohmic

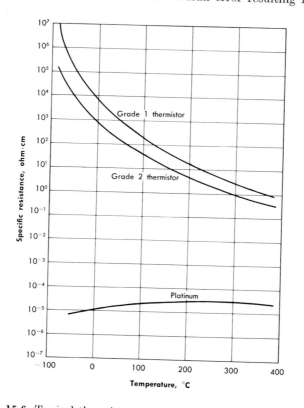

Fig. 15.6 Typical thermistor temperature-resistance relations.

heating may be had by making readings at different current values and extrapolating to zero current.

c) *Thermistors.* The thermistor is a thermally sensitive variable resistor made of a ceramiclike semiconducting material. Unlike metals, thermistors respond negatively to temperature. As the temperature rises, the thermistor resistance decreases. Figure 15.6 shows typical temperature-resistance relations.

Thermistors are composed of oxides of manganese, nickel, and cobalt in formulations having resistivities of 100 to 450,000 ohms·cm. They are

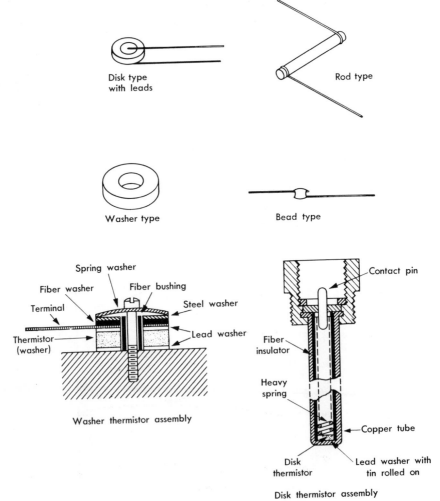

Disk type
with leads

Rod type

Washer type

Bead type

Spring washer

Fiber washer Fiber bushing

Terminal

Steel washer

Thermistor (washer)

Lead washer

Washer thermistor assembly

Contact pin

Fiber insulator

Heavy spring

Copper tube

Disk thermistor

Lead washer with tin rolled on

Disk thermistor assembly

Fig. 15.7 Various thermistor forms available commercially.

TABLE 15.3

PROPERTIES OF SELECTED THERMISTORS*

Code No.†	Description	Temperature coefficient, % resistance change per °C	Nominal resistance, ohms			Maximum continuous ambient temperature, °C	Primary application‡
			At 32°F	At 78°F	At 122°F		
1A	Bead in glass bulb	−3.0	140,000	60,000	38,000	—	1
3A	$\frac{3}{4}$″ dia. washer	−3.9	28.3	10	4.1	150	2
5A	0.4″ dia. washer	−3.9	283	100	40.7	150	2
7A	0.2″ dia. washer	−4.4	3270	1000	360	150	2
9A	$0.11 \times \frac{13}{16}$″ rod	−4.4	103,000	31,500	11,300	150	2
13A	$0.055 \times \frac{1}{2}$″ rod	−4.4	327,000	100,000	36,000	150	2
15A	0.4″ dia. disk	−3.9	283	100	40.7	150	2
D168391	Disk on plate	−4.4	2350	725	255	125	3
D171933	Rod	−4.4	145,000	45,000	17,000	100	4
20A	Glass coated bead, 0.030″ dia.	−3.8	8800	3100	1270	300	5

* Courtesy of Western Electric Co., New York, N.Y.

† Western Electric Co.

‡ 1. Time delay.   2. Temperature measurement and control.   3. Temperature compensation.   4. Temperature control.   5. Temperature measurement.

available in various forms, such as shown in Fig. 15.7. Table 15.3 lists the properties of certain commercially available thermistors.

The temperature-resistance function for a thermistor is given by the relationships [6]

$$R = R_0 e^k \tag{15.4}$$

and

$$k = \beta \left(\frac{1}{T} - \frac{1}{T_0} \right), \tag{15.4a}$$

in which

R = resistance at any temperature T, °K,

R_0 = resistance at reference temperature T_0, °K,

e = base of Naperian logarithms,

β = a constant.

The constant β usually has a value between 3400 and 3900, depending on the thermistor formulation or grade.

When a thermistor is used in an electrical circuit, current normally flows through it, and ohmic heating is generated by its resistance. This will raise the temperature of the element, the amount depending upon the rate with which the heat is dissipated. For given ambient conditions, a temperature equilibrium will occur at which a definite resistance value will exist. Through proper application of thermistor and electrical circuit characteristics, the devices may be used for temperature measurement or control. In addition, they are quite useful for compensating electrical

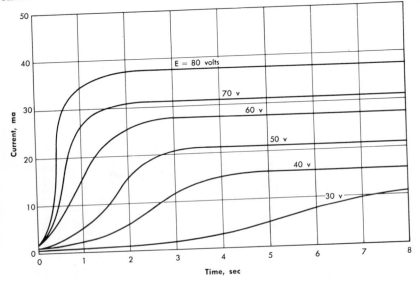

Fig. 15.8 Typical current-time relations for thermistors.

circuitry for changing ambient temperature. This is largely possible because of the *negative* temperature characteristics of the thermistor, in contrast to *positive* characteristics possessed by most electrical components. Also, time-delay actions over large ranges are possible through proper balancing of electrical and heat-transfer conditions. Figure 15.8 shows characteristic thermistor current-time relations.

The inherently high sensitivity possessed by thermistors permits use of very simple electrical circuitry for measurement of temperature. Ordinary ohmmeters (Article 8.2d) may be used within the limits of accuracy of the meter itself. More often one of the various forms of resistance bridge is used (Article 7.18), either in the null-balance form or as a deflection bridge. Simple ballast circuits (Article 7.15) are also usable.

Through use of the thermistor's temperature-resistance characteristics alone, or in conjunction with controlled heat transfer, thermistors have been used for measurement of many quantities, including pressure, liquid level, power, and others. They are also used for temperature control, timing (through use of their delay characteristics in combination with relays), overload protectors, warning devices, etc.

15.6 THERMOCOUPLES

In 1821, T. J. Seebeck discovered that an electromotive force exists across a junction formed of two unlike metals [7]. Later it was shown [8, 9] that the potential actually comes from two different sources: that resulting solely from *contact* of the two dissimilar metals and the *junction temperature*, and that due to *temperature gradients* along the conductors in the circuit. These two effects are named the Peltier and Thomson effects after their respective discoverers. In most cases the Thomson emf is quite small relative to the Peltier emf, and with proper selection of materials may be disregarded. These effects form the basis for a very important temperature-measuring element, the *thermocouple*.

If a circuit is formed including a thermocouple, as shown in Fig. 15.9, a minimum of two conductors will be necessary, unavoidably resulting in two junctions, p and q. Disregarding the Thomson effect, the net emf will

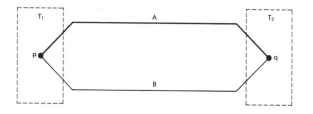

Fig. 15.9 Elementary thermocouple circuit.

be the result of the difference between the two Peltier emf's occurring at the two junctions. If the temperatures T_1 and T_2 are equal, the two emf's will be equal but opposed, and no current will flow. However, if the temperatures are different, the emf's will not balance, and a current *will* flow. The net emf is a function of the two materials used to form the circuit and the *temperatures* of the two junctions. The actual relations, however, are empirical, and the temperature-emf data must be based on experiment. An important fact is that the results are reproducible and therefore provide a reliable method for measuring temperature.

Note particularly that *two* junctions are *always* required. In general, one senses the desired or unknown temperature; this one we shall call the *hot* or *measuring* junction. The second will usually be maintained at a known fixed temperature; this one we shall refer to as the *cold* or *reference* junction.

a) Application laws for thermocouples. In addition to the Seebeck effect, there are certain laws which thermoelectric circuits abide by, as follows:

Law of intermediate metals [10]. *Insertion of an intermediate metal into a thermocouple circuit will not affect the net emf, provided the two junctions introduced by the third metal are at identical temperatures.*

Applications of this law are shown in Fig. 15.10. As shown in part (a) of the figure, if the third metal C is introduced and if the new junctions r and s are both held at temperature T_3, the net potential for the circuit will remain unchanged. This, of course, permits insertion of a measuring device or circuit, without upsetting the temperature function of the thermocouple circuit. In Fig. 15.10(b) the third metal may be introduced at either a *measuring* or *reference junction*, as long as couples p_1 and p_2 are maintained at the same temperature T_1. This makes possible the use of joining materials, such as soft or hard solder, in fabricating the thermocouples, provided a temperature gradient does not exist through the junction. In addition, the thermocouple may be actually embedded directly into the surface or interior of either a conductor or nonconductor without altering the thermocouple's usefulness.

Law of intermediate temperatures [10]. *If a simple thermocouple circuit develops an emf, e_1, when its junctions are at temperatures T_1 and T_2, and an emf, e_2, when its junctions are at temperatures T_2 and T_3, it will develop an emf, $e_1 + e_2$, when its junctions are at temperatures T_1 and T_3.*

This makes possible direct correction for secondary junctions whose temperatures may be known but are not directly controllable. It also makes possible the use of thermocouple tables based on a "standard" reference temperature (say 32°F) although neither junction may actually be at the "standard" temperature.

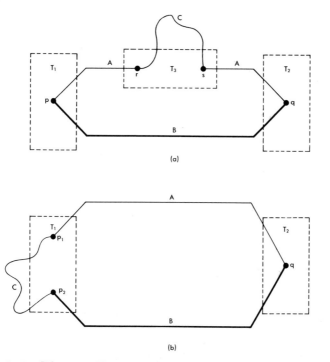

Fig. 15.10 Diagrams illustrating the law for intermediate metals.

b) Thermocouple materials and installation. Theoretically any two unlike conducting materials could be used to form a thermocouple. Actually, of course, certain materials and combinations are better than others, and some have practically become standard for given temperature ranges. Materials commonly used in thermocouples may be listed as follows:

> Copper–an element
> Iron–an element
> Platinum–an element
> Rhodium–an element
> Iridium–an element
> Constantan–60% copper, 40% nickel
> Chromel–10% chromium, 90% nickel
> Alumel–2% aluminum, 90% nickel,
>   remainder silicon and manganese

Combinations of the above materials and their characteristics are listed in Table 15.4.

Size of wire is of some importance. Usually the higher the temperature to be measured, the heavier should be the wire. As the size is increased,

TABLE 15.4

CHARACTERISTICS OF COMMERCIAL THERMOCOUPLES OF
VARIOUS MATERIALS

Type	Temperature limits, °F	Comments
Copper–constantan	−300 to 650	Inexpensive; high output; copper oxidizes above 650°F.
Chromel–constantan	0 to 1000	Highest output of common couples; good stability.
Iron–constantan	0 to 1500	Inexpensive; high output; iron oxidizes rapidly above 1500°F; temperature gradients along leads should be avoided to minimize Thomson emf's.
Chromel–alumel	600 to 2000 (continuous) Up to 2300 (intermittent)	Very resistant to oxidation within specified temperature ranges; emf tends to reduce with use.
Platinum–10% rhodium	1300 to 2700 (continuous) Up to 2850 (intermittent)	Expensive, with low output, hence should be used only for high-temperature applications; very stable and resistant to oxidation; specified for reproducing International Temperature Scale between 1166 and 1945°F (Article 3.5).
Platinum–6% rhodium, Platinum–30% rhodium	Up to 3100	Not common.
Iridium, iridium–rhodium	Up to 3500 +	Not common.

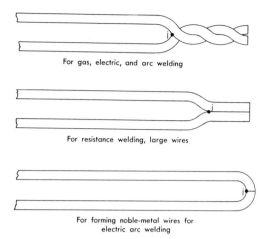

For gas, electric, and arc welding

For resistance welding, large wires

For forming noble-metal wires for
electric arc welding

Fig. 15.11 Common forms of thermocouple construction. [See Article 15.6(d) for significance of point j.]

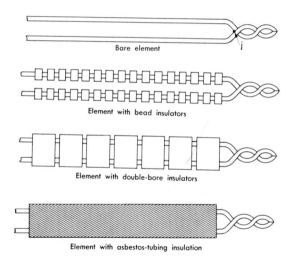

Bare element j

Element with bead insulators

Element with double-bore insulators

Element with asbestos-tubing insulation

Fig. 15.12 Methods for insulating thermocouple leads.

however, the time response of the couple to temperature change increases. Therefore some compromise between response and life may be required.

Thermocouples may be prepared by twisting the two wires together and brazing, or preferably welding, as shown in Fig. 15.11. Low-temperature couples are often used bare; however, for higher temperatures, some form of protection is usually required. Figure 15.12 illustrates common methods

TABLE 15.5

VALUES OF THERMAL EMF IN ABSOLUTE MILLIVOLTS FOR
SELECTED METAL COMBINATIONS [25]
(Based on reference junction temperature at 32°F)

Temper-ature, °F	Copper–constantan	Chromel–constantan	Iron–constantan	Chromel–alumel	Platinum–10% rhodium
−300	−5.284	−8.30	−7.52	−5.51	
−200	−4.111	−6.40	−5.76	−4.29	
−100	−2.559	−3.94	−3.49	−2.65	
0	−0.670	−1.02	−0.89	−0.68	
100	1.517	2.27	1.94	1.52	0.221
200	3.967	5.87	4.91	3.82	0.595
300	6.647	9.71	7.94	6.09	1.017
400	9.525	13.75	11.03	8.31	1.474
500	12.575	17.95	14.12	10.57	1.956
700	19.100	26.65	20.26	15.18	2.977
1000		40.06	29.52	22.26	4.596
1500		62.30		33.93	7.498
2000				44.91	10.662
2500				54.92	13.991
3000					17.292

for separating the wires, and Fig. 15.4 shows a section through a typical protective tube.

c) *Measurement of thermal emf.* The actual magnitude of electrical potential developed by thermocouples is quite small when judged in terms of many standards. Table 15.5 indicates the range of values for different types of couples, and Table 15.6 lists expanded (although still very much abbreviated) values for copper–constantan thermocouples.

Either of two measuring devices are commonly employed for determining thermocouple output: some form of ordinary galvanometer, or the voltage-balancing potentiometer (see Article 7.17). Probably the latter is the most commonly used, either as a simple manually balanced instrument (Fig. 15.13) or as an automatically balancing recording system (Fig. 15.14). The potentiometer possesses an important advantage over the galvanometer in that by basing its operation on a bucking emf, no current flows in the thermocouple circuit at balance. Hence any resistance problem in the leads is largely eliminated.

Figure 15.15 shows a simple temperature-measuring system employing a thermocouple as the sensing element and a potentiometer for indication. In this illustration, the thermoelectric circuit consists of a measuring junction, p, and a somewhat less obvious reference junction, q, at the potentiometer.

TABLE 15.6

VALUES OF THERMAL EMF IN ABSOLUTE MILLIVOLTS FOR COPPER—
CONSTANTAN THERMOCOUPLES [25]

(Based on reference junction temperature at 32°F)

°F	0	10	20	30	40	50	60	70	80	90	100
−300	−5.284	−5.379									
−200	−4.111	−4.246	−4.377	−4.504	−4.627	−4.747	−4.863	−4.974	−5.081	−5.185	−5.284
−100	−2.559	−2.730	−2.897	−3.062	−3.223	−3.380	−3.533	−3.684	−3.829	−3.972	−4.111
(−)0	−0.670	−0.872	−1.072	−1.270	−1.463	−1.654	−1.842	−2.026	−2.207	−2.385	−2.559
(+)0	−0.670	−0.463	−0.254	−0.042	+0.171	0.389	0.609	0.832	1.057	1.286	1.517
100	1.517	1.751	1.987	2.226	2.467	2.711	2.958	3.207	3.458	3.712	3.967
200	3.967	4.225	4.486	4.749	5.014	5.280	5.550	5.821	6.094	6.370	6.647
300	6.647	6.926	7.208	7.491	7.776	8.064	8.352	8.642	8.935	9.229	9.525
400	9.525	9.823	10.123	10.423	10.726	11.030	11.336	11.643	11.953	12.263	12.575
500	12.575	12.888	13.203	13.520	13.838	14.157	14.477	14.799	15.122	15.447	15.773
600	15.773	16.101	16.429	16.758	17.089	17.421	17.754	18.089	18.425	18.761	19.100
700	19.100	19.439	19.779	20.120	20.463	20.463					

Fig. 15.13 A manually balanced thermocouple potentiometer. (Courtesy Leeds and Northrup Company, Philadelphia, Pennsylvania.)

Fig. 15.14 An automatically balanced recording thermocouple potentiometer. (Courtesy of The Bristol Company, Waterbury, Connecticut.)

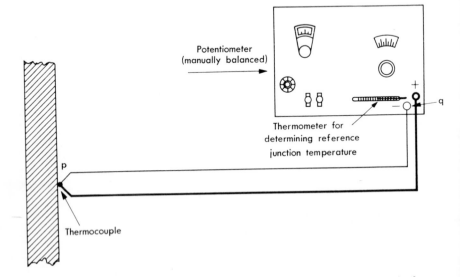

Fig. 15.15 Schematic diagram illustrating use of potentiometer terminals as a reference junction.

Comparison with Fig. 15.10(b) indicates that the instrument box may be considered an intermediate conductor in the same sense as C in the figure. If we assume the two instrument binding posts to be at identical temperature, the cold junction will then be formed by the ends of the two thermocouple leads as they attach to the posts. If a reference temperature is determined by use of a good liquid-in-glass thermometer placed near the binding posts, application of the law of intermediate metals and use of the tables referred to 32°F permits determination of the hot-junction temperature.

Example. Let us assume a copper-constantan couple, a reference temperature of 70°F determined as described above, and a potentiometer reading of 2.878 millivolts. We may express the law of intermediate temperatures as follows:

$$e_{x-32} = e_{x-70} + e_{70-32}.$$

Substituting,

$$e_{x-32} = 2.878 + 0.832 = 3.710.$$

Reference to Table 15.6 indicates a *hot* temperature of 190°F.

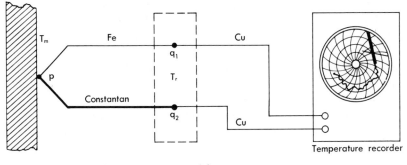

(a)

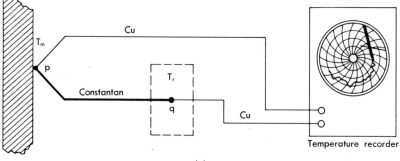

(b)

Fig. 15.16 Diagrams showing use of extension leads.

Thermocouple wire is relatively expensive compared with most common materials, such as ordinary copper. It is therefore often desirable to minimize the use of the more costly materials by employing extension leads. Arrangements of this are shown in Fig. 15.16. In these cases the measuring junction is shown at p and the reference junction(s), at q. Comparison with Fig. 15.10(a) indicates the similarity. Of course, a requirement for accuracy is that q_1 and q_2 be maintained at the same temperature, and further, that the temperature be accurately known.

An iron–constantan couple is indicated in Fig. 15.16(a), although any of the common materials could be used. Should a copper–constantan couple be employed, with copper extension leads, the arrangement would become that shown in Fig. 15.16(b). As before, temperature T_r must be accurately known. This corresponds to the simplified version shown in Fig. 15.10(a).

As indicated above, reference temperature T_r should be known. This may not always be strictly true. In certain commercial equipment, particularly of the recording type, electrical or electromechanical compensation is sometimes built in. When this is done, T_r may not actually be indicated, but its effect is nevertheless recognized and compensated.

Laboratory methods for using thermocouples often employ reference junctions at accurately controlled temperatures. The common arrangements make use of ice baths, such as shown in Fig. 15.17. These systems correspond to simplified circuits shown in Figs. 15.10(a) and (b), respectively. (One circuit employs extension leads, while the other does not.) For the most accurate work, distilled water with ice made therefrom is advisable to eliminate shifts in the freezing point caused by contaminants. Some form of Dewar flask is convenient to use to reduce melting rate. Although the ice bath supplies an easily obtained, accurately controlled reference temperature, any other controlled source could be employed using the same procedures.

d) Effective junction. In certain instances it may be highly desirable to know, as precisely as possible, the location of the effective thermocouple junction, i.e., where, within the dimensional extent of the couple, the indicated temperature occurs. This becomes of greater importance as both the temperature gradient and the size of the couple are increased. In general terms, the *effective location* is at the point of junction symmetry nearest the leads (points j in Figs. 15.11 and 15.12). Baker, Ryder, and Baker [27] define the area $D \times d$ (Fig. 15.18) as the region of uncertainty in a "bead-type" couple, when there is a temperature gradient through the junction.

e) Thermopiles and thermocouples connected in parallel. Thermocouples may be connected electrically in series or parallel, as seen in Fig. 15.19. When connected in series, the combination is usually called a *thermopile*, while parallel-connected couples have no particular name.

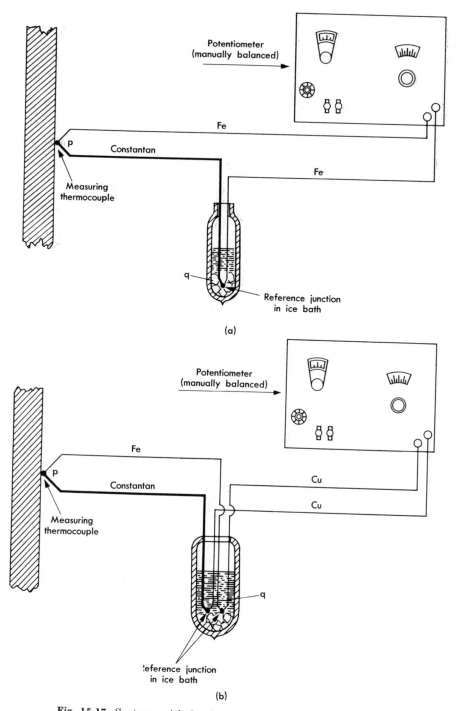

Fig. 15.17 Systems with fixed reference temperature (ice bath).

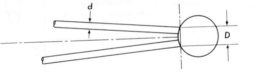

Fig. 15.18 Region of uncertainty for a bead-type thermocouple.

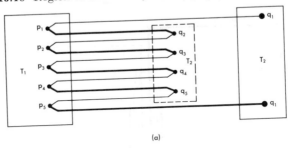

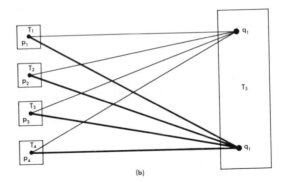

Fig. 15.19 (a) Series-connected thermocouples forming a thermopile.   (b) Parallel-connected thermocouples.

The total output from n thermocouples connected to form a thermopile [Fig. 15.19(a)] will be equal to the sums of the individual emf's, and if the thermocouples are identical, the total output will equal n times the output of a single couple. The purpose of using a thermopile rather than a single thermocouple is, of course, to obtain a more sensitive element.

When the couples are combined in the form of a thermopile, it is usually desirable to cluster them together as closely as possible in order to measure the temperature at an approximate point source. It is obvious, however, that when thermocouples are combined in series, the law for intermediate metals, as illustrated in Fig. 15.10(b), cannot be applied to combinations of thermocouples, for the individual thermocouple emf's would be shorted. Care must therefore be used to insure that the individual couples are electrically insulated one from the other.

Parallel connection of thermocouples provides an averaging, which in certain cases may be advantageous. This form of combination is *not* usually referred to as a thermopile.

15.7 THE LINEAR-QUARTZ THERMOMETER

The interrelationship between temperature and the resonating frequency of a quartz crystal has long been recognized. In general, the relationship is nonlinear, and for many applications very considerable effort has been expended in attempts to minimize the frequency drift caused by temperature variation. Recently, however, Hammond [30] discovered a new orientation termed the "LC" or "linear cut" which provides a temperature-frequency relationship of 1000 Hz/°C with a deviation from the best straight line of less than 0.05 % over a range of −40 to 230°C (−40 to 446°F). This linearity may be compared with a value of 0.55 % for the platinum-resistance thermometer.

Nominal-resonator frequency is 28 MHz and the sensor output is compared to a reference frequency of 28.208 MHz supplied by a reference oscillator. The frequency difference is detected, converted to pulses, and passed to an electronic counter which provides a digital display of the temperature magnitude. Various probes are available, all with time constants of 1 sec. Resolution is dependent upon repetitive readout rate, with a value of 0.0001°C attainable in 10 sec. Readouts as fast as four per second may be obtained. Remote sensing to 10,000 ft is possible.

15.8 PYROMETRY

The term *pyrometry* is derived from the Greek words *pyros*, meaning "fire," and *metron*, "to measure." Literally the term means general temperature measurement. However, in engineering usage, the word normally refers to the measurement of temperatures in the range extending upward from about 1000°F. Although certain of the thermocouples and resistance-type thermometers are usable above 1000°F, pyrometry is generally thought of as consisting primarily of the various forms of thermal-radiation measurement.

There are two distinct instruments referred to as pyrometers. They are the *total-radiation pyrometer* and the *optical pyrometer*. The first one, as the name implies, accepts a controlled sample of *total* radiation and, through determination of the heating effect of the sample, obtains a measure of temperature. Again as the name implies, the optical pyrometer employs optical means for estimating the change in average wavelength of *visual* radiation with temperature. Neither instrument is dependent upon direct contact with the source and, within reason, neither is dependent on distance from the source.

TABLE 15.7

TOTAL EMISSIVITY FOR CERTAIN SURFACES [11]

Surface	Temperature, °F	Emissivity
Polished silver	440–1160	0.0198–0.0324
Platinum filament	80–2240	0.036–0.192
Polished nickel	74	0.045
Aluminum foil	212	0.087
Concrete	70	0.63
Roofing paper	69	0.91
Plaster	50–190	0.91
Rough red brick	70	0.93
Asbestos paper	100–700	0.93–0.945
Smooth glass	72	0.937
Water	32–212	0.95–0.963
Blackbody	—	1.00

a) Pyrometry theory. All bodies above absolute zero temperature radiate energy. Not only do they radiate or emit energy, but they also receive and absorb it from other sources. We all know that when a piece of steel is heated to about 1000°F it begins to glow, i.e., we become conscious of visible light being *radiated* from its surface. As the temperature is raised, the light becomes brighter or more intense. In addition, there is a change in color; it changes from a dull red, through orange to yellow, and finally approaches an almost white light at the melting temperature (2600 to 2800°F).

We know, therefore, that through the range of temperatures from approximately 1000 to 2800°F, energy in the form of *light* is radiated from the body. We can also sense that at temperatures below 1000°F and almost down to room temperature, the piece of steel is still radiating energy in the form of *heat*, for if the mass is large enough we can feel it even though we may not be touching it. We know, then, that energy is radiated through certain temperature ranges because our senses provide the necessary information. Although our senses are not as good at lower temperatures, on occasion one can actually "feel" the presence of cold walls in a room because heat is being radiated from one's body *to* the walls. Energy transmission of this sort does not require an intervening medium for conveyance; in fact, intervening substances actually interfere with transmission.

The energy of which we are speaking is transmitted as electromagnetic waves traveling at the speed of light. Although the processes are not completely understood, it is known that all substances emit and absorb radiant energy at a rate depending on the absolute temperature and physical properties of the substance. Waves striking the surface of a substance are partially absorbed, partially reflected, and partially transmitted. These portions are measured in terms of *absorptivity*, α, *reflectivity*, ρ, and *transmissivity*, τ, where:

$$\alpha + \rho + \tau = 1. \tag{15.5}$$

For an ideal reflector, a condition approached by a highly polished surface, $\rho \to 1$. Many gases represent substances of high transmissivity, for which $\tau \to 1$, and a *blackbody* approaches the ideal absorber, for which $\alpha \to 1$.

Before a body can emit energy it must have first absorbed it. It follows, therefore, that a good absorber is also a good radiator, and it may be concluded that the *ideal radiator* is one for which the value of α is equal to unity. When we refer to radiation as distinguished from absorption, the term *emissivity*, ε, is used rather than absorptivity, α. However, from Kirchhoff's law

$$\epsilon = \alpha.$$

Table 15.7 lists values of emissivities for certain materials.

According to the Stefan-Boltzmann law [11], the net rate of exchange of energy between two ideal radiators A and B is

$$q = \sigma(T_A^4 - T_B^4). \tag{15.6}$$

This may be modified for the nonideal case to read

$$q = \sigma \epsilon C_A(T_A^4 - T_B^4), \tag{15.7}$$

in which

q = radiant-heat transfer, Btu/hr·ft²,

C_A = configurational factor to allow for relative position and geometry of bodies,

T_A and T_B = absolute temperatures of bodies A and B, respectively, °R,

σ = Stefan-Boltzmann constant

= 0.173×10^{-8} Btu/hr·ft² °R⁴.

The foregoing supplies the theoretical basis for total-radiation pyrometry.

In application, recognition must be made of the fact that the radiators are nonideal. They are of nonoptimum geometry and position; absorption takes place in the intervening media, and the bodies themselves never possess emissivities equal to unity. Account of such things as these must be made through calibration.

As mentioned earlier, we all know that the color changes and the wavelength of radiation decreases with temperature increase, which accounts for the change in color of the piece of steel as its temperature is raised. If we heat an ideal radiator and determine the relative intensities at each wavelength, we will obtain the data for a characteristic energy-distribution

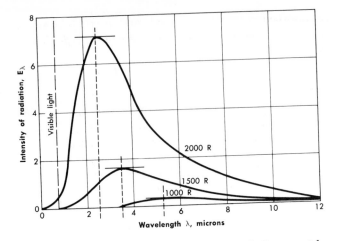

Fig. 15.20 Graphical representation of Wien's displacement law.

curve. For different temperatures, curves such as shown in Fig. 15.20 will be obtained. It will be noted that not only is the radiation intensity of the higher-temperature body increased, but also that there is a shift in the wavelength of maximum emission toward the shorter waves (from red toward blue). The rule governing this latter effect, referred to as the *Wien displacement law*, is the primary basis for the optical pyrometer. The intensity relation may be expressed as follows [12]:

$$E_\lambda = C_1 \lambda^{-5}/[e^{(C_2/\lambda T)} - 1] \qquad (15.8)$$

in which

E_λ = energy emitted at wavelength λ,

C_1 and C_2 = constants,

e = base of Naperian logarithms,

T = absolute temperature of blackbody.

This forms the basis for temperature measurement by optical means.

b) Total-radiation pyrometry. Figure 15.21 shows, in simplified form, the method of operation of the total-radiation pyrometer. Essential parts of the device consist of some form of directing means, shown here as baffles but which is more often a lens, and an approximate blackbody receiver with means for sensing temperature. Although the sensing element may be any of the types discussed earlier in this chapter, it is generally one of the thermoelectric kinds, such as the thermocouple or resistance thermometer. Usually some form of thermopile is used. A balance is quickly established between the energy absorbed by the receiver and that dissipated by conduction through leads and emission to surroundings. The receiver equilibrium

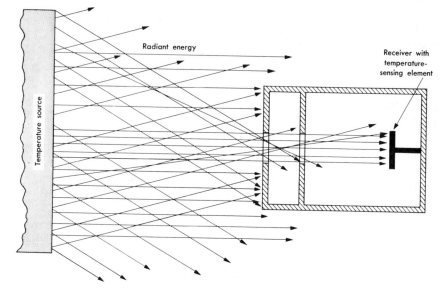

Fig. 15.21 A simplified form of total-radiation pyrometer.

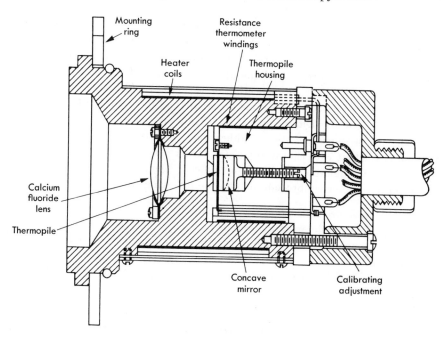

Fig. 15.22 Section through a commercially available low-temperature, total-radiation pyrometer. (Courtesy Minneapolis-Honeywell Regulator Company, Brown Instruments Division, Philadelphia, Pennsylvania.)

temperature then becomes the measure of source temperature, with the scale established by calibration.

Figure 15.22 shows a sectional view of a commercially available pyrometer. Although total-radiation pyrometry is primarily used for temperatures above 1000°F, the pyrometer shown is selected to illustrate an instrument sensitive to very low-level radiation (125–700°F). The arrangement, however, is typical of general radiation-pyrometry practice. A lens-and-mirror system is used to focus the radiant energy on a thermopile, whose output is measured by a voltage-balancing potentiometer. Thermocouple reference temperature is supplied by maintaining the assembly at constant temperature through use of a heater controlled by a resistance thermometer. In many cases compensation is obtained through use of temperature-compensating resistors in the electrical circuit.

Particular attention must be given to the optical system of a radiation pyrometer, and appropriate optical glasses must be selected to pass the necessary range of wavelengths. Pyrex glass may be used for the range of from 0.3 to 2.7 microns, fused silica for 0.3 to 3.8 microns, and calcium fluoride for 0.3 to 10 microns. While pyrex glass may be used for high-temperature measurement, it is practically opaque to low-temperature radiation, say below 1000°F.

Radiation pyrometers are used ideally in applications where the sources approach blackbody conditions, that is, where the source has an emissivity ϵ approaching unity. In general, however, the radiated energy is a good measure of temperature only if the applicable value of ϵ is accounted for. This may be made clear by inspection of Eq. (15.7). Although pyrometers calibrated for blackbody conditions are available, in general they must be calibrated for the particular application. They are not normally considered general-purpose instruments. Calibration consists of comparing the pyrometer readout with that of some standardized device, such as a thermocouple. Often single-point calibration suffices. Devices for adjusting total-radiation pyrometer calibration include the following [13]:

1. Movable aperture in front of the thermopile.
2. Variable thermopile aperture area or pyrometer lens or window area.
3. Movable metal plug screwed into the thermopile housing adjacent to the hot junction.
4. Movable concave mirror reflecting varying amounts of energy back to the thermopile of a lens-type pyrometer (see Fig. 15.22).
5. Variable shunt resistor in the electrical circuit.

Although radiation pyrometers may theoretically be used at any reasonable distance from a temperature source, there are practical limitations that should be mentioned. First, the size of target will largely determine the degree of temperature averaging, and in general, the greater the distance from the source, the greater the averaging will be. Secondly, the nature of the intervening atmosphere will have a decided effect on the pyrometer indication. If smoke or dust is present or certain optically transparent gases or solids are in the path, considerable energy absorption may occur. This will be a particularly troublesome problem if such absorbents are not constant, but are varying with time. For these reasons, minimum practical distance is advisable, along with careful selection of pyrometer sighting methods.

There are three common arrangements used to obtain a sample of radiated energy for the thermopile to sense. They are:

1. A lens system which has the power to concentrate the sampled energy over a smaller area, but which is subject to aberrations and must be kept carefully cleaned.

2. An open-end tube (illustrated schematically in Fig. 15.21). In effect, this is a selective "baffle" arranged to transmit energy from a selected area only.

3. A closed-end sighting tube, usually of ceramic, which may be inserted into a furnace or immersed in a liquid bath.

Of course, the primary purpose of these sighting methods and devices is to obtain a true sample from the area or point of interest, uninfluenced by any surrounding conditions.

c) *Optical pyrometry.* Optical pyrometers use a method of matching as the basis for their operation. In general, a reference temperature is provided in the form of an electrically heated lamp filament, and a measure of temperature is obtained by optically comparing the visual radiation from the filament with that from the unknown source. In principle, the radiation from one of

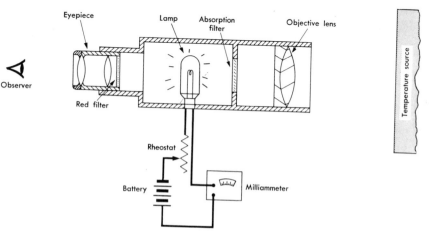

Fig. 15.23 Schematic diagram of an optical pyrometer.

the sources, as viewed by the observer, is adjusted to match that from the other source. Two methods are employed: (1) the current through the filament may be controlled electrically, through a resistance adjustment, or (2) the radiation accepted by the pyrometer from the unknown source may be adjusted optically by means of some absorbing device such as an optical wedge, polarizing filter, or iris diaphragm. The two methods are referred to, respectively, as the method using the variable-intensity comparison lamp and the method using the constant-intensity comparison lamp [14]. In both cases the adjustment required is used as the means for temperature readout.

Figure 15.23 illustrates schematically an arrangement of a variable-intensity pyrometer. In use, the pyrometer is sighted at the unknown temperature source at a distance such that the objective lens focuses the source in the plane of the lamp filament. The eyepiece is then adjusted so that the filament and the source appear superimposed and in focus to the observer. In general, the filament will appear either hotter than or colder than the unknown source, as shown in Fig. 15.24. By adjusting the battery current, the filament (or any prescribed portion such as the tip) may be made to disappear, as indicated in Fig. 15.24(c). The current indicated by the milliammeter to obtain this condition may then be used as the temperature readout. Other readout arrangements use the rheostat setting, in which case battery standardization is required, or some form of potentiometric null-balancing system.

A red filter is usually employed to obtain approximately monochromatic conditions, and an absorption filter is used so that the filament may be operated at reduced intensity, thereby prolonging its life.

The constant-intensity comparison-lamp method employs the same basic principle of operation as just described for the variable-intensity type, but a different method of adjustment is used. As the name implies, the lamp filament is maintained at constant intensity, while the comparative

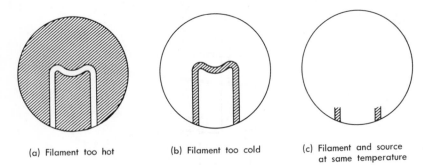

(a) Filament too hot   (b) Filament too cold   (c) Filament and source
                                                       at same temperature

Fig. 15.24 Appearance of filament when (a) filament temperature is too high, (b) filament temperature is too low, and (c) filament temperature is correct.

radiation from the unknown source is attenuated by methods mentioned earlier. The required adjustment is used for indication of temperature. It is obvious that some means must be employed to standardize filament current. This is accomplished through use of a milliammeter and rheostat in the manner shown in Fig. 15.23.

15.9 OTHER METHODS OF TEMPERATURE INDICATION

Two methods of temperature measurement listed in the introduction to this chapter have not been referred to in the intervening pages. They are: the application of changes in physical state and of changes in chemical state. Several devices based on these principles should be mentioned.

Seger cones have long been used in the ceramic industry as a means of checking temperatures [15]. These devices are simply small cones made of an oxide and glass. When a predetermined temperature is reached, the tip of the cone softens and curls over, thereby providing the indication that the temperature has been attained. Seger cones are made in a standard series covering a range from 1100 to 3600°F in steps of 20–70°F.

Somewhat similar temperature-level indicators are available under the registered trademarks of Tempilstiks,* Tempilaq,* and Tempil Pellets.* These are crayonlike sticks, lacquer, and pill-like pellets, respectively. Each is available at temperature intervals of about 12–50°F through a range of from 110–2000°F. Pellets may be had to a temperature of 2500°F. The crayon or lacquer is stroked or brushed on the part whose temperature is to be indicated. After the lacquer dries, it and the crayon marks appear dull and chalky. When the calibrated temperature is reached, the marks become liquid and shiny. This provides indication that the particular temperature has been reached. The pellets are used in a similar manner, except that they simply melt and assume a shiny liquidlike appearance at stated temperature. With the use of crayons, lacquer, or pellets covering a range of temperatures, the maximum temperature attained during a test may be rather closely determined.

Other similar temperature-level indicators are called Thermocolor† and Thermochrome.† The former is a paint and the latter a crayon. Both indicate attainment of a predetermined temperature by changing color. For example, Thermocolor No. 40 makes four color changes at specified temperatures, as follows: from pink to light blue at 149°F, to yellow at 293°F, to black at 347°F, and finally to olive green at 644°F. Other formulations have single, double, or triple changes at various temperatures.

---

* Registered trademarks, Tempil Corporation, New York.

† Registered trademarks. Distributed by Curtiss-Wright Corporation, Research Division, Clifton, N.J.

15.10 SPECIAL PROBLEMS

The number of special problems associated with temperature measurement is unlimited. However, several are significant enough to warrant special note. These will be discussed in the next several pages.

a) Errors resulting from conduction and radiation. In considering this item, let us think in terms of using a thermocouple to measure the gas temperature in a furnace, bearing in mind, however, that the principles discussed apply to other temperature *probes* and to many other situations.

Basically, any temperature element senses temperature because heat is transferred between the surroundings and the element until some kind of equilibrium condition is reached. When a bare thermocouple is inserted through the wall of a furnace (assume it to be gas- or coal-fired), heat is transferred to it from the immersing gases by convection. Heat also reaches the element through radiation from the furnace walls and from incandescent solids such as a fuel bed or those carried along by the swirling gases. Finally, heat will flow from the element through any connecting leads by conduction. The temperature indicated by the probe therefore will be a function of all these environmental factors, and consideration must be given to their effects in order to intelligently interpret or control the results.

First of all, in the more common case, the major heat flow will occur directly by forced convection between the gases and the probe. This may be expressed by the relation [16]

$$q_1 = h_c A (T_g - T_t),$$ (15.9)

in which

q_1 = heat transferred, Btu/hr,

h_c = coefficient of heat transfer, Btu/hr·in^2·°F,

A = surface area of probe, in^2,

T_g = gas temperature, °F,

T_t = probe temperature, °F.

Although the transfer coefficient, h_c, is a function of a number of things, including viscosity, density, and specific heat of the gas, of particular importance in the application under discussion is the fact that it also is a function of a power of the velocity of gas over the probe.

Radiation effects. Radiation between the probe and any source or sink of different temperature is a function of the difference in the fourth powers of the (absolute) temperatures [see Eq. (15.7)]. It is generally true, therefore, that radiation becomes an increasingly important source of temperature error as the temperatures and their differences increase. Increased temperature differences generally result from temperature extremes, either high or low, and in either of these cases, particular attention must be given to radiation effects.

As discussed in Article 15.8a, radiant-heat transfer is also a function of the emissivities of the members involved. For this reason, a bright, shiny probe is less affected by thermal radiation than is one tarnished or covered with soot.

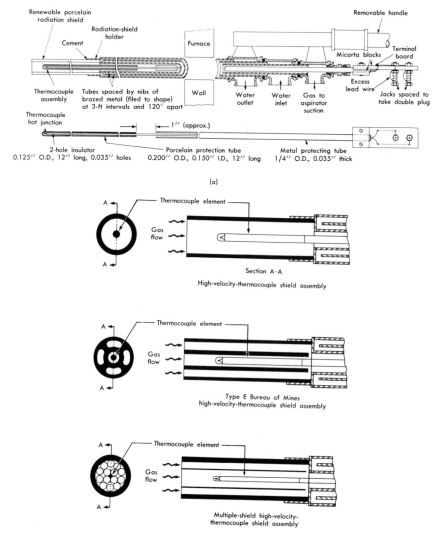

Fig. 15.25 (a) Section through an aspirated high-velocity thermocouple, HVT. (b) Various tips used on high-velocity thermocouples. (Both figures courtesy of The Babcock and Wilcox Company, N.Y.)

Radiation error may be largely eliminated through proper use of thermal shielding. This consists of placing barriers to thermal radiation around the probe, which prevent the probe from "seeing" the radiant source or sink, as the case may be. For low-temperature work, such shields may simply be made of sheet metal appropriately formed to provide the necessary protection. At higher temperatures, metal or ceramic sleeves or tubes may be employed. In applications where gas temperatures are desired, however, care must be exercised in placing radiation shields so as not to cause stagnation of flow around the probe. As pointed out earlier, desirable convection transfer is a function of gas velocity.

Consideration of these factors led quite naturally to the development of an aspirated high-temperature probe known as the *high-velocity thermocouple* (HVT) [17]. Figure 15.25 illustrates an aspirated probe with several types of tips. Gas is induced through the end, over the temperature-sensing element, and is exhausted either to the exterior or, if it will not alter process or measuring functioning, may be returned to the source. A renewable shield provides radiation protection for the element, and through use of aspiration, convective transfer is enhanced. Gas mass-flow over the element should be not less than 15,000 lb/hr/ft² for maximum effectiveness [17].

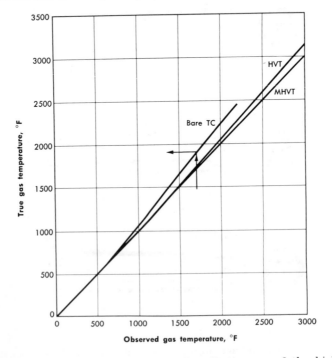

Fig. 15.26 Graphical representation of the effectiveness of the high-velocity thermocouple. (Courtesy of The Babcock and Wilcox Company, N.Y.)

When a single shield is used, as shown in Fig. 15.25(a), the shield temperature is largely controlled by convective transfer from the aspirated gas through it. Its exterior, however, is subject to thermal-radiation effects, and thus its equilibrium temperature, and hence that of the sensing element, will still be somewhat influenced by radiation. Maximum shielding may be obtained through use of multiple shields, as shown in the lower two sections of Fig. 15.25(b). Thermocouples using multiple shielding are known as *multiple high-velocity thermocouples* (MHVT) [17]. The effectiveness of both the HVT and MHVT relative to a bare thermocouple is graphically illustrated in Fig. 15.26.

Our discussion here of radiation effects has been centered largely on high-temperature application of thermocouples. However, once again it should be made clear that the principles involved apply to *any* temperature-measuring system or situation to one extent or another. Radiation may introduce errors at low temperatures as well as at high ones, and will present similar problems to all types of sensing elements. When the fluids are liquid rather than gaseous, the problem is considerably reduced, however, because most liquids, and even water vapor in air, act as effective thermal-radiation filters.

Errors caused by conduction. All temperature-measuring elements of the probe type must have mechanical support, and, in general, some connection must be made to external indicating apparatus. Such connections provide conduction paths through which heat may be transferred to or away from the sensing element. Such transfer of heat will result in a discrepancy between the indicated temperature and that desired, namely, the temperature that would exist where the instrument not present. Factors influencing such errors may be itemized as follows:

1. Conductivity of lead or support material.
2. Lead and element sizes.
3. Properties of surrounding media.
4. Flow conditions over the probe.
5. Presence of lead insulation or protective well.
6. Configuration of the immersed leads.
7. Temperature magnitudes and the form of temperature gradient along the leads or support.
8. Depth of immersion of the probe.

Johnson, Weinstein, and Osterle [18] have found that except for extreme conditions, variation in the last two factors may usually be ignored. In addition, they show that increasing the ratio of element to lead size reduces the error, or more specifically, a large value of ηL minimizes error. ($\eta^2 = 2h/kd$, where h = convection coefficient, k = conductivity of the lead wire, L = a length indicating gradient intensity, and d = diameter of the lead

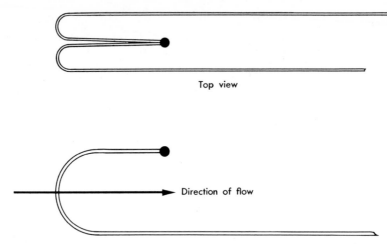

Top view

Direction of flow

Fig. 15.27 Reversed lead configuration, which may be used to reduce lead error.

wire.) It is also shown that the lead error may be reduced to zero by employ-ing a reversed lead configuration of proper proportions, as illustrated in Fig. 15.27, wherein the leads are bent back and brought downstream.

b) Measurement of temperature in rapidly moving gas. When a temperature probe is placed in a stream of gas, the flow will be partially stopped by the presence of the probe. The lost kinetic energy will be converted to heat, which will have some bearing on the indicated temperature. Two "ideal" states may be defined for such a condition. A *true* state would be that observed by instruments moving with the stream, and a *stagnation* state would be that obtained if the gas were brought to rest and its kinetic energy completely converted to heat, resulting in a temperature rise. A fixed probe inserted into the moving stream will indicate conditions lying between the two states. For exhaust gases from internal combustion engines, we find that temperature differences between the two states may be as great as 400°F [19].

An expression relating stagnation and true temperatures for a moving gas, assuming adiabatic conditions, may be written as follows [20]:

$$(T_t - T_s) = \frac{V^2}{2g_c J C_p} \, . \tag{15.10}$$

This relation may also be written

$$\frac{T_s}{T_t} = 1 + \frac{(k-1)M^2}{2} \, , \tag{15.11}$$

in which

T_s = stagnation or total temperature, °F,

T_t = true or static temperature, °F,

V = velocity of flow, ft/sec,

g_c = proportionality factor for expressing pounds force in terms of pounds mass, 32.2 lbf/lbm × ft/sec²,

J = mechanical equivalent of heat, ft·lb/Btu,

C_p = mean specific heat at constant pressure, Btu/lb·°F,

k = ratio of specific heats,

M = Mach number.

A measure of the effectiveness of a probe in bringing about kinetic energy conversion may be expressed by the relation

$$r = \frac{T_i - T_t}{T_s - T_t},$$ (15.12)

where

T_i = temperature indicated by the probe, °F,

r = a term called the "recovery factor," which is proportional to the energy conversion.

If $r = 1$, the probe would measure the stagnation temperature, and if $r = 0$, it would measure the true temperature. Experiment has shown that for a given instrument, the recovery factor is essentially a constant and is a function of the probe configuration. It changes little with composition, temperature, pressure, or velocity of the flowing gas [20].

Combining Eqs. (15.10) and (15.12), we obtain

$$T_t = T_i - \frac{rV^2}{2g_c J C_p}$$ (15.13)

or

$$T_s = T_i + \frac{(1 - r)V^2}{2g_c J C_p}$$ (15.13a)

The recovery factor, r, for a given probe may be determined experimentally [19]. However, this does not generally provide sufficient information to determine either true or stagnation temperature. Inspection of Eqs. (15.13) and (15.13a) indicates that in addition to knowing the indicated temperature T_i and the recovery factor r, the stream velocity and certain properties of the fluid must be known. When these values are known, the relations yield the desired temperatures directly. In many cases, however,

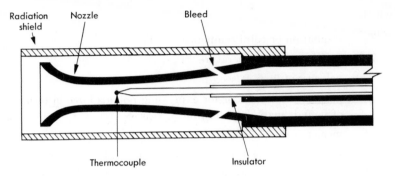

Fig. 15.28 Schematic of a sonic-flow pyrometer. (Courtesy of the National Bureau of Standards.)

it is particularly difficult to determine the flow velocity, and further theoretical consideration of the situation is required.

It has been shown [20] that for sonic velocities ($M = 1$),

$$T_s = \phi T_i, \tag{15.14}$$

in which

$$\phi = \frac{k + 1}{2 + r(k - 1)}. \tag{15.15}$$

One solution to the temperature measurement of high-velocity gases has been to make the measurement at Mach 1, through use of an instrument termed a *sonic-flow pyrometer*. Such a device is shown in Fig. 15.28. The basic instrument comprises a temperature-sensing element (thermocouple) located at the throat of a nozzle. Gas whose temperature is to be measured is aspirated (or pressurized by the process) through the nozzle to produce critical or sonic velocity at the nozzle throat. Under these conditions, Eqs. (15.14) and (15.15) apply, and in this manner determination of flow velocity need not be made. It is still necessary to know the ratio of specific heats, but these can usually be determined or estimated with sufficient accuracy. (It may be observed that the dependence of ϕ upon k reduces as r is increased.)

c) Temperature-element response. An ideal temperature transducer would faithfully respond to fluctuating inputs regardless of the time rate of temperature change; however, the ideal is not realized in practice. A time lag exists between cause and effect, and the system seldom, if ever, actually indicates true temperature input. Figure 15.29 illustrates quite graphically the magnitude of errors that may result from poor response.

The time lag that exists is determined by the particular heat-transfer circumstances that apply, and the complexity of the situation depends to a large extent on the relative importance of the convective, conductive,

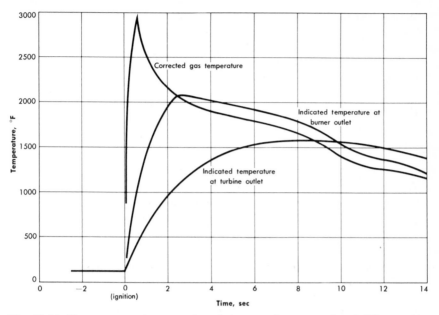

Fig. 15.29 Temperature-time record made by two thermocouples of different size and location during starting cycle of a large jet engine. (Courtesy of the Instrument Society of America.)

and radiative components. If we assume that radiation and conduction are minimized by design and application, we may equate the heat accepted by the probe per unit time to the rate of heat transferred by convection [21]:

$$Wc\frac{dT_p}{dt} = \frac{hA}{3600}(T_g - T_p) \qquad (15.16)$$

or

$$\tau\frac{dT_p}{dt} + T_p = T_g, \qquad (15.16a)$$

in which

T_p = temperature of probe, °F,

T_g = temperature of surrounding gases, °F,

W = weight of probe, lb,

c = specific heat of probe, Btu/lb·°F,

t = time, sec,

h = convective heat-transfer coefficient, Btu/hr·ft² °F,

A = surface area of probe exposed to gases, ft²,

$\tau = \dfrac{3600Wc}{hA}$.

We may write Eq. (15.16a) as follows:

$$\int_0^t dt = \tau \int_{T_{p_0}}^{T} \frac{dT_p}{T_g - T_p} . \qquad (15.16b)$$

Solving,

$$T_p = T_g - (T_g - T_{p_0})e^{-t/\tau}.$$

If we let

$$\Delta T_p = T_p - T_{p_0},$$

then

$$\Delta T_p = (T_g - T_{p_0})(1 - e^{-t/\tau}). \qquad (15.17)$$

This relation corresponds to suddenly exposing the probe at temperature T_{p_0} to a gas temperature T_g. This would be approximated if the probe were quickly inserted through the wall of a furnace or immersed in a liquid bath.

In response to a steady sinusoidal variation in temperature of angular frequency ω, indicated temperature will oscillate with reduced amplitude and will lag in phase and time [21] [see also, Example in Article 2.8(b)].

The value of τ will be recognized as the *time constant* or *characteristic time* [22] for the probe, or the time in seconds required for 63.2% of the maximum possible change $(T_g - T_p)$ [see Article 2.8(a)]. Obviously, τ should be as small as possible, and inspection shows, as should be expected, that this corresponds to low mass, low heat capacity, high transfer coefficient, and large area. Probes with low time constant provide fast response, and vice versa.

Even under idealized conditions (convective transfer only), as assumed, the time constant for a given probe is not determined by the probe alone. The convective heat-transfer coefficient is also dependent upon the character of the gas flow. For this reason, a given probe may show different time constants when subjected to different conditions.

In general, two parameters, total temperature (Article 15.10b) and mass velocity, are sufficient to describe the flow. Moffat [22] gives the following empirical equation for evaluating the time constant for bare wire thermocouples:

$$\tau = \frac{3500\rho c\, d^{1.25}}{T} G^{-15.8/\sqrt{\tau}} \qquad (15.18)$$

where

d = wire diameter, in.,

G = mass velocity, lb/sec·ft²,

T = total temperature, °R,

ρ = *average* density for the two wires, lb/ft³,

c = *average* specific heat for the two wires, Btu/lb·°F.

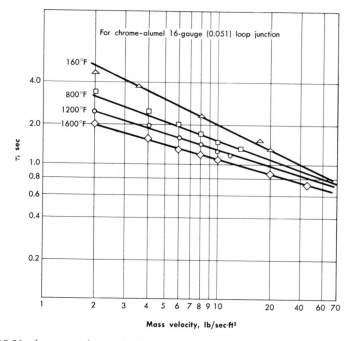

Fig. 15.30 A comparison of time constants calculated by Eq. (15.18) and determined from test data. (Courtesy of the Instrument Society of America.)

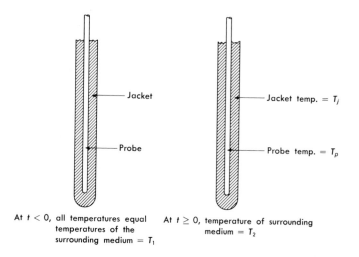

Fig. 15.31 Temperature probe in jacket subjected.

A comparison between time constants calculated by Eq. (15.18) and determined from test data is shown in Fig. 15.30.

Although practical probe-response characteristics may, in many cases, be closely approximated by the application of Eq. (15.17), in many other cases more complicated situations exist. Other elements, in addition to the actual temperature-sensing element may be involved, resulting in the multiple-time-constant problem.

The case of the common thermometer in a well, or a thermocouple or resistance thermometer in a protective sheath (Fig. 15.4) may be better approximated by a two-time-constant model. Both probe and jacket will have characteristic time constants. Let us analyze this situation as follows: we will assume that a probe-jacket assembly (Fig. 15.31) at temperature T_1 is suddenly inserted into a medium at temperature T_2. In the manner of Eq. (15.16), we may write two relationships, as follows:

$$W_j c_j \frac{dT_j}{dt} = \frac{h_j A_j}{3600}(T_2 - T_j) - \frac{h_p A_p}{3600}(T_j - T_p) \qquad (15.19\text{a})$$

and

$$W_p c_p \frac{dT_p}{dt} = \frac{h_p A_p}{3600}(T_j - T_p), \qquad (15.19\text{b})$$

where subscripts j and p refer to the protective jacket and the probe, respectively.

The relationships may be rewritten as

$$\tau_j \frac{dT_j}{dt} = T_2 - T_j - \frac{h_p A_p}{h_j A_j}(T_j - T_p) \qquad (15.20\text{a})$$

and

$$\tau_p \frac{dT_p}{dt} = T_j - T_p. \qquad (15.20\text{b})$$

Simplification may be had if we assume that the last term in Eq. (15.20a) may be neglected. This will be legitimate if:

$$A_p \ll A_j \qquad \text{and/or} \qquad T_j - T_p \ll T_2 - T_j.$$

If the above assumption is made, Eqs. (15.20a) and (15.20b) may be combined, yielding

$$\tau_j \tau_p \frac{d^2 T_p}{dt^2} + (\tau_j + \tau_p)\frac{dT_p}{dt} + T_p = T_2. \qquad (15.21)$$

A solution to this relationship is

$$\frac{T_2 - T_p}{T_2 - T_1} = \frac{\Delta T}{\Delta T_{\max}} = \left(\frac{\zeta}{\zeta - 1}\right)e^{-t/\zeta \tau_p} - \left(\frac{1}{\zeta - 1}\right)e^{-t/\tau_p} \qquad (15.22)$$

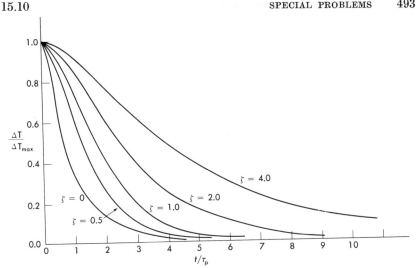

Fig. 15.32 Two-time constant problem. Plot of $\Delta T/\Delta T_{max}$ versus t/τ_p for various ratios of $\zeta = \tau_j/\tau_p$.

where

$$\Delta T = \text{momentary difference between the indicated and actual temperatures,}$$

$$\Delta T_{max} = \text{difference between temperature of medium and probe temperature at } t = 0,$$

$$\zeta = \tau_j/\tau_p.$$

Characteristics for various values of ζ are shown in Fig. 15.32. It is seen that for $\zeta = 0$, Eq. (15.22) reverts to Eq. (15.17). In addition, as the time constant for the well is increased, the overall lag is increased, as one would suspect it should.

Still more refined methods are sometimes applied, employing multiple-time-constants and dead times [28, 29]. For example, careful consideration of the simple mercury-in-glass thermometer indicates that the glass envelope, in addition to functioning as a necessary part of the differential-expansion pair, also acts as a thermal shield for the mercury.

d) *Electrical compensation.* Lag in electrical temperature-sensing elements (thermocouples and resistance thermometers) may be approximately compensated by use of appropriate electrical networks. This is done by selecting a type of filter (Article 7.25c) whose electrical-time characteristics complement those of the sensing element [23, 24]. Figure 15.33 illustrates a simple form of such a compensator. In the example illustrated, thermocouple response drops off with increased input frequency. (This is shown in terms of multiples of time constant reciprocals.) By proper choice of resistors and

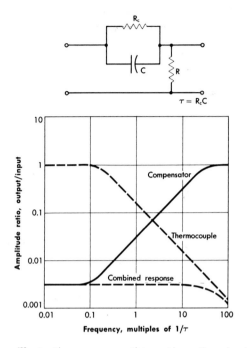

Fig. 15.33 Curves illustrating compensating action of a simple R-C network. (Courtesy of the Instrument Society of America.)

capacitance, satisfactory combined response may be extended approximately one hundred times.

15.11 CALIBRATION OF TEMPERATURE-MEASURING DEVICES

As stated in Article 2.4, for the results to be meaningful, measuring procedure and apparatus must be provable. This is true of all areas of measurement, but for some reason the impression seems prevalent that it is less true for temperature-measuring systems than for others. For example, it is very generally thought that the only limitation in the use of thermocouple tables is in satisfying the requirement for metal combination indicated in the table heading. Mercury-in-glass scale divisions and resistance-thermometer characteristics are commonly accepted without question. And it is assumed that once proved, the calibrations will hold indefinitely.

Of course, we know that these ideas are incorrect. Thermocouple output is very dependent on purity of elementary metals and consistency and homogeneity of alloys. Alloys of supposedly like characteristics but manufactured by different companies may have temperature-emf relations sufficiently at variance to require different tables. In addition, aging with

TABLE 15.8

SECONDARY REFERENCE POINTS [26]

Source	Temperature	
	°C	°F
Sublimation point of carbon dioxide	−78.5	−109.3
Freezing point of mercury	−38.87	−37.97
Triple point of water*	0.0100	32.0180
Triple point of benzoic acid*	122.36	252.25
Boiling point of naphthalene	218.0	424.4
Freezing point of tin	231.9	449.4
Boiling point of benzophenone	305.9	582.6
Freezing point of cadmium	320.9	609.6
Freezing point of lead	327.3	621.1
Freezing point of zinc	419.5	787.1
Freezing point of antimony	630.5	1166.9
Freezing point of aluminum	660.1	1220.2
Freezing point of copper	1083	1981
Freezing point of palladium	1552	2826
Freezing point of platinum	1769	3216

* Temperature of equilibrium between solid, liquid, and vapor.

use will alter thermocouple outputs. Resistance-thermometer stability is very dependent on the degree of freedom from residual strains in the element, and comparative results from like elements require very careful use and control of the metallurgy of the materials.

Methods used to calibrate temperature-measuring systems fall into two general classifications: (1) comparison with the primary standards, the fixed temperature points, or interpolated points as specified by the International Practical Temperature Scale of 1968, and (2) comparison with reliably calibrated secondary standards. (See Article 4.6.)

Basically, the primary temperature standard consists of the fixed physical conditions discussed in Article 4.6. Arbitrary scales used to relate the following fixed points:

Oxygen point	−297.346°F,
Ice point	32.0°F,
Steam point	212.0°F,
Sulfur point	832.28°F,
Silver point	1761.4°F,
Gold point	1945.4°F.

Intermediate points are established by specified interpolation procedures.

Therefore, for primary calibration, the problems are those of technique attending reproduction of these fundamental points and reproduction of interpolation methods.

Certain secondary references are also specified by the International Temperature Scale (see listings in Table 15.8). For many of these points, commercial "standards" are available. These consist of a sealed container enclosing the reference materials. Glass is used for the container material for the lower temperatures and graphite is used for the higher temperatures. Integral heating coils are employed. To use the standard, the element to be calibrated is placed in a well extending into the center of the container. The heater is then turned on, and the temperature carried above the melting point of the reference substance and held until melting is completed. It is then permitted to cool, and when the freezing point is reached, the temperature stabilizes and remains constant at the specified value as long as liquid and solid are both present (several minutes). It is claimed that accuracies of approximately 0.2°F may be easily attained and that 0.02°F may be attained if care is exercised.

SUGGESTED READINGS

American Institute of Physics, *Temperature; Its Measurement and Control in Science and Industry*, New York: Reinhold Publishing Corp., V. 1, 1941; V. 2, 1962, V. 3, Pt. 1, 1963; Pt. 2, 1962; Pt. 3, 1963.

ASME Performance Test Code, PTC 19.3-1961, *Temperature Measurement*.

Baker, H. D., E. A. Ryder, and N. H. Baker, *Temperature Measurement in Engineering*. V. 1, 1963; V. 2, 1961; New York. John Wiley & Sons, Inc.

National Bureau of Standards Monograph 27, *Bibliography of Temperature Measurement*, January, 1953 to June, 1960, U.S. Department of Commerce, 1961.

National Bureau of Standards Monograph 27, Supp. 2, *Bibliography of Temperature Measurement*, July 1960 to December 1965, U.S. Department of Commerce.

National Bureau of Standards Special Publication 300, Vol. 2, *Temperature*. U.S. Department of Commerce, 1969.

PROBLEMS

15.1 The temperature of a fluid is 155°F. What is the temperature in degrees C? In degrees K? In degrees R? What are the various temperatures corresponding to 10°F?

15.2 At what temperature °F do the Fahrenheit and centigrade scales coincide?

15.3 The temperature indicated by a "total-immersion" mercury-in-glass thermometer is 155°F. Actual immersion is to the 40-degree mark. What correction should be applied to account for the partial immersion? Assume ambient temperature is 70°.

15.4 A 20 Ga (0.032-in. thick) $\times$ 6-in.-long strip of phosphor bronze is brazed to a similarly dimensioned strip of Invar to form a bimetal temperature-sensing element as shown in Fig. 15.1 (see Table 7.1 for physical properties). Recalling that for a beam in bending $1/r = d^2y/dx^2$, determine the incremental deflection of the free end of the strip per degree change in temperature.

15.5 A resistance-thermometer element is to be constructed of nickel wire (see Table 11.1 for properties). The thermometer resistance at 68°F is to be 50 ohms. What length of 0.001-in.-diameter wire should be used? Assuming that the temperature coefficient of resistance is constant over the common range of atmospheric temperatures, what will be the element resistances at 0 and 120°F?

15.6 If platinum is substituted for nickel in Problem 15.5, what would be the calculated values?

15.7 Search the literature for the range of values for the constants A and B in Eq. (15.3) for commonly used resistance-thermometer materials.

15.8 Devise a simple thermistor-calibration facility consisting of a variable-temperature environment, an accurate resistance-measuring means which avoids significant ohmic heating, and a reliable temperature-measuring system to use as the "standard." Calibrate several thermistors and evaluate their degree of adherence to Eq. (15.4). (Avoid the problems inferred in Problem 15.18.)

15.9 Investigate the schemes employed by the various automobile manufacturers for measuring and indicating engine-block temperatures. What accuracies do you think are obtained by the various systems?

15.10 An ice-bath reference junction is employed in conjunction with a copper-constantan thermocouple. For four different conditions millivolt outputs are read as follows: -4.334, 0.00, 8.133, and 11.130. What are the respective measuring-junction temperatures?

15.11 Copper-constantan thermocouples are employed for measuring the temperatures at various points in an air-conditioning unit. A reference-junction temperature of 73°F is recorded. If the following emf outputs are supplied by various couples, what are the corresponding temperatures? -1.623, -1.088, -0.169, and $+3.250$.

15.12 The temperature difference between two points on a heat exchanger is desired. The measuring and reference junctions of a copper-constantan thermocouple are embedded within the inlet and outlet tubes A and B, respectively, and an emf of 0.381 millivolts is read. Why does this provide insufficient data to determine the differential temperature accurately? What additional information must be obtained before the answer may be obtained?

15.13 Plot the ratio of T_s/T_t versus Mach number for velocities ranging from zero to Mach 3 and for $k = 1.3$, 1.4, and 1.5.

15.14 Plot T_i/T_s versus r for the range $r = 0$ to 1 and for $k = 1.3$, 1.4, and 1.5.

15.15 If Problems 2.9, 2.10, 2.11, 2.19, 2.20, 2.21, and 2.22 were not worked previously, consider them at this time.

15.16 A two-time-constant temperature transducer possesses time constants in the ratio $\zeta = 4/1$, where $\tau_p = 1.5$ sec.

   If the transducer, at an initial temperature of 80°, is suddenly immersed in a 500° environment, what will be the temperature indicated after 3 sec?

15.17 If the transducer in Problem 15.16 is initially at 500° and is suddenly immersed in an 80° environment, what temperature will be indicated after 3 sec? Although definition of the actual thermodynamic situation would require specification of temperature scale (e.g., Fahrenheit or centigrade), why is such specification unnecessary for the stated problem?

15.18 A small insulated box is constructed for the purpose of obtaining temperature-calibration data for thermistors. Provision is made for mounting a thermistor within the box and bringing suitable leads out for connection to a commercial Wheatstone bridge. The bulb of a standardized mercury-in-glass thermometer is inserted into the box for the purpose of determining reference temperatures. A small heating element (a miniature soldering iron tip) is employed as a heat source.

   After the heater is turned on, thermistor resistances and thermometer readings are periodically made as the temperature rises from ambient to a maximum. The heater is then turned off and further data are taken as the temperature falls.

   It is quickly noted, however, that there is a very considerable discrepancy in the "heating" resistance-temperature relationship compared with the corresponding "cooling" data. Why should this have been expected? Criticize the design of the arrangement described above when used for the stated purpose. How would you make a *simple* laboratory setup for obtaining reasonably accurate calibration data for a thermistor over a temperature range of, say, 80–400°F?

CHAPTER 16

VIBRATION AND SHOCK:
MEASUREMENT AND TEST METHODS

16.1 INTRODUCTION

For purposes of discussion we may divide dynamic displacement-time relationships into two broad categories, namely, vibration and shock. Very broadly, if the displacement-time variation is of a generally continuous form with some degree of repetitive nature, it is thought of as being a *vibration*. On the other hand, if the action is of a single-event form, a transient, with the motion generally decaying or damping-out before further dynamic action takes place, then it may be referred to as *shock*. Obviously, shock action may be repetitive and in any case the displacement-time relationships will normally contain vibratory characteristics. To be so termed, however, shock must in general possess the property of being discontinuous. Additionally, steep wave-fronts are often associated with shock action, although this is not a necessary characteristic.

In any event, both mechanical shock and mechanical vibration involve the parameters of frequency, amplitude, and wave form. Basic measurement normally consists of applying the necessary instrumentation to obtain a time-based record of displacement, velocity, or acceleration. Subsequent analysis can then provide such additional information as the frequencies and amplitudes of harmonic components and derivable displacement-time relationships not directly measured.

In many respects, instrumentation employed for vibration measurements are directly applicable to shock measurement. On the other hand, testing procedures and methods are quite different. The chapter is therefore divided into three sections, I. Instrumentation for Vibration and Shock Measurements, II. Vibration Testing, and III. Shock Testing.

I. INSTRUMENTATION FOR VIBRATION AND SHOCK MEASUREMENTS

16.2 VIBROMETERS AND ACCELEROMETERS

Current nomenclature applies the term *vibration pickup* or *vibrometer* to detector-transducers yielding an output, usually a voltage, that is proportional to either displacement or velocity. Whether displacement or velocity is sensed is determined primarily by the secondary transducing element. For example, if a differential transformer (Article 6.14) or a voltage-dividing potentiometer (Articles 6.9 and 7.16) is used, the output will be proportional to a displacement. On the other hand, if a variable-reluctance element (Article 6.15) is used, the output will be a function of velocity.

The term *accelerometer* is applied to those pickups whose outputs are functions of acceleration. There is a basic difference in design and application between vibration pickups and accelerometers.

16.3 ELEMENTARY VIBROMETERS AND VIBRATION DETECTORS

In spite of the tremendous advances made in vibration-measuring instrumentation, one of the most sensitive vibration detectors is the human touch. Tests conducted by a company specializing in balancing machines determined that the average person can detect, by means of his fingertips, sinusoidal vibrations having amplitudes as low as 12 microinches [1]. When the vibrating member was tightly gripped, the average minimum detectable amplitude was only slightly greater than *one* microinch. In both cases, by fingertip touch and by gripping, greatest sensitivity occurred at a frequency of about 300 Hz.

When amplitudes of motion are greater than say $\frac{1}{32}$ of an inch, a simple and useful tool is the *vibrating wedge*, shown in Fig. 16.1(a). This is simply a wedge of paper or other thin material of contrasting tone, often black, attached to the surface of the vibrating member. The axis of symmetry of the wedge is placed at right angles to the motion. As the member vibrates, the wedge successively assumes two extreme positions, as shown in Fig. 16.1(b). The resulting double image is quite well defined, with the center portion remaining the color of the wedge and the remainder of the images a compromise between dark and light. By observing the location of the point where the images overlap, marked X, one may obtain a measure of the amplitude. At this point the width of the wedge is equal to the double amplitude of the motion. This device does not yield any information as to the waveform of the motion.

Another simple amplitude-measuring device is an adaptation of the ordinary dial indicator. In this case the spindle of the indicator is held against the vibrating member while a mass attached to the case keeps the indicator relatively stationary. Motion from the member is transmitted

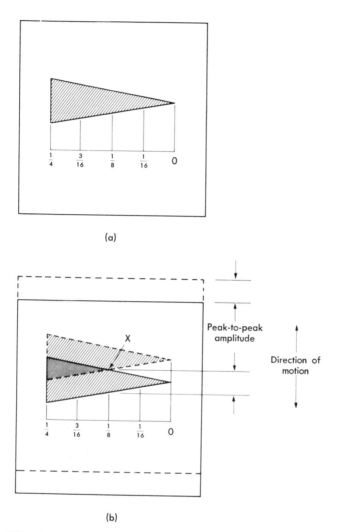

Fig. 16.1 Vibrating-wedge amplitude indicator.   (a) Stationary wedge.   (b) Extreme positions of moving wedge.

through the indicator linkage to the indicator hand, which whips back and forth, causing a blur. Mechanical inertia, friction, and backlash makes this instrument of somewhat limited use.

Devices described so far have been used to measure vibration amplitudes. A simple apparatus for measuring frequency involves a small cantilever beam whose resonant frequency may be varied by changing its effective length. In use, the instrument case is held against the member whose frequency is to be measured, and the beam length is slowly adjusted, searching for the length of beam at which resonance will occur. When this condition is found, the end of the beam whips back and forth with considerable amplitude. The device is quite sensitive, with the accuracy limited only by the resolution of the scale.

16.4 ELEMENTARY ACCELEROMETERS

Probably the most elementary acceleration-*measuring* device is the acceleration-level indicator. There are different forms of this instrument, but they are all of the yes-or-no variety, indicating that a predetermined level of acceleration has or has not been reached. Figure 16.2 is a schematic of one such instrument, which makes use of a preloaded electrical contact [2, 3]. In theory, when the effect of the inertia forces acting on the spring and mass exceed the preload setting, contact will be broken, and this action may then be used to trip some form of indicator. Rather elaborate forms of this arrangement have been devised.

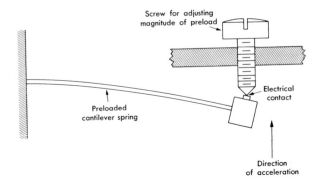

Fig. 16.2 Preloaded spring-type accelerometer.

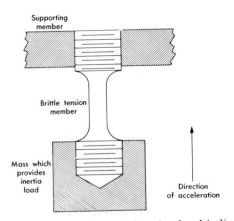

Fig. 16.3 Brittle-member acceleration-level indicator.

A second acceleration-level indicator of the *one-shot* type is illustrated in Fig. 16.3. Acceleration level is determined by whether or not the tension member fractures. Strictly brittle materials should be used for the tension member, otherwise cold working caused by previous acceleration history will change the physical properties and hence the calibration. Since such materials do not exist, this limitation is an important one.

Both the instruments described above can only be considered "rough" indicators, whose primary value lies in their simplicity.

16.5 THE SEISMIC INSTRUMENT

Vibration pickups and accelerometers are usually of the "seismic mass" form illustrated schematically in Fig. 16.4. A spring-supported weight is mounted in a suitable housing, with a sensing element provided to detect the relative motion between the mass and the housing. As will be seen later, damping must also be provided. In the figure this is represented by a dashpot mounted between the mass and the housing.

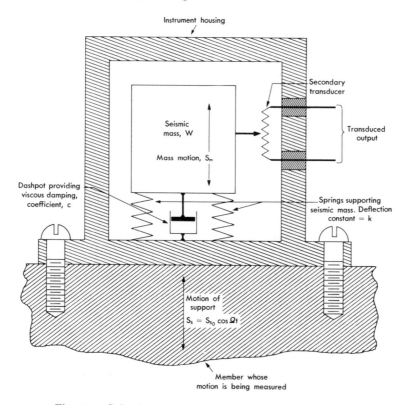

Fig. 16.4 Seismic type of motion-measuring instrument.

Basically, the action of the seismic instrument is a function of acceleration through the inertia of the mass. The output, however, is determined by the relative motion between the mass and the housing. This results in two varieties of seismic mass instruments, the *vibrometer* and the *accelerometer*. Several of the more commonly used types of vibration pickups employ a variable-reluctance transducer, in which the relative motion between a coil and the flux field from a permanent magnet is used. In this case, the instrument is velocity-sensitive because the output is proportional to the rate at which the lines of flux are cut.

By proper selection of natural frequency and damping, it is possible to design the seismic instrument so that the relative displacement between mass and housing is a function of acceleration. The output from such an instrument could therefore be calibrated in terms of acceleration, and the instrument would be an accelerometer. The fundamental requirements for the two types of instrument will be developed in the following articles.

16.6 GENERAL THEORY OF THE SEISMIC INSTRUMENT

A seismic-type motion-measuring instrument, Fig. 16.4, consists of a one-degree-of-freedom system with viscous damping, excited by a harmonic motion applied to the support. Special notice should be taken of the fact that *simple harmonic* excitation is assumed, which, strictly speaking, restricts the relations to be developed to a rather limited case. As will be seen, however, much can be learned about the seismic instruments by studying this special case. Let

W = weight of the seismic mass,

g = acceleration due to gravity,

k = deflection constant for the spring support,

c = damping coefficient,

S_m = absolute displacement of mass W, measured from the static equilibrium condition,

S_{m_0} = displacement amplitude of mass W,

S_{s_0} = displacement amplitude of supporting member,

S_s = absolute displacement of supporting member,

    = $S_{s_0} \cos \Omega t$,

$S_r = S_m - S_s$,

    = relative displacement between mass and the support, which is the displacement the secondary transducer will detect,

S_{r_0} = relative displacement amplitude between mass and supporting member,

t = any instant of time measured from $t = 0$,

Ω = exciting frequency,

ω = undamped natural frequency of the system,

ϕ = phase angle.

Applying Newton's second law to the free body of mass W, we find that the differential equation for the motion of the mass will be

$$\frac{W}{g}\frac{d^2 S_m}{dt^2} + c\frac{dS_r}{dt} + kS_r = 0. \tag{16.1}$$

Each term represents a force: the first is the inertia force, the second is the damping force, and the third is the spring force. Substituting,

$$S_m = S_r + S_s, \tag{16.2}$$

$$\frac{W}{g}\frac{d^2 S_r}{dt^2} + c\frac{dS_r}{dt} + kS_r = -\frac{W}{g}\frac{d^2 S_s}{dt^2}. \tag{16.3}$$

However,

$$S_s = S_{s_0}\cos\Omega t.$$

Then,

$$\frac{W}{g}\frac{d^2 S_r}{dt^2} + c\frac{dS_r}{dt} + kS_r = \frac{W}{g}S_{s_0}\Omega^2\cos\Omega t. \tag{16.4}$$

This equation is a linear differential equation of the second order, with constant coefficients, and is very similar to Eq. (2.18). Therefore, by comparison, the solution may be written as:

$$S_r = e^{-t/\tau}[A\cos\sqrt{1 + (c/c_c)^2}\,\omega t + B\sin\sqrt{1 + (c/c_c)^2}\,\omega t]$$

$$+ \frac{(W/kg)S_{s_0}\Omega^2\cos(\Omega t - \phi)}{\sqrt{[1 - (\Omega/\omega)^2]^2 + [2(c/c_c)(\Omega/\omega)]^2}}, \tag{16.5}$$

where

$$\phi = \tan^{-1}\left[\frac{2(c/c_c)(\Omega/\omega)}{1 - (\Omega/\omega)^2}\right]. \tag{16.6}$$

The first term on the right-hand side of Eq. (16.5) provides the transient component and the second term, the steady-state component. If we assume a time interval that is several time constants in length, the transient term may be ignored. We may then write

$$S_{r_0} = \frac{S_{s_0}(\Omega/\omega)^2}{\sqrt{[1 - (\Omega/\omega)^2]^2 + [2(c/c_c)(\Omega/\omega)]^2}}. \tag{16.7}$$

a) *The vibration pickup.* Let us now consider just what we have in Eq. 16.7 by recalling that S_{r_0} is the relative displacement amplitude between the seismic mass, W, and the support, and that S_{s_0} is the displacement

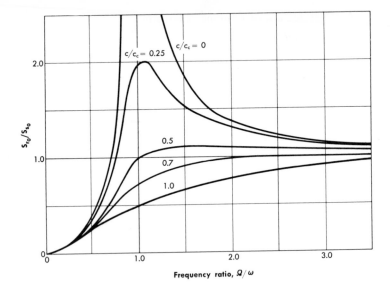

Fig. 16.5 Response of a seismic instrument to harmonic displacement.

amplitude of the instrument housing and hence of the supporting member to which it is attached. We should also recall that the instrument output will be some function of S_r, the relative displacement, and not a direct function of the quantity we wish to measure, S_s.

An inspection of Eq. (16.7) forces us to the conclusion that if the support amplitude is to be a direct linear function of the relative amplitude, it will be necessary that the total coefficient of S_{s_0} be a constant. For a given instrument, the natural frequency and damping will be *built in*, hence the only variable will be the forcing frequency, Ω. Let us see, then, how the function behaves by plotting the ratio S_{r_0}/S_{s_0} versus Ω/ω. This can be done for various damping ratios, thereby obtaining a family of curves. Fig. 16.5 is the result. Inspection of the curves shows that for values of Ω/ω considerably greater than 1.0, the amplitude ratio is indeed near unity, which is as desired. It may also be observed that the value of the damping ratio is not important for high values of Ω/ω. However, in the region near a frequency ratio of 1.0, the amplitude ratio varies considerably and is quite dependent upon damping. Below $\Omega/\omega = 1.0$, the ratios of amplitude break widely from unity. It may also be observed by inspection of Fig. 16.5 that for certain damping ratios, the amplitude ratio does not stray very far from unity, *even in the vicinity of resonance*.

We may conclude from our inspection that *damping in the order of 65 to 70% of critical is desirable* if the instrument is to be used in the frequency region just above resonance. We also see that, in any case, damping of a

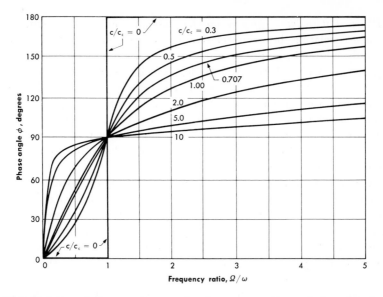

Fig. 16.6 Relations between phase angle, frequency ratio, and damping for a seismic instrument.

general-purpose instrument is a compromise, and inherent errors resulting from the principle of operation will be present. To these would be added errors that may be introduced by the secondary transducer and the second- and third-stage instrumentation.

As an example, let us check the discrepancy for the following conditions:

$$c/c_c = \text{damping ratio} = 0.68,$$
$$S_{s_0} = 0.015 \text{ in.,}$$
$$f_n = \text{natural frequency of instrument} = 4.75 \text{ Hz,}$$
$$f_e = \text{exciting frequency} = 7 \text{ Hz.}$$

Then

$$\frac{f_e}{f_n} = \frac{7}{4.75} = 1.472, \qquad \left(\frac{f_e}{f_n}\right)^2 = 2.17.$$

Using Eq. (16.7),

$$S_{r_0} = \frac{2.17 \times 0.015}{\sqrt{[1 - (2.17)]^2 + (2 \times 0.68 \times 1.472)^2}}$$

$$= 0.01403,$$

$$\text{Inherent error} = \left(\frac{0.01403}{0.015} - 1\right)100 = -6.14\%.$$

b) Phase shift in the seismic vibrometer. Let us now turn our attention to the phase relation between relative amplitude and support amplitude. Naturally it would be very desirable to have a fixed phase relation for all frequencies. A plot of Eq. (16.6) is shown in Fig. 16.6. This indicates that for zero damping the seismic mass moves exactly in phase with the support (but not with the same amplitude), as long as the forcing frequency is below resonance. Above resonance the mass motion is completely out of phase (180°) with the support motion. At resonance there is a sudden shift in phase. For other damping values, a similar shift takes place, except there is a gradual change with frequency ratio.

A simple experiment verifies this phase-shift relation. A crude support-excited seismic mass can be constructed by tying together five or six rubber bands in series to form a long, soft *spring*, and attaching a weight, of say one-half pound, to one end, while holding the other end in the hand. When the hand is moved up and down, a relative motion is obtained and the natural frequency of the system can easily be found, it being the frequency which provides greatest amplitudes with least effort. Now try moving the hand up and down at a frequency considerably below the natural frequency. It will be observed that the weight moves up and down at very nearly the same time as the hand. The motion of the seismic mass is approximately *in phase* with the motion of the supporting member, the hand. Now move the hand up and down at a frequency considerably above the natural frequency. It will be observed that the weight now moves downward as the hand moves up, and the weight moves up as the hand moves down. The motions are *out of phase*.

Our observations indicate that from the standpoint of phase shift, a "best" solution would be to design an instrument with zero damping (if that were possible). However, the amplitude relation near resonance would then be in serious error.

Perhaps any amplitude and phase-shift effects near resonance, such as we have been discussing, could be accounted for by a calibration! It would seem at first glance that such a possibility would be good, and indeed it would be feasible if single-frequency harmonic motions were always encountered. In fact, if simple sinusoidal motion were always to be measured, phase shift would not be of consequence. We would not care particularly whether the peak relative motion coincided exactly with the peak support motion as long as the waveform and measured amplitudes were correct. The difficulty arises when the input is in the complex waveform, made up of the fundamental and many other harmonics, with each harmonic simultaneously experiencing a different phase shift.

As we saw in Article 3.8c, if certain harmonic terms in a complex waveform shift relative to the remaining terms, the shape of the resulting wave is distorted, and an incorrect output results. On the other hand, bodily shifts without *relative* changes retain the true shape, and in most applications no problem results.

We find that there are three possible ways in which distortion from phase shift may be minimized. First, if there is no lag for any of the terms, there will be no distortion. Secondly, if all components lag by 180°, their relative values remain unchanged. And finally, if the shifts are in proportion to the harmonic orders, i.e., there is a linear shift with frequency, correct relative relations will be retained.

Zero shift requires no further comment, other than to suggest that it rarely, if ever, exists. When 180° shift takes place, all sine and cosine terms will simply have their signs reversed, and their relative magnitudes will remain unaffected.

A phase shift linear with frequency would be of the type in which the first harmonic lagged by say, ϕ degrees, the second by 2ϕ, the third by 3ϕ, and so on. Let us consider this situation by means of the following relation:

$$f(t) = A_0 + A_1 \cos \omega t + A_2 \cos 2\omega t + A_3 \cos 3\omega t \dots . \quad (16.8)$$

Linear phase shifts would alter this equation to read

$$\begin{aligned} f(t) &= A_0 + A_1 \cos (\omega t - \phi) + A_2 \cos (2\omega t - 2\phi) \\ &\quad + A_3 \cos (3\omega t - 3\phi) \dots \\ &= A_0 + A_1 \cos \beta + A_2 \cos 2\beta + A_3 \cos 3\beta \dots , \quad (16.8a) \end{aligned}$$

where $\beta = \omega t - \phi$. It is seen, then, that the whole relation is retarded uniformly, and that each term retains the same relative harmonic relationship with the other terms. Therefore there will be no phase distortion. As we shall see, the vibration pickup approximates the second situation, i.e., 180° phase shift, while the accelerometer is of the linear phase-shift type.

Figure 16.6 shows that in the frequency region above resonance, used by a seismic-type displacement or velocity pickup, phase shift approaches 180° as the frequency ratio is increased. The swiftness with which it does so, however, is determined by the damping. For zero damping, the change is immediate as the exciting frequency passes through the instrument's resonant frequency. At higher damping rates, the approach to 180° shift is considerably reduced. It is seen, therefore, that damping requirements for good amplitude and phase response in this frequency area *are in conflict*, and some degree of compromise is required. In general, however, amplitude response is more of a problem than phase response, and commercial instruments are often designed with 60 to 70% of critical damping, although in some cases the damping is kept to a minimum. In any case, the greater the frequency ratio above unity, the more accurately will the relative motion to which the vibrometer responds represent the desired motion.

c) *General rule for vibrometers.* We may say, therefore, that in order for a vibration pickup of the seismic mass type to yield satisfactory motion information, use of the instrument must be restricted to input forcing frequencies above its own natural frequency. Hence the lower the instrument's natural frequency, the greater will be its range. In addition, in the frequency region immediately above resonance, compromised amplitude and phase response must be accepted.

16.7 THE SEISMIC ACCELEROMETER

We shall now turn our attention to a very similar type of seismic instrument, the accelerometer. Basically the construction of the accelerometer is the same as that of the vibrometer (Fig. 16.4), except that its design parameters are adjusted so that its output is proportional to the applied acceleration.

Let us rewrite Eq. (16.7) as follows:

$$S_{r_0} = \frac{S_{s_0}\Omega^2}{\omega^2\sqrt{[1 - (\Omega/\omega^2)^2 + [2(c/c_c)(\Omega/\omega)]^2}} \qquad (16.9)$$

or

$$S_{r_0} = \frac{a_{s_0}}{\omega^2\sqrt{[1 - (\Omega/\omega)^2]^2 + [2(c/c_c)(\Omega/\omega)]^2}}, \qquad (16.10)$$

in which a_{s_0} is the acceleration amplitude of the supporting member.

Inspection of Eq. (16.10) makes the problem of properly designing and using an accelerometer clear. In order that the relative displacement between the supporting member and the seismic mass may be used as a measure of the support acceleration, the radical in the equation should be a constant. The term ω^2 in the denominator is fixed for a given instrument and does not change with application. Hence if the radical is a constant, the relative displacement will be directly proportional to the acceleration.

Let

$$K = \frac{1}{\sqrt{[1 - (\Omega/\omega)^2]^2 + [2(c\Omega/c_c\omega)]^2}}. \qquad (16.11)$$

By plotting K versus Ω/ω for various damping ratios, Fig. 16.7 is obtained. Inspection of the plot indicates that the only possibility of maintaining a reasonably constant-amplitude ratio as the forcing frequency changes is over a range of frequency ratio between 0.0 and about 0.40 and for a damping ratio of around 0.7. The extent of the usable range will depend upon the magnitude of error that may be tolerated.

a) *Phase lag in the accelerometer.* Referring again to Fig. 16.6 and to the limited accelerometer operating range just indicated, that is, $\Omega/\omega = 0.0$ to about 0.4 and $c/c_c \approx 0.70$, we see that the phase changes very nearly

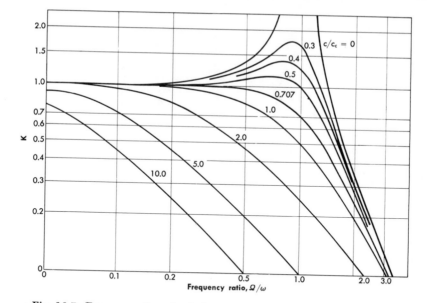

Fig. 16.7 Response of a seismic instrument to sinusoidal acceleration.

linearly with frequency. This is fortunate, for as we have seen, it results in good phase response.

b) General rule for accelerometers. We may now say that in order for a seismic instrument to provide satisfactory acceleration data, it must be used at forcing frequencies *below* approximately 40% of its own natural frequency and the instrument damping should be in the order of 70% of critical damping.

It may be observed that both vibration pickups and accelerometers may use about the same damping; however, the range of usefulness of the two instruments lies on opposite sides of their natural frequencies. The vibration pickup is made to a low natural frequency, which means that it uses a "soft" sprung mass. On the other hand, the accelerometer must be used well below its own natural frequency; therefore it uses a "stiff" sprung mass. This makes the accelerometer an inherently less sensitive but more rugged instrument than the vibration pickup.

16.8 PRACTICAL VIBROMETERS

Some form of variable-reluctance instrument is commonly used as a vibration pickup. Figure 16.8 shows the internal construction of a commercial instrument, consisting of a seismic mass in the form of a spring-restrained coil, *A*, which moves in a magnetic field supplied by the permanent magnets,

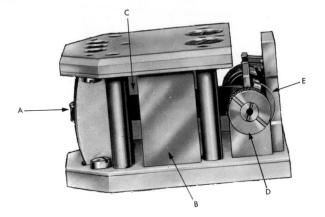

Fig. 16.8 Construction of a vibrometer which employs a reluctance-type secondary transducer. (Courtesy of MB Manufacturing Company, New Haven, Connecticut.)

B. The coil is mounted at the end of arm C, which is pivoted on jeweled bearings at points D. Weight of arm and coil is counterbalanced by hair-springs, E. As the seismic mass moves relative to the magnetic field, lines of flux are cut and a voltage is produced at the pickup terminals. Damping for this instrument is obtained electrically through the movement of coil and arm in the magnetic field, which produces a reversed emf.

Another type of variable-reluctance instrument employs a permanent magnet as the seismic mass, and the pickup coil is attached to the instrument housing. Silicon fluid surrounding the seismic mass is used to provide damping. In both cases the instruments are *velocity*-sensitive, because it is the rate at which the lines of flux are cut that determines the output from the secondary transducers.

16.9 PRACTICAL ACCELEROMETERS

A popular form of accelerometer, shown in Fig. 16.9, makes use of an *unbonded* strain-gage bridge. The seismic mass, A, is constrained to single-degree motion by small flexure springs (not shown). The unbonded strain-gage element, B, behaves in the same manner as the bonded type discussed in Chapter 11. Damping is accomplished by use of a silicon fluid surrounding the moving mass, and a small diaphragm is employed to provide expansion room required by temperature changes.

An advantage enjoyed by this type of accelerometer is the ease with which the secondary transducer, the strain-sensitive elements, may be calibrated in the field by paralleling calibration resistors (see Article 11.13). Initial calibration of the complete accelerometer is performed by the manu-

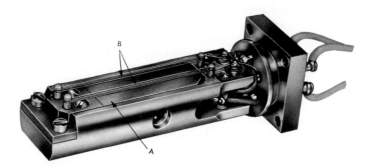

Fig. 16.9 Internal construction of the Statham Instruments Model A-6 accelerometer. (Courtesy of Statham Instruments, Inc., Los Angeles, California.)

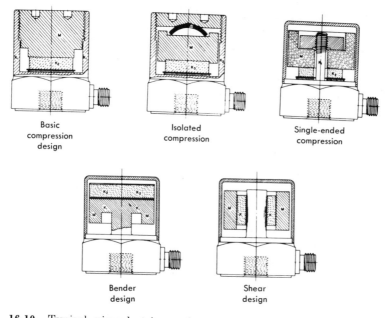

Basic
compression
design

Isolated
compression

Single-ended
compression

Bender
design

Shear
design

Fig. 16.10 Typical piezoelectric accelerometer designs. (Courtesy Endevco Corporation, Pasadena, California.)

facturer. Instruments of this kind are available covering an acceleration range from 0.5 g to 200 g.

Variable-differential transformers and voltage-dividing potentiometers are also used as secondary transducing elements in accelerometers. When these devices are employed, proper damping is often obtained by means of viscous fluids or through use of eddy-current damping provided by a permanent magnet incorporated into the design.

Probably the most popular type of accelerometer employs a piezoelectric element in some form* as shown in Fig. 16.10 (see Article 6.17). Polycrystalline ceramics including barium titanate, lead zirconate, lead titanate, and lead metaniobate are among the piezoelectric materials that have been used [7]. Various design arrangements are also employed as shown in Fig. 16.10, depending upon the characteristics desired, such as frequency range and sensitivity.

Important advantages enjoyed by the piezoelectric type are high sensitivity, extreme compactness, and ruggedness. Although the damping ratio is relatively low (0.002 and 0.25), the useful linear frequency ranges that may be attained are still large because of the high natural frequencies (to 100,000 Hz) inherent in the designs.

The output impedance of a piezoelectric device is quite high, and presents certain problems associated with proper matching, noise, and connecting cable motion and length. Either a cathode follower or a charge amplifier (Article 7.23) is normally required for proper signal conditioning. Each device has both advantages and limitations. To be effective, the cathode follower must be inserted near the accelerometer, but this may not be convenient. Although the charge amplifier minimizes the effects of noise and cable length, in combination with the pickup it is more sensitive to temperature [8]. Proper selection of instrumentation will, therefore, depend upon the application.

16.10 CALIBRATION

To be useful as amplitude-measuring instruments, both vibration pickups and accelerometers must be calibrated. This consists of determining the units of output signal (usually voltage) per unit of input (displacement, velocity, or acceleration). For example, if a displacement-sensitive vibration pickup is to be calibrated, volts per 0.001-inch amplitude may be the desired quantity. For the velocity pickup, volts per inch per second would probably be desired. For the accelerometer, volts per g could be determined. Also, the calibration should indicate how such "constants" vary over the useful frequency range.

There are two basic approaches to the calibration of seismic-type transducers: (a) by absolute methods (based directly on the physical concepts of mass, length, and time), and (b) by comparative techniques.

---

* A simple crystal phonograph cartridge, which may be purchased for a dollar or two, serves very nicely as a vibration detector, or even as a frequency pickup [5]. It should not be relied upon for amplitude information, however, because of its nonlinear response. Another simple frequency pickup that may be used where sound is involved is the microphone [6].

The latter approach employs a "standard" against which the subject transducer is compared. It is clear that the standard must have highly reliable characteristics whose own calibration is not questioned. "Identical" motions are then imposed on subject and standard and the two outputs compared. Although this method would appear to be quite simple and is undoubtedly the one most commonly used, there are many pitfalls which must be avoided. Error-free results depend upon a number of factors [9, 10].

a) the impressed motions must indeed be identical,

b) readout apparatus associated with the standard should preferably be and remain a part of the standard, and the *entire* system have traceable calibration,

c) associated readout apparatus in both circuits must have identical responses, and

d) the standard must have long-term reliability.

These are all requirements that are not easily achieved.

The motion source employed for this purpose is usually some form of exciter system as described in Article 16.17.

The following two articles discuss some of the more fundamental methods applied to calibration of seismic-type transducers.

16.11 CALIBRATION OF MOTION TRANSDUCERS

Vibration pickups are often calibrated by subjecting them to steady-state harmonic motion of known amplitude and frequency. The output of the pickup is then a sinusoidal voltage which is measured either by a reliable vacuum-tube voltmeter or a cathode-ray oscilloscope. The primary problem, of course, is in obtaining a harmonic motion of *known* amplitude and frequency.

Occasionally mechanical tables driven by cams or Scotch yoke are used. Frequency limitation inherent in this method is one problem; however, another important one is the mechanical "noise" that exists. Electromechanical exciters are commonly used [11]. Devices of this sort are described in detail in Article 16.17. Exciters of this type are capable of producing usable amplitudes at frequencies to several thousand cycles per second.

The only really positive method for determining actual instrument amplitude in a calibration test of this sort is to measure the excursion directly by means of some form of displacement-measuring device. Measuring microscopes (Article 10.15) are very useful for this purpose [12]. Either the filar type or the graduated reticule type, having a magnification of about 40–100 times, may be used. It is necessary that the microscope be mounted on a rigid support so that the pickup will be credited with no more motion

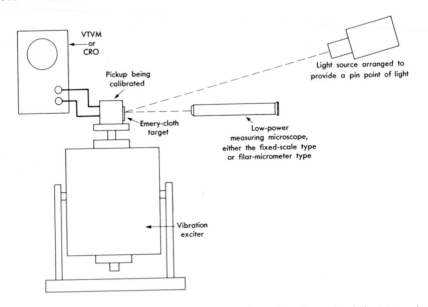

Fig. 16.11 Schematic diagram of arrangement for calibrating seismic instruments by use of steady-state harmonic motion.

than it is actually experiencing. Figure 16.11 shows schematically the general arrangement that may be used for this method of calibration.

A convenient target to observe is a small patch of #320 grit emery cloth cemented to the exciter table or directly on the pickup. A pin point of light is then directed on the emery-cloth patch. The light reflected from the emery cloth appears through the microscope as a myriad of small light sources reflected from the rough sides of the individual crystals, as shown in Fig. 16.12(a). As the exciter table moves, the individual points of light each become bright lines, Fig. 16.12(b), having lengths equal to the double amplitude of the motion. The lengths of these lines may easily be measured through use of the microscope.

One of the requirements of this method is that the center of gravity of the pickup and any mounting fixture must be placed directly on the force axis of the exciter. Otherwise lateral motion may also occur, which must be avoided. Lateral motion will not go unnoticed, however, because if it exists, Lissajous traces (Article 9.5) will be described by the light points and the condition will immediately become obvious.

The following example illustrates the procedure and data for calibrating a velocity pickup. A pickup was shaken sinusoidally by a small electromechanical shaker, and the amplitude was measured through use of a filar microscope by the procedure outlined immediately above. Data were

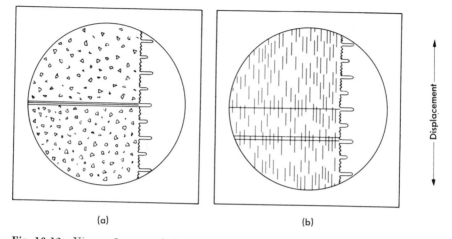

(a)   (b)

Fig. 16.12 Views of emery-cloth target observed through a microscope. (a) View when exciter table is stationary. (b) View when table is vibrated.

obtained as follows:

$$f = \text{frequency} = 120 \text{ Hz},$$
$$A_0 = \text{amplitude} = 0.0030 \text{ in. } (0.0060 \text{ in. peak to peak}),$$
$$e = \text{rms voltage measured by VTVM} = 0.150 \text{ V}.$$

Calculations:

$$e_0 = \text{voltage amplitude} = 0.150 \times 1.414 = 0.212 \text{ V},$$
$$V_0 = \text{velocity amplitude} = 2\pi \times 120 \times 0.0030 = 2.26$$
$$\text{in./sec,}$$
$$\text{Sensitivity} = e_0/V_0 = 0.212/2.26 = 0.0938 \text{ V/in./sec.}$$

(The manufacturer's nominal rating for the sensitivity of this model was 0.0945 V/in./sec.)

16.12 CALIBRATION OF ACCELEROMETERS

Accelerometer-calibration methods may be classified as follows:

I. Static

    A. Plus or minus 1-g turnover method

    B. Centrifuge method

II. Steady-state periodic

    A. Rotation in gravitational field

    B. Using a sinusoidal shaker or exciter

III. Pulsed
    A. One-g step, using free fall
    B. Multiple spring-mass device
    C. High-g methods

a) Static calibration. Plus and minus 1-g turnover method. Low-range accelerometers may be given a 2-g step calibration by simply rotating the sensitive axis from one vertical position 180° through to the other vertical position, i.e., by simply turning the accelerometer upside down. This method is positive but is, of course, limited in the magnitude of acceleration that may be applied. A simple fixture is described in Reference 13.

Centrifuge method. Practically unlimited values of static acceleration may be had by means of a centrifuge or rotating table. The normal component of acceleration toward the center of rotation is expressed by the equation:

$$a_n = \frac{\rho}{386}\left(\frac{2\pi \times \text{rpm}}{60}\right)^2 = 0.000285\rho(\text{rpm})^2, \qquad (16.12)$$

where:
    a_n = acceleration of seismic mass, g's,
    ρ = radius of rotation measured from the center
        of rotation of the table to the center of
        gravity of the seismic mass, in.,
    rpm = turntable speed, revolutions per minute,
    386 = acceleration due to gravity, in./sec².

One of the problems in this method, although not serious, is that the electrical connections must be brought out through slip rings.

b) Steady-state periodic calibration. Rotation in gravitational field. This is simply a variation of the centrifuge method, in which the turntable is rotated about a horizontal axis [14]. To the average static component as determined by Eq. (16.12), a sinusoidal 1-g gravitational component is superimposed.

Using a sinusoidal vibration exciter. A very satisfactory procedure for obtaining a steady-state periodic calibration is that described in Article 16.11 for calibrating a vibration pickup. The primary difference lies in the fact that the input for the accelerometer is the harmonic acceleration,

$$a = S_{s_0}\Omega^2 \cos \Omega t \qquad (16.13)$$

and

$$a_0 = S_{s_0}\Omega^2. \qquad (16.14)$$

c) Pulsed calibration. The free-fall method. A 1-g stepped acceleration may be obtained by suspending an accelerometer with something like a string. When the support is suddenly cut, the accelerometer is subjected to an acceleration change of 1 g.

High-g methods. Calibration of accelerometers in the high-g range (up to 40,000 g or higher) presents special problems which can be discussed only briefly at this point. Calibration methods are usually based on velocity measurements [15] and use of the following relation:

$$V_2 - V_1 = \int_{t_1}^{t_2} a \, dt. \tag{16.15}$$

The integration covers the time duration of velocity change.

Various arrangements are used for obtaining the necessary acceleration pulse, including use of ballistic pendulums, drop testers, air guns, and inclined troughs [7, 15, 16]. The problem consists in obtaining the accelerometer calibration factor, expressed by the following relation:

$$K = \frac{e}{a}, \tag{16.15a}$$

in which

K = calibration factor, volts·sec²/in.,

e = accelerometer output, V,

a = acceleration input to accelerometer, in./sec².

To obtain this value the accelerometer is excited by in impact pulse through some means such as a ballistic pendulum, one arrangement of which is shown in Fig. 16.13, and a record is made of the resulting accelerometer

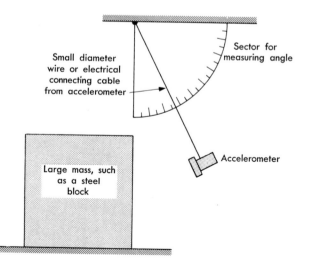

Fig. 16.13 A simple form of ballistic pendulum for calibrating accelerometers.

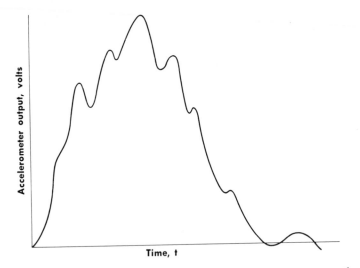

Fig. 16.14 Typical acceleration-time curve obtained by impact method.

output, Fig. 16.14. Substituting the value of a from Eq. (16.15a) into Eq. (16.15) yields

$$K = \frac{1}{(V_2 - V_1)} \int_{t_1}^{t_2} e \, dt. \tag{16.16}$$

It will be observed that the integral represents the area under the output curve, which can be obtained graphically. By experimentally determining the velocity change resulting from the impact, $(V_2 - V_1)$, the value of the calibration factor K may then be calculated.

16.13 DETERMINATION OF NATURAL FREQUENCY AND DAMPING RATIO IN A SEISMIC INSTRUMENT

On occasion it may be desirable to determine the dynamic characteristics, including damping ratio, for an *existing* vibration pickup or an accelerometer. This will be necessary, for example, when a special-purpose instrument is constructed by the user, or perhaps to check an instrument before a particularly important testing job, or when instrument damage is suspected.

Two quantities must be determined, the damping ratio and the natural frequency. It should be clear at this point that by *natural frequency* we mean the frequency of oscillation that would occur if damping were zero; we do not mean the *damped natural frequency* that would result if the seismic mass were released from an initially displaced condition. Natural frequency cannot actually be directly measured, because we cannot completely eliminate damping.

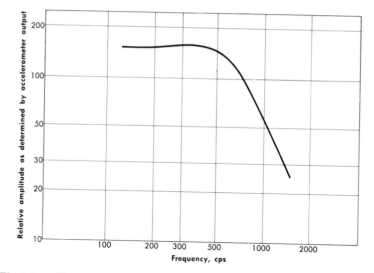

Fig. 16.15　Experimental response curve obtained from test equipment.

Theoretically it is possible to determine the damping ratio and natural frequency of a seismic instrument by subjecting it to a step input and measuring the readout "overshoot" on the first cycle [17]. Practical limitations associated with this method make its application difficult.

A very workable and accurate solution to the problem may be had by determining the instrument's response over a range of driving frequencies [18]. This may be done by sinusoidally exciting the instrument through a frequency range including the instrument's estimated natural frequency. The output in terms of relative amplitude is obtained (Fig. 16.15).

The frequency-amplitude data thus obtained is then compared with theoretical or "master" curves. If the instrument is a vibration pickup, the master curves would be those shown in Fig. 16.5. If it is an accelerometer, Fig. 16.7 would be used. The comparison is most easily done by plotting the measured relative amplitudes as ordinates and the corresponding actual frequencies as abscissas on transparent paper having the same logarithmic scales as the master curves. The experimental curve thus plotted is then superimposed on the theoretical curves and adjusted to take its proper place in the family, as determined by its shape. The damping ratio is determined by interpolation, and the experimental frequency corresponding to the dimensionless frequency ratio of 1.0 is the natural frequency. Figure 16.16 illustrates how this would be done for an accelerometer. The theoretical curves are shown as dashed lines, and the experimental curve, with corresponding scales, is shown as a solid line.

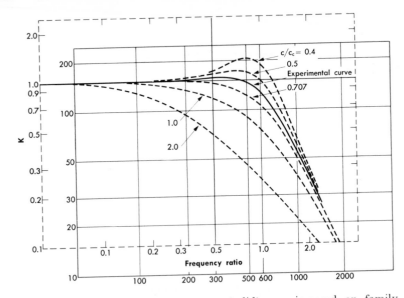

Fig. 16.16 Experimental response curve (solid) superimposed on family of theoretical curves (dashed). This shows that the experimental instrument has a natural frequency of 600 Hz and a damping ratio of 0.6.

16.14 RESPONSE OF THE SEISMIC INSTRUMENT TO TRANSIENTS

Our discussion of seismic instruments, to this point, has largely been in terms of simple harmonic motion. How will they respond to complex waveforms and transients? As stated at the beginning of the chapter, it would seem that a seismic instrument capable of responding faithfully to a range of simple harmonic frequencies should also respond faithfully to complex inputs made up of frequency components within that range. Of course, if such an assumption is true, the inherent nature of the accelerometer places it initially in a much more restricted area of operation than the vibration pickup. Whereas the accelerometer is limited to a frequency range up to roughly 40% of its own natural frequency, the only frequency restriction on the vibration pickup is that it be operated above its own natural frequency. In comparing the relative merits of the vibrometer and the accelerometer, we must not forget, however, that in terms of phase response the advantage is definitely on the side of the accelerometer (Articles 16.6b and 16.7a). This is an important advantage.

By constructing the vibration pickup with a lightly sprung seismic mass, almost any frequency range can be satisfactorily covered. For the accelerometer, frequency components above the approximate 40% value may be unavoidable. In general, however, higher frequency inputs are

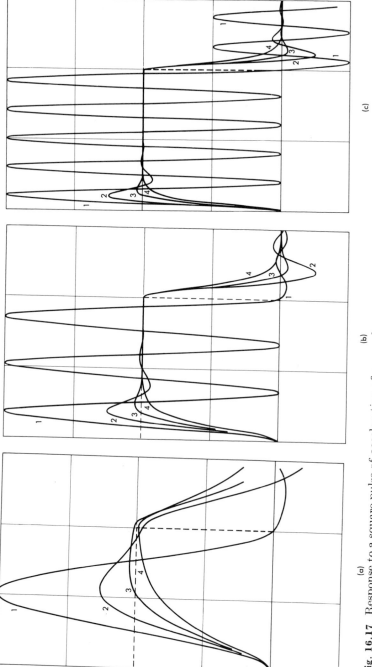

(a)   (b)   (c)

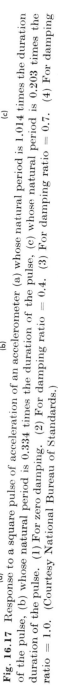

Fig. 16.17 Response to a square pulse of acceleration of an accelerometer (a) whose natural period is 1.014 times the duration of the pulse, (b) whose natural period is 0.334 times the duration of the pulse, (c) whose natural period is 0.203 times the duration of the pulse. (1) For zero damping. (2) For damping ratio = 0.4. (3) For damping ratio = 0.7. (4) For damping ratio = 1.0. (Courtesy National Bureau of Standards.)

attenuated. Figure 16.17 shows the theoretical accelerometer response to square-wave pulses [17, 19, 20]. Results are shown for different damping ratios and natural frequencies.

First of all, these curves confirm our previous conclusion that a damping ratio of about 0.7 is optimum. They also show that insofar as mass response is concerned, use of an instrument with as high a natural frequency as possible is desirable. This is not surprising, because our previous investigation has pointed to this conclusion. A high natural frequency, however, requires a stiff suspension. This means that an extra burden will be placed on the secondary transducer because of the small relative motions between mass and instrument housing. Hence, what is gained in response may be immediately lost in resolution.

The situation emphasizes even more the fact that accelerometer selection must be based on compromise. It means that the accelerometer cannot be selected entirely on its own merits, but that the whole system must be considered and then the accelerometer selected with the highest natural frequency consistent with satisfactory over-all response.

16.15 MEASUREMENT OF VELOCITY WITH SEISMIC INSTRUMENTS

Through proper instrumentation, seismic instruments may be used to measure either periodic or aperiodic velocities. Integration of acceleration-time data will provide the necessary velocity information. Such integration may be performed either by subsequent analytical treatment or by electrical integration at the time of measurement.

To begin with, however, we should understand the real meaning of the term *velocity-sensitive* as applied to seismic instruments. This designation refers to the action of the secondary transducer that may be used, and in general applied to the self-generating types. All seismic-type instruments are *basically* acceleration-sensitive. It is an acceleration or inertia force that produces the relative motion between the seismic mass and the instrument housing. Most so-called velocity-sensitive instruments use some form of variable-reluctance secondary transducer, sensitive to *relative velocity* of mass to housing, which may have only short-term relation to the absolute velocity of the member to which the instrument is attached. Assuming a perfectly responding instrument, we can readily see that if it is subjected to a constant acceleration, only a temporary output will be produced while the initial relative velocity takes place. With constant acceleration, however, the velocity of the supporting member will continue to increase.

Strain-gage, inductive, or capacitive secondary transducers usually provide an output proportional to relative displacement rather than relative velocity of mass to housing. They may be used to determine the characteristics of prolonged motion in a single direction in addition to the characteristics of periodic motions. Absolute velocity may be determined by single integration, while displacement requires double integration. The repeated operation required for double integration, however, will unavoidably introduce greater error. Properly designed and used equipment can perform the integration process with errors of less than 1 % [21].

16.16 VIBRATION AND SHOCK TESTING

Vibration and shock test systems are particularly important in relation to numerous R&D contracts. Many specifications require that equipment perform satisfactorily at definite levels of steady-state or transient dynamic conditions. Such testing requires the use of special test facilities, often unique for the test at hand but involving principles basic to all.

Numerous items for civilian consumption require dynamic testing as part of their development. All types of vibration-isolating methods require testing to determine their effectiveness. Certain material-fatigue testing employs vibration test methods. Specific examples of items subjected to dynamic tests include many automobile parts, such as car radios, clocks, head lamps, radiators, ignition components, and larger parts like fenders and body panels. Also, many aircraft components and other items for use by the armed services must meet definite vibration and shock specifications. Missile components are subjected to extremely severe dynamic conditions of both mechanical and acoustic origin.

It might be assumed that dynamic testing should exactly simulate field conditions. However, this is not always necessary or even desirable. First of all, field conditions themselves are often nonrepetitive; situations at one time are not duplicated at another time. Conditions and requirements today differ from those of yesterday. Hence, to define a set of *normal* operating conditions is often difficult if not impossible. Dynamic testing, on the other hand, may be employed to pinpoint particular areas of weakness under accurately controlled and measurable conditions. For example, such factors as accurately determined resonant frequencies, destructive amplitude-frequency combinations, and the like, may be uncovered in the development stage of a design. With such information the design engineer then may judge whether corrective measures are required, or perhaps determine that such conditions lie outside operating ranges and are therefore unimportant. Another factor making dynamic testing attractive is that the accelerated testing is possible. Field testing, in many cases, would require inordinate lengths of time.

Our discussion of dynamic testing is divided into two parts, namely vibration testing and shock testing.

II. VIBRATION TESTING

16.17 EXCITER SYSTEMS

In order to submit a test item to a specified vibration, a source of motion is required.   Devices used for supplying vibrational excitation are usually referred to simply as *shakers* or *exciters*.   In most cases, simple harmonic motion is provided, but systems supplying complex waveforms are also available.

There are various forms of shakers, the variation depending on the source of driving force.   In general, the primary source of motion may be electromagnetic, mechanical, or hydraulic-pneumatic or, in certain cases, acoustical.   Each is subject to inherent limitations, which usually dictate the choice.

a) Electromagnetic systems [22, 23, 24].   A section through a small electromagnetic exciter is shown in Fig. 16.18.   This consists of a field coil which supplies a fixed magnetic flux across the air-gap *h* and a driver coil supplied

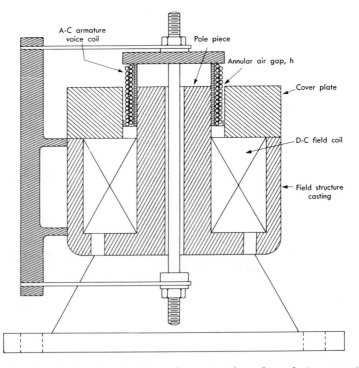

Fig. 16.18 Sectional view showing internal construction of an electromagnetic shaker head.

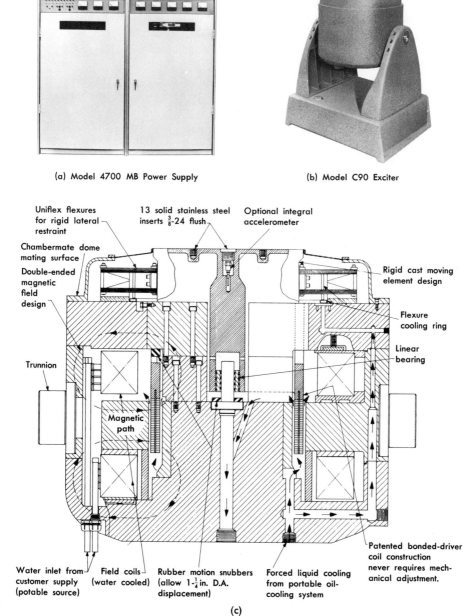

(a) Model 4700 MB Power Supply

(b) Model C90 Exciter

Uniflex flexures for rigid lateral restraint

13 solid stainless steel inserts $\frac{3}{8}$-24 flush

Optional integral accelerometer

Chambermate dome mating surface

Double-ended magnetic field design

Rigid cast moving element design

Flexure cooling ring

Linear bearing

Trunnion

Magnetic path

Water inlet from customer supply (potable source)

Field coils (water cooled)

Rubber motion snubbers (allow $1\frac{1}{4}$ in. D.A. displacement)

Forced liquid cooling from portable oil-cooling system

Patented bonded-driver coil construction never requires mechanical adjustment.

(c)

Fig. 16.19 (a) and (b) Modern power supply and 9000 lb vector force exciter. (c) Section through Model C90 exciter. (Courtesy of MB Electronics, Division of Textron Co., New Haven, Connecticut.)

from a variable-frequency source. Permanent magnets are also sometimes used for the fixed field. Support of the driving coil is by means of flexure springs, which permit the coil to reciprocate when driven by the force interaction between the two magnetic fields. It is seen that the electromagnetic driving head is very similar to the field and voice coil arrangement in the ordinary radio loudspeaker.

Figure 16.19a shows a modern system of 9000-lb capacity. An electronic power supply is used in this case. Figure 16.19(b) shows a section through the exciter head.

An electromagnetic shaker is rated according to its vector-force capacity, which in turn is limited by the current-carrying ability of the voice coil. Temperature limitations of the insulation basically determine the shaker force capacity. Note that the head illustrated in Fig. 16.19(b) is water-cooled. The driving force is commonly simple harmonic (complex waveforms are also employed) and may be thought of as a rotating vector in the manner of harmonic displacements discussed in Article 3.3. The force used for the rating is the vector force exerted between the voice and field coils.

Rated force, however, is never completely available for driving the test item. It is the force developed within the system, from which must be subtracted the force required by the moving portion of the shaker system proper. This may be expressed as follows:

$$F_n = F_t - F_a, \tag{15.1}$$

in which

F_n = net usable force available to shake the test item, lb,

F_t = manufacturer's rated capacity, or total force provided by the magnetic interaction of the voice and field coils, lb,

F_a = force required to accelerate the moving parts of the shaker system, including the voice coil, table, and appropriate portions of the voice-coil flexure beams, lb.

In practice, it is often convenient to think in terms of the total vector force, F_t, and to simply add the weight of the shaker's moving parts to that of the test item and any required accessories such as mounting brackets, etc. Table 16.1 lists the specifications for several typical commercially available electromagnetic shaker systems.

b) Mechanical exciters. There are two basic kinds of *mechanical* shakers, the directly driven type and the inertia type. The directly driven shaker consists simply of a test table which is caused to reciprocate by some form of mechanical linkage.   Crank and connecting rod mechanisms, Scotch yokes, or cams may be used for this purpose. Mechanical arrangement is provided for varying the stroke, and the frequency is controlled by use of infinitely variable drive system placed between the motor and the shaker.

TABLE 16.1

SPECIFICATIONS OF TYPICAL COMMERCIALLY AVAILABLE
EXCITER SYSTEMS

Manufacturer and model	Max. rated Force, pounds	Freq. range, Hz	Max. Disp., inches	Max. accel., g's	Wgt. Moving Armature, pounds
Ling Electronics*					
A280S/TP300	100 (sine)	0–10,000	1	80	1.25
330/CP5/6-4	3200 (sine) 2000 RMS (random)	5–3000	0.55	100	25
340/PP200/350	30,000 (sine) 34,000 RMS (random)	5–2000	1	100	240
MB Electronics†					
PM-25/2125	25 (sine)	5–10,000	0.5	50	0.5
C90/4700	9000 (sine) 6000 RMS (random)	5–3000	1	150	39
C220/5140	35,000 (sine) 35,000 RMS (random)	5–2000	1		395
Unholtz-Dickie‡					
83CD	2000 (sine) 1100 RMS (random)	5–3000	$1\frac{1}{4}$	100	20
201/A105	2500 (sine) 1600 RMS (random)	5–5000	$1\frac{1}{4}$	100	

* Anaheim, Calif.

† New Haven, Conn.

‡ Hamden, Conn.

(*Note:* All systems have electronic-type power supplies. All systems listed have electromagnetic field supplies excepting MB-25 which has a permanent magnet.)

A primary disadvantage of these shakers is that the shaking forces are fundamentally supplied by the tie-down connections between the machine and the floor to which it is attached. These forces may be considerable in some cases, resulting in shaking more than the test item. The advantage of the system, of course, lies in its simplicity and also in its ability to shake relatively larger masses with comparatively small machinery.

Figure 16.20 illustrates the method used for the second type of mechanical exciter. In this case the normal acceleration forces resulting from counterrotating masses are used as the source of excitation. In the plane of rotation, the centrifugal forces add in one direction (the vertical, as shown) and cancel in the right-angled direction.

Here again, as in the case of the electromagnetic type, the exerted force is not completely available for driving the test item. The weight of the

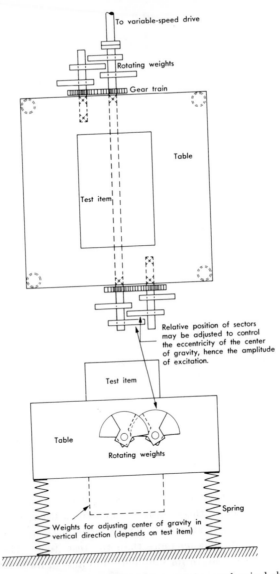

Fig. 16.20 Schematic of a rotating-weight mechanical shaker.

shaker mechanism and table must be added to the weight of the test item in determining the shaked mass.

Frequency is controlled by means of a variable-speed motor, while force amplitude is made adjustable through use of split weights, as illustrated. By adjustment of the relative angular position of the two weight halves, the center of gravity of the combination may be varied radially, thereby

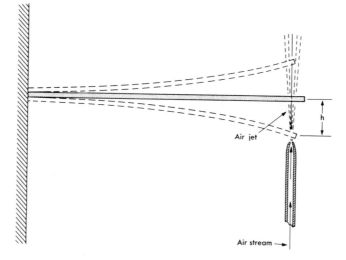

Fig. 16.21 How an air jet may be used to excite a vibration.

permitting force adjustment for a given speed. The weights may also be adjusted to provide horizontal shaking.

There are two primary advantages in an inertia system. In the first place, high force capacities are not difficult to obtain, and secondly, the shaking amplitude of the system remains unchanged by frequency cycling. This means that if a system is set up to provide a 0.05-inch amplitude at 20 Hz, changing the frequency to 50 Hz will not alter the amplitude. The reason for this will be understood if it is remembered that both the *available* exciting force and the *required* accelerating force are harmonic functions of the *square* of the exciting frequency; hence as the requirement changes with frequency, so also does the available force.

c) Hydraulic and pneumatic systems. Important disadvantages of electro-magnetic and mechanical shaker systems are limited load capacity and limited frequency, respectively. As a result, the search for other sources of controllable excitation has led to investigation in the areas of hydraulics and pneumatics.

A very simple system is shown in Fig. 16.21. This shows how an air jet may be used to excite a mass-spring system at its resonant frequency. By adjusting the air gap *h* for a given nozzle and pressure, a condition may be reached wherein the natural frequency of the beam and that of the pneumatic system coincide, with each being excited by the pulsations resulting from the periodically varying air gap. Although a cantilever beam is shown in the figure, any other simple vibrating system could be used [25]. For the most part, however, a system of this sort is only practical for small, single-purpose applications.

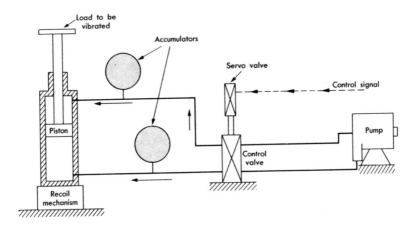

Fig. 16.22 Block diagram of a hydraulically operated shaker.

Figure 16.22 illustrates in block form a hydraulic system that has been used for vibration testing [26, 27]. In this arrangement an electrically actuated servo valve operates a main control valve. This in turn regulates flow to each end of the main driving cylinder. Capacities as great as 500,000 lb, frequencies as high as 400 Hz, and double amplitudes as great as 18 in. have been obtained.* As would be expected, a primary problem in designing a satisfactory system of this sort has been in developing valving with sufficient capacity and response to operate at the required speeds.

In an arrangement of the type shown, force, hence acceleration, is determined by the differential pressure across the piston, and the velocity is proportional to the flow capacity of the control valve. The integrated product of the force at the piston and the velocity determines the indicated power required. Pump capacity and pressure requirements are determined by the indicated power plus flow losses.

d) Relative merits and limitations of each system. Frequency range. The upper frequency ranges are available only through use of the electromagnetic shaker. In general, the larger the force capacity of the electromagnetic exciter, the lower its upper frequency will be. However, even the 35,000-lb shaker listed in Table 16.1 boasts an upper useful frequency of 2000 Hz. To attain this value with a mechanical exciter would require rotative speeds of 120,000 rpm. The maximum frequency available from the smaller mechanical units is limited to approximately 120 Hz (7200 rpm) and for the larger machines to 60 Hz (3600 rpm). Hydraulic units are presently limited to about 400 Hz.

---

* MB Electronics, New Haven, Conn.

Force limitations. Electromagnetic shakers have been built with maximum vector-force ratings of 35,000 lb, and larger units have been planned. Variable-frequency power sources for shakers of this type and size are nearly prohibitive in cost. Within the frequency limitations of mechanical and hydraulic systems, corresponding or higher force capacities may be obtained at lower costs by the latter shakers. Careful design of mechanical and hydraulic types is required, however, or maintenance costs become an important factor. Mechanical shakers are particularly susceptible to bearing and gear failures, while valve and packing problems are inherent in the hydraulic ones.

Maximum excursion. One inch, or slightly more, may be considered the upper limit of peak-to-peak displacement for the electromagnetic exciter. Mechanical types may provide displacements as great as 5 or 6 in.; however, total excursions as great as 18 in. have been provided by the hydraulic-type exciter.

Magnetic fields. Because the electromagnetic shaker requires a relatively intense fixed magnetic field, special precautions are sometimes required in testing certain items such as solenoids or relays, or any device in which induced voltages may be a problem. Although the flux is rather completely restricted to the magnetic field structure, relatively high stray flux is nevertheless present in the immediate vicinity of the shaker. Operation of items sensitive to magnetic fields may therefore be affected. Degaussing coils are sometimes used around the table to reduce flux level.

Nonsinusoidal excitation. Shaker head motions may be sinusoidal or complex, periodic or completely random. While sinusoidal motion is by far the most common, other waveforms and random motions are sometimes specified [28, 29]. In this area, the electromagnetic shaker enjoys almost exclusive franchise. Although the hydraulic type may produce nonharmonic motion, precise control of a complex waveform is not easy. Here again, future development of valving may alter the situation.

The voice coil of the ordinary loudspeaker normally produces a complex random motion, depending on the sound to be reproduced. Complex random shaker head motions are obtained in essentially the same manner. Instead of using a fixed-frequency harmonic oscillator as the signal source, either a strictly random or a predetermined random signal source is used. Electronic *noise* sources are available, or a record of the motion of the actual end use of the device may be recorded on magnetic tape and used as the signal source for driving the shaker. As an example, electronic gear may be subjected to combat-vehicle motions by first tape-recording the output of motion transducers, then using the record to drive a shaker. In this manner, controlled repetition of an identical program is possible.

16.18 VIBRATION TEST SPECIFICATIONS [30]

In many cases tests are run to simulate certain operational characteristics or conditions. Ideally, the specification should control all the parameters significant to producing the desired condition, thereby permitting repeated testing as often as required. Such tests are desirable to determine various things, primarily satisfactory life and correct functioning of the device.

A typical vibration test specification may include the following:

1. Determination of any mechanical resonant frequencies (either desirable or undesirable).
2. Determination of durability of an item (a) when cycled through a frequency range, or (b) at any resonant condition or other specified frequency.

Resonant conditions may be found by simply attaching the test item to a shaker and exciting vibration through a range of frequencies. Frequency limits and either minimum amplitudes or peak accelerations are usually prescribed, along with the axes of motion. Often tests are specified for each of three mutually perpendicular axes through the item. In many cases resonances may be detected visually or aurally. Another method makes use of the fact that *exciting-power requirements are reduced at resonance*. Most shaker power supplies provide an output-current meter, and a dip in the meter reading may be employed to indicate resonance. Of course, this is only usable when the resonating system forms a significant portion of the total shaker system. It is especially useful in connection with resonant test systems (Article 16.19b).

In other cases, critical conditions must be identified by use of vibration or acceleration transducers, or through appropriate observation of their function (for example, resonance of an electronic tube element could be detected by observing change of functioning in, say, an oscillator circuit).

Durability testing may be performed in two different ways. If there is a possibility that the test item may be operated indefinitely at a resonant condition, it would be well to test the item under such a condition. The length of test may be based on the expected life of the item.

If the application of the item makes a spectrum of frequencies probable, then either carefully specified random excitation or *cycled* sinusoidal excitation may be used. When cycling is employed, upper and lower frequency limits are specified, along with the period of cycle and duration of test and either the displacement or acceleration amplitude.

Typical specifications for cycling an item may be stated as follows: "A vibration test shall be conducted with a constant double displacement amplitude of 0.060 in. and the frequency cycled uniformly between 10 and 55 Hz in one-minute cycles for a total period of 60 min."

A typical specification for determination of resonant frequencies may read: "Resonant frequencies of the test specimen shall be determined by varying the frequency of applied vibration slowly through a range of 10 to 500 Hz at 0.06-in. double displacement amplitude or ± 10 g's, whichever is less severe. Individual resonant-frequency surveys shall be conducted with vibration applied along each of any set of three mutually perpendicular axes of the test specimen. Functioning of the test specimen shall be checked concurrently with scanning for resonant frequencies. The test specimen shall be vibrated at any indicated resonant condition for 60 min at the displacement or acceleration amplitude specified above. When no resonant condition is observed, the test item is to be subjected to a frequency of 55 Hz at a double amplitude of 0.060 in. for a period of 120 min. The test item shall function correctly at all times."

16.19 VIBRATION TEST METHODS

Two basic methods are used in applying a sinusoidal force to the test item. They are: the *brute-force method* and the *resonant method*. In the first case the item is attached or mounted on the shaker table, and the shaker supplies sufficient force to literally drive the item back and forth through its motion. The second method makes use of a mechanical spring-mass fixture having the desired natural frequency. The test item is mounted as a part of the system which is excited by the shaker. The shaker simply supplies the energy dissipated by damping.

a) The brute-force method. Brute-force testing requires that the exciter supply all the accelerating force to drive the item through the prescribed motion. Such motion is usually sinusoidal, although complex waveforms may be used. The problems inherent in an arrangement of this sort may be shown by the following example.

Suppose a vibration test specification calls for sinusoidally shaking a 20-lb test item at 100 Hz with a displacement amplitude of 0.05 in. What force amplitude will be required?

Maximum force will correspond to maximum acceleration, and maximum acceleration may be calculated as follows:

$$\text{Circular frequency } \Omega = 2\pi \times 100 = 628 \text{ rad/sec,}$$
$$\text{Maximum acceleration} = \text{amplitude } (\Omega)^2$$
$$= (0.05/12)(628)^2$$
$$= 1640 \text{ ft/sec}^2$$
$$= 51 \text{ } g,$$
$$\text{Maximum force} = \text{force amplitude} = 51 \times 20 = 1020 \text{ lb.}$$

This, of course, is the force amplitude required to shake the test item only. If support fixtures are required, they too must be shaken along with the moving coil of the shaker itself. Suppose these (the fixture and voice coil assembly) weigh 12 lb; then an additional vector force of 12 × 51, or 612 lb, would be required. The rated capacity of the shaker would therefore have to be at least 612 + 1020 = 1632 lb at 100 Hz. Thus a shaker capable of producing roughly a *2000-lb vector force* would be required to shake the *20-lb item* at the specified frequency and amplitude.

The advantage of the brute-force method is that it is not limited to any particular frequency. The disadvantage, of course, is that the shaker must supply *all* the force necessary to provide the required acceleration as dictated by frequency and amplitude. As a result, outsized shakers are sometimes required.

b) Resonant methods. The resonant system uses a spring-mass arrangement to which the test item is attached. As a simple example, a table arrangement as shown in Fig. 16.23 may be used. Leaf springs *S* support a table *T* on which is attached a test item *X*. With the shaker attached to the table through a "fuse,"* the system will have a definite natural frequency of vibration. If the frequency of the exciting force supplied by the shaker is adjusted to the natural frequency of the system, then the amplitude of the system will be determined by the condition when input energy from the exciter just balances dissipated energy due to damping.

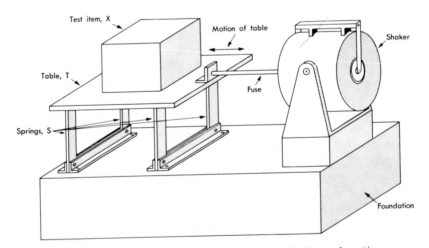

Fig. 16.23 Exciter-driven resonant table for horizontal motion.

---

* The tie linking a resonant test table and shaker has generally been given the term "fuse." It is made quite flexible in all directions, except the direction of driving, thus minimizing any reflection of transverse forces back to the exciter tending to destroy coil alignment. Since the tie serves as a protective device, the term "fuse" seems very appropriate.

c) *The sprung table.* As an example of a resonant system, an experimental setup such as shown in Fig. 16.23 used a 5-lb vector-force electromagnetic exciter which drove a horizontally resonant table. The system was designed to shake a test item weighing 7.7 lb at 50 Hz by proportioning the moving-weight and spring-deflection constants to provide a resonant frequency of 50 Hz. It was realized, however, that flexibility at the end connections to the springs would not be accounted for by simple calculations and that the actual natural frequency would be somewhat less than the calculated. The small amount of damping present would also tend to reduce the resonant frequency. For this reason, the calculations were based on a frequency of 60 rather than 50 Hz. The natural frequency of the final system turned out to be 51 Hz. This was reduced to 50 Hz by the addition of a small amount of weight to the table. The final weight distribution was as follows:

Test item	7.7
Table with mounting accessories	4.75
Moving weight of exciter	0.35
$\frac{1}{3}$ weight of leaf springs	0.50
Total	13.3 lb

As a test of the maximum capacity of the system at 50 Hz, it was found that the 5-lb shaker could actually move the table with test load through an amplitude of $\pm 0.17 \cdot$ inch at 50 Hz. The force required to accomplish this may be calculated as follows:

Maximum acceleration $= S_0\Omega^2 = (0.17)(2\pi \times 50)^2 = 16{,}750$ in/sec^2,

Necessary accelerating force $= ma = (16{,}750/386) \times 13.3 = 577$ lb.

Obviously the 5-lb shaker did not supply the force: the necessary accelerating forces were supplied almost in their entirety by the springs.

It will be readily realized that resonant systems of the types just described are limited to *one frequency.* Although a limited range of application might be designed into such a system through use of adjustable springs and masses, in general the system must be designed for the problem at hand and for that only. A wide-range, mechanically tunable, resonant system would be indeed a welcome addition in the vibration test field.

d) *The free-free beam.* Other forms of mechanically resonant systems include the free-free beam and a tuning fork arrangement [31]. Figure 16.24 shows a free-free beam setup. The beam is supported at the nodal points for the first mode of vibration. It is excited by means of a shaker, as shown. The test item is usually mounted in the center, and the resonant frequency of the beam may be tuned downward either by adjusting the mass at the center by adding to the test weight, or by applying masses at the ends. Therefore, the designer should aim for a frequency that is slightly high

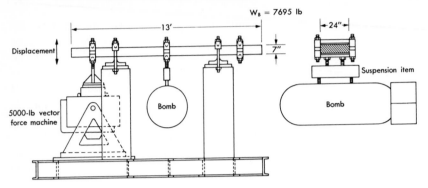

Fig. 16.24 Significant dimensions of a free-free beam system for shaking 1000-lb test items.

rather than low. The nodal points occur at approximately 22.5% of the total length from each end, their exact location depending on the relative values of end loads.

Design relations for uniformly sectioned free-free beams may be found in References [32] and [33].

The beam shown in the figure has a natural frequency of 57 Hz. It is supported on rubber-type vibration mounts located at 23% of the beam length from the ends. Exact nodal locations were determined by sprinkling a fine sand on the top of the vibrating beam, which immediately sought the location of minimum disturbance.

III. SHOCK TESTING

16.20 INTRODUCTION

Mechanical engineers are called upon to design machinery to operate at higher and higher speeds. As speed goes up, accelerations increase, for the most part, not in direct proportion, but as the square of the speed. Both the magnitude of acceleration and acceleration gradients are increased. Resulting body loads often become much greater than applied loading, therefore becoming very significant factors in the design. The complexity of many problems has led to an area of investigation generally referred to as *shock testing* [34].

Actually, shock testing is only one of two phases of a broader classification which might better be called *acceleration testing*. Acceleration testing would include any test wherein acceleration loading is of primary significance. This would include tests involving static or relatively slowly changing accelerations of any magnitude. Shock testing, on the other hand, is usually thought of as involving acceleration transients of moderate to high magnitude. In both cases the basic problem is to determine the ability of the test item to continue functioning properly either during or after application of such loading.

The more passive type of acceleration testing involves static or relatively slowly changing accelerations, which, however, may be of high magnitude. It involves the use of centrifuges, rocket sleds, maneuvering aircraft, etc., for the purpose of testing the capabilities of system components, including the human body, to withstand sustained or slowly changing high-level accelerations. Such tests are usually of quite specialized nature, generally applied to the study of performance in high-speed aircraft and missiles. Therefore, we shall only note this phase in passing and shall devote our primary attention to the first type of acceleration testing, namely shock testing.

Most military apparatus must satisfactorily pass specified shock tests before acceptance. Equipment aboard ship, for example, is subject to shock from the ship's own armament, noncontact mine explosions, and the like. Aircraft equipment must withstand sharp maneuvering and landing loads, and artillery and communication equipment are subject to severe handling in crossing rough terrain. In addition, many items of industrial and civilian application are also subject to shock, often simply caused by normal handling during distribution, such as railroad-car humping, mail chuting, etc.

As a result, shock testing has become accepted as a necessary step in determining the usefulness of many items. It is becoming generally recognized, however, that to be meaningful, considerably more than magnitude of acceleration must be considered. In addition to magnitude, the rate and duration of application, along with the dynamic characteristics of the test item, must all be studied in setting up a useful shock test.

16.21 SHOCK RIGS

Several different methods are employed for producing the necessary motion relations for shock testing. The approach generally used is to store the required energy as some form of potential energy until needed, then to release it at a rate supplying the desired acceleration-time relation. Methods for doing this include the use of compressed air in air guns, loaded springs,

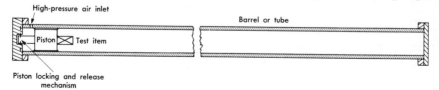

Fig. 16.25 A high-energy air-gun shock-producing system.

TABLE 16.2*

CAPACITIES OF AIR-GUN SHOCK TESTING MACHINES

Bore, in.	Length, ft	Piston weight, lb	Maximum† accelerating force, lb	Maximum net energy, ft-lb	Maximum working pressure, psi
5	58	7	295,000	270,000	15,000
5.6	43	10	74,000	90,000	3000
15	58	65	175,000	250,000	1000
21	93	120	345,000	1,300,000	1000

* Ref. 35.

† Maximum acceleration in g's is equal to the maximum accelerating force divided by the total weight accelerated.

hydraulic methods, and application of gravity. The latter is the most commonly used.

a) *Air-gun shock testing machines.* Figure 16.25 shows schematically the principle of operation of an air-gun shock testing machine. Basically, the test item is accelerated by a piston moved in a tube or barrel by the action of high-pressure air on one face. Energy is stored by pressurizing air in an accumulator. The high pressure is introduced against the left face of the piston, while the piston is restrained by a mechanical latching mechanism. When released, the piston and test item are accelerated sharply to the right. Air trapped in the right-hand portion of the cylinder serves to decelerate the piston and finally bring it to rest. Acceleration-time data may be obtained by means of an accelerometer mounted on the piston. Electrical connections are made through moving brushes which slide against fixed runners. A typical record is shown in Fig. 16.26 [35]. Machines based on this principle of operation have been made with the capacities shown in Table 16.2.

As can be readily perceived, this type of testing device requires rather elaborate design and is relatively costly. In addition, it is limited to the

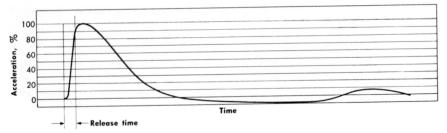

Fig. 16.26 Acceleration-time relation for a system such as shown in Fig. 16.25.

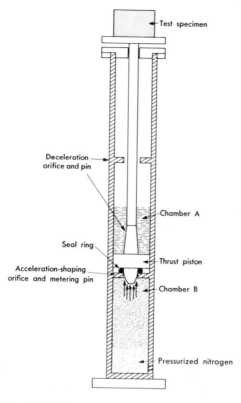

Fig. 16.27 The "Hyge" shock-producing device.   (Courtesy of Consolidated Vacuum Corporation, Rochester, N.Y.)

testing of items that can be fitted into the bore. For these reasons it is used to a considerably lesser extent than the drop-test type to be discussed later.

b) Spring-loaded test rigs. As the name indicates, these machines employ some form of mechanical spring for storing the energy required for acceleration.   One machine designed to provide vertical accelerations uses helical tension springs attached at one end to a test carriage and at the other end to anchors that may be moved to put various initial tensions in the springs. With the springs initially tensioned, the test carriage (with test item) is released by means of a mechanical triggering mechanism and is accelerated suddenly upward. After the carriage has traveled a predetermined distance, the carriage ends of the spring strike stationary hooks. The spring's working stroke is thereby limited; however, the carriage continues upward until stopped by gravity.

c) A hydraulic-pneumatic shock tester. Figure 16.27 illustrates a shock-testing machine employing both hydraulic and pneumatic methods.   A

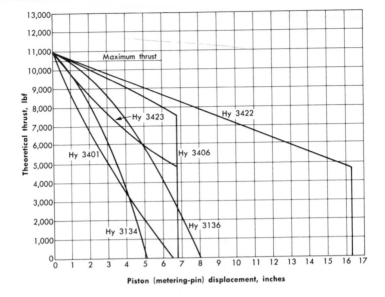

Fig. 16.28 Curves showing the output characteristics of various "Hyge" machines. (Courtesy of Consolidated Vacuum Corporation, Rochester, N.Y.)

main cylinder is divided into two chambers by an orifice plate, positioned to seal the orifice as shown in the figure. The piston rod passes through a stuffing box at the upper end of chamber A and is attached to the test table. A small pressure is applied in chamber A, which serves to provide positive seal at the orifice. The orifice seal ring is so proportioned that only a relatively small piston area communicates with chamber B.

When chamber B is pressurized, a very much higher pressure than that in chamber A will be required to lift the piston from the seal. At the instant it is lifted, however, the higher pressure in chamber B immediately acts over the full piston area, causing a high-level transient thrust to be developed. This forces the piston, piston rod, table, and test item to move upward with high acceleration.

A primary advantage claimed for this type of machine is the comparative ease with which the acceleration-time relation may be shaped. This is accomplished through use of *acceleration metering pins*. A similar method has long been used to control hydraulic buffer action [36, 37]. By properly formed metering pins, acceleration-time relations may be controlled to provide a wide range of waveforms.

The same idea is also used to control deceleration through use of a second orifice and pin, as shown. While the total stroke is fixed, the percentages devoted to acceleration and deceleration may be controlled both by varying metering-pin proportions and initial pressure in chamber A. An alternative deceleration means consists of a rail and carriage system and pneumatically controlled brakes. This system can be employed to provide lower-level decelerations than are normally obtained by hydraulic or pneumatic methods.

An idea of the maximum output of these devices may be had by reference to Fig. 16.28. This shows a plot of maximum gross thrust versus piston position over the maximum strokes for several different commercial models. Gross thrust is that available to accelerate all moving parts and is greater than that available at the test item by the force required to accelerate the piston, table, fixtures, etc.

d) Gravity rigs. There are two commonly used gravity-type shock rigs: the drop type and the hammer type. The hammer type is often referred

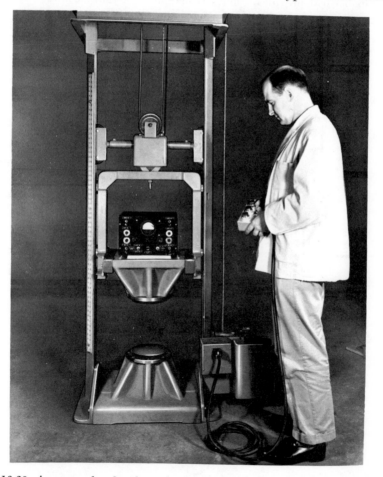

Fig. 16.29 An example of a drop shock testing machine. (Courtesy of Barry Controls, Inc., Watertown, Mass.)

to as high-g machine and normally provides higher values of acceleration than the drop machine does.

Basically the drop machine consists of a platform to which the test item is attached, an elevating system for raising the platform, a releasing device which allows the platform to drop, and an impact pad or arrester against which the platform strikes. Guides are provided for controlling the fall. The machine shown in Fig. 16.29 is an example.

Acceleration-time relations are adjusted by controlling the height of drop and type of arrester pad. Pad selection is of great importance in determining the exact shock characteristics. If the pad is very rigid, for example, an acceleration pulse of very short duration results. On the other hand, a more flexible pad provides a longer time base. Actual *magnitude* of peak acceleration is controlled by adjusting the height of drop.

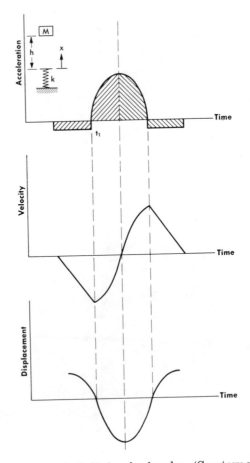

Fig. 16.30 Representation of a half-sine shock pulse. (Courtesy of Barry Controls, Inc., Watertown, Mass.)

Neoprene pads are employed in the machine shown in Fig. 16.29. These are shaped to provide a half-sine wave acceleration-time pulse such as illustrated in Fig. 16.30. With the use of different pad combinations, a range of pulse magnitudes and durations may be obtained, as shown in Fig. 16.31, and other pulse *shapes* may be obtained. Sand pits* have been used with variously shaped impacting surfaces employed on the underside of the test platform to provide the desired characteristics. As another example, the test platform may be equipped with shaped pins or punches which strike lead blocks to provide controlled deceleration.

In single-application rigs, that is, rigs set up for the purpose of testing one item and one item only, the retarding members may be determined by trial and error simply by experimenting with rubber pads, blast mats, etc.,

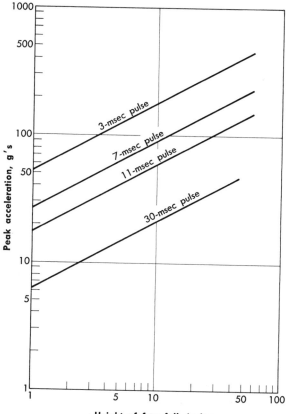

Fig. 16.31 Range of peak acceleration-time pulses obtainable with different pad arrangements and heights of fall. (Courtesy of Barry Controls, Inc., Watertown, Mass.)

---

* Ottawa River sand is used because of its remarkable uniformity.

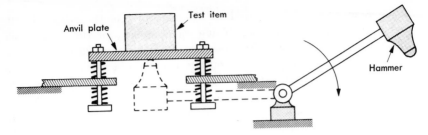

Fig. 16.32 A hammer-and-table machine for vertical impacts.

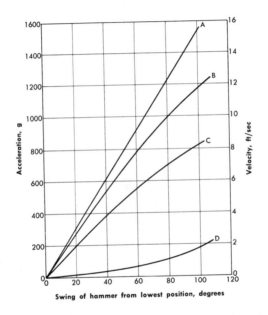

Fig. 16.33. Maximum values of acceleration and velocity of a small hammer machine. *A*, acceleration of anvil plate. *B*, Hammer velocity at impact. *C*, anvil plate velocity just after impact. *D*, hammer velocity just after impact. (Courtesy of the Society for Experimental Stress Analysis.)

until the desired acceleration-time is obtained as shown by an accelerometer record.

Figure 16.32 illustrates the operation of hammer shock testers. The test item is mounted on a platform, or "anvil plate," which is struck by a "tup" swinging as a pendulum. The character of the impact is governed by the hammer height, relative position of the anvil plate, and the weight of the anvil and test item. Figure 16.33 shows the characteristics of a relatively small machine for testing electronic items [38]. This machine has a hammer weight of 89 lb, anvil plate of 80 lb, and is usable for test items weighing up to 50 lb.

TABLE 16.3

RANGES OF SHOCK-TESTING MACHINES

Type of machine	Range of accelerations, g's	Range of total velocity change, ft/sec
Air gun	5 to 25,000	20 to 800
Spring	4 to 40	1 to 30
Drop	80 to 70,000	1 to 60
Hammer	0 to 4000	0 to 15

e) Relative merits and limitations of each shock rig. Each of the shock testing machines discussed in this article possesses certain distinctive characteristics. The air-gun type produces what may be called a *high-energy* shock. Generally speaking, high energy is synonomous with *high velocity*, and to reach a high velocity, considerable displacement of the test item is required. High velocity can only be acquired by relatively large accelerations, or relatively long time intervals, or a combination of the two. In either case, the test item will be displaced a considerable distance.

On the other hand, the drop and hammer machines are of the low-energy category. High acceleration levels are possible, but only for short time intervals. This results in comparatively low test-item velocities, and hence low energies The hydraulic-pneumatic machine would be classified as a medium-energy machine. A rough idea of the ranges of the various machines may be obtained from Table 16.3.

Idealized relative acceleration-time relations for these categories of shock may be visualized as shown in Fig. 16.34. The energy absorbed by the test item will be a function of the area under the curves.

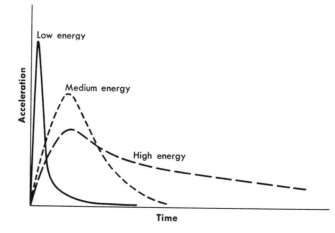

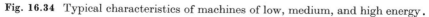

Fig. 16.34 Typical characteristics of machines of low, medium, and high energy.

We may now ask just what parameters are important in shock testing, and immediately conclude that we have a problem comparable to that of determining the applicable theory of failure for a given combined stress situation. Although test conditions may be set up providing correlative results, probably only an acceleration-time relation identical with that of the simulated field condition can be considered completely adequate. In general, both energy and acceleration levels comparable to those of the simulated condition must be supplied.

16.22 ELEMENTARY SHOCK-TESTING THEORY

An impact or shock test may be simplified as shown in Fig. 16.35. This diagram may be thought of as representing a drop machine of the spring-retarding type. Here M_1 represents the table mass; k_1 is the modulus of the retarding spring; M_2 is the test item, or a portion thereof, supported by a linear spring of modulus, k_2. A viscous damping coefficient, c_2, may be assumed acting between the table and the test mass. If we assume c_2 to be comparatively small, we may write

$$M_2 \frac{d^2 S_2}{dt^2} + k_2(S_2 - S_1) = 0. \tag{16.18}$$

The relative displacement between the masses may be expressed as

$$S = S_2 - S_1.$$

Hence,

$$M_2 \frac{d^2 S}{dt^2} + k_2 S = - \frac{M_2 \, d^2 S_1}{dt^2}. \tag{16.19}$$

It has been shown [39] that under conditions similar to those depicted, the table acceleration may be approximated by

$$\frac{d^2 S_1}{dt^2} = -v_1 \omega_1 \sin \omega_1 t, \qquad \text{for} \qquad 0 \leq t \leq \frac{\pi}{\omega_1}, \tag{16.20}$$

and

$$\frac{d^2 S_1}{dt^2} = 0, \qquad \text{for} \qquad t \geq \frac{\pi}{\omega_1},$$

where

$v_1 = $ striking velocity of the table $= \sqrt{2gh}$,

$h = $ height of drop.

This means that the table experiences a half-sine wave acceleration pulse during its contact with spring k_1. Assuming

$$S = \frac{dS}{dt} = 0 \qquad \text{at} \qquad t = 0,$$

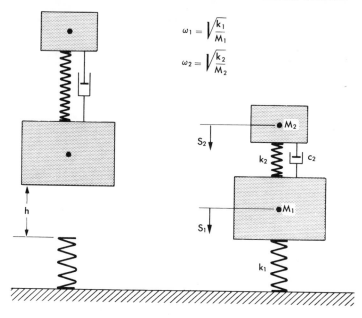

Fig. 16.35 Idealized shock situation.

and substituting Eqs. (16.20), the solution to Eq. (16.19) may be written as

$$S = \frac{v_1\omega_1}{\omega_1^2 - \omega_2^2}\left(\frac{\omega_1}{\omega_2}\sin\omega_2 t - \sin\omega_1 t\right), \qquad \text{for} \qquad 0 \le t \le \frac{\pi}{\omega_1},$$

and

$$S = \frac{2v_1\omega_1^2 \cos(\pi\omega_2/2\omega_1)}{\omega_2(\omega_1^2 - \omega_2^2)}\sin\omega_2\left(t - \frac{\pi}{2\omega_1}\right), \qquad \text{for} \qquad t \ge \frac{\pi}{\omega_1}.$$

Maximum displacements are as follows:

$$S_{\max} = \frac{v_1}{\omega_2[(\omega_2/\omega_1) - 1]}\sin\frac{2n\pi}{(\omega_2/\omega_1) + 1}, \qquad \text{for} \qquad 0 \le t \le \frac{\pi}{\omega_1},$$

and

$$S_{\max} = \frac{2v_1\omega_1^2 \cos(\pi\omega_2/2\omega_1)}{\omega_2(\omega_1^2 - \omega_2^2)}, \qquad \text{for} \qquad t \ge \frac{\pi}{\omega_1}.$$

Here n is a positive integer, chosen to make the sine term as large as possible, while

$$\frac{2n}{(\omega_2/\omega_1) + 1} < 1.$$

If the gradient of acceleration (d^2S_1/dt^2) is very low so that the peak value $(v\omega)$ is reached very slowly, the maximum disturbance experienced by M_2 will be very nearly proportional to (d^2S_1/dt^2). In other words, under

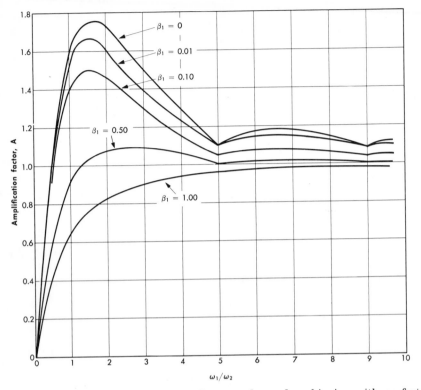

Fig. 16.36 Amplification factors for linear undamped cushioning with perfect rebound.

such conditions secondary vibrations would not be excited in the supported system. Under these conditions an equivalent static displacement, S_{st} may be obtained from Eq. (16.19) by dropping the time-dependent terms, or

$$S_{st} = \frac{v\omega_1}{\omega_2^2}. \tag{16.23}$$

Mindlin, Stubner, and Cooper [40] call the ratio (S_{max}/S_{st}) the *amplification factor*, A. Solving for A, using Eqs. (16.22) and (16.23), we obtain

$$A = \frac{\omega_2/\omega_1}{(\omega_2/\omega_1) - 1} \sin\left[\frac{2n\pi}{(\omega_2/\omega_1) + 1}\right], \quad \text{for} \quad 0 \le t \le \frac{\pi}{\omega_1},$$

and

$$A = \frac{2(\omega_2/\omega_1) \cos(\pi\omega_2/2\omega_1)}{1 - (\omega_2/\omega_1)^2}, \quad \text{for} \quad t \ge \frac{\pi}{\omega_1}. \tag{16.24}$$

It will be observed that the amplification factor is only dependent upon the ratio ω_2/ω_1. By plotting Eqs. (16.24), the upper curve of Fig. 16.36 is obtained.   Damping ratios β reduce the magnitude of the amplification factor as shown.

The foregoing theoretical consideration of the shock problem is quite elementary from the standpoint that a very simple pulse excitation was assumed and a simple single-degree-of-freedom test situation was considered. In the field, and also when most shock test machines are used, the exciting pulse or disturbance consists of a complex acceleration-time relation including many frequency components of different amplitude, and the test item will normally possess many degrees of freedom, hence many modes of vibration. The theoretical results are quite useful, however, in providing information on the limiting maxima that may be expected.

Readers interested in pursuing the theoretical aspects of this problem further are directed particularly to Reference [40] for this chapter.

SUGGESTED READINGS

Brennan, J. N., ed., *Bibliography on Shock and Shock Excited Vibrations*. Eng. Res. Bull. No. 68, University Park, Penna.: The Pennsylvania State University, Vol. 1, 1957, Vol. 2, 1958.

Doeblin, E. O., *Measurement Systems: Application and Design*, New York: McGraw-Hill Book Co., 1966.

Dove, R. C., and P. H. Adams, *Experimental Stress Analysis and Motion Measurement*, Columbus, Ohio: Charles E. Merrill Books, Inc., 1964.

Harris, C. M., and C. E. Crede, *Shock and Vibration Handbook*, 3 vol., New York: McGraw-Hill Book Co., 1961.

Hetenyi, M., *Handbook of Experimental Stress Analysis*. New York: John Wiley and Sons, Inc., 1950.

Keast, D. N., *Measurements in Mechanical Dynamics*, New York: McGraw-Hill Book Co., 1967.

PROBLEMS

16.1 The displacement amplitude of a sinusoidal vibration is measured as 0.40 in. The frequency is 420 Hz. Calculate velocity and acceleration amplitudes.

16.2 The velocity amplitude of a sinusoidal vibration is determined to be 50 in./sec. The frequency is 600 Hz. Calculate the maximum displacement and the maximum acceleration of the vibrating member.

16.3 Sketch the general characteristic curves for each of the following type of seismic instruments, indicating ranges and requirements for satisfactory operation:
   a) The accelerometer
   b) The vibrometer

16.4 A weight of 40 lb is mounted on springs whose total deflection constant is 640 lb/in. Damping equal to 12% of critical exists. If the support has an amplitude of 0.002 in. at 15 Hz, what will be the absolute steady-state displacement amplitude of the weight?

16.5 A vibrometer has a natural frequency of 3.75 Hz and the critical damping factor has been estimated to be 0.092 lb-sec/in. Approximately what should be the normal damping factor and over what frequency range would it be useful?

16.6 A vibrometer has a natural frequency of 5 Hz and a damping ratio of 0.4.
   a) What inherent error in percentage would be expected at a frequency of 7 Hz?
   b) What will be the theoretical phase lag?

16.7 Determine the percentage of inherent amplitude error and phase shift for an accelerometer subjected to a sinusoidal input having a frequency of 1200 Hz. Accelerometer natural frequency is 2000 Hz and it is viscously damped at 55% of critical.

16.8 An accelerometer has a natural frequency of 1000 Hz and the critical damping factor for the device has been calculated as 0.092 lb-sec/in. Approximately what should be the damping factor for the instrument and over what frequency range would it be useful?

16.9 An accelerometer has a natural frequency of 2500 Hz and a damping ratio of 0.65.
   a) What would normally be considered the upper frequency limit for this instrument? Is the damping ratio what would normally be considered satisfactory?
   b) What inherent error would be expected if the instrument is used at a frequency of 1200 Hz?
   c) What is the theoretical phase lag at 1200 Hz?

16.10 A vibration pickup has a natural frequency of 1 Hz and it is essentially undamped. When subjected to a vibration of 4 Hz the instrument reads an amplitude of 0.050 in. What is the true amplitude of vibration and the percent error?

16.11 An accelerometer has the following specifications: $f_n = 2000$ Hz and $c/c_c = 0.65$ (liquid damped). Because of unauthorized tampering, the damping fluid has been lost and for practical purposes the damping ratio is now very nearly zero.
       If the instrument is used at a harmonic frequency of 600 Hz, what multiplying factor (correction) would be required to obtain correct amplitude and what would be the phase lag?

16.12 A system, as shown in Fig. 16.37, has a period of free vibration of 2 sec. It is attached to a machine frame which is vibrating with a frequency of 1 Hz. If the pointer has an amplitude of 0.3 in., relative to the scale of the instrument, what is the amplitude of vibration of the machine? Neglect the weight of the arm. Assume there is no damping.

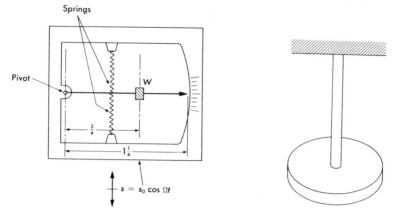

Figure 16.37

Figure 16.38

16.13 A torsional system consists of a disk and shaft as shown in Fig. 16.38. The mass moment of inertia of the disk is 10 in.-lb-sec^2 and the deflection constant of the shaft is 200 in.-lb/rad. There is no damping. If a harmonic torque of 50 in.-lb amplitude at 2 Hz is applied, what will be the displacement amplitude of the disk? What is the phase relation between the displacement and the torque?

16.14 Obtain an inexpensive crystal-type phonograph pickup designed for replaceable phonograph needles. The least expensive will be quite satisfactory. Place a small overhanging mass in the needle chuck, connect the pickup output to an oscilloscope, and use it to investigate various sources of mechanical vibration. [*Note:* This device will not be adequate to determine vibrational amplitudes. Its response will display too many resonances. It is useful, however, as a "frequency" pickup.]

16.15 Mount the pickup described in Problem 16.14 on a small vibration exciter and investigate its response characteristics,

a) by comparison with a "quality" motion transducer, and

b) by the methods described in Article 16.11.

16.16 Calibration data for a velocity-type vibrometer are given in Article 16.11 (sensitivity = 0.938 V/in./sec). Assume that the identical vibrometer is mounted on a vibrating beam and that its output is connected to the vertical input of a CRO.

When a sine-wave oscillator set at 82 Hz supplies the horizontal CRO input, a Lissajous diagram corresponding to the one shown in the lower-right illustration of Fig. 9.13 is the result. If the calibrated vertical height of the diagram indicates 5.42 V input, what are the amplitudes for velocity (inches per second), displacement (inches), and acceleration (g's)?

16.17 Figure 16.39(a) illustrates an arrangement for a laboratory experiment in dynamic-stress analysis. Pendulum OA was caused to swing freely and strike a steel cantilever beam BC on the end of which is mounted a small

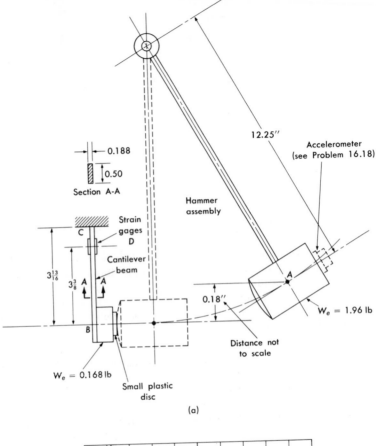

0.188

0.50

Section A-A

12.25″

Accelerometer
(see Problem 16.18)

Hammer
assembly

Strain
gages
D

Cantilever
beam

C

$3\frac{13}{16}$

$3\frac{3}{8}$

A   A

B

$W_e = 0.168$ lb

Small plastic
disc

0.18″

A

Distance not
to scale

$W_e = 1.96$ lb

(a)

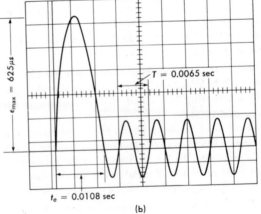

$\epsilon_{max} = 625\,\mu s$

$T = 0.0065$ sec

$t_e = 0.0108$ sec

(b)

Figure 16.39

mass. A small plastic disk was placed at the point of impact to promote inelastic impact. Strain gages on each side of the beam are placed at D. Gage output was appropriately amplified and fed to a CRO for readout. Figure 16.39(b) shows the CRO trace obtained when the pendulum swings from rest through an arc equivalent to $h = 0.18$ in. Calibration-based readout values are shown on the figure.

Applying theoretical relationships, compare analytical and experimental values for (a) maximum strain, (b) period of free vibration, and (c) time of contact between pendulum tup and beam. Pertinent dimensions and weights are shown on the sketch. Equivalent weight w_e is taken as the sum of small weight at B plus one-third the weight of the beam. Equivalent weight of the pendulum W_e is taken as the weight of the tup plus one-third the weight of the arm.

Note that the initial potential energy of the pendulum will be completely converted into kinetic energy at impact (assuming negligible losses). On the basis of the inelastic conditions between weights w_e and W_e, the law of conservation of momentum will apply, from which the impacting energy loss may be evaluated. The remaining energy must be absorbed by the beam.

Time of contact may be estimated through the assumption that the strain-time relationship is a half-sine wave, corresponding to the free vibration of the beam with both w_e and W_e rigidly attached.

16.18 Referring to Problem 16.17, suppose an accelerometer is placed on the rear of the tup of the pendulum, as indicated by the dotted outline, Fig. 16.39(a). Sketch the shape of the readout plot that you would expect covering the time interval from the instant of initial release until the pendulum returns to zero velocity after rebound. From the data given in Problem 16.17, calculate the expected peak acceleration.

CHAPTER 17

ACOUSTICAL MEASUREMENTS

17.1 INTRODUCTION

Sound may be described on the basis of two considerably different points of view: (a) from the standpoint of the physical phenomenon itself, or (b) in terms of the "psychoacoustical" effect sensed through the human process of hearing. It is very important that these basically different aspects be kept continually in mind. To measure the particular physical parameters associated with a specific sound, either of simple or complex waveform, is a much simpler assignment than to attempt to evaluate the effects of the parameters as sensed by human hearing.

Occasionally, the reasons for measuring a sound may not be associated with hearing. For example, sound pressure variations accompanying high-thrust rocket-motor or jet-engine operation may be of sufficient magnitude to endanger the structural integrity of the missile or aircraft [1]. Structural fatigue failures have been induced by sound excitation. In such cases measurement of the parameters that are involved does not directly include the psychoacoustical relationship. But in the great majority of cases the effect of the measured sound *is* directly related to human hearing and this added complication is therefore unavoidable.

Noise may be defined as unwanted sound. Noise affects human activities in many ways. Excessive noise may make communication by direct speech difficult or impossible. Noise may be a factor in marketing appliances or other equipment. Prolonged ambient noise levels may eventually cause permanent damage to hearing or, of course, it may simply impair efficiency of workers because of the annoyance factor. All of these aspects of sound are unavoidably coupled with human hearing.

Physically, airborne sound is, within a certain range of frequencies, a periodic variation in air pressure about the atmospheric mean. The air particles oscillate back and forth along the direction of propagation, and for this reason the wave-form is said to be longitudinal. For a single tone or frequency (as opposed to a sound of complex form), the oscillation is simple harmonic and may be expressed as [2]:

$$S = S_0 \cos \frac{2\pi}{\lambda} (x \pm ct), \tag{17.1}$$

where

S = displacement of a particle of the transmitting medium,

S_0 = displacement amplitude,

λ = wavelength = c/f,

x = distance from some origin (e.g., the source), in the direction of propagation,

c = velocity of propagation,

t = time,

f = frequency.

For a gaseous medium [2]:

$$p = -B \frac{\partial s}{\partial x} = -B \frac{2\pi}{\lambda} S_0 \sin \frac{2\pi}{\lambda} (ct - x), \tag{17.2}$$

where

p = pressure variation about an ambient pressure, P,

B = the adiabatic bulk modulus.

17.2 BASIC ACOUSTICAL PARAMETERS

a) *Sound pressure.* In the presence of a sound wave, the instantaneous difference in pressure at a point and the static pressure is termed *sound pressure*. The common unit is the *microbar* (μ bar), which is a pressure of one dyne per square centimeter. Standard atmospheric pressure (14.696 psia) corresponds to 1,013,250 μ bars. It is seen then that a pressure of one bar very closely approximates one standard atmosphere.

b) *Sound pressure level.* The ratio between the greatest sound pressure that a person with normal hearing may tolerate without pain and that of the softest discernible sound is roughly 10 million to 1 (Fig. 17.1). This tremendous range cannot be covered practically with a linear unit only, and suggests the use of some form of logarithmic scale. It will be recalled that the decibel is based on the logarithm of a ratio (Article 7.21) and it is this unit that is most commonly employed in sound or noise measurements. Basically the decibel is a measure of power ratio,

$$\text{db} = 10 \log_{10} (\text{power}_1/\text{power}_0). \tag{17.3}$$

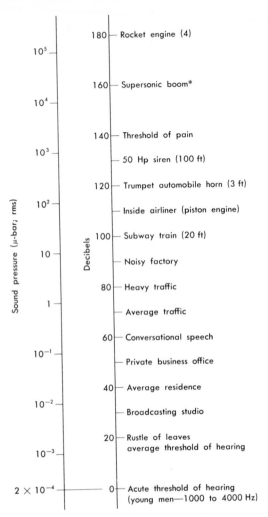

Fig. 17.1 Typical sound pressures and sound pressure levels.

Because sound power is proportional to the *square* of sound pressure [3], Eq. (17.3) may be written as

$$\mathrm{db} = 20 \log_{10}(p_1/p_0). \tag{17.4}$$

It is seen that the decibel is not an absolute quantity but a comparative one which, however, can be employed *in the manner* of an absolute quantity if referred to some generally accepted base. This is done and the *rms* value, $p_0 = 0.0002 \, \mu$ bar, is widely accepted as the standard reference for sound pressure. This value takes on additional significance when we note that it corresponds closely to the acute threshold of hearing.

Therefore,

$$SPL = 20 \log_{10}(p/0.0002) \; re \; 0.0002, \tag{17.5}$$

where

SPL = sound pressure level, db,

p = rms pressure of sound in question, μ bar.

"re 0.0002" is sometimes appended to make clear that the reference pressure is 0.0002 μ bar; however, throughout the remainder of this chapter it will be omitted.

Example. What is the sound pressure level corresponding to a sound pressure of 10 μbar rms?

$$SPL = 20 \log_{10} (10/0.0002) = 94 \; db$$

Special attention should be directed at this point to the use of the word "level." Various terms to be discussed later employ the word, e.g., "sound level," "loudness level," "noise level," etc. Use of the word "level" implies a logarithmic ratio expressed in decibels. Remembering this fact helps in keeping units straight.

c) Sound power level. As suggested by Eq. (17.3), sound involves energy, which can be expressed in terms of power. The common unit is the watt (W), and when the quantity, *sound power level* (PWL) is employed, the usual

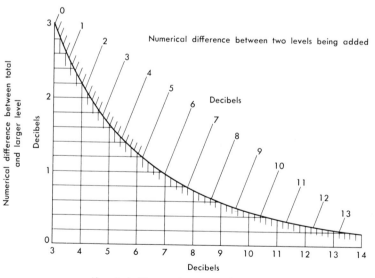

Fig. 17.2 Diagram for adding or subtracting decibels. (Courtesy of General Radio Co., W. Concord, Mass.)

reference power is 10^{-12} W. Power level is therefore defined as

$$PWL = 10 \log (W/10^{-12}) \text{ db } re \text{ } 10^{-12} \text{ W.} \tag{17.6}$$

There is no instrument for measuring power level directly. However, the quantity can be calculated from sound pressure measurements [6].

d) Combination of sounds. It should be noted at this point that when two sound sources are combined the resulting sound-pressure level is *not* the algebraic sum of the two sound-pressure levels. For example, the SPL from a twin-engined airplane will not be twice that of one of the engines alone.

The proper procedure for determining the combined result is to make the addition on the basis of power. Figure 17.2 provides a convenient method for either adding or subtracting sound-pressure levels.

Example. What is the combined level of two sound sources having SPL's of 66 db and 70 db?

The difference is 4 db. The 4-db line intersects the curved line at 1.45 db on the vertical scale. Thus the total value is 70 + 1.45 or 71.45 db. (Note that the same result would be obtained if the lower scale were used.)

Example. The combined pressure level from two sources is 85 db. One of the sources alone has a sound-pressure level of 80 db. What would be the resulting SPL if the 80-db source were removed?

The difference between the 80-db source and the total is 5 db. The 5-db vertical line intersects the curve at 1.6 db on the vertical scale. Thus the unknown level is 83.4 db.

17.3 PSYCHOACOUSTIC RELATIONSHIPS

As mentioned earlier, in only a relatively few situations is human hearing disassociated from sound measurements. Measuring systems and techniques are therefore unavoidably greatly influenced by the physiological and psychological makeup of the human ear as a transducer and the brain as the final evaluator. The ear is a nonlinear device, in terms of both input magnitudes and frequencies. Figure 17.3 shows average thresholds of hearing and tolerance for young persons. It will be noted that the greatest sensitivity occurs at about 4000 Hz and that a considerably greater sound-pressure level (SPL) is required at lower frequencies. Figure 17.4 shows the free-field *equal-loudness* contours for pure tones as determined by Robinson and Dadson at the National Physical Laboratory, Teddington, England.

Loudness is the measure of relative sound magnitudes or strengths *as judged by a listener.* It is measured in terms of *loudness level* and the unit is called the *phon.* One phon is numerically equal to the sound-pressure level in decibels at a frequency of 1000 Hz. (Inspection of Fig. 17.4 reveals that at 1000 Hz the values of loudness and SPL correspond.)

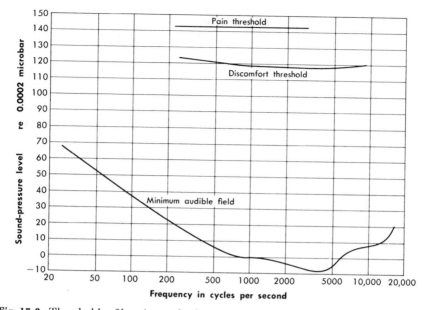

Fig. 17.3 Thresholds of hearing and tolerance for young people with good hearing. (Courtesy of General Radio Co., W. Concord, Mass.)

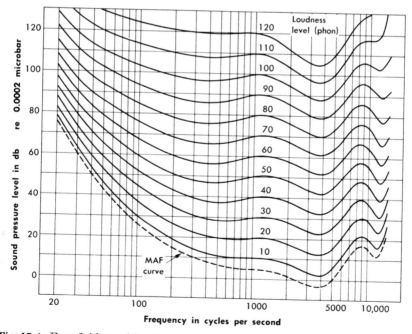

Fig. 17.4 Free-field equal-loudness contours for pure tones. (Courtesy of General Radio Co., W. Concord, Mass.)

To illustrate, if the magnitude of a sound-pressure level at 1000 Hz is 30 db (loudness level equal to 30 phon), for an equal loudness level at 100 Hz the average person will require a SPL of 44 db.

It is clear that loudness level based on the phon is a logarithmic quantity. Although this is quite useful, still another measure of strength is employed. *Loudness* (note the absence of the word "level") is also measured with a *linear unit*, the *sone*. One sone is the loudness of a 1000-Hz tone with a SPL of 40 db *re* 0.0002 bar (note that this also corresponds to 40 phons). A tone that sounds twice as loud has a loudness of 2 sone, etc.

17.4 SOUND-MEASURING APPARATUS AND TECHNIQUES

Measurement of the parameters associated with sound employ a basic system made up of a detector-transducer (the microphone), intermediate modifying devices (amplifiers and filtering systems), and readout means (a meter, CRO, or recording apparatus). Most sound-measuring systems are used to obtain psychoacoustically related information. It is therefore necessary to build into the apparatus nonlinearities approximating those of the average human ear. Elaborate filtering networks also provide the basis for analyzers, devices for separating and identifying the various frequency components or ranges of components forming a complex sound.

a) Microphones. Most microphones incorporate a thin diaphragm as the primary transducer, which is moved by the air acting against it. The mechanical movement of the diaphragm is converted to an electrical output by means of some form of secondary transducer which provides an analogous electrical signal.

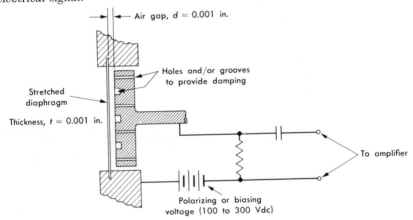

Fig. 17.5 Schematic of the condenser-type microphone.

Common microphones may be classified on the basis of the secondary transducer, as follows:

1. Capacitor or condenser
2. Crystal
3. Electrodynamic (moving coil or ribbon)
4. Carbon

The *capacitor* or *condenser microphone* is probably the most respected microphone for sound-measurement purposes. It is arranged with a diaphragm forming one plate of an air-dielectric capacitor (Fig. 17.5). Movement of the diaphragm caused by impingement of sound pressure results in an output voltage [3],

$$E \approx Qd, \tag{17.7}$$

where

E = voltage,

Q = charge provided by the polarizing voltage (relatively constant),

d = separation of the plates.

The capacitive microphone is widely employed as the primary transducer for sound measurement purposes.

The *crystal microphone* employs a piezoelectric-type element (Article 6.17), usually activated by bending. For greatest sensitivity, a cantilevered element is mechanically linked to the diaphragm. Other constructions employ direct contact between diaphragm and element, either by cementing (element placed in bending) or by direct bearing (element in compression). Crystal microphones are extensively used for serious sound measurement.

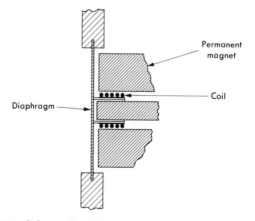

Fig. 17.6 Schematic of the electrodynamic microphone.

The *electrodynamic microphone* employs the principle of the moving conductor in a magnetic field. The field is commonly provided by a permanent magnet thereby placing the transducer in the variable-reluctance category (Article 6.15). As the diaphragm is moved, voltage is induced proportional to the *velocity* of the coil relative the the magnetic field, thereby providing an analogous electrical output. Two different constructions are employed, the "moving coil" (Fig. 17.6) and the "ribbon" type. The inductive member of the latter type consists of a single element in the form of a ribbon which serves the dual purpose of "coil" and diaphragm.

The carbon microphone. The secondary transducer of the carbon microphone consists of a capsule of carbon granules, the resistance of which varies with change in sound pressure sensed by a diaphragm. Its limited high- and low-frequency response precludes its use for serious sound measurements and it is mentioned here merely to round out the list. The limited-frequency characteristics of the carbon microphone, coupled with its ruggedness, make it ideal for use as the transmitter in the ordinary telephone handset.

b) Microphone selection factors. An ideal microphone employed for measurement purposes would possess the following characteristics:

1) flat frequency response over the audible range,

2) nondirectivity,

3) predictable, repeatable sensitivity over the complete dynamic range,

4) at the lowest sound level to be measured, output signal that is several times the system's internal noise level,

5) minimum dimensions and weight,

6) output that is unaffected by all environmental conditions except sound pressure.

The capacitor-type microphone undoubtedly enjoys the top position for sound measurement use, while the crystal type runs a close second. As with all measurement, the presence of the sensor (microphone) unavoidably alters (loads) the signal to be measured (see Article 6.2). In this application the microphone should be as physically small as possible. It is obvious, however, that size, particularly diaphragm diameter, must have an important influence on both sensitivity and response. Microphones are therefore available in a range of sizes, and final selection must be based on a balance of the requirements for the specific application. Table 17.1 summarizes microphone characteristics.

c) The sound-level meter. The basic sound-level meter is a measuring system which senses the input sound pressure and provides a meter readout yielding a measure of the sound magnitude. The sound may be wideband,

TABLE 17.1

SUMMARY OF MICROPHONE CHARACTERISTICS

Type of Microphone	Principle of operation	Sensitivity (db re 1 V (dyne/cm²)	Relative impedance	Linearity	Advantages	Disadvantages
Capacitor	Capacitive	−50 to −100	Very High	Excellent	Stable: holds calibration. Low sensitivity to vibration. Wide range.	Sensitive to temperature and pressure variations. Relatively fragile. Requires high polarizing voltage. Requires impedance coupling device near microphone.
Crystal	Piezoelectric	−55	High	Good to excellent	Self-generating. May be hermetically sealed. Relatively rugged. Relatively inexpensive.	Requires impedance-matching device. Relatively sensitive to vibration.
Electrodynamic	Reluctive	−80	Low	Good	Self-generating. Very rugged. Inexpensive.	Physically large.
Carbon	Resistive		Moderate	Poor		Severely limited frequency range.

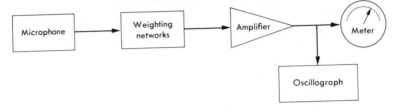

Fig. 17.7 Block diagram of a typical sound-level meter, or sound-level recorder.

Fig. 17.8 Sound level meter,
Type 2203. (Courtesy of
B & K Instruments, Inc.,
Cleveland, Ohio.)

it may be of random frequencies, or it may contain definite discrete tones. Each of these factors will, of course, affect the readout.

Usually the system includes weighting networks (filters) to roughly match the instrument response to that of human hearing. The readout, therefore, is of a psychoacoustic nature, a number ranking the sound magnitude in terms of the ability of the human measuring system to do so.

A block diagram of the basic sound-level meter is shown in Fig. 17.7, and Fig. 17.8 shows a commercially available system. When used in conjunction with an oscillograph the system is commonly referred to as a *sound-level recorder.*

Figure 17.9 displays the internationally standardized weighting characteristics selectable by a panel switch. It is seen that the filter responses selectively discriminate against low and high frequencies, much as the human ear does. It is customary to use characteristic *A* for sound levels

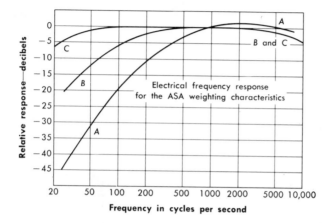

Fig. 17.9 Frequency-response characteristics in the American Standard for Sound-Level Meters, S1.4, 1961.

below 55 db, B between 55 and 85 db, and C for levels above 85 db. Certain broad generalizations as to the frequency makeup of a sound may be made by taking separate readings with each network. A resulting reading is said to be the *sound level*. The sound-*pressure* level is obtained only for an instrument with a flat response.

d) Frequency spectrum analysis. Although determination of a value of sound pressure or sound level provides a measure of sound intensity, it yields no indication of frequency distribution. For noise abatement purposes, for example, it is very desirable to know the predominant frequencies involved. This can often point directly to the prominent noise sources.

Determination of intensities versus frequency is referred to as *spectrum analysis* and is accomplished through the use of pass-band filters (Article 7.25c). Various combinations of filters may be employed, determined by their relative band-pass widths. Probably the most commonly used are the "full-octave" filters having center frequencies as follows: 31.5, 63, 125, 250, 500, 1000, 2000, 4000, 8000, 16,000 and 31,500 Hz. Figure 17.10 shows such an instrument, and Fig. 17.11 depicts the frequency characteristics of the filters.

In addition to full-octave spectrum division, $\frac{1}{2}$, $\frac{1}{3}$, $\frac{1}{10}$ and other fractional octave divisions are also employed. Although the ear may be capable of distinguishing pure tones in the presence of other tones, it does tend to integrate complex sounds over roughly $\frac{1}{3}$ octave intervals [7], thus lending some additional importance to $\frac{1}{3}$ octave analyzers.

Simple analyzers are used by taking a separate reading for each pass-band. It is seen therefore that an appreciable period of time is required to scan the range. Not only does the bulk of data increase with reduced band

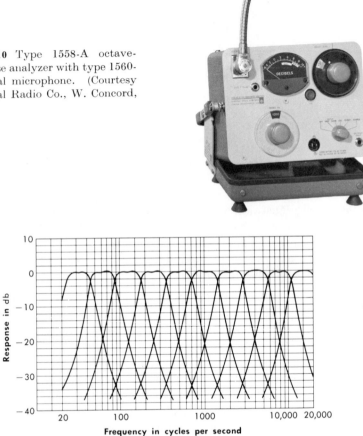

Fig. 17.10 Type 1558-A octave-band noise analyzer with type 1560-P6 crystal microphone. (Courtesy of General Radio Co., W. Concord, Mass.)

Fig. 17.11 Response characteristics, Type 1558 octave-band noise analyzer. (Courtesy of General Radio Co., W. Concord, Mass.)

width, but the constancy of the sound source becomes of greater importance also as the necessary time for measurement increases. A partial solution to the latter problem is to employ a tape recorder for sampling and then analyze the recorded sound at leisure.

To this point most of the discussion has assumed sound sources that are of constant or relatively constant amplitude and character. Examples would be sound from a blower, an electric motor, or an internal combustion engine running at constant speed and load. Noise from open factory spaces, offices with business machines, or street traffic may also be of relatively

Fig. 17.12 Hewlett-Packard Model 8051A. Loudness Analyzer. This instrument is capable of measuring loudness of narrow-band or wide-band, continuous, or impulsive sounds. (Courtesy of Hewlett-Packard Co., Palo Alto, Calif.)

constant magnitudes. On the other hand, there may be noise sources of discrete time-wise characteristics. Sources of this type would include the impulse of a forging hammer, almost any form of explosion, or the single stroke of a typewriter key.

One type of impact-measuring system* employs a temporary storage of certain sound parameters for later readout. Three different sound characteristics may be simultaneously sampled: maximum instantaneous sound level, average sound level, and a continuous indication of sound level. The electrical analogs of the first two quantities are temporarily stored in capacitors for later reading on the meter. The duration of the impact may be estimated through a comparison of the average and the instantaneous maximum levels.

A more sophisticated instrument for automatic analysis of either constant-level or impact-type sounds is shown in Fig. 17.12. In this case the frequency spectrum is scanned automatically and the readout is displayed on a storage-type CRO tube [8].

17.5 CALIBRATION METHODS

As with most electromechanical measuring systems, sound measurement involves a detector-transducer (the microphone) followed by an intermediate electrical/electronic signal-conditioning stage and some sort of readout stage.

---

* General Radio Type 1556-B

Calibration of the complete system involves introducing a known sound-pressure variation and comparing it with the readout. In practice such a calibration may be made or more commonly, the various component stages are "calibrated" separately. For example, sound-level meters commonly provide a simple check of the electrical and readout stages by substituting a carefully measured voltage for the microphone output and adjusting the readout to a predetermined value for the particular instrument. This procedure ignores the microphone altogether; however, it does provide a convenient method for checking the remainder of the system.

Comparative methods may also be employed wherein a microphone or a complete system is directly compared with a "standard." Of course, as with all comparative methods, it is imperative that great care be exercised to assure that a true comparison is made. It is necessary that both the standard and the test microphone "hear" the identical sound.

Standard sound sources involve mechanical loudspeaker-type drivers for producing the necessary pressure fluctuations. One system* employs two battery-driven pistons moving in opposition in a cylinder at 250 Hz. A precalibrated pressure level of 124 ± 0.2 db, $re\ 2 \times 10^{-4}\ \mu$-bar is produced. A somewhat similar system† employs a small, rugged loudspeaker as the driver. In both cases proper calibration is maintained only if the proper coupling cavity is employed between driver and microphone.

A more basic calibration method is known as *reciprocity calibration*. In addition to the microphone under test, a *reversible*, linear transducer and a sound source with a proper coupling cavity are required. Calibration procedure is divided into two steps. First, both the test microphone and the reversible transducer are subjected to the common sound source, and the two outputs, hence the ratio between them, is determined. Second, the reversible transducer is employed as a sound source with known input (current), and the output (voltage) of the test microphone is measured. It can be shown [9, 10] that this step provides a relationship which is a function of the product of the two sensitivities, that of the test microphone and that of the reversible transducer. Results of the two steps yield sufficient information to determine the absolute sensitivity of the tested microphone.

17.6 FINAL REMARKS

Sound is a complex physical quantity. Its evaluation through measurement becomes doubly difficult in comparison with other engineering quantities because of the necessary involvement of the process of human hearing.

---

* Pistonphone Type 4220, B&K Instruments, Inc., Cleveland, Ohio.

† Type 1552-B Sound-Level Calibrator, General Radio Co., West Concord, Mass.

It must, therefore, be recognized that the foregoing sections merely serve as a relatively superficial introduction to the subject. The student is directed to the appended general references for further study.

Little distinction has been made in this chapter between sound and noise. The latter has been considered merely as "undesirable" sound. Noise, however, is becoming an increasingly important problem to the mechanical engineer due to controls that are set up by ordinance and statute [11] as well as to company-union agreements and court decisions [12] intended to protect the employee. As a result the mechanical engineer will be called upon to design "quiet" machinery and processes. He will be increasingly criticized for his part in "silence pollution." It is quite necessary, therefore, that he not only be knowledgeable in the area of measurement of noise (sound) but also in the theory and art of noise abatement.

SUGGESTED READING

Beranek, L. L., *Noise Reduction*. New York: McGraw-Hill Book Co., 1960.

Broch, J. T., *Acoustic Noise Measurements*. Cleveland, O.: B & K Instruments, Inc., 1967.

Harris, C. M., *Handbook of Noise Control*. New York: McGraw-Hill Book Co., 1957.

Journal of the Acoustical Society of America, New York.

Kinsler, L. E., and A. R. Frey, *Fundamentals of Acoustics*. New York: John Wiley & Sons, Inc., 1962.

Peterson, A. P. G., and E. E. Gross, Jr., *Handbook of Noise Measurement*. 5th. Ed., W. Concord, Mass.: General Radio Co., 1963.

Randall, R. H., *An Introduction to Acoustics*. Reading, Mass.: Addison-Wesley Publishing Co., 1951.

Sound: (a journal formerly titled, *Noise Control*, published by the American Institute of Physics).

United States of America Standards Institute, New York. (Formerly American Standards Association.) Various standards including those pertaining to sound measurement.

CHAPTER 18

APPLICATION OF RADIOACTIVE ISOTOPES TO MECHANICAL MEASUREMENTS

18.1 INTRODUCTION

Because of their radioactivity, radioisotopes, or unstable isotopes, provide an increasingly useful tool in mechanical measurements. Although unstable isotopes have always occurred in nature to some extent, development of the nuclear reactor was necessary to supply them in generally usable quantities. They are used in two major areas of mechanical engineering, as tracers and for density measurement. However, many other applications are being found, and their use will undoubtedly be expanded very rapidly.

18.2 WHAT ARE RADIOISOTOPES?

The Greek word from which "atom" is derived means "indivisible." But although the atom *does* remain unchanged in ordinary chemical reactions, we now know that it can be broken down in a nuclear reaction.

An atom consists of a nucleus (obviously the source of the word "nuclear"), surrounded by orbiting electrons. The nucleus possesses positive electrical charges which are balanced by the negative charges of the electrons. The number of positive electrical charges determines the particular element, each element having a different number. In addition to the positive charges, the nucleus contains *neutrons* having no charge but contributing to the total mass of the atom. While a given element is identified by the number of positive charges in the nucleus, and a change in this number will produce a different element, the number of neutrons *may* be varied without causing major changes in the chemical properties. Such a variation yields what are termed *isotopes** of an element.

---

* The word "isotope" is derived from Greek words meaning "the same place," referring to the fact that element's position in the periodic table remains unchanged in spite of the variation in the number of neutrons in the nucleus.

Isotopes of an element have identical chemical properties (which means, incidentally, that they cannot be separated chemically), but they may possess quite different physical or nuclear properties. For one thing, since the neutrons contribute to the mass of the atom, various isotopes of an element will have somewhat different mass numbers. Isotopes may be either stable or unstable. Although both occur naturally, stable isotopes are much more common. Seventy-three of the natural elements occur in nature as mixtures of two or more isotopes. For example, naturally occurring carbon consists of 98.9% C^{12} and 1.1% C^{13}. The difference between the two is that C^{13} contains an additional neutron. Both are stable. Naturally occurring uranium, on the other hand, possesses three isotopes, all of which are radioactive.

18.3 THE NATURE OF RADIOACTIVITY

There are two general forms of nuclear radiation: (a) particles, (alpha, beta, and neutrons), and (b) extremely short electromagnetic waves (gamma and x-rays). There are certain other radiations such as protons, deuterons, etc., but because these do not apply to the purpose of this chapter they will not be mentioned further.

An *alpha particle* is a fast-moving helium nucleus. It is emitted by unstable isotopes, primarily of the heavier elements such as uranium, thorium, and radium. Particles emitted from different sources will have different ranges, speeds, and energies, but all particles from a given homogeneous source will possess identical values. Alpha particles are of low energy and range compared with beta and gamma sources; their maximum range is about 8.6 cm in air, and a sheet of ordinary paper will stop them. Their velocity range is between about 4.6 to 7% of the velocity of light. Alpha particles exist only until they pick up two electrons, at which time they become normal helium atoms. An important property of alpha particles is that they cause intense ionization of the air along their paths; it is this factor which makes them valuable in measurements.

Beta emissions are of fast-moving negatively-charged particles or electrons. They are not necessarily of the same velocity, range, or energy, even though emitted from a single source. Their velocities may be only slightly below that of light, and their penetration limit is about 0.08 in. of brass or 0.04 in. of lead. This means that beta radiations may be shielded or stopped by a relatively thin layer of metal. Ionization caused by their passage is very much less than that of the alpha particles.

Gamma rays are electromagnetic radiations from the nucleus of an unstable isotope and are very similar to x-rays. They have the velocity of light, and are *not* in the form of discrete particles. Their range is much greater than that of alpha or beta particles, and the radiation is highly

TABLE 18.1

IMPORTANT CHARACTERISTICS OF ALPHA, BETA, AND
GAMMA RADIATION

Type of radiation	Form	Ionizing power	Penetrating power	Deflected by electrostatic and magnetic fields	Charge
Alpha	Helium nuclei	Very strong	Quite weak	Yes	Positive
Beta	High-speed electrons	Less strong	Stronger	Yes	Negative
Gamma	High-frequency electromagnetic radiation (similar to x-rays)	Weak	Very strong	No	—

dangerous. Unlike radiated particles, they are not completely stopped by any material, but diminish exponentially with shielding thickness. This makes adequate shielding difficult in many cases.

The important characteristics of the above three types of radiation are given in Table 18.1.

Radiation from a radioactive source reduces or decays with time. This is true only on an aggregate basis, for there is no way of knowing precisely when particular nuclei will disintegrate. It may be said, however, that a certain proportion or fraction will decay during a given time interval, say one second. This quantity, the proportionate reduction per second, is

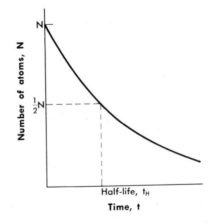

Fig. 18.1 Decay of a radioactive substance with time.

TABLE 18.2

PROPERTIES OF SELECTED ISOTOPES HAVING USEFUL
APPLICATION TO MECHANICAL MEASUREMENTS

Isotope	Half-life*	Useful radiation mode and energies, Mev†			Application
		α	β	γ	
Calcium-45	152 d		0.25		Tracer of soluble salts
Carbon-14	5500 y		0.155		Thickness gages and chemical tracer
Cerium-144	290 d		0.3		Thickness gages
Cesium-134	2.3 y		0.658	1.35	Liquid level
Cesium-137	33 y		1.2	0.662	Thickness gages
Cobalt-60	5.2 y		0.31	1.33	Radiography
Helium-6	0.85 s		3.7		
Iridium-192	70 d			0.651	Radiography
Iron-59	46.3 d		0.46	1.3	Tracer in iron and steel
Phosphorus-32	14.3 d		1.712		Tracer
Polonium-210	140 d	5.3			Air ionizer
Radium-226	1620 y	4.79		0.19	Thickness and level gaging
Strontium-89	53 d		1.50		Thickness gages and tracer
Strontium-90	25 y		0.537		Thickness gages
Thallium-204	2.7 y		0.783		Thickness gages
Yttrium-90	61 h		2.18		Thickness gages

* h = hours, d = days, y = years.
† When complex modes exist, maximum energies only are listed.

known as the *decay constant*, λ. If the number of radioactive nuclei present at a given time is N, then it can be shown that the following relation holds [1]:

$$N = N_0 e^{-\lambda t}, \qquad (18.1)$$

where

t = time, seconds,

λ = decay constant, 1/sec (see time constant, Article 2.8a),

N_0 = number of nuclei present at $t = 0$,

e = base of Naperian logarithms.

If this relation is plotted, it appears as shown in Fig. 18.1.

Half-life is the common measure for the time-activity relation for a given isotope. It may be defined as the time required for an initial number of radioactive nuclei to be reduced to one-half. By substituting $N = \frac{1}{2}N_0$ in Eq. (18.1),

$$t_h = \frac{0.693}{\lambda}, \qquad (18.1a)$$

where t_h is the half-life.

Half-lives of different isotopes vary over a wide range. Table 18.2 lists values for a few representative materials. It will be noted that half-life extremes of 0.85 sec for helium-6 to 5500 years for carbon-14 are given. These are not limit values, however, for half-lives much shorter than 0.85 sec and much greater than 5500 years occur.

A quantitative measure of a radioisotope is the *curie*. Early measurements indicated that one gram of radium-226 (half-life, 1620 years) disintegrated at the rate of about 3.7×10^{10} disintegrations per second. Although the more exact value is 3.61×10^{10} disintegrations per second, the former has been retained as a measure of isotope activity. One curie of a radioactive substance is the *quantity* in grams which decays at the rate of 3.7×10^{10} atoms per second. This is a factor of considerable interest to the user, for in one sense, it is a measure of "quality" or concentration of product. Another way of expressing relative amounts of radioactive source is with the term *specific activity*, which is the number of curies per gram.

Relative energies of particles or rays are measured in terms of the electron volt. Because of the magnitudes encountered, millions of electron volts, or Mev, is the unit commonly employed. One Mev may be defined as the energy acquired by an electron falling through a potential of one million volts (1.603×10^{-6} erg, or about 4.26×10^{-10} ft·lb).

18.4 SOURCES OF RADIOISOTOPES

As mentioned earlier, radioisotopes of certain types are available in limited quantities from natural sources. Examples of these are radium-226 and uranium-235. The natural occurrence of these isotopes is very limited, practically prohibiting their use for most of the industrial applications to be discussed in later articles. However, development of the cyclotron, or other particle accelerators, and of the nuclear reactor, have made many isotopes, not occurring naturally, available in sufficient quantities to make their use reasonably economical.

Radioisotopes may be produced or manufactured in three different ways: (a) by neutron radiation in a nuclear reactor, (b) in particle accelerators such as the cyclotron, and (c) by nuclear fission.

Various nuclear reactions may take place in a reactor. Neutrons may be added to a stable isotope by simply placing the element in a pile and irradiating it with the neutron flux. Other unstable isotopes may be produced by more complicated reactions. When a material is irradiated by a neutron flux such as occurs in a reactor, a dual action takes place. As radioactive nuclei are formed, those in existence will decay by the process previously described. Hence, the buildup takes place at a decreasing rate. This relation may be expressed as follows [2]:

$$N_t = N_s(1 - e^{-\lambda t}), \tag{18.2}$$

where

N_t = number of unstable atoms at time t,

N_s = number of unstable atoms when rate of formation
and rate of decay are equal, a condition referred
to as *saturation*.

Certain isotopes are presently obtainable only from a cyclotron. The cyclotron employs two hollow, semicircular chambers (called D's, or Dees, because of their shape), similar to the halves of a shallow, enclosed tank. These are located in an evacuated container, and the entire assembly is placed in an intense magnetic field. A controlled high-frequency voltage is applied alternately to the two Dees. Particles introduced at the center from sources not unlike a cathode-ray tube gun are accelerated in a spiral path, increasing in energy with each turn. The beam is finally directed to a target area containing the material to be irradiated. While this method is limited to the production of only one type of isotope at a time and in small quantities, the flexibility of the method makes it a very important source.

Fission-produced isotopes are obtained as by-products from the operation of nuclear reactors, which makes this source more prolific than the others. Splitting of the uranium-235 atom produces fragments near the center of the periodic table, many of which are radioactive. Examples of these are strontium-90, cesium-137, and cerium-144.

18.5 MEASUREMENT OF RADIOACTIVITY

There are a number of means for sensing or measuring radioactivity. The more common ones applicable to mechanical measurements are: photographic emulsions, ionization chambers, proportional counters, Geiger-

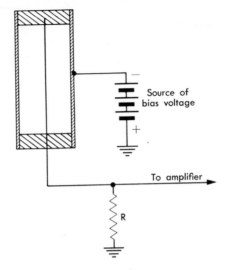

Fig. 18.2 Schematic section through a simple ionization chamber.

Mueller counters, and scintillation counters. Most of the methods are either directly or indirectly dependent on the ionizing effect of radiation. The ionization may be the result of a primary radiation, such as obtained from alpha or beta particles, or it may result from a secondary ionized particle produced by a gamma ray or a neutron.

a) Use of photographic emulsions. The earliest and still a practical method for sensing and measuring radioactive radiation is by means of photographic plates. This was used by Becquerel in 1896 when he discovered the radiation from uranium. All forms of radiation may be detected either directly or indirectly by this method.

Radiation affects the photographic plate in much the same manner as light does. The chemically developable blackening or density is proportional to intensity, time, and type of radiation—an effect which can be calibrated. For quantitative determinations, measurement of emulsion density is necessary, requiring photometric apparatus and very careful control of the photographic processing.

Qualitative results can be determined through use of special film and microscopic inspection of tracks resulting from radiations. Straightness, number of developed grains per unit length, relative angles of tracks, etc., may be used to establish the character of the radiation. Photographic means are quite useful for monitoring of personnel and areas. Devices for this general purpose are called "dosimeters."

b) Devices depending on ionization. Ionization chambers, proportional counters, and Geiger-Mueller (G-M) counters are all similar in general

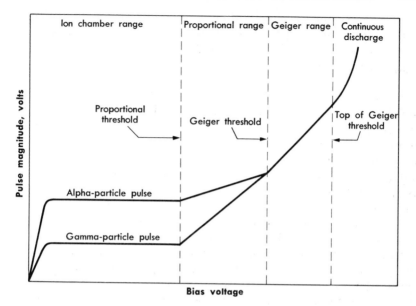

Fig. 18.3 Characteristics of an ionization chamber with change in bias voltage.

construction, and all depend on ionization for their operation. The basic form is shown schematically in Fig. 18.2. An enclosure or chamber is employed, the shell of which serves as an electrode. A second electrode in the form of a wire passes coaxially through the center. The chamber is filled with a gas, and an electrical potential, termed the *bias voltage*, is applied across the electrodes. Any negative ions (electrons) in the chamber are quickly drawn to the positively charged center electrode, and the positive ions are drawn to the shell. This causes a current to flow, which produces a voltage drop across resistance R.

The three devices mentioned above are all of this basic form. They differ, however, in the type and pressure of gas employed and the bias voltage. Variation in the latter quantity has the most marked influence, which is illustrated in Fig. 18.3. Three general ranges of sensitivity are obtained, referred to as the *plateau, proportional,* and *Geiger ranges.*

In the first, or plateau range, the cell behaves as an integrating device producing a current which is *proportional to the ionization* for a given time. A small number of high-energy particles can cause the same current as a large number of low-energy particles. The device is said to be *intensity-sensitive,* and is called an *ion chamber.*

In the proportional range, the cell is *energy-sensitive.* Since individual voltage pulses are of considerably greater energy, we can make a discrete pulse count, and the *pulse magnitudes are proportional to particle energy.*

The Geiger range is characterized by greater sensitivity and still larger pulses. However, the pulse magnitudes are identical and *independent of the original particle energy;* hence they make it impossible to determine the energy-distribution spectra.

c) Ionization chambers. The simple ionization or ion chamber may be used to detect all three types of radiation mentioned previously, alpha particles, beta particles, and gamma rays. The very weak penetrating power of alpha particles practically requires the source to be placed directly in the chamber or between electrodes, while the moderate penetrating power of beta radiation permits the use of a foil window, usually of stainless steel, through which the particles may be introduced to the chamber. Gamma rays, on the other hand, permit chambers of more sturdy construction.

Ionization is directly proportional to the mass of gas in the chamber; hence pressures are employed as high as are consistent with the chamber design. When subjected to steady beams of radiation, the output, although fundamentally pulsed, may be considered d-c. Amplification is required, and for this purpose some form of d-c or chopper amplifier may be used. In general, the simple ionization chamber is used when the radiation level is relatively large. It is, therefore, of limited application to mechanical measurement problems.

A common form of dosimeter is based on the idea of the simple ionization chamber. Basically, it is a high-leakage capacitor charged with a potential of between 20 and 200 V. Radiation reduces the charge in approximate proportion to the energy of radiation. When the charge is measured with an electrometer, a measure of radiation exposure is obtained. A personnel dosimeter of this type is approximately the shape and size of the ordinary fountain pen.

d) The Geiger-Mueller and proportional counters. Geiger-Mueller (G-M) and proportional counters are quite similar in construction to the ionization chamber. However, they are operated to take advantage of what is known as the Townsend avalanche phenomenon, also known as *gas amplification* [3]. Under certain conditions it is found that the ions formed directly by the radiation (primary ionization) may be accelerated to velocities producing a secondary ionization. This results in a "snow-balling" of ion movement, and hence a very much increased output. Practically, it produces a considerably increased sensitivity.

Gas amplification depends upon a number of factors, including the nature and pressure of the gas charge and the dimensions of the chamber and bias voltage, the latter having the greatest effect. At relatively low voltages (specific values will not be given, for they are dependent on the other factors mentioned), primary ions caused by radiation are simply drawn to the positive electrode, providing more-or-less continuous current flow. This is the range of the ionization cell discussed in the previous article. At somewhat higher voltages, gas amplification begins, providing bursts or pulses of ion flow. In the intermediate range of bias voltage, it is found that the pulse amplitudes are approximately *proportional* to the number of primary ions triggering the pulse. Alpha particles produce large pulses (because they are highly ionizing), while beta particles produce much smaller pulses. Because gamma rays do not directly produce ionization, they are scarcely detectable. Proportional counters, therefore, are detectors of the ionization type, operating with an intermediate bias voltage.

If the bias voltage is increased still further, *Geiger-Mueller counters* are obtained. In this range the pulses are much larger than obtainable from the proportional counter; however, they are very nearly uniform in size regardless of primary ionization. Alpha and beta particles and gamma rays are practically indistinguishable.

It is seen, then, that the basic difference in simple ionization chambers, proportional counters, and Geiger-Mueller counters lies in the relative magnitudes of bias voltages.

Although the construction of the Geiger-Mueller tube is quite similar to that of the ionization chamber, the gas composition and pressure and electrical potential are adjusted to take fullest advantage of gas amplification discussed in the previous section. Improved performance is obtained by replacing air, as used in ionization chambers, with a mixture of vapor and certain gases under reduced pressure. Common types employ either mixtures of alcohol vapor and argon or a neon-argon combination with a halogen gas. Once initiated, the ionizing continues until a state of complete discharge results. This requires a time interval of a few thousandths of a second, and if a second triggering particle arrives during this time, its effect will be ignored. For this reason the counting rate of the Geiger-Mueller tube is limited.

In spite of its relatively slow counting rate and the fact that it cannot easily distinguish the different types of radiation, the G-M tube is popular as a radiation-sensing device. This is largely due to its relatively low cost and high output, requiring only simple intermediate and indicating circuitry.

e) Scintillation counters. When radioactive radiation strikes certain fluorescing materials, tiny flashes of visible light are produced. The scintillation counter employs this phenomenon along with a photomultiplier tube for sensing the flashes. The light flashes are directed to a photosensitive surface in the multiplier tube, from which electrons are emitted. The multiplier tube has the characteristic of being able to amplify the electron flow many times (in the order of 10^6), which may be further amplified by conventional means and indicated or recorded.

18.6 INTERMEDIATE AND READOUT DEVICES

Radiation-sensing elements discussed in the preceding paragraphs require some form of readout device or devices. Density of *photographic film* may be measured by some form of densitometer or photometer. This value may then be compared with the film density resulting from standardized conditions, and the amount of radiation which reached the test film may be evaluated by comparison.

Ion chambers provide a current output proportional to the radiation reaching the chamber. Either direct-coupled amplifiers (Article 7.23b) or some form of chopper or carrier-type amplifier (Articles 7.23c and d) may be used to provide sufficient current or voltage output to actuate simple meter-type indicators or recorders.

Geiger-Mueller, proportional, and scintillation tubes sense radiation as a series of single events. When detectors of this kind are employed, EPUT counters may be used (Articles 8.5b and 9.2b). The term *rate meter* is often applied to the latter. Usually the number of events is of such magnitude that some form of *scaler* is employed. This is an electronic circuit which divides the number of distinct input events by a constant factor, often 10 or 100.

18.7 DIRECT USE OF IONIZING CHARACTERISTICS OF RADIOACTIVE ISOTOPES

Although not an example of measurement, probably the simplest application of radioactive isotopes is for the dissipation of static electrical charges. For this purpose, the low-energy alpha particles may be employed. This is useful for the removal of dust and lint over small areas or in small volumes. An example of widespread application of this technique is the use of a small amount of polonium-210 attached to phonograph pickups for dissipation of static charges on recordings. Similar applications have been employed in the manufacture and use of paper, textiles, rubber, and plastic items.

18.8 APPLICATION TO DENSITY AND THICKNESS MEASUREMENTS

An important mechanical-measurement application of isotopes involves radiation absorption by the measured substance. Different materials, or different thicknesses or densities of a given material, impede the passage of radiation to different degrees. Through proper application and calibration, this shielding effect may be employed to measure such quantities as material density, thickness of materials, thickness of coatings, extent of corrosion, fluid levels, imperfections in castings, welds, etc.

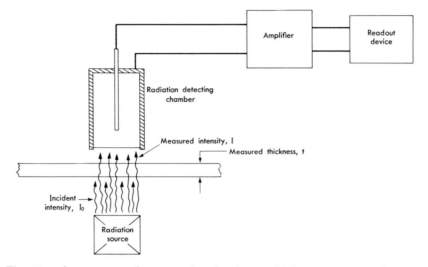

Fig. 18.4 Arrangement for measuring density or thickness using a radioactive isotope and radiation detector.

Basically, beta and gamma radiation is absorbed in accordance with the relation [4, 5]

$$I = I_0 e^{-\mu w}, \qquad (18.3)$$

where

I = radiation intensity,

I_0 = intensity of incident radiation,

w = density of absorbing material per unit area,

μ = absorption coefficient based on density.

If the absorbing material is *homogeneous*, Eq. (18.3) may be expressed in terms of the material thickness, t,

$$I = I_0 e^{-\mu' t}, \qquad (18.4)$$

where μ' is the absorption coefficient based on thickness. When various radiations are present, then

$$I = I_0(e^{-\mu_1 w} + e^{-\mu_2 w} + \cdots + e^{-\mu_n w}). \qquad (18.5)$$

It is seen, then, that if we place a radiation source on one side of a test item or specimen and an intensity-sensing device such as an ion chamber, or a G-M, or proportional, or scintillation counter on the other side, a means for measuring variations in intervening thickness or density is obtained. This assumes that such a device may be calibrated. An arrangement of this kind is shown schematically in Fig. 18.4.

Thickness-measuring devices of these types are particularly attractive for two reasons: the sensing elements are noncontacting, and continuous measuring or monitoring is possible. The latter permits use of the system for control of such apparatus as metal rolling mills, paper mills, calendering machines, and the like.

In most applications, however, compensation must be employed to eliminate the effect of two variables: decay of the radiating source and pickup of ambient radiation from extraneous sources such as cosmic rays. Compensation is accomplished through use of two sensing cells and ratio circuitry. One cell is arranged to sense direct radiation from the source, unaffected by the specimen; the other senses the radiation only after passing through the specimen. The two cells are affected equally by decay and extraneous pickup, with the only factor governing the ratio being the absorption caused by the specimen material.

Various radiation sources may be used, but for industrial applications a reasonably long half-life is mandatory. In addition, beta radiations are more desirable than gamma rays for safety reasons. A common radiation source is strontium-90, which has a half-life of 25 years. This isotope is usable for measuring thicknesses up to two inches in plastics and 0.030 in. for steel [4].

18.9 USE OF GAMMA RADIATION FOR RADIOGRAPHY

As stated previously, gamma rays are not completely absorbed by any thickness of shielding, but are absorbed exponentially. One of the more important applications taking advantage of this characteristic is the use of gamma-ray sources for radiographic inspection. Photographic methods are employed to obtain "shadowgraphs" of the internal structure of such things as welds and castings. Internal defects in the form of inclusions and cracks result in different opacities on the film because of density differences or discontinuities. A primary advantage of gamma-ray radiography as compared with x-ray inspection is reduced cost. It has been stated that radioisotopes and associated equipment costing about $3000 can perform many functions as well as a $50,000 x-ray source [6].

18.10 RADIOACTIVE TRACER TECHNIQUES

Radioactive isotopes may be employed in diverse ways for the purpose of following or tracing process characteristics. Probably the first applications were concerned with medical and biological research, but many of the techniques are applicable to engineering studies. Applications may be concerned only with qualitative investigations, although quantitative results are also possible in many cases. Some of the methods that may be employed will be discussed in terms of the following examples.

Tracer techniques may be quite easily applied to many fluid-flow problems. For example, the flow through a complex system such as a heat exchanger or boiler may be studied by introducing a small quantity of soluble radioactive material, then following it through the system by means of strategically placed pickups. Relative flow through various branches of the system may be determined by means of relative counts. This method has been applied to the measure of river flow [7].

Leak detection is a closely related fluid-flow problem. As an example, "tightness" between the tube and shell sides of a heat exchanger may be determined by introducing a radioactive material at the inlet of one side, then monitoring the flow from the opposite side with a radiation detector. A measure of leakage magnitude should also be possible.

Determination of wear. Before radioactive isotopes were generally available, wear studies were usually based on change in either weight or dimension of the subject part. The techniques required measurements of great sensitivity and accuracy because of the extremely small increments involved. This alone placed severe limitations on the methods. Use of radioactive isotopes have largely eliminated this problem. The following outlines a radioactive procedure applied to a study of gear wear [8].

Two identical gears were prepared: one as the subject for direct study and the other as a control for calibration purposes. Material samples were obtained from the second gear by grinding, using a very fine grinding wheel. The subject gear and grinding dust were then irradiated by neutron bombardment in an atomic reactor for 28 days.

When they were removed from the reactor, the radioactivity of both were monitored by means of a Geiger-Mueller tube and appropriate counting means. At first the decay rate was quite rapid because of the influence of short half-life elements. After about ten days the decay rate approached a straight line, thereby providing convenient time corrections during the actual test.

Calibration was supplied by adding known weights of the irradiated grinding dust to a predetermined amount of lubricating oil. After thorough mixing, radioactive counts were made providing correlation with the known iron-particle concentration.

The subject gear was then loaded and run under controlled conditions, and the radioactivity of the known weight of circulating oil was monitored. By comparing the resulting radioactivity count with the calibration results, corrected for difference in quantities of oil and decay rate, the actual weight of material removed by wear from the gear teeth could be determined.

Although the primary advantage of a method such as described is its extreme sensitivity (50 micrograms per minute could be detected), there are also other advantages. Particles worn from any other gears or bearings supplied by the same oil source have no influence because they have not been radioactively tagged, and hence are ignored by the counter. Another advantage lies in the possibility of continuous measurement during the process. This, of course, supplies information on *wear rate* in addition to total wear.

In this example, the radioactive isotopes were formed subsequent to manufacture of the gear. Another method is to add appropriate isotopes in the part material before manufacture. This has been done in the study of tire wear [9]. Phosphorus isotope P^{32} was added to the tread composition when molded. Tread material deposited on various roadways was then measured by counting techniques and compared with calibration based on known simulated conditions. Through use of similar methods, erosion of jet-engine fuel nozzles has been measured at rates as low as one microgram per hour.

18.11 SAFETY CONSIDERATIONS

Certain hazards accompany the handling and use of radioactive materials, and safety precautions must be taken to prevent contamination or excessive exposure. Although this form of hazard may be somewhat unfamiliar to the average engineer, he should accept it with the same attitude as when he works with extreme pressures or temperatures or handles highly toxic or explosive materials, all of which may be controlled with correct technique, and all of which may be deadly when treated with improper respect. One general difference may be observed, however: the effects of radioactive exposure may be *cumulative*.

Known hazards have been classified in the order of their importance as follows [10]: (a) deposition of radioisotopes in the body, (b) exposure of the whole body to gamma radiation, (c) exposure of the body to beta radiation, and (d) exposure of the hands or other limited parts to beta or gamma radiation.

Basically, as with any hazard, complete knowledge of the nature of the hazard along with proper and adequate handling equipment and techniques are the keys to safe and successful application. Involved are, (a) adequate shielding, (b) proper handling equipment, (c) continuous monitoring, and (d) avoidance of all unnecessary radiation, regardless of any arbitrary "safe" levels.

With proper precautions and equipment, all the above problems may be safely solved. However, the details are many and varied, depending on the application and source of radioactivity employed, the discussion of which is beyond the scope of this book. For further details the reader is directed to the references for this chapter.

STRESS AND STRAIN RELATIONSHIPS

A.1 THE GENERAL PLANE STRESS SITUATION

Suppose an element dx wide by dy high is selected from a general plane stress situation in equilibrium, as shown in Fig. A.1. Assume that the element is of uniform thickness, t, normal to the paper.

From strength of materials it will be remembered that for equilibrium, τ_{xy} must be equal to τ_{yx}. We will therefore employ the symbol τ_{xy} for both. We will also assume that not only must the complete element be in equilibrium, but so also must be all its parts. Therefore, if the element is bisected by a diagonal ds long, each half of the element must also be in equilibrium.

As shown in Fig. A.2, there must be a normal stress σ_θ and a shear stress τ_θ acting on the diagonal area. If the various stresses shown on the partial element, Fig. A.2, are multiplied by the areas over which they act, forces are obtained as shown in Fig. A.3. Let all the directions be considered positive, as shown. If the force components in the direction normal to the

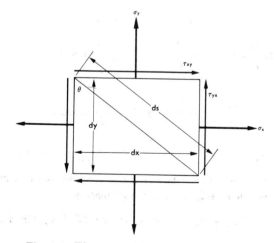

Fig. A.1 Element subject to plane stresses.

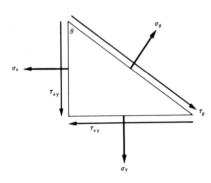

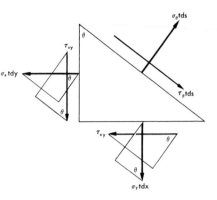

Fig. A.2 Element used to define positive stress directions.

Fig. A.3 Element illustrating the requirement for force equilibrium.

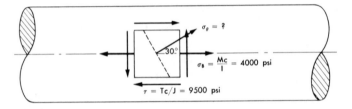

Fig. A.4 Stresses acting on an element on the outer surface of a shaft subject to torsion and bending.

diagonal plane are added, then for equilibrium

$$\sigma_\theta t\, ds - \sigma_x t\, dy \cos\theta - \sigma_y t\, dx \sin\theta - \tau_{xy} t\, dx \cos\theta - \tau_{xy} t\, dy \sin\theta = 0.$$

$$(A.1)$$

Now if Eq. (A.1) is divided by $t\, ds$ and solved for σ_θ,

$$\sigma_\theta = \sigma_x \frac{dy}{ds} \cos\theta + \sigma_y \frac{dx}{ds} \sin\theta + \tau_{xy} \frac{dx}{ds} \cos\theta + \tau_{xy} \frac{dy}{ds} \sin\theta.$$

However,

$$\frac{dy}{ds} = \cos\theta \quad \text{and} \quad \frac{dx}{ds} = \sin\theta.$$

Therefore

$$\sigma_\theta = \sigma_x \cos^2\theta + \sigma_y \sin^2\theta + 2\tau_{xy} \sin\theta \cos\theta.$$

A more convenient form of this equation may be had in terms of double angles. By substituting trigonometric equivalents,

$$\sigma_\theta = \tfrac{1}{2}(\sigma_x + \sigma_y) + \tfrac{1}{2}(\sigma_x - \sigma_y) \cos 2\theta + \tau_{xy} \sin 2\theta. \qquad (A.2)$$

Using this equation, the stress on any plane may be determined if values of σ_x, σ_y, and τ_{xy} are known.

Example. A shaft is subject to a torque, T, which results in a shear stress, $Tc/J = 9500$ psi (Fig. A.4), and at the same time and at the same point a bending moment due to gear loads causes an outer fiber stress, $Mc/I = 4000$ psi. What will be the normal stress on the outer surface in a direction $30°$ to the shaft center line?

Solution.

$$\sigma_{30°} = \tfrac{1}{2}(\sigma_x + \sigma_y) + \tfrac{1}{2}(\sigma_x - \sigma_y)\cos 2\theta + \tau_{xy}\sin 2\theta$$
$$= \tfrac{1}{2}(4000 + 0) + \tfrac{1}{2}(4000 - 0)\cos 60° + 9500\sin 60°$$
$$= 11{,}240 \text{ psi.}$$

Of course, the normal stress, 11,240 psi, is not necessarily the maximum normal stress, because the angle $30°$ was chosen at random; undoubtedly some other angle may result in a larger normal stress.

A.2 DIRECTION AND MAGNITUDES OF PRINCIPAL STRESSES

To calculate the maximum normal stress, the particular angle θ_1 determining the plane over which it will act must be found. This may be done by differentiating Eq. (A.2), with respect to θ, setting the derivative equal to zero, and solving for the angle θ_1. This should also give us the plane over which the normal stress is a minimum.

$$\frac{d\sigma_\theta}{d\theta} = -(\sigma_x - \sigma_y)\sin 2\theta + 2\tau_{xy}\cos 2\theta = 0$$

or

$$\tan 2\theta_{1,2} = \frac{\pm 2\tau_{xy}}{\pm(\sigma_x - \sigma_y)}. \tag{A.3}$$

Two angles, $2\theta_{1,2}$, are determined by Eq. (A.3), and consideration of the trigonometry involved shows that the two angles are $180°$ apart. This would mean, then, that the two angles $\theta_{1,2}$ are $90°$ apart, and leads to a very important fact: *The planes of maximum and minimum normal stress are always at right angles to each other.*

The maximum and minimum normal stresses are called the *principal stresses*, and the planes over which they act are called the *principal planes*. We have just found, therefore, that the principal planes are at right angles to each other. If we know the direction of the maximum normal stress, we automatically know the direction of the minimum normal stress.

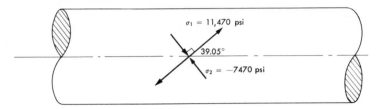

Fig. A.5 The principal stresses corresponding to the situation shown in Fig. A.4.

Now we would like to find an expression for the principal stresses. From Eq. (A.3) we may write

$$\sin 2\theta_{1,2} = \frac{2\tau_{xy}}{\sqrt{(\sigma_x - \sigma_y)^2 + (2\tau_{xy})^2}},$$

$$\cos 2\theta_{1,2} = \frac{(\sigma_x - \sigma_y)}{\sqrt{(\sigma_x - \sigma_y)^2 + (2\tau_{xy})^2}}.$$

(A.4)

Substituting these values in Eq. (A.2) gives us the principal stresses, which we shall designate σ_1 and σ_2, and

$$\sigma_{\theta max} = \sigma_1 = \tfrac{1}{2}(\sigma_x + \sigma_y) + \tfrac{1}{2}\sqrt{(\sigma_x - \sigma_y)^2 + (2\tau_{xy})^2},$$

$$\sigma_{\theta min} = \sigma_2 = \tfrac{1}{2}(\sigma_x + \sigma_y) - \tfrac{1}{2}\sqrt{(\sigma_x - \sigma_y)^2 + (2\tau_{xy})^2}.$$

(A.5)

Example. Referring to the example in Article A.1, determine the magnitudes of the principal stresses and the positions of the principal planes relative to the shaft center line.

Solution. Using Eqs. (A.5),

$$\sigma_1 = \tfrac{1}{2}(4000 + 0) + \tfrac{1}{2}\sqrt{(4000 + 0)^2 + (2 \times 9500)^2}$$
$$= 11,470 \text{ psi}$$

and

$$\sigma_2 = \tfrac{1}{2}(4000 + 0) - \tfrac{1}{2}\sqrt{(4000 + 0)^2 + (2 \times 9500)^2}$$
$$= -7470 \text{ psi.}$$

From Eq. (A.4),

$$\tan 2\theta = \frac{2 \times 9500}{4000 - 0} = \frac{19,000}{4000} = 4.75,$$

$$2\theta = 78.1°,$$

$$\theta = 39.05°.$$

The orientation is as shown in Fig. A.5.

A.3 VARIATION IN SHEAR STRESS WITH DIRECTION

Following the same procedure used for normal stresses, and again referring to Fig. A.3, if the forces parallel to the diagonal plane are summed, the following shear relations are obtained. The equation for the shear stress on any plane in terms of σ_x, σ_y, and θ is

$$\tau_\theta = \tfrac{1}{2}(\sigma_x - \sigma_y) \sin 2\theta - \tau_{xy} \cos 2\theta. \tag{A.6}$$

The angle determining the planes over which the shear stresses are maximum and minimum may be determined by the relation

$$\tan 2\theta_s = \frac{\mp(\sigma_x - \sigma_y)}{\pm 2\tau_{xy}}. \tag{A.7}$$

By substituting the angles determined by Eq. (A.7) in Eq. (A.6), relations for maximum and minimum shear stress may be obtained.

$$\begin{aligned} \text{Maximum shear stress} &= \tau_{\theta\text{max}} = \tfrac{1}{2}\sqrt{(\sigma_x - \sigma_y)^2 + (2\tau_{xy})^2}, \\ \text{Minimum shear stress} &= \tau_{\theta\text{min}} = -\tfrac{1}{2}\sqrt{(\sigma_x - \sigma_y)^2 + (2\tau_{xy})^2}. \end{aligned} \tag{A.8}$$

We must be careful, however, because the shear stress extremes given by Eq. (A.7) only account for the x- and y-directions. Consideration of the three-dimensional condition often shows the greatest shear stress to occur on yet another plane. See Article A.6.

A.4 SHEAR STRESS ON PRINCIPAL PLANES

Equations (A.8) allow us to determine the shear stress on any plane defined by θ. Therefore, let us substitute the expressions for $\theta_{1,2}$, Eq. (A.3), in Eq. (A.6), and thereby determine the shear stresses acting over the principal planes. Doing this gives

$$\begin{aligned} \tau_{1,2} &= \frac{-\tfrac{1}{2}(\sigma_x - \sigma_y)(2\tau) + \tau_{xy}(\sigma_x - \sigma_y)}{\sqrt{(\sigma_x - \sigma_y)^2 + (2\tau_{xy})^2}} \\ &= 0. \end{aligned}$$

This proves a very important fact about any plane stress situation: *The shear stresses on the principal planes are zero.* This in itself often provides the necessary clue to determine the orientation of the principal planes by inspection. In any case where it can be said, "there can be no shear on this plane," then the fact we have just established tells us that the plane we are referring to is a principal plane. Or, often just as important, if it can be said that shear stresses *do* exist on a plane, we know the plane *cannot* be one of the principal planes.

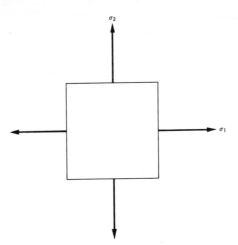

Fig. A.6 An element subjected to principal stresses.

In strain-gage applications, knowing the directions of the principal planes at the point of interest provides a very decided advantage. With this information, gages may be aligned in the principal directions, and usually only two gages are required. More important, however, is the fact that the calculations become much simpler and less time consuming (Article 11.18b).

A.5 GENERAL STRESS EQUATIONS IN TERMS OF PRINCIPAL STRESSES

Checking back on our original assumptions in Article A.1, we see that we assumed a simple element subject to two orthogonal normal stresses, σ_x and σ_y, and shear stresses, τ. Using the information since developed, namely that the principal stresses are at right angles to each other and that the shear stresses are zero on the principal planes, we may now rewrite certain of our equations in terms of principal stresses σ_1 and σ_2.

By selecting just the right element orientation, i.e., aligning it with the principal planes, our basic element could be made to appear as it does in Fig. A.6. σ_1 and σ_2 are orthogonal stresses, and we know also that there will be no shear on the planes over which they act. Therefore, any of our equations written so far may be modified by substituting σ_1 and σ_2 for σ_x and σ_y, respectively, and making the shear stress equal to zero.

Substitution in Eqs. (A.2) and (A.6) yields particularly useful relations. Substitution in most of the others simply confirms our definitions.

Substituting in Eqs. (A.2) and (A.6) gives us

$$\sigma_\theta = \tfrac{1}{2}(\sigma_1 + \sigma_2) + \tfrac{1}{2}(\sigma_1 - \sigma_2) \cos 2\theta, \qquad \text{(A.9a)}$$

$$\tau_\theta = \tfrac{1}{2}(\sigma_1 - \sigma_2) \sin 2\theta. \qquad \text{(A.9b)}$$

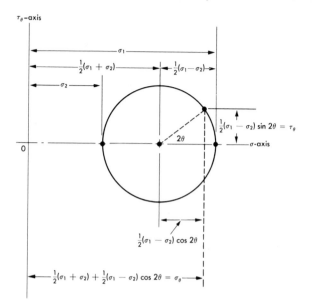

Fig. A.7 Mohr's circle for plane stresses.

These equations are particularly useful in helping us visualize the overall stress condition as shown in the following section.

A.6 MOHR'S CIRCLE FOR STRESS

Let us establish a coordinate system with σ_θ plotted as the abscissa and τ_θ as the ordinate (Fig. A.7). The shear stress corresponding to the principal stresses is zero, hence σ_1 and σ_2 will be plotted along the σ_θ-axis.

If a circle is drawn passing through the σ_1 and σ_2 points and having its center on the σ_θ-axis, the construction shown in Fig. A.7 will result. It will be noted that for any point on the circle, the distance along the abscissa represents σ_θ, and the ordinate distance represents τ_θ. This construction, which is very useful in helping to visualize stress situations, is known as Mohr's stress circle.

At this point we should consider the third or z-direction. In general three orthogonal stresses σ_x, σ_y, and σ_z, along with corresponding shear stresses τ_{xy}, τ_{yz} and τ_{zx}, will occur on an element, as shown in Fig. A.8a. In this case a third principal plane exists, over which, as for the two-dimensional case, *shear is zero*. Also, it may be shown* that the three principal planes, along with the three principal stresses σ_1, σ_2, and σ_3, are at right angles to one another. By considering the three directions in combinations

---

* e.g., in most intermediate solid mechanics books.

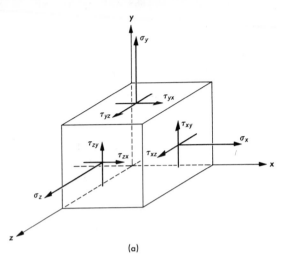

(a)

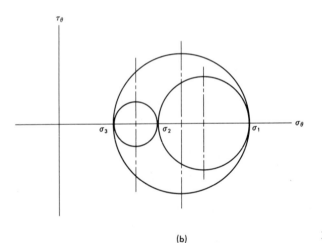

(b)

Figure A.8

of two, we may reduce the problem to three related two-dimensional situations. The resulting combined Mohr's diagrams are illustrated in Fig. A.8(b).

In the majority of cases, strain gages are applied to free, unloaded surfaces and the condition thought of as being two-dimensional. It is well, however, to consider every condition in terms of three dimensions, even though the third stress may be zero, and to plot or merely sketch the three-circle Mohr's diagram. This procedure often reveals a maximum shear which might otherwise be overlooked.

A few examples will demonstrate the power of the Mohr diagram.

Example 1. Figure A.9(a) shows a simple tension member. We know there is no shear stress on a transverse section; hence we know that this must be a principal plane. Since the other principal plane must be normal to the first

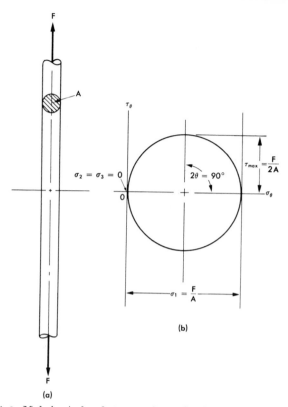

Fig. A.9 Mohr's circle of stresses for a simple tension member.

principal plane and hence be aligned with the axis of the specimen, the normal stress on this plane must be zero. Therefore

$$\sigma_1 = \frac{F}{A} \quad \text{and} \quad \sigma_2 = \sigma_3 = 0.$$

Plotting Mohr's circle for this situation gives us Fig. A.9(b). One of the Mohr circles degenerates to a point in this case.

By inspection we see that the maximum shear stress is equal to $F/2A$ at an angle $2\theta = 90°$ or $\theta = 45°$ measured relative to the axis of the specimen. This confirms our previous knowledge of the stress condition for this simple situation.

Example 2. Figure A.10(a) shows a thin-walled cylindrical pressure vessel. From elementary theory, the hoop, longitudinal and normal stresses may be calculated by the following relations:

$$\sigma_H = \frac{PD}{2t}, \quad \sigma_L = \frac{PD}{4t}, \quad \text{and} \quad \sigma_N = 0.$$

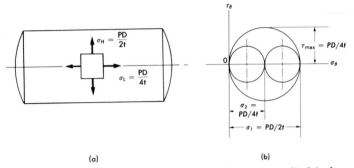

(a)   (b)

Fig. A.10 Mohr's circle of stresses for the free surface of a cylindrical pressure vessel. P = pressure, psi; D = shell diameter, inches; and t = shell wall thickness, inches.

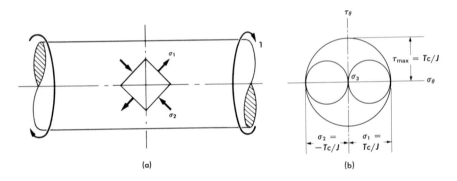

(a)   (b)

Fig. A.11 Mohr's circle for a shaft subject to "pure" torsion. T = torque, pounds; J = polar moment of inertia of section, in^4; c = distance from neutral axis to fiber of interest, inches.

Consideration of the nature of the stress field makes it difficult to imagine shear stresses on planes parallel to the hoop, longitudinal, or normal directions. Assuming this to be correct, then the circumferential, the longitudinal, and the normal directions must be the principal directions and σ_H, σ_L, and σ_N must be the principal stresses. Mohr's circle for this situation is shown in Fig. A.10b. The maximum shear is seen to be $PD/4t$, over a plane inclined 45° to the circumferential and normal directions.

Example 3. Figure A.11(a) shows a shaft in simple torsion. From strength of materials we know that the shear stress on the outer fiber acting on a plane normal to the shaft center line is equal to Tc/J. The fact that shear stress exists on this plane eliminates it from consideration as a principal plane. Since principal planes must be normal to each other, we immediately think of the one other symmetrical possibility—the two planes inclined 45°

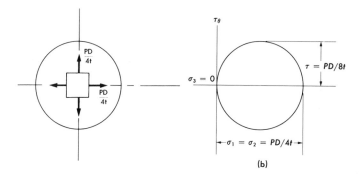

Fig. A.12 Mohr's circle for the free surface of a spherical pressure vessel.

to the shaft center line. Careful consideration of the stresses that may exist on these two planes leads us to conclude that tension would exist on one and compression on the other. The wringing of a wet towel is often used as an example of this situation. In the one 45° direction, the threads of the towel are obviously in tension, while in the other 45° direction, compression is employed to squeeze out the water.

Because of symmetry, we are led to the conclusion that the magnitudes of the two stresses are equal. Plotting equal tensile and compressive principal stresses, using Mohr's circle construction, gives us Fig. A.11(b).

The third principal direction is normal to the shaft surface and we see that $\sigma_N = 0$. Although the above discussion can hardly be considered rigorous proof, Fig. A.11(b) does represent the actual stress situation for a shaft subject to pure torsion. We know that maximum shear stress is equal to Tc/J. Therefore, inspection shows us that the principal stresses σ_1 and σ_2 must also have the same magnitude, Tc/J, tension and compression respectively.

Example 4. Figure A.12(a) shows a thin-wall spherical pressure vessel, for which elementary theory shows that the stress in the wall abides by the following relation:

$$\sigma = \frac{PD}{4t}.$$

In this case it is difficult to see how direction has significance. At any point on the outside of the shell, the normal stresses must be equal in all directions, simply because of symmetry. We must therefore conclude that

$$\sigma_1 = \sigma_2 = \frac{PD}{4t} \qquad \text{and} \qquad \sigma_3 = \sigma_N = 0.$$

Mohr's diagram for this condition is shown in Fig. A.12(b), from which it is seen that $\tau_{\max} = PD/8t$.

From the preceding discussion and consideration of Mohr's circle construction, we may now make the following general observations:

1. A stress state involving shear without normal stress is impossible.

2. Maximum shear stress always occurs on planes oriented 45° to the principal stresses, and is equal to one-half the algebraic difference of the principal stresses.

3. The shear stresses on any mutually perpendicular planes are of equal magnitude.

4. The sum of the normal stresses on any mutually perpendicular planes is a constant.

5. The maximum ratio of shear stress to principal stress occurs when the principal stresses are of equal magnitude but opposite sign.

A.7 STRAIN AT A POINT

Through use of Hooke's law and the stress relations developed in the preceding pages, the following relations for strain at a point may be derived:

$$\epsilon_\theta = \tfrac{1}{2}(\epsilon_x + \epsilon_y) + \tfrac{1}{2}(\epsilon_x - \epsilon_y)\cos 2\theta + \frac{\gamma_{xy}}{2}\sin 2\theta, \qquad \text{(A.10a)}$$

$$\frac{\gamma_\theta}{2} = \tfrac{1}{2}(\epsilon_x - \epsilon_y)\sin 2\theta - \frac{\gamma_{xy}}{2}\cos 2\theta. \qquad \text{(A.10b)}$$

Comparison of the above two equations with Eqs. (A.2) and (A.6), respectively, indicates that with a minor exception [the shear strains γ are divided by 2, whereas their counterparts are not], the stress and the strain relations at a point are functionally alike. It follows, therefore, that we can draw a Mohr's diagram for strain, provided the ordinate is made $\gamma_\theta/2$. This is sometimes useful in treating strain-rosette data.

Example 5. Power piping is subject to a combination of loading whose complexity will serve as an interesting example of a combined stress situation. In addition to pressure loading, differential expansion between the hot and cold conditions may superimpose bending, torsional, and axial loading.

Of course, the primary problem involved in piping design is the determination of the loading brought about by pipe expansion, end movements, and movement-limiting stops. In the simple situations good estimates of these loads may be determined analytically, but experimental methods using models are often required (Article 12.9). More recently, the various computer techniques have been applied to the problem.

In this example it will be assumed such preliminary work has been finished and the critically stressed location found. The remaining problem, then, is to combine the stress components and to determine the net stress condition. The problem is as follows:

Pipe data (14 inch, Schedule 100)

Outside diameter = 14 in.,
Wall thickness = 0.937 in.,
Inside diameter = 12.125 in.,
Bending moment of inertia = 825 in^4,
Bending section modulus = 117.9 in^3,
Torsional moment of inertia = 1650 in^4,
Torsional section modulus = 235.8 in^3,
Cross-sectional area = 38.47 in^2,
Young's modulus = 23 $\times$ 10^6 psi,
Poisson's ratio = 0.29.

Loading Data

Internal pressure = 620 psi,
Bending moment = 700,000 in·lb,
Torsional moment = 480,000 in·lb,
Axial load = 35,000-lb tension.

Problem. For the outer surface of the pipe, calculate (a) the maximum shear stress, (b) the principal stresses, (c) the direction of the stress σ_1 relative to the axis of the pipe, (d) the axial and circumferential unit strains, and (e) the principal strains. Also, (f) sketch Mohr's diagrams for stress and for strain.

Solution. The stress components are found as follows:

$$\text{The ratio} \frac{\text{I.D.}}{\text{O.D.}} = \frac{12.125}{14} = 0.87,$$

which is the range usually termed "thin wall." Hence,

$$\sigma_H = \text{hoop stress} = \frac{PD}{2t} = \frac{620 \times 12.125}{2 \times 0.937} = 4011 \text{ psi},$$

$$\sigma_L = \text{longitudinal stress} = \frac{PD}{4t} = \frac{1}{2}\sigma_H = 2005 \text{ psi},$$

$$\sigma_B = \text{bending stress} = \frac{Mc}{I} = \frac{700,000}{117.9} = \pm 5937 \text{ psi},$$

$$\tau = \text{torsional stress} = \frac{Tc}{J} = \frac{480,000}{235.8} = 2035 \text{ psi},$$

$$\sigma_A = \text{axial stress} = \frac{F}{A} = \frac{35,000}{38.47} = 910 \text{ psi}.$$

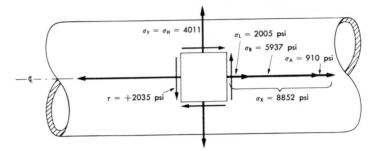

Fig. A.13 Axial and hoop stresses acting in the pipe of Example 5.

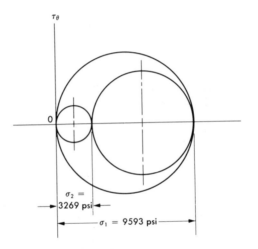

Fig. A.14 Principal stresses and principal stress directions for the pipe of Example 5.

The above conditions are illustrated in Fig. A.13.

a) Using Eq. (A.8), we obtain

$$\tau = \tfrac{1}{2}\sqrt{(8852 - 4011)^2 + (2 \times 2035)^2} = 3162 \text{ psi.}$$

From Fig. A.14 we see, however, that the true maximum shear stress is

$$\tau_{\text{max}} = (9593/2) = 4796 \text{ psi.}$$

b) Using Eq. (A.5) (see also Fig. A.15), we get

$$\sigma_1 = \tfrac{1}{2}(8852 + 4011) + 3162 = 9593 \text{ psi,}$$

$$\sigma_2 = \tfrac{1}{2}(8852 + 4011) - 3162 = 3269 \text{ psi.}$$

Also,

$$\sigma_3 = \sigma_N = 0.$$

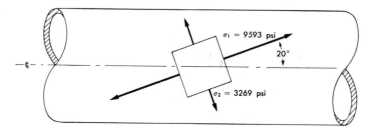

Fig. A.15 Principal stress element for Example 5.

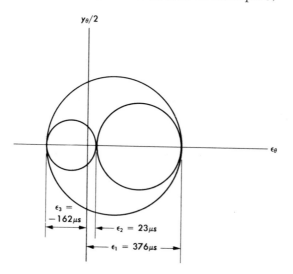

Fig. A.16 Mohr's strain diagram (outer surface) for Example 5.

c) Using Eq. (A.3), we obtain

$$\tan 2\theta_{\sigma_1} = \frac{2 \times 2036}{(8852 - 4011)} = \frac{4072}{4841} = 0.8411,$$

$$2\theta_{\sigma_1} = 40°4', \qquad \theta_{\sigma_1} = 20°2'.$$

d) Using Eqs. (11.16), we get

$$\epsilon_x = \frac{1}{23 \times 10^6} [8852 - 0.29(4011 + 0)] = 334 \,\mu s,$$

$$\epsilon_y = \frac{1}{23 \times 10^6} [4011 - 0.29(0 + 8852)] = 63 \,\mu s,$$

$$\epsilon_z = \frac{1}{23 \times 10^6} [0 - 0.29(8852 + 4011)] = -162 \,\mu s.$$

e) Again, from Eqs. (11.16)

$$\epsilon_1 = \frac{1}{23 \times 10^6} [9593 - 0.29(3269 + 0)]$$

$$= 376 \, \mu s,$$

$$\epsilon_2 = \frac{1}{23 \times 10^6} [3269 - 0.29(0 + 9539)]$$

$$= 23 \, \mu s,$$

$$\epsilon_3 = \frac{1}{23 \times 10^6} [0 - 0.29(9593 + 3269)]$$

$$= -162 \, \mu s.$$

f) Mohr's diagrams for stress and for strain are shown in Figs. A.14 and A.16, respectively.

[*Student assignment:* Modify the above calculations and diagrams for conditions on the inner pipe surface ($\sigma_3 = \sigma_N = -620$ psi).]

THEORETICAL BASIS FOR THE HARMONIC-ANALYSIS PROCEDURE

The theoretical basis for the harmonic-analysis procedure may be described as follows: any single valued function $f(x)$ which is continuous (except for a finite number of finite discontinuities) in the interval $-\pi$ to π, and which has only a finite number of maxima and minima in that interval, can be represented by a series in the form:

$$f(x) = \frac{A_0}{2} + A_1 \cos x + A_2 \cos 2x + \cdots + A_n \cos nx$$

$$+ B_1 \sin x + B_2 \sin 2x + \cdots + B_n \sin nx. \quad \text{(B.1)}$$

If each term in Eq. (B.1) is multiplied by dx and integrated over any interval of 2π length, all sine and cosine terms will drop out, leaving

$$\int_a^{2\pi+a} f(x)\, dx = \int_a^{2\pi+a} \frac{A_0}{2} = A_0 \pi$$

or

$$A_0 = \frac{1}{\pi} \int_a^{2\pi+a} f(x)\, dx. \quad \text{(B.2)}$$

The factor A_n may be determined if we multiply both sides of Eq. (B.1) by $\cos mx\, dx$ and integrate each term over the interval of 2π.

In general there are the following terms:

$$\int_a^{2\pi+a} \sin nx \cos mx\, dx = 0$$

and

$$\int_a^{2\pi+a} \cos nx \cos mx\, dx = 0, \qquad \text{except for} \quad m = n.$$

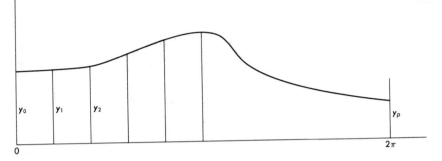

Fig. B.1 Graphical representation of Eq. (B.5.) (Ordinates taken at equal intervals.)

For the special case $m = n$,

$$\int_a^{2\pi+a} \cos^2 nx\, dx = \frac{1}{2n}\,[nx + \sin nx \cos nx]_{-\pi}^{\pi}$$

$$= \pi,$$

hence

$$\int_a^{2\pi+a} f(x) \cos nx\, dx = A_n \pi$$

or

$$A_n = \frac{1}{\pi}\int_a^{2\pi+a} f(x) \cos nx\, dx. \tag{B.3}$$

[*Note:* for $n = 0$, Eq. (B.3) reduces to Eq. (B.2).]

In like manner, if we multiply both sides of Eq. (B.1) by $\sin nx\, dx$ and integrate term by term over the interval 2π, we may obtain

$$B_n = \frac{1}{\pi}\int_a^{2\pi+a} f(x) \sin nx\, dx. \tag{B.4}$$

If Eqs. (B.2), (B.3) or (B.4) are plotted separately, but in general as

$$y = \frac{1}{\pi}\int_a^{2\pi+a} \phi(x)\, dx,$$

and the area under the resulting curve divided into p equal intervals along the x-axis, as shown in Fig. B.1, then

$$A_n(\text{or } B_n) = \frac{1}{\pi}\left(\frac{2\pi}{p}\right)\left[\left(\frac{y_0 + y_1}{2}\right) + \left(\frac{y_1 + y_2}{2}\right) + \cdots + \left(\frac{y_{p-1} + y_p}{2}\right)\right]$$

$$= \frac{1}{(p/2)}\sum_0^p y_n. \tag{B.5}$$

HARMONIC SINES AND COSINES

606　　APPENDIX C

APPENDIX C
HARMONIC SINES AND COSINES

Deg.		Fundamental	2nd	3rd	4th	5th	Deg.
0	cos	1.00000 –	1.00000	1.00000 –	1.00000	1.00000 –	180
	sin	0.00000	0.00000	0.00000	0.00000	0.00000	
5	cos	0.99619 –	0.98481	0.96593 –	0.93969	0.90631 –	175
	sin	0.08716	0.17365 –	0.25882	0.34202	0.42262	
10	cos	0.98481 –	0.93969	0.86603 –	0.76604	0.64279 –	170
	sin	0.17365	0.34202 –	0.50000	0.64279 –	0.76604	
15	cos	0.96593 –	0.86603	0.70711 –	0.50000	0.25882 –	165
	sin	0.25882	0.50000 –	0.70711	0.86603 –	0.96593	
20	cos	0.93969 –	0.76604	0.50000 –	0.17365	-0.17365	160
	sin	0.34202	0.64279 –	0.86603	0.98481 –	0.98481	
25	cos	0.90631 –	0.64279	0.25882 –	-0.17365	-0.57358	155
	sin	0.42262	0.76604 –	0.96593	0.98481 –	0.81915	
30	cos	0.86603 –	0.50000	0.00000	-0.50000	-0.86603	150
	sin	0.50000	0.86603 –	1.00000	0.86603 –	0.50000	
35	cos	0.81915 –	0.34202	-0.25882	-0.76604	-0.99619	145
	sin	0.57358	0.93969 –	0.96593	0.64279 –	0.08716	
40	cos	0.76604 –	0.17365	-0.50000	-0.93969	-0.93969 –	140
	sin	0.64279	0.98481 –	0.86603	0.34202 –	-0.34202	

		135 / 45	130 / 50	125 / 55	120 / 60	115 / 65	110 / 70	105 / 75	100 / 80	95 / 85	90 / 90
	cos	−0.70711	−0.34202	0.08716	0.50000	0.81915	0.98481	0.96593	0.76604	0.42262	0.00000
	sin	−0.70711	−0.93969	−0.99619	−0.86603	−0.57358	−0.17365	0.25882	0.64279	0.90631	1.00000
	cos	−1.0000	−0.93969	−0.76604	−0.50000	−0.17365	0.17365	0.50000	0.76604	0.93969	1.00000
	sin	0.00000	−0.34202	−0.64279	−0.86603	−0.98481	−0.98481	−0.86603	−0.64279	−0.34202	0.00000
	cos	−0.70711	−0.86603	−0.96593	−1.0000	−0.96593	−0.86603	−0.70711	−0.50000	−0.25882	0.00000
	sin	0.70711	0.50000	0.25882	0.0000	−0.25882	−0.50000	−0.70711	−0.86603	−0.96693	−1.00000
	cos	0.00000	−0.17365	−0.24002	−0.50000	−0.64279	−0.76604	−0.86603	−0.93969	−0.98481	−1.00000
	sin	1.00000	0.98481	0.93969	0.86603	0.76604	0.64279	0.50000	0.34202	0.17365	0.00000
	cos	0.70711	0.64279	0.57358	0.50000	0.42262	0.34020	0.25882	0.17365	0.08716	0.00000
	sin	0.70711	0.76604	0.81915	0.86603	0.90631	0.93969	0.96593	0.98481	0.99619	1.00000
		45	50	55	60	65	70	75	80	85	90

(cont.)

HARMONIC SINES AND COSINES (cont.)

Deg.		6th	7th	8th	9th	10th	Deg.
0	cos	1.00000	1.00000 −	1.00000	1.00000 −	1.00000	180
	sin	0.00000	0.00000	0.00000	0.00000 −	0.00000	
5	cos	0.86603	0.81915 −	0.76604	0.70711 −	0.64279 −	175
	sin	0.50000 −	0.57358	0.64279 −	0.70711 −	0.76604 −	
10	cos	0.50000	0.34202 −	0.17365	0.00000 −	−0.17365 −	170
	sin	0.86603 −	0.93969	0.98481 −	1.0000	0.98481 −	
15	cos	0.00000	−0.25882	−0.50000	−0.70711 −	−0.86603 −	165
	sin	1.00000 −	0.96593	0.86603 −	0.70711 −	0.50000 −	
20	cos	−0.50000	−0.76604	−0.93969	−1.0000 −	−0.93969 −	160
	sin	0.86603 −	0.64279 −	0.34202 −	0.00000	−0.34202 −	
25	cos	−0.86603	−0.99619	−0.93969	−0.70711 −	−0.34202 −	155
	sin	0.50000 −	0.08716	−0.34202 −	−0.70711 −	−0.93969 −	
30	cos	−1.00000	−0.86603	−0.50000	0.00000 −	0.50000	150
	sin	0.00000	−0.50000 −	−0.86603	−1.00000 −	−0.86603 −	
35	cos	−0.86603	−0.42262	0.17365	0.70711 −	0.98481	145
	sin	−0.50000	−0.90631 −	−0.98481	−0.70711 −	−0.17365	
40	cos	−0.50000	0.17365 −	0.76604	1.00000 −	0.76604	140
	sin	−0.86603	−0.98481 −	−0.64279	0.00000	0.64279 −	

angle	angle						
45	135	cos	0.00000	0.70711	1.00000	0.70711	0.00000
		sin	−1.00000	−0.70711	0.00000	0.70711	1.00000
50	130	cos	0.50000	0.98481	0.76604	0.00000	−0.76604
		sin	−0.86603	−0.17365	0.64279	1.00000	0.64279
55	125	cos	0.86603	0.90631	0.17365	−0.70711	−0.98481
		sin	−0.50000	0.42262	0.98481	0.70711	−0.17365
60	120	cos	1.00000	0.50000	−0.50000	−1.00000	−0.50000
		sin	0.00000	0.86603	0.86603	0.00000	−0.86603
65	115	cos	0.86603	−0.08716	−0.93969	−0.70711	0.34202
		sin	0.50000	0.99619	0.34202	−0.70711	−0.93969
70	110	cos	0.50000	−0.64279	−0.93969	0.00000	0.93969
		sin	0.86603	0.76604	−0.34202	−1.00000	−0.34202
75	105	cos	0.00000	−0.96593	−0.50000	0.70711	0.86603
		sin	1.00000	0.25882	−0.86603	−0.70711	−0.50000
80	100	cos	−0.50000	−0.93969	0.17365	1.00000	0.17365
		sin	0.86603	−0.34202	−0.98481	0.00000	0.98481
85	95	cos	−0.86603	−0.57358	0.76604	0.70711	−0.64279
		sin	0.50000	−0.81915	−0.64279	0.70711	0.76604
90	90	cos	−1.00000	0.00000	1.00000	0.00000	−1.00000
		sin	0.00000	−1.00000	0.00000	1.00000	0.00000

(cont.)

HARMONIC SINES AND COSINES (*cont.*)

Deg.		16th	15th	14th	13th	12th	11th	Deg.
180	cos	1.0000	1.0000 —	1.0000	1.0000 —	1.0000	1.0000 —	0
	sin	0.0000	0.0000	0.0000	0.0000	0.0000	0.0000	
175	cos	0.17365	0.25882 —	0.34202	0.42262 —	0.50000	0.57358 —	5
	sin	0.98481 —	0.96593	0.93969	0.90631	0.86603 —	0.81915	
170	cos	−0.93969	−0.86603	−0.76604 —	−0.64279	−0.50000 —	−0.34202	10
	sin	0.34202	0.50000	0.64279 —	0.76604	0.86603	0.93969	
165	cos	−0.50000	−0.70711	−0.86603 —	−0.96593	−1.00000 —	−0.96593	15
	sin	−0.86603	−0.70711 —	−0.50000 —	−0.25882 —	−0.00000	0.25882	
160	cos	0.76604	0.50000	0.17365 —	−0.17365	−0.50000 —	−0.76604	20
	sin	−0.64279 —	0.86603 —	−0.98481 —	−0.98481 —	−0.86603	−0.64279 —	
155	cos	0.76604	0.96593	0.98481	0.81915 —	0.50000	0.08716 —	25
	sin	0.64279 —	0.25882	−0.17365 —	−0.57358 —	−0.86603	−0.99619 —	
150	cos	−0.50000	0.00000	0.50000	0.86603 —	1.00000	0.86603 —	30
	sin	0.86603 —	1.0000	0.86603 —	0.50000	0.00000	−0.50000 —	
145	cos	−0.93969	−0.96593	−0.64279 —	−0.08716 —	0.50000	0.90631 —	35
	sin	−0.34202	0.25882	0.76604 —	0.99619	0.86603 —	0.42262	
140	cos	0.17365	−0.50000 —	−0.93969 —	−0.93969	−0.50000 —	0.17365 —	40
	sin	−0.98481	−0.86603 —	−0.34202 —	0.34202	0.86603 —	0.98481	

45	cos	−0.70711	−1.00000 −	−0.70711	0.00000	0.70711 −	1.00000	135
	sin	0.70711	0.00000	−0.70711 −	−1.00000	−0.70711 −	0.00000	
50	cos	−0.98481	−0.50000 −	0.34202 −	0.93969	0.86603 −	0.17365	130
	sin	−0.17365 −	−0.86603	−0.93969 −	−0.34202	0.50000	0.98481 −	
55	cos	−0.42262	0.50000	0.99619 −	0.64279	−0.25882	−0.93969 −	125
	sin	−0.90631 −	−0.86603	−0.08616 −	0.76604 −	0.96593	0.34202 −	
60	cos	0.50000 −	1.00000	0.50000 −	−0.50000 −	−1.00000	−0.50000	120
	sin	−0.86603 −	0.00000	0.86603	0.86603 −	0.00000	−0.86603	
65	cos	0.99619 −	0.50000	−0.57358	−0.98481 −	−0.25882	0.76604	115
	sin	−0.08716 −	0.86603 −	0.81915	−0.17365	−0.96593 −	−0.64279	
70	cos	0.64279 −	−0.50000 −	−0.98481	−0.17365 −	0.86603 −	0.76603	110
	sin	0.76604	0.86603 −	−0.17365 −	−0.98481	−0.50000 −	0.64279 −	
75	cos	−0.25882	−1.00000 −	−0.25882	0.86603	0.70711 −	−0.50000 −	105
	sin	0.96593	0.00000	−0.96593 −	−0.50000	0.70711	0.86603 −	
80	cos	−0.93969	−0.50000 −	0.76604 −	0.76604	−0.50000	−0.93969 −	100
	sin	0.34202	−0.86603	−0.64279 −	0.64279 −	0.86603	−0.34202	
85	cos	−0.81915	0.50000	0.90631 −	−0.34202 −	−0.96593	0.17365	95
	sin	−0.57358 −	−0.86603	0.42262	0.93969 −	−0.25882 −	−0.98481	
90	cos	0.00000	1.00000	0.00000	−1.00000 −	0.00000	1.00000	90
	sin	−1.00000 −	0.00000	1.00000	0.00000	−1.00000 −	0.00000	

(*cont.*)

TABLES

THE INTERNATIONAL SYSTEM OF UNITS

The International System of Units (designated SI, for *Système International d'Unités*) was given official status in a resolution of the 11th General Conference on Weights and Measures, Paris, 1960. SI terms, units, and symbols are employed by the National Bureau of Standards in all its publications except where use of the units would reduce the usefulness of the communication. The system is given below because of its obvious relationship to mechanical measurements.

Quantity	Unit	Symbol	
	Elemental units		
Length	meter	m	
Mass	kilogram	kg	
Time	second	s	
Electric current	ampere	A	
Temperature	degree Kelvin	°K	
Luminous intensity	candela	cd	
	Supplementary Units		
Plane angle	radian	rad	
Solid angle	steradian	sr	
	Derived Units		
Area	square meter	m²	
Volume	cubic meter	m³	
Frequency	hertz	Hz	(s⁻¹)
Density	kilogram per cubic meter	kg/m³	
Velocity	meter per second	m/s	
Angular velocity	radian per second	rad/s	
Acceleration	meter per second squared	m/s²	
Angular acceleration	radian per second squared	rad/s²	
Force	newton	N	(kg·m/s²)
Pressure	newton per sq meter	N/m²	

TABLE D.1 (cont.)

Quantity	Unit	Symbol	
Kinematic viscosity	sq meter per second	m²/s	
Dynamic viscosity	newton-second per sq meter	N·s/m²	
Work, energy, quantity of heat	joule	J	(N·m)
Power	watt	W	(J/s)
Electric charge	coulomb	C	(A·s)
Voltage, potential difference, electromotive force	volt	V	(W/A)
Electric field strength	volt per meter	V/m	
Electric resistance	ohm	Ω	(V/A)
Electric capacitance	farad	F	(A·s/V)
Magnetic flux	weber	Wb	(V·s)
Inductance	henry	H	(V·s/A)
Magnetic flux density	tesla	T	(Wb/m²)
Magnetic field strength	ampere per meter	A/m	
Magnetomotive force	ampere	A	
Luminous flux	lumen	lm	(cd·sr)
Luminance	candela per sq meter	cd/m²	
Illumination	lux	lx	(lm/m²)

TABLE D.2

The National Bureau of Standards recommends the following prefixes and symbols for multiples and sub-multiples of units.

Multiples and submultiples	Prefixes	Symbols	Pronunciations
10^{12}	tera	T	tĕr′ȧ
10^{9}	giga	G	jĭ′ gȧ
10^{6}	mega	M	mĕg′ȧ
10^{3}	kilo	k	kĭl′ô
10^{2}	hecto	h	hek′tô
10	deka	da	dĕk′ȧ
10^{-1}	deci	d	dĕs′ĭ
10^{-2}	centi	c	sĕn′tĭ
10^{-3}	milli	m	mĭl ĭ
10^{-6}	micro	μ	mī′ krô
10^{-9}	nano	n	năn′ô
10^{-12}	pico	p	pē′ cô
10^{-15}	femto	f	fĕm′ tô
10^{-18}	atto	a	ăt′ tô

Tech. News Bull., Oct. 1949, v. 43, No. 10

TABLE D.3

PROPERTIES OF WATER

Temperature, °F	Absolute viscosity, $(lb_f\text{-sec}/ft^2) \times 10^5$	Specific weight, lb_f/ft^3
40	3.23	62.42
50	2.72	62.41
60	2.33	62.37
70	2.02	62.30
80	1.77	62.22
90	1.58	62.11
100	1.43	61.99
110	1.30	61.86
120	1.15	61.71

TABLE D.4

PROPERTIES OF DRY AIR AT ATMOSPHERIC PRESSURE

Temperature, °F	Absolute viscosity, $(lb_f\text{-sec}/ft^2) \times 10^{6}$*	Specific weight, lb_f/ft^3†
40	0.362	0.0794
50	0.368	0.0779
60	0.374	0.0764
70	0.379	0.0749
80	0.385	0.0735
90	0.390	0.0722
100	0.396	0.0709
110	0.401	0.0697
120	0.407	0.0685

* Over the range from atmospheric pressure to 1000 psia, the viscosity of dry air increases at a rate of approximately 1% for each 100 psi increase in pressure.

† For pressures other than atmospheric, use $\gamma/\gamma_{atm} = P/14.7$.

REFERENCES

Chapter 1

1. *SAE Quarterly Transactions*, **4**, 4 (Oct. 1950).

Chapter 2

1. G. M. Clemence, "Time and Its Measurement," *Am. Scientist*, **40**, 2 (April 1952), p. 260.
2. H. C. Roberts, "Universal Characteristics of Measuring Methods," *ISAJ.*, **1**, 8 (Aug. 1954), p. 48.
3. M. F. Behar, "Handbook of Measurement and Control," *Inst. and Automation*, **27**, 12, Part 2 (Dec. 1954), pp. 187, 199.
4. J. L. Murphy, "Electronic Instruments for Mechanical Measurements," *Prod. Engr.*, **22**, 11 (Nov. 1951), p. 153.
5. A. H. Church, *Elementary Mechanical Vibrations.* New York: Pitman Publishing Corp. 1948, p. 35.
6. J. P. Den Hartog, *Mechanical Vibrations*, 4th ed. New York: McGraw-Hill Book Co., 1956, p. 38.
7. Ref. 6, p. 26.
8. E. B. Pearson, *Technology of Instrumentation.* Princeton, N.J.: D. Van Nostrand Co., 1957.
9. C. S. Draper and G. P. Bentley, "Design Factors Controlling the Dynamic Performance of Instruments," *Trans. ASME*, **62** (1940), pp. 421, 432.
10. C. S. Draper, W. McKay, and S. Lees, *Instrument Engineering*, 3 vols. New York: McGraw-Hill Book Co., 1952.
11. R. H. Bube, *Photoconductivity of Solids*, New York: John Wiley & Sons, Inc., 1960.
12. "Control Designer's Guide to Solid State Photosensors," *Control Engineering*, Oct. 1964, p. 71.
13. J. A. Jamieson, "Detectors for Infrared Systems," *Electronics*, Dec. 9, 1960, p. 82.
14. W. O. Bennett, "Accutron—A Chronometric Micro-Powerplant," *S.A.E.* Paper 711C, June, 1963.

Chapter 3

1. W. C. Johnson, *Mathematical and Physical Principles of Engineering Analysis.* New York: McGraw-Hill Book Co., 1944, p. 241.

2. R. E. Doherty and E. G. Keller, *Mathematics of Modern Engineering,* Vol. 1. New York: John Wiley & Sons, Inc., 1936, p. 83.

3. A. R. Knight and G. H. Fett, *Introduction to Circuit Analysis.* New York: Harper and Brothers, 1943, p. 413.

References not cited

A. Eagle, *Fourier's Theorem and Harmonic Analysis.* London: Green and Co., 1925.

H. Von Sanden, *Practical Mathematical Analysis,* New York: E. P. Dutton and Co., 1924.

A. G. Worthing and J. Geffner, *Treatment of Experimental Data.* New York: John Wiley & Sons, Inc., 1943.

F. A. Willers, *Practical Analysis.* New York: Dover Publications, Inc., 1948.

Chapter 4

1. "Units of Weight and Measure," *NBS Miscel. Pub.* 214, July 1, 1955.

2. U.S. Coast and Geodetic Survey Bulletin 26, April 5, 1893.

3. American Standards Association (ASA b 48.1–1933).

4. F. B. Silsbee, "Fundamental Units and Standards," *Instruments,* **26** (Oct. 1953), p. 1520.

5. G. M. Clemence, "Time and its Measurement," *Am. Scientist,* **40,** 2 (April 1952), p. 260.

6. J. D. Ryder, *Electronic Fundamentals and Applications.* Englewood Cliffs, N.J., Prentice-Hall, Inc., 1950, p. 440.

7. "Improvements in the NBS Primary Standard of Frequency," *NBS Tech. News Bull.,* **37,** 1 (Jan. 1953).

8. "Time Signals Available from Canadian Sources," *Radio and Television News* (Feb. 1955), p. 154.

9. "The International Practical Temperature Scale of 1968," *Metrologia,* v. 5, n. 2, p. 35, April 1969.

10. J. de Boer, "Temperature as a Basic Physical Quantity," *Metrologia,* v. 1, n. 4, p. 158, Oct. 1965.

11. F. B. Silsbee, "Extension and Dissemination of the Electrical and Magnetic Units by the National Bureau of Standards," *NBS Circular* 531 (1952).

12. C. Snow, "Formulas for Computing Capacitance and Inductance," *NBS Circular* 544 (1954).

13. R. L. Driscoll and R. D. Cutkosky, "Measurement of Current with the National Bureau of Standards Current Balance," *NBS J. Res.,* **60,** 4 (April 1958).

14. H. W. Simpson, "A Method of Comparing Transducers for Instrumentation Applications" *ISA J.*, **2**, 7 (July 1955), p. 251.

15. J. Terrien, "Scientific Metrology on the International Plane and the Bureau International des Poids et Mesures," *Metrologia*, v.1, n.2, p. 15, Jan. 1965.

16. R. D. Cochrane, *Measures for Progress, A History of the National Bureau of Standards*, U.S. Dept. of Commerce, 1966, p. 47.

17. Martin Gardner, "Can Time Go Backward," *Scientific American*, Jan. 1967, p. 98.

18. Lewis V. Judson, "Weights and Measures of the United States," *NBS Miscel. Pub.* 247, October, 1963.

19. John Cohen, "Psychological Time," *Scientific American*, 211, 5 (Nov. 1964), p. 116.

20. *NBS Tech. News Bull.*, v. 52, n. 1 (Jan. 1968), p. 10.

21. C. L. Stong, *Sci. Amer.*, v. 203, n. 1 (July 1960), p. 165 and n. 2 (Aug. 1960), p. 158.

22. *NBS Tech. News Bull.*, v. 48, n. 4 (April 1964), p. 61.

23. A. H. Morgan, "Time and Frequency Broadcasting," *I.S.A. Jour.*, p. 49 (June 1963).

24. "WWV to be Relocated," *NBS Tech. News Bull.*, p. 218 (Dec. 1965).

25. "New Standard Frequency Broadcasts," *NBS Tech. News Bull.*, p. 120 (July 1960).

26. R. H. Muller, "New Precise Temperature Standard Accurate to 5×10^{-4} deg. C.," *Anal. Chem.*, v. 32 (Nov. 1950), p. 103A.

27. "Standardization of the Inch," *NBS Tech. News Bull.*, 42, 3 (March 1958), p. 42.

28. "Frequency and Time Standards," *Application Note* 52, Hewlett-Packard Co., 1965.

Chapter 5

1. T. N. Whitehead, *Instruments and Accurate Mechanism*. New York: Dover Publications, Inc., 1953.

2. H. W. Simpson, "A Method of Comparing Transducers for Instrumentation Applications," *ISA J.*, 2, 7 (July 1955), p. 251.

3. Hugh D. Young, *Statistical Treatment of Experimental Data*. New York: McGraw-Hill Book Co., 1962.

4. B. Austin Barry, *Engineering Measurements*. New York: John Wiley & Sons, Inc. 1964.

5. Yardley Beers, *Introduction to the Theory of Error*. Reading, Mass: Addison Wesley Publishing Co., 1957.

6. D. C. Baird, *Experimentation: An Introduction to Measurement Theory and Experiment Design*. Englewood Cliffs, N.J.: Prentice-Hall, Inc., 1962.

7. A. M. Neville and J. B. Kennedy, *Basic Statistical Methods for Engineers and Scientists*. Scranton, Pa.: International Textbook Co., 1964.

8. B. W. Lindgren and G. W. McElrath, *Introduction to Probability and Statistics*. New York: The Macmillan Co., 1959.

9. *RADC Reliability Notebook*, U.S. Dept. of Commerce, PB 161894, October 1959, p. 4–2.

10. S. J. Kline and F. A. McClintock, "Describing Uncertainties in Single-Sample Experiments," *Mech. Engr.*, **75** (Jan. 1953), p. 3.

11. W. A. Wilson, "Design of Power Plant Tests to Insure Reliability of Results," *Trans. ASME*, **77** (May 1955), p. 405.

Chapter 6

1. H. W. Simpson, "A Method of Comparing Transducers for Instrumentation Application," *ISA J.*, **2**, 7 (July 1955), p. 251.

2. W. Kneen, "A Review of the Electric Displacement Gages Used in Railroad Car Testing," *ISA Proc.*, **6** (1951), p. 74.

3. H. L. Gray, Jr., "A Guide to Applying Resistance Pots," *Control Engr.*, **3**, 7 (July 1956), p. 80.

4. K. P. Dowell, "Thermistors as Components Open Product Design Horizons," *Electrical Manufacturing*, **42**, 2 (Aug. 1948).

5. *The Radio Amateur's Handbook*, 31st ed. American Radio Relay League, West Hartford, Conn., 1954, p. 27.

6. M. Hetenyi, *Handbook of Experimental Stress Analysis*. New York: John Wiley & Sons, Inc., 1950, p. 239.

7. "Electronic Micrometer Uses Dual Coils," *Prod. Engr.*, **19**, 1 (Jan. 1948), p. 134.

8. A. Brenner and E. Kellogg, "An Electric Gage for Measuring the Inside Diameter of Tubes," *NBS J. Res.*, **42** (May 1949), p. 461.

9. A. G. Boggis, "Design of Differential Transformer Displacement Gauges," *SESA Proc.*, **9**, 2 (1952), p. 171.

10. *Notes on Linear Variable Differential Transformers*, Bulletin AA–1a, Schaevitz Engineering Co., Camden, N.J., 1955.

11. Ref. 5, p. 24.

12. "Low Temperature Liquid Level Indicator for Condensed Gases," *NBS Tech. News Bull.*, **38**, 1 (Jan. 1954).

13. Ref. 6, p. 287.

14. Y. T. Sihvonen, G. M. Rassweiler, A. F. Welch, and J. W. Bergstrom, "Recent Improvements in a Capacitor Type Pressure Transducer," *ISA J.*, **2**, 11 (Nov. 1955).

15. J. W. Leggat, G. M. Rassweiler, and Y. T. Sihvonen, "Engine Pressure Indicators, Application of a Capacitor Type," *ISA J.*, **2**, 9 (Aug. 1955).

16. Welch, Weller, Hanysz, and Bergstrom, "Auxiliary Equipment for the Capacitor-Type Transducer," *ISA J.*, **2**, 12 (Dec. 1955).

17. *Piezotronic Technical Data*, Brush Electronics Company, Cleveland, Ohio, 1953.

18. L. T. Fleming, "A Ceramic Accelerometer of Wide Frequency Range," *ISA Proc.*, **5** (1950), p. 62.

19. J. R. McDermott, "Control Designer's Guide to Solid State Photosensors," *Control Engineering*, Oct. 1960, p. 71.

20. C. W. Gadd and T. C. Van Degrift, "A Short-Gage-Length Extensometer and Its Application to the Study of Crankshaft Stresses," *J. App. Mech.*, **9** (March 1942), p. A.15.

21. *Dew Point Equipment to Measure Moisture in Gases*, General Electric Co., Bulletin GEC–588. Schenectady, N.Y., 1950.

22. J. O. Campbell, "Special Electrical Applications in the Steel Industry," *Iron and Steel Engr.*, **19** (Feb. 1942), pp. 78–89.

23. R. Gunn, "A Convenient Electrical Micrometer and Its Use in Mechanical Measurements," *J. App. Mech.*, **7**, 6 (June 1940), p. A–49.

24. A. F. Underwood and J. B. Bidwell, "New Instruments for Roughness Measurement," *Brush Strokes*, Brush Electronics Company, **3**, 3 (Dec. 1953).

25. R. C. Lewis, "The Design and Characteristics of a High Sensitivity Direct Current Operated Accelerometer," *J. Acoust. Soc. Am.*, **5**, 28 (May 1950), pp. 357–361.

26. B. P. McKay, "A Mechanical-Electronic Transducer for Low Range Pressure and Force," *ISA J.*, **3**, 6 (June 1956).

27. "A Gas Discharge Phenomenon Leads to Development of New Transducer," *Industrial Lab.*, **5**, 9 (Sept. 1954).

28. K. S. Lion, "Mechanic-Electric Transducer," *Rev. Sci. Inst.*, **27**, 2 (April 1956), pp. 222–225.

29. E. V. Hardway, Jr., "Electrokinetic Transducers," *Instruments*, **26** (Aug. 1953), p. 1186.

30. "Electrokinetic Measurement of Dynamic Pressures," *GEC Recordings*, (Nov.–Dec. 1956), p. 6, Consolidated Electrodynamics Corporation, Pasadena, California.

31. John A. Jamieson, "Detectors for Infrared Systems," *Electronics*, Dec. 9, 1960, p. 82.

32. Richard H. Bube, *Photoconductivity of Solids*. New York: John Wiley & Sons, Inc., 1960.

33. Ante Lujic, "Semiconductor Sensors," *Machine Design*, July 22, 1965, p. 149.

Chapter 7

1. C. W. Ham, E. J. Crane, and W. L. Rogers, *Mechanics of Machinery*, 4th ed. New York: McGraw-Hill Book Co., 1958.

2. A. R. Holowenko, *Dynamics of Machinery*. New York: John Wiley & Sons, Inc., 1955.

3. F. A. Willers, *Practical Analysis.* New York: Dover Publications, Inc., 1947, p. 191.

4. A. M. Wahl, *Mechanical Springs.* Cleveland, Ohio: Penton Publishing Co., 1944, pp. 29, 36.

5. R. Gitlin, "How Temperature Affects Instrument Accuracy," *Control Engr.,* **2,** 4 (April 1955).

6. R. Gitlin, "Compensating Instruments for Temperature Changes," *Control Engr.,* **2,** 5 (May 1955).

7. F. A. Laws, *Electrical Measurements,* 2nd ed.   New York: McGraw-Hill Book Co., Inc., 1938, p. 217.

8. "Improvements in the NBS Primary Standard of Frequency," *NBS Tech. News Bull.,* **37,** 1 (Jan. 1953).

9. R. C. Geldmacher, "Ballast Circuit Design," *SESA Proc.,* **12,** 1 (1954), p. 27.

10. J. H. Meier, Discussion of Ref. 9, in same source, p. 33.

11. F. R. Bradley and R. D. McCoy, "Computing with Servo-Driven Potentiometers," *Tele-Tech and Electronic Industries* (Sept. 1952).

12. L. Goldsmith, "Electro-Mechanical Position Transducers," *Product Engineering Annual Handbook.* New York: McGraw-Hill Book Co., 1955.

13. *Encyclopedia Britannica,* **23** (1957) p. 566.

14. C. Wheatstone, "An Account of Several New Instruments and Processes for Determining the Constants of a Voltaic Circuit," *Phil. Trans. Roy. Soc.,* (*London*), **133** (1843), p. 303 ff.

15. F. A. Laws, in Ref. 7, p. 179.

16. B. Hague, *Alternating Current Bridge Methods.* London: Pitman Publishing Corporation, 1938.

17. W. D. Oliphant, "New Method for Recording Minute Changes in Capacitance," *J. Sci. Inst.* (*London*), **14,** 937 (1937), pp. 173–177.

18. J. D. Ryder, *Electronic Fundamentals and Applications.* New York: Prentice-Hall, Inc., 1950, p. 242.

19. O. G. Villard, "Tunable A-F Amplifier," *Electronics,* **22,** 7 (July 1949), pp. 77–79.

20. R. J. Gunderman, "Electronic Audio Bandpass Filter," *Radio-Electronic Engr.* (Nov. 1953), p. 16.

21. J. N. Mandalakas, "A Temperature Compensated Light Sensor," *I.S.A. Journal,* September, 1966, p. 37.

22. Anon., "Basic Course in Solid-State Electronics," *Machine Design,* in 12 parts beginning Nov. 24, 1966.

23. H. D. Arlowe and R. C. Dove, "Signal Conditioning Sensor Outputs," *Instrumentation Technology,* April, 1967, p. 39.

24. C. A. Bowes, "Variable Resistance Sensors Work Better with Constant Current Excitation," Instrument Technology, March 1967.

25. N. Sion, "Bridge Networks in Transducers," Inst. & Control Systems, August 1968.

Chapter 8

1. D. Christian, "BL–310 Strain Analyzer," *SESA Proc.*, **7**, 1 (1949), p. 21.

2. M. A. LeGette, *The Theory of Recording, Galvanometers*, Consolidated Electrodynamics Corp., Pasadena, Calif., July 1958.

Chapter 9

1. *Encyclopedia Britannica*, **6** (1957), p. 892.

Chapter 10

1. C. G. Peters and W. B. Emerson, "Interference Methods for Producing and Calibrating End Standards," *NBS J. Res.*, **44** (April 1950), p. 427.

2. *The Science of Precision Measurement*, The DoAll Company, Des Plaines, Illinois, 1953.

3. "Metrology of Gage Blocks," *NBS Circular* 581 (April 1, 1957), p. 67.

4. M. Graneek, "A Pneumatic Comparator of High Sensitivity," *The Engineer*, (July 13, 1951).

5. W. F. Meggers and F. O. Westfall, "Lamps and Wavelengths of Mercury 198," *NBS J. Res.*, **44** (May 1950), p. 447.

6. "Atomic Standards of Length Now Available to Industry," *Prod. Engr.*, **22**, 5 (May 1951), p. 127.

7. A. Kastelowitz, "Optical Airframe Tooling, A New Assembly Technique," *Automotive Industries*, **113**, 10 (April 15, 1951), p. 38.

8. "The World's First Man-Made Satellite," *Automotive Industries*, **104**, 8 (Nov. 15, 1956).

9. I. Goldman, "Which Method to Evaluate Surface Roughness?" *Materials and Methods*, **36**, 6 (Dec. 1952), p. 89.

10. E. J. Abbott and E. Goldschmidt, "Surface Quality," *Mech. Engr.*, **59**, 11 (Nov. 1937), p. 813.

11. R. E. Reason, "Surface Finish," *Prod. Engr.*, **28**, 10 (Sept. 16, 1957), p. 77.

12. D. E. Williamson, "Tracer-Point Sharpness as Affecting Roughness Measurements," *Trans. ASME*, **69**, 5 (May 1947), p. 319.

13. L. P. Tarasov, "Relation of Surface Roughness Readings to Actual Surface Profile," *Trans. ASME*, **67**, 4 (April 1945), p. 189.

14. W. Kneen, "A Review of Electric Displacement Gages used in Railroad Car Testing," *ISA Proc.*, **6** (1951), p. 74.

15. K. E. Gillilland, H. D. Cook, K. D. Mielenz, and R. B. Stephens, "Use of a Laser for Length Measurement by Fringe Counting," *Metrologia*, vol. 2, no. 3, p. 96, July 1966.

16. Amer. Soc. of Tool & Mfg. Engineers, *Handbook of Industrial Metrology*, Chapter 7, Englewood Cliffs, N.J., Prentice-Hall, Inc., 1967.

17. C. Hubbard, *Recent Advances in Long-path Metrology*, Airborne Instruments Laboratory, Division of Cutler-Hammer, Inc. Paper presented to the Amer. Ordinance Soc., Philadelphia, March 1965.

18. American Standard, *Surface Texture* (ASA B46. 1-1962), ASME, New York, 7, N.Y.

Chapter 11

1. L. B. Tuckerman, "Optical Strain Gages and Extensometers," *ASTM Proc.*, Part II, **23** (1923), p. 602.

2. B. L. Wilson, "Characteristics of the Tuckerman Strain Gage," *ASTM Proc.*, **44** (1944), p. 1017.

3. A. S. Kobayashi and E. E. Day, "Recording Interferometer for Strain Gage Calibration," *SESA Proc.*, **11**, 1 (1953), p. 235.

4. R. V. Vose, "An Application of the Interferometer Strain Gage in Photo-Elasticity," *J. App. Mech.*, **2** (Sept. 1935), p. 99.

5. B. F. Langer, "Design and Application of a Magnetic Strain Gage," *SESA Proc.*, **1**, 2 (1943), p. 82.

6. C. M. Hathaway, "Electrical Instruments for the Measurement of Strain," *ISA Proc.*, 8 (1953), p. 108.

7. C. H. W. Brookes–Smith and J. A. Colls, "Measurement of Pressure, Movement, Acceleration and Other Mechanical Quantities by Electrostatic Systems," *J. Sci. Inst. (London)*, **14** (1939), p. 361.

8. B. C. Carter, J. F. Shannon, and J. R. Forshaw, "Measurement of Displacement and Strain by Capacity Methods," *Proc. Instn. Mech. Engrs.*, **152** (1945), p. 215.

9. B. F. Langer, "Measurement of Torque Transmitted by Rotating Shafts," *J. App. Mech.*, **67**, 3 (March 1945), p. A.39.

10. E. A. Ripperger, "A Piezoelectric Strain Gage," *SESA Proc.*, **12**, 1 (1954), p. 117.

11. J. W. Mark and W. Goldsmith, "Barium Titanate Strain Gages," *SESA Proc.*, **13**, 1 (1955), p. 139.

12. Thompson (Lord Kelvin), "On the Electro-Dynamic Qualities of Metals," *Phil. Trans. Roy. Soc. (London)*, **146** (1856), pp. 649–751.

13. E. C. Eaton, "Resistance Strain Gage Measures Stresses in Concrete," *Engr. News Record*, **107** (Oct. 1931), pp. 615–616.

14. A. Bloach, "New Methods for Measuring Mechanical Stresses at Higher Frequencies," *Nature*, **136** (Aug. 19, 1935), pp. 223–224.

15. D. S. Clark and G. Datwyler, "Stress-Strain Relations under Tension Impact Loadings," *Proc. ASM*, **38** (1938), pp. 98–111.

16. E. W. Krammer and T. E. Pardue, "Electric Resistance Changes of Fine Wires During Elastic and Plastic Strains," *SESA Proc.*, **7**, 1 (1949), p. 7.

17. E. W. Krammer and I. Vigness, "Lattice Defects and Strain Gage Factors," *SESA Proc.*, **15**, 1 (1957), p. 179.

18. I. Vigness, "Magnetostrictive Electricity in Strain Gages," *Rev. Sci. Inst.*, **27** (Dec. 1956), p. 1012.

19. I. Vigness, "Magnetostrictive Effects in Wire Stain Gages," *SESA Proc.*, **14**, 2 (1957).

20. J. Gunn and E. Billinghurst, "Magnetic Fields Affect Strain Gages," *Control Engr.*, **4**, 8 (Aug. 1957).

21. Bulletin 279–B, 1951, Baldwin-Lima-Hamilton Corporation, Waltham 54, Mass.

22. "Characteristics and Applications of Resistance Wire Strain Gages," *NBS Circular* 528 (1954), p. 12.

23. E. Frank, "Series Versus Shunt Bridge Calibration," *Instruments*, **31** (March 1958).

24. V. J. McDonald and H. C. Roberts, "Practical Strain Gage Circuit Assembly," ISA Paper 54-23-1.

25. W. R. Campbell and R. F. Suit, Jr., "A Transistorized AM-FM Radio-Link Torque Telemeter for Large Rotating Shafts," *SESA Proc.*, **14**, 2 (1957), p. 55.

26. J. J. Rebeske, Jr., "Investigation of an NACA High-Speed Strain Gage Torquemeter," *NACA Tech. Note* 2003 (Jan. 1950).

27. J. H. Meier, "Some Aspects of Observing the Performance of Large Machinery Under Operating Load," *SESA Proc.*, **1**, 2 (1944), p. 11.

28. R. A. Berger and A. W. Brunot, "Dynamic Stress Measurements in Gas Turbines," *SESA Proc.*, **12**, 2 (1955), p. 45.

29. A. Goloff, "Determination of Operating Loads and Stresses in Crankshafts," *SESA Proc.*, **2**, 2 (1945), p. 139.

30. D. A. Drew, "The Measurement of Turbine Stresses in Aircraft Engines in the Laboratory; on the Test Bed and in Flight," *SESA Proc.*, **10**, 1 (1952), p. 187.

31. D. K. Wright, Jr., and J. R. Jeromson, Jr., "Application of Silver-Painted Slip Rings for Strain Gage Circuits," *SESA Proc.*, **11**, 2 (1954), p. 139.

32. Baumberger and Hines, "Practical Reduction Formulas for Use on Bonded Wire Strain Gages in Two-Dimensional Stress Fields," *SESA Proc.*, **2**, 1 (1944), p. 113.

33. C. C. Perry and H. R. Lissner, *The Strain Gage Primer*, 2nd ed. New York: McGraw-Hill Book Company, Inc., 1962.

34. G. H. Lee, *An Introduction to Experimental Stress Analysis*. New York: John Wiley and Sons, Inc., 1950.

35. M. Hetenyi, *Handbook of Experimental Stress Analysis*. New York: John Wiley and Sons, Inc., 1950.

36. W. M. Murray and P. K. Stein, *Strain Gage Techniques* (in two parts). Massachusetts Institute of Technology, Cambridge, Mass., 1957.

37. A. U. Huggenberger, "Abstract of a New Strain Gage Without Transverse Sensitivity," and G. V. A. Gustafsson, "How to Use G–H Gages," *NBS Circular* 528, *Characteristics and Applications of Resistance Strain Gages*, Feb. 15, 1954.

38. *Testing Topics*, **8**, 3 (July, Aug., Sept. 1953), Baldwin-Lima-Hamilton Corp., Waltham 54, Mass.

39. *Testing Topics*, **4**, 2 (1949), p. 5. Baldwin-Lima-Hamilton Corp.

39a. *Testing Topics*, **9**, 3 (1954), p. 6. Baldwin-Lima-Hamilton Corp.

40. E. G. Coker and L. N. G. Filon, *A Treatise on Photoelasticity*, Cambridge: Cambridge University Press, 1931.

41. M. M. Frocht, *Photoelasticity*, New York: John Wiley & Sons, Inc., Vol. 1, 1941, Vol. II, 1948.

42. A. J. Durelli and W. F. Riley, "Introduction to Photomechanics," Englewood Cliffs, N.J.: Prentice-Hall, Inc., 1965.

43. L. N. G. Filon, *A Manual of Photoelasticity for Engineers*, Cambridge: Cambridge University Press, 1936.

44. H. T. Jessop and F. C. Harris, *Photoelasticity, Principles and Methods*' New York: Dover Publications, Inc., 1950.

45. G. U. Oppel, "Photoelastic Strain Gages," *Exp. Mech.*, v. 1, n. 3, p. 65, Mar. 1961.

46. J. Schaighofer, "Application of Photoelastic Strain Gages," *Exp. Mech.*, v. 1, n. 12, Dec. 1961.

47. Felix Zandman, "Concepts of the Photoelastic Stress Gage," *Exp. Mech.*, v. 2, n. 8 Aug. 1962.

48. L. J. Weymouth, J. E. Starr, and J. Dorsey, "Bonded Resistance Strain Gages," *Exp. Mech.*, v. 6, n. 4, p. 19A, April 1966.

49. M. J. Lebow: "Extensometers," *Exp. Mech.* v. 6, n. 6, p. 21A, June 1966.

50. C. S. Smith, "Piezoresistive Effect in Germanium and Silicon," *Phys. Rev.*, v. 94, n. 42, Apr. 1, 1954.

51. W. P. Mason and R. N. Thurston, "Use of Piezoresistive Materials in the Measurement of Displacement, Force and Torque," *Jour. of the Acoustical Soc. of America*, v. 29, 1957.

52. P. M. Palermo, "Methods of Waterproofing SR-4 Strain Gages," *S.E.S.A. Proc.*, v. XIII, n. 2, p. 79, 1956.

53. Mills Dean III, "Strain Gage Waterproofing Methods and Installation of Gages on Propeller Strut of USS Saratoga," *S.E.S.A. Proc.*, v. XVI, n. 1, p. 137, 1958.

54. A. J. Bush, "Soldered-Cap Method of Waterproofing Protection for Strain Gages," *19th Annual I.S.A. Conf. Proc.*, v. 19, Part II, 1964.

55. T. G. Beckwith and N. L. Buck: "*Mechanical Measurements*," 1st. Ed., Reading, Mass.: Addison-Wesley Publishing Co., p. 276, 1961.

56. C. S. Ades and L. H. N. Lee, "Strain Gage Measurements in Regions of High Stress Gradient," *Exp. Mech.*, v. 1, n. 6, p. 199 June 1961.

57. M. C. Mucuoglu, "Solution of Rectangular Strain Rosettes by Vectorial-layout Method," *Exp. Mech.*, p. 302, Dec. 1963.

58. J. H. Meier, "On the Transverse Sensitivity of Foil Gages," *Exp. Mech.*, July 1961.

59. C. T. Wu, "Transverse Sensitivity of Bonded Strain Gages," *Exp. Mech.*, p. 338, Nov. 1962.

60. T. H. H. Pian, "Reduction of Strain Rosettes in the Plastic Range," *Jnl. Aerospace Sci.*, 26, No. 12., 842, Dec. 1959.

61. C. S. Ades, "Reduction of Strain Rosettes in the Plastic Range," *Exp. Mech.*, p. 345, Nov. 1962.

62. John C. Telinde, "Investigation of Strain Gages at Cryogenic Temperatures," *Douglas Paper No.* 3835, Douglas Missile and Space Systems Division, Huntington Beach, Calif., 1966.

63. S. W. Leszynski, "The Development of Flame Sprayed Sensors," *ISA Jour.* p. 35, July 1962.

64. S. P. Wunk, Jr., "Innovations in Flame-spraying Techniques," *Two Parts, BLH Measurement Topics*, Vol. 3 Nos. 2 and 3, 1965.

65. V. Rastogi, K. D. Ives, and W. A. Crawford, "High-temperature Strain Gages for Use in Sodium Environments," *Exp. Mech.*, v. 7, n. 12, Dec. 1967.

66. A. J. Karnie and E. E. Day, "A Laser Extensometer for Measuring Strain at Incandescent Temperatures," *Exp. Mech.*, v. 7, n. 11, Nov. 1967.

Chapter 12

1. T. W. Lashof and L. B. Macurdy, "Precision Laboratory Standards of Mass and Laboratory Weights," *NBS Circular* 547 (1954).

2. R. E. Bell and J. A. Fertle, "Electronic Weighing on the Production Line," *Electronics*, 28, 6 (June 1955), p. 152.

3. *Encyclopedia Britannica*, 23 (1957), p. 483.

4. *Instruments*, 25 (Sept. 1952), p. 1300.

5. B. L. Wilson, D. R. Tate, and G. Borkowski, "Proving Rings for Calibrating Testing Machines," *NBS Circular* C454 (1946).

6. S. Timoshenko, *Strength of Materials*, Part II, 2nd ed. New York: D. Van Nostrand, Inc., 1941, p. 88.

7. Ref. 6, p. 82.

8. "High Capacity Load Calibrating Devices," *NBS Tech. News Bull.*, 37, 9 (Sept. 1953).

9. P. K. Stein, "Strain Gage Transducers," *Prod. Engr.*, 27, 3 (March 1956), p. 196.

10. "Industrial Load Cells," *Prod. Engr.*, 20, 3 (March 1949), p. 111.

11. C. S. Moore, A. E. Biermann, and F. Voss, "The NACA Balanced-Diaphragm Dynamometer-Torque Indicator," NACA RB-4C28 (WR E–139), March 1944.

12. J. E. Voytilla, "Pneumatic Weighing in Process Control Systems," *ISA Proc.*, 5, 18 (1950).

13. R. G. Bosseler, "Air Operated Thrus-Torq System," *Trans. ASME*, **70**, 5 (May 1948), p. 271–273.

14. R. J. Sweeney, *Measurement Techniques in Mechanical Engineering.* New York: John Wiley & Sons, Inc., 1953.

15. A. C. Ruge, "The Bonded Wire Torquemeter," *SESA Trans.*, **1**, 2 (1943), p. 68.

16. J. J. Rebeske, Jr., "Investigation of a NACA High-Speed Strain-Gage Torquemeter," *NACA Tech. Note* 2003, Jan. 1950.

17. B. F. Langer, "Measurement of Torque Transmitted by Rotating Shafts," *J. App. Mech.*, **67**, 3 (March 1945), p. A.39.

18. B. F. Langer and K. L. Wommack, "The Magnetic-Coupled Torquemeter, *SESA Trans.*," **2**, 2 (1944), p. 11.

19. M. Hetenyi, *Handbook of Experimental Stress Analysis.* New York: John Wiley & Sons, Inc., 1950, Chapters 6 and 7.

20. D. C. Andrews, "Piping Flexibility Analysis by Model Test," *ASME Trans.*, **74**, 1 (Jan. 1952), p. 123.

21. Handbook 77, Vol. III, *Precision Measurement and Calibration, Optics, Metrology and Radiation*, National Bureau of Standards, pp. 588 and 615, 1961.

Chapter 13

1. E. E. Ambrosius, R. D. Fellows and A. D. Brickman, *Mechanical Measurement and Instrumentation*, New York: The Ronald Press, 1966.

2. J. S. Doolittle, *Mechanical Engineering Laboratory.* New York: McGraw-Hill Book Co., 1957.

3. C. W. Messersmith, C. F. Warner, and R. A. Olsen, *Mechanical Engineering Laboratory*, 2nd ed. New York: John Wiley & Sons. Inc., 1958.

4. A. Wolfe, "An Elementary Theory of the Bourdon Gage," *J. Appl. Mech.*, **68**, 9 (Sept, 1946), p. A–207.

5. C. Sunatani, "Theory of the Bourdon Tube Pressure Gage and an Improvement in Its Mechanism," *Tech. Rpts.*, Tohoku Imperial University, Japan, **4**, 1 (1924–1925), pp. 69–110 (in English).

6. E. Wenk, Jr., "A Diaphragm-Type Gage for Measuring Low Pressures in Fluids," *SESA Proc.*, **8**, 2 (1951), p. 90.

7. R. J. Roark, *Formulas for Stress and Strain*, 4th ed. New York: McGraw-Hill Book Co., 1965.

7a. C. K. Stedman, "The Characteristics of Flat Annular Diaphragms," *Instrument Notes*, No. 31 (Jan. 1957), Statham Laboratories, Los Angeles 64, Calif.

8. J. L. Patterson, "A Miniature Electrical Pressure Gage Utilizing a Stretched Flat Diaphragm," *NACA Tech. Note* 2659 (April 1952).

9. H. J. Grover and J. C. Bell, "Some Evaluations of Stresses in Aneroid Capsules," *SESA Proc.*, **5**, 2 (1948), p. 125.

10. F. D. Werner, "The Design of Diaphragms for Pressure Gages Which use the Bonded Wire Resistance Strain Gage," *SESA Proc.*, **11**, 1 (1953), p. 137.

11. J. Delmonte, "A Versatile Miniature Flush-Diaphragm Pressure Transducer," *ISA Proc.*, **7** (1952), p. 174.

12. M. F. Spotts, *Design of Machine Elements*, 3rd ed. New York: Prentice-Hall, Inc., 1961, p. 470.

13. W. H. Howe, "What's Available for High Pressure Measurement and Control," *Control Engr.*, **2**, 4 (April 1955), p. 53.

14. W. H. Howe, "The Present Status of High Pressure Measurement," *ISA J.*, **2**, 3–4 (March, April 1955), pp. 77 and 109.

15. W. A. Wildhack, "Pressure Drop in Tubing in Aircraft Instrument Installations," *NACA Tech Note* 593 (1937).

16. A. S. Iberall, "Attenuation of Oscillatory Pressures in Instrument Lines," *NBS J. Res.*, **45** (July 1950), p. 85.

17. J. C. Moise, "Pneumatic Transmission Lines," *ISA Proc.*, **8** (1953), p. 152.

18. C. G. Hylkema and R. B. Bowersox, "Experimental and Mathematical Techniques for Determining the Dynamic Response of Pressure Gages," *ISA Proc.*, **8** (1953), p. 115, or *ISA J.*, **1**, 2 (Feb. 1954), p. 27.

19. C. K. Stedman, "Alternating Flow of Fluids in Tubes," *Instrument Notes*, No. 30 (Jan. 1956), Statham Laboratories, Los Angeles 65, Calif.

20. Lord Rayleigh, *The Theory of Sound*, Vol. II, 2nd ed. New York: Dover Publications, Inc., reprinted 1945, p. 188.

21. R. D. Mylius and R. J. Reid, "Acoustical Filters Protect Pressure Transducers," *Control Engr.*, **4**, 1 (Jan. 1957), p. 115.

22. G. White, "Liquid Filled Pressure Gage Systems," *Instrument Notes*, No. 7 (Jan.–Feb. 1949), Statham Laboratories, Los Angeles 64, Calif.

23. I. Taback, "The Response of Pressure Measuring Systems to Oscillating Pressure," *NACA Tech. Note* No. 1819 (Feb. 1949).

24. A. Badmaieff, "Techniques of Microphone Calibration," *Audio Engr.*, **38**, 12 (Dec. 1954).

25. R. Reid, "Use Standard Functions to Test Pneumatic Systems," *Control Engr.*, **5**, 1 (Jan. 1958), p. 117.

26. R. C. Baird, R. L. Solnich, and J. R. Amiss, "Calibrator for Dynamic Pressure Transducers," *Inst. and Automation*, **27** (July 1954), p. 1074.

27. R. D. Meyer, "Dynamic Pressure Transmitter Calibrator," *Rev. of Sci. Inst.*, **17**, 5 (May 1946), p. 199.

28. D. P. Eckman and J. C. Moise, "A Pneumatic Sine-Wave Generator for Process Control Study," *ISA Proc.*, **7** (1952), p. 13.

29. W. R. Davis, "Measuring High Pressure Transients," *Auto. Control*, **4**, 1 (Jan. 1956), p. 24.

30. National Bureau of Standards Monograph 67, *Methods for the Dynamic Calibration of Pressure Transducers*, Washington: U.S. Dept. of Commerce 1963.

31. R. Bowersox, "Calibration of High Frequency Response Pressure Transducers," *ISA J.*, **5** 11 (Nov. 1958).

32. T. G. Beckwith and N. L. Buck, *Mechanical Measurements*, 1st Ed., Reading, Mass.: Addison-Wesley Publishing Co., 1961.

33. W. E. Teale, "Pressure Measurement in a Closed Vessel," *Inst. & Control Sys.*, March 1966, p. 127.

References not cited

R. E. Duff, *The Use of Real Gases in a Shock Tube*, Report 51–3, Engineering Research Institute, University of Michigan, Ann Arbor, Mich., March 5, 1951.
P. W. Huber, C. E. Fitton, Jr., and F. Delpino, "Experimental Investigation of Moving Pressure Disturbances and Shock Waves and Correlation with One-Dimension Unsteady-Flow Theory," *NACA Tech. Note* 1903 (June 1949).
Y. T. Li, "Pressure Transducers for Missile Testing and Control," *ISA J.*, **5**, 11 (Nov. 1958), p. 81.
W. G. Brombacher and T. W. Lashof, "Bibliography and Index on Dynamic Pressure Measurement," *NBS Circular* 558 (1955).

Chapter 14

1. R. C. Binder, *Fluid Mechanics*, 3rd ed. Englewood Cliffs, N.J.: Prentice-Hall, Inc., 1955, p. 113.

2. Ref. 32, p. 306.

3. *ASME Performance Test Code*, "Flow Measurement," Part 5, Chapter 4, 1959, p. 68.

4. E. M. Schoenborn and A. P. Colburn, "The Flow Mechanism and Performance of the Rotameter," *Trans. AI Ch.E.*, **35**, 3 (1939), p. 359.

5. W. Gracey, "Measurement of Static Pressure on Aircraft," *NACA Tech. Note* 4184 (Nov. 1957).

6. L. N. Krause and C. C. Gettelman, "Effect of Interaction Among Probes, Supports, Duct Walls and Jet Boundaries on Pressure Measurements in Ducts and Jets," *ISA Proc.*, **7** (1952), p. 138.

7. C. C. Gettelman and L. N. Krause, "Considerations Entering into the Selection of Probes for Pressure Measurement in Jet Engines," *ISA Proc.*, **7** (1952), p. 134.

8. F. A. Nagler, "Use of Current Meters for Precise Measurement of Flow," *ASME Trans.*, **57** (1935), p. 59.

9. J. Grey, "Transient Response of the Turbine Flowmeter," *Jet Propulsion* **26**, 2 (Feb. 1956).

10. L. V. King, "On the Convection of Heat from Small Cylinders in a Stream of Fluid, with Applications to Hot-Wire Anemometry," *Phil. Trans. Roy. Soc. (London)* **214**, 14, Ser. A (1914), pp. 373–432.

11. J. C. Laurence and L. G. Landes, "Auxiliary Equipment and Techniques for Adapting the Constant Temperature Hot-Wire Anemometer to Specific Problems in Air-Flow Measurements," *NACA Tech. Note.* 2843 (Nov. 1952).

12. J. C. Laurence and L. G. Landes, "Application of the Constant Temperature Hot-Wire Anemometer to the Study of Transient Air Flow Phenomena," *ISA J.*, **1**, 12 (Dec. 1953), p. 128.

13. A. E. Knowlton, *Standard Handbook for Electrical Engineers*, 8th ed. New York: McGraw-Hill Book Co., Inc., 1949, pp. 36–40.

14. W. G. James, "An A-C Induction Flow Meter," *ISA Proc.*, **6** (1951), p. 5.

15. W. C. Gray and E. R. Astley, "Liquid Metal Magnetic Flowmeters," *ISA J.*, **1**, 6 (June 1954).

16. *NBS Tech. News Bull.* **37**, 3 (March 1953).

17. A. G. Hansen, *Fluid Mechanics*, New York: John Wiley & Sons, Inc., p. 422, 1967.

18. W. T. Collins and T. W. Selby, "A Gravimetric Flow Standard," Flow Measurement Symposium, ASME, p. 290, 1966.

19. D. H. Liebenberg, R. W. Stokes, and F. J. Edeskuty, "The Calibration of Flowmeters with Liquid Hydrogen in the Region between 1000 and 7000 GPM," Flow Measurement Symposium, ASME, p. 155, 1966.

20. R. P. Bowen, "Designing Portability into a Flow Standard," *I.S.A. Jour.*, p. 40, May 1961.

21. R. V. Giles, *Fluid Mechanics and Hydraulics*, 2nd. Ed. New York: Schaum Publishing Co., p. 137, 1962.

Chapter 15

1. *Handbook of Chemistry and Physics.* Chemical Rubber Publishing Company, Cleveland, Ohio, 37th ed., 1955.

2. H. T. Wenzel, *Temperature.* New York: Reinhold Publishing Corporation, p. 3, 1941.

3. M. L. Smith and K. W. Stinson, "Fuels and Combustion." New York: McGraw-Hill Book Co., 1952, p. 157.

4. M. F. Behar, *The Handbook of Measurement and Control*, 2nd ed. The Instruments Publishing Company, Inc., Pittsburgh, 1954, p. 26.

5. S. G. Eskin, and J. R. Fritze, "Thermostatic Bimetals," *ASME Trans.*, **62**, 7 (July 1910).

6. K. P. Dowell, "Thermistors as Components Open Product Design Horizons," *Electrical Manufacturing*, **42**, 2 (August 1948).

7. T. J. Seebeck, *Evidence of the Thermal Current of the Combination Bi-Cu by Its Action on Magnetic Needle.* Abt. d. Konigl, Akak. d. Wiss Berlin, 1822–23, p. 265.

8. M. Peltier, "Investigation of the Heat Developed by Electric Currents in Homogeneous Materials and at the Junction of Two Different Conductors," *Ann. Chim. Phys.*, **56** (1834), p. 371.

9. W. Thomson, "Theory of Thermoelectricity . . . in Crystals," *Trans. Edinb. Soc.* **21**, 153 (1847). Also, *Math. Phys. Papers*, **1** (1882), p. 232, 266.

10. P. H. Dike, *Thermoelectric Thermometry.* Philadelphia: Leeds and Northrup Company, 1954.

11. J. F. Lee and F. W. Sears, *Thermodynamics*. Reading, Mass.: Addison-Wesley Publishing Co., 1955, p. 292.

12. *ASME Power Test Code*, "Information on Instruments and Apparatus," Part 3, Temperature Measurement, Chapter 8, Optical Pyrometers, PTC. 3.8–1940.

13. T. J. Schlottenmier, "Radiation Pyrometry," *Proc. ISA.*, **7** (1952), p. 102.

14. Ref. 12.

15. L. S. Marks, *Mechanical Engineer's Handbook*, 5th ed. New York: McGraw-Hill Book Co., 1951, p. 2006.

16. Ref. 11, p. 281

17. *Steam, Its Generation and Use*, 37th ed. New York: The Babcock and Wilcox Company, 1955.

18. N. R. Johnson, A. S. Weinstein, and F. Osterle, "The Influence of Gradient Temperature Fields on Thermocouple Measurements," *ASME Paper* No. 57–HT–18.

19. H. C. Hottel and A. Kalitinsky, "Temperature Measurement in High-Velocity Air Streams," *J. App. Mech.*, **67**, 3 (March 1945), p. A25.

20. G. T. Lalos, "A Sonic-Flow Pyrometer for Measuring Gas Temperatures, *NBS J. Res.*, **47**, 3 (Sept. 1951), p. 179.

21. M. D. Scadron and I. Warshawsky, "Experimental Determination of Time Constants and Nusselt Numbers for Bare-Wire Thermocouples in High-Velocity Air Streams and Analytic Approximation of Conduction and Radiation Errors," *NACA Tech. Note* 2599 (Jan. 1952).

22. R. J. Moffat, "How to Specify Thermocouple Response," *ISA J.*, **4**, 6 (June 1957), p. 219.

23. C. E. Shepard and I. Warshawsky, "Electrical Techniques for Compensation of Thermal Time Lag of Thermocouples and Resistance Thermometer Elements," *NACA Tech. Note* 2703 (May 1952).

24. C. E. Shepard and I. Warshawsky, "Electrical Techniques for Time Lag Compensation of Thermocouples used in Jet Engine Gas Temperature Measurements," *ISA J.*, (Nov. 1953), p. 119.

25. "Reference Tables for Thermocouples," *NBS Circular* 561, April 1955.

26. W. F. Roeser and S. T. Lonberger, "Methods of Testing Thermocouples and Thermocouple Materials," *NBS Circular* 590, Feb. 1958.

27. H. D. Baker, E. A. Ryder, and N. H. Baker, *Temperature Measurement in Engineering*, Vol. 1, p. 49. New York: John Wiley & Sons, Inc., 1953.

28. L. Lefkowitz, "Methods of Dynamic Analysis," *ISA Jour* (June 1955), p. 203.

29. J. R. Louis, and W. E. Hartman, The Determination and Compensation of Temperature Sensor Transfer Functions, *ASME* Paper 64-WA/AUT-13.

30. D. L. Hammond, and A. Benjaminson, "Linear Quartz Thermometer," *Instruments and Control Systems* (Oct. 1965), p. 115.

Chapter 16

1. R. J. Fibikar, "Touch and Vibration Sensitivity," *Prod. Engr.* **27**, 11 (Nov. 1956), p. 177.

2. D. E. Hudson and O. D. Terrell, "A Preloaded Spring Accelerometer for Shock and Impact Measurements," *SESA Proc.*, **9**, 1 (1951), p. 1.

3. C. B. Phillips and M. J. Anderson, "Development of an Automatic Counting Accelerometer," *Civil Aeronautics Admin. Tech. Devel. Report* No. 166 (April 1952).

4. L. T. Fleming, "A Ceramic Accelerometer of Wide Frequency Range," *ISA Proc.*, **5** (1952), p. 62.

5. G. O. Sankey, "Plastic Models for Vibration Analysis," *Mach. Des.*, **25**, 5 (May 1953), p. 125.

6. D. D. Rosard, "Natural Frequencies of Twisted Cantilever Beams," *ASME Paper* 52–A–15 (1952).

7. Dale Pennington, *Piezoelectric Accelerometer Manual.* Pasadena: Endevco Corporation, 1965.

8. R. I. Butler, and M. McWhirter, "Current Practices in Shock and Vibration Sensors," *Instrumentation Technology*, March 1967, p. 41.

9. Walter P. Kistler, "Precision Calibration of Accelerometers for Shock and Vibration," *Test Engineering*, May 1966, p. 16.

10. Seymour Edelman, "Additional Thoughts on Precision Calibration of Accelerometers," *Test Engineering*, Nov. 1966, p. 17.

11. R. C. Lewis, "Electro-Dynamic Calibrators for Vibration Pickups," *Prod. Engr.*, **22**, 9 (Sept. 1951).

12. K. Unholtz, "The Calibration of Vibration Pickups to 2000 CPS," *ISA Proc.*, **7** (1952), p. 325.

13. N.B.S. Tech. News Bull. "Easily made Device Calibrates Accelerometer," June 1966, p. 94.

14. W. A. Wildhack and R. O. Smith, "A Basic Method of Determining the Dynamic Characteristics of Accelerometers by Rotation," *ISA Paper* No. 54–40–3.

15. R. W. Conrad and I. Vigness, "Calibration of Accelerometers by Impact Techniques," *ISA Proc.*, **8** (1953), p. 166.

16. T. A. Perls and C. W. Kissinger, "High-g Accelerometer Calibrations by Impact Methods with Ballistic Pendulum, Air Gun and Inclined Trough," *ISA Paper* No. 54–40–2.

17. D. E. Weiss, "Design and Application of Accelerometers," *SESA Proc.*, **4**, 2 (1947).

18. J. Burns and G. Rosa, "Calibration and Test of Accelerometers," *Instrument Notes* No. 6 (Dec. 1948), Statham Laboratories, Los Angeles 64, Calif.

19. S. Levy and Wilhemina D. Kroll, "Response of Accelerometers to Transient Accelerations," *NBS J. Res.* **45**, 4 (Oct. 1950).

20. W. P. Welch, "A Proposed New Shock Measuring Instrument," *SESA Proc.*, **5**, 1, p. 39, 1947.

21. A. B. Kaufman, "Accelerometer Integration," *Radio-Electronic Engr.*, (June 1952).

22. J. A. Dickie, "Vibration Testing Technique and its Use in Improving Designs," *Prod. Engr.*, **21**, 11 (Nov. 1950).

23. R. C. Lewis, "Applying Shakers for Product Vibration Analysis," *Prod. Engr.*, **21**, 11 (Nov. 1950).

24. R. C. Lewis, "Test Techniques Using Electro-Dynamic Shakers," *Prod. Engr.* **21**, 1 (Jan. 1951).

25. J. C. Truman, J. R. Martin, and R. V. Hunt: "Pulsed-Air Vibration Technique for Testing High-performance Turbomachinery Blading," *Exp. Mech.*, June, 1961.

26. J. A. Adler, "Hydraulic Shakers," *Test Engineering*, April 1963.

27. J. A. Dickie, "Hydraulic Vibrations," *Prod. Engr.*, **22**, 1 (Dec. 9, 1957).

28. G. B. Booth, "Random Motion," *Prod. Engr.*, **27**, 11 (Nov. 1956).

29. S. H. Crandall, *Random Vibration*. New York: John Wiley and Sons, Inc. 1959.

30. K. Unholtz, "Factors to Consider in Setting up Vibration Test Specifications," *Mach. Des.*, **28**, 6 (Mar. 22, 1956).

31. G. P. Wozney, "Resonant Vibration Fatigue Testing," *Exp. Mech.*, Jan. 1962.

32. "Application and Design Formulae for Free-Free Resonant Beams," *MB Vibration Notebook*, 1, 1, MB Manufacturing Co., New Haven, Conn., March 1955.

33. T. G. Beckwith and N. L. Buck, *Mechanical Measurements*, 1st. Ed., Reading, Mass.: Addison Wesley Publishing Co., 1961, p. 485.

34. Maxwell Lazarus, "Shock Testing, a Design Guide," *Machine Design*, Oct. 12, 1967.

35. J. H. Armstrong, "Shock Testing Technology at the Naval Ordnance Laboratory," *SESA Proc.*, **6**, 1 (1948), p. 55.

36. J. Brown, "Selection Factors for Mechanical Buffers," *Prod. Engr.*, **21**, 11 (Nov. 1950), p. 156.

37. J. Brown, "Further Principles of Buffer Design," *Prod. Engr.*, **21**, 12 (Dec. 1950), p. 125.

38. I. Vigness, "Some Characteristics of Navy High Impact Type Shock Machines," *SESA Proc.*, **5**, 1, p. 101.

39. R. D. Mindlin, F. W. Stubner, and H. L. Cooper, "Response of Damped Elastic Systems to Transient Disturbances," *SESA Proc.*, **5**, 2, p. 69.

40. R. D. Mindlin, "Dynamics of Package Cushioning," *Bell System Tech. J.*, **24** (July–Oct. 1945), or *Bell Telephone System Monograph* 13–1369, p. 70.

Chapter 17

1. D. C. Skilling, "Acoustical Testing at Northrup Aircraft," *SESA Proceedings*, Vol. 16, No. 2., p. 121, 1959.

2. Robert H. Randall, *An Introduction to Acoustics*. Reading Mass., Addison-Wesley Publishing Co., 1951.

3. David N. Keast, *Measurements in Mechanical Dynamics*. New York: McGraw-Hill Book Co., 1967.

4. *Scientific American*, December, 1966, p. 76.

5. *Product Engineering*, December 22, 1958, p. 58.

6. Arnold P. G. Peterson, and Ervin E. Gross, Jr., *Handbook of Noise Measurement*, Fifth Edition, West Concord, Massachusetts, General Radio Company.

7. James R. Ranz, "Noise Measurement Methods," *Machine Design*, November 10, 1966.

8. Blasser, Heinz and Helmut Finckh, "Automatic Loudness Analysis," *Hewlett-Packard Journal*, November, 1967, p. 12.

9. Various papers, Handbook 77, Vol. II, *Precision Measurement and Calibration, Heat and Mechanics*, National Bureau of Standards, 1961.

10. *American Standard Method for Calibration of Microphones*, S1.10–1966, American Standards Associations, New York.

11. *Wall Street Journal*, February 7, 1968, p. 1.

12. *Time Magazine*, January 11, 1954, p. 59.

Chapter 18

1. R. L. Murray, *Introduction to Nuclear Engineering*. Englewood Cliffs, N.J.: Prentice-Hall, Inc., 1954, p. 27.

2. S. Jefferson, *Radioisotopes, A New Tool for Industry*. New York: Philosophical Library, Inc., 1958, p. 72.

3. L. P. Curtiss, "The Geiger-Mueller Counter," *NBS Circular* 490, Jan. 23, 1950.

4. G. B. Foster, "Beta Radiation Gauging Methods," *Radio-Electronic Engr.* (April 1953).

5. P. E. Ohmart and H. L. Cook, Jr., "Applications of Radioactive Density Gaging to Process Measurement and Control," *ISA J.*, **2**, 1 (Jan. 1955).

6. H. A. Tuttle and G. E. Noakes, "The Applications of Radioactivity for the Control and Testing of Automotive Materials," *SAE Preprint No.* 775 for presentation June 3–8, 1956, at Atlantic City, N.J.

7. "Gold Measures River Flow," *Electronics*, **31**, 3 (Jan. 17, 1958), p. 112.

8. F. L. Schwartz and R. H. Eaton, "Wear Rates of Gears by the Radioactive Method," *SAE Preprint No.* 351 for presentation Sept. 13–16, 1954, at Milwaukee, Wisconsin.

9. "Radioactive Tires Used in Wear Tests," *Chem. and Eng. News*, **29** (1951), p. 2914.

10. "Safe Handling of Radioactive Isotopes," *NBS Handbook* 42.

References not cited

L. F. Curtiss, "Measurements of Radioactivity," *NBS Circular* 476, Oct. 1949.

W. B. Mann and H. H. Seliger, "Preparation, Maintenance and Application of Standards of Radioactivity, *NBS Circular* 594, June 1958.

G. D. Calkins and R. L. Belcher, "A Primer on Radioisotopes," *Prod. Engr.*, **23**, 10 (Oct. 1952), p. 177.

C. C. Gambill, "An Application of Radioisotope Wear Study Techniques," *General Motors Engr. J.* (April, May, June 1958), p. 21.

L. B. Shore and K. F. Ockert, "Combustion Chamber Deposits—A Radiotracer Study," *SAE Paper* No. 145, for presentation June 1957.

"Radiological Monitoring Methods and Instruments," *NBS Handbook* 51, April 1952.

V. N. Smith, "Radioactivity Analyzes Liquids and Gases," *Control Engr.*, **4**, 9 (Aug. 1957).

P. J. Stewart and G. J. Leighton, "Use Radioactive Instruments," *Control Engr.*, **2**, 3 (March 1955).

A. M. Smith, "Radioisotopes as Design Tools," *Prod. Engr.* (May 1955).

"Nuclear Data," *NBS Circular* 499.

F. L. Brannigan, *Living with Radiation*, No. 1. Office of Industrial Relations, U.S. Atomic Energy Commission, Washington, D.C., no date, 65 pp.

F. L. Brannigan and G. S. Miles, *Living with Radiation*, No. 2. (Same source as above.) 199 pp.—*Radioisotopes in Science and Industry.* Special Report, U.S. Atomic Energy Commission, Washington, D.C., Jan. 1960, 176 pp.— *Radiation: A Tool for Industry* (AL152) *A Survey of Current Technology.* U.S. Atomic Energy Commission, Washington, D.C., Jan. 1959, 392 pp.

ANSWERS TO SELECTED PROBLEMS

Chapter 2

2.6   0.43, 0.83, 1.93 sec

2.9   95 deg. F

2.13 129.3, 88.7, 73.7, 65.6 units

2.16 0.0216 sec   2.17 1.225 sec

2.26 $\tau = W/cg$ 2.30 17.6%, 28 deg.

2.7   0.036, 0.089, 0.177 sec

2.11 1.39 sec

2.14 5000, 250,000 μs

2.19 12.32 sec

Chapter 3

3.2   c) 1.256 sec   3.3   a) 50 Hz

Chapter 5

5.1 40 mph

5.4   For group A: 15, 13, 20.2, 4.73, 1.49, 1.06

5.5   1.449, 1.865, 26.4%, 0.33

5.6   b) 0.62613, 0.0093 (based on $\bar{x} = 0.626$), 0.0119

Chapter 7

7.5   2.75%, 4.8%   7.9   0.0734 v   7.11 0.0732 v

7.12 1292 ohms   7.13 600, 1.5 v

Chapter 8

8.1   1860 ohms   8.2   0.1401 ohms   8.3   1.357 v

Chapter 10

10.4 1.718368 in.   10.7   7.1   10.10 54.3   10.11 3.789070 in.

10.13 48.8   10.16 177

Chapter 11

11.3 1.2   11.5   15,980 psi   11.6   15.64 × 10^6 psi, 0.281

11.7 74.5μs

11.10 8660 psi

11.11 2475, —415μs, 38,565, 4965 psi, 16,800 psi, —40.82, 49.17 deg.
11.15 620, —155, 1080μs   11.17 30.8 hp (max)
11.19 29μs, 664 psi   11.20 21,178 psi

Chapter 13

13.1   2.31, 2.04   13.5   68.6 rpm   13.6   9.46 deg.
13.10 18.02, 85%

Chapter 14

14.4   99 in. H_2O 14.5 $h > 27.1$ in. H_2O 14.9   13.2 lb/sec
14.12 3.27 ft/sec   14.13 193 mph   14.15 0.76R
14.16 27.1 ft/sec

Chapter 15

15.2   —40 deg.   15.4 $0.39 \times 10^{-3} \, \Delta T$
15.5   61.2, 41.4 ohms   15.11 —5, 25, 65, 205 deg. F
15.16 388 deg. F   **15.17 192 deg. F**

Chapter 16

16.4   0.0037 in.   16.5   0.06 lb-sec/in., $5 < f < \infty$
16.6   36%, 65.4 deg.   16.9   +0.9%, 39 deg.
16.10 0.047 in., 6.25%   16.11 0.89
16.16 0.369 in., 2.89 in/sec, 22.62 in/sec^2
16.17 (Theoretical values) 654μs, 0.0062 sec, 0.011 sec

INDEX